Springer Proceedings in Mathematics & Statistics

Volume 245

Springer Proceedings in Mathematics & Statistics

This book series features volumes composed of selected contributions from workshops and conferences in all areas of current research in mathematics and statistics, including operation research and optimization. In addition to an overall evaluation of the interest, scientific quality, and timeliness of each proposal at the hands of the publisher, individual contributions are all refereed to the high quality standards of leading journals in the field. Thus, this series provides the research community with well-edited, authoritative reports on developments in the most exciting areas of mathematical and statistical research today.

More information about this series at http://www.springer.com/series/10533

James W. Cogdell · Günter Harder
Stephen Kudla · Freydoon Shahidi
Editors

Cohomology of Arithmetic Groups

On the Occasion of Joachim Schwermer's 66th Birthday, Bonn, Germany, June 2016

Springer

Editors
James W. Cogdell
Department of Mathematics
Ohio State University
Columbus, USA

Günter Harder
Mathematical Institute
University of Bonn
Bonn, Germany

Stephen Kudla
Department of Mathematics
University of Toronto
Toronto, Canada

Freydoon Shahidi
Department of Mathematics
Purdue University
West Lafayette, USA

ISSN 2194-1009 ISSN 2194-1017 (electronic)
Springer Proceedings in Mathematics & Statistics
ISBN 978-3-030-07056-4 ISBN 978-3-319-95549-0 (eBook)
https://doi.org/10.1007/978-3-319-95549-0

Mathematics Subject Classification (2010): 11F75, 11F70, 11G18, 22E40, 57Q45, 12F10, 11R23

This Springer imprint is published by the registered company Springer Nature Switzerland AG
The registered company address is: Gewerbestrasse 11, 6330 Cham, Switzerland

Preface

In June of 2016, a conference was held at the Max Planck Institute for Mathematics, Bonn, to celebrate the 66th birthday of Joachim Schwermer. All such conferences offer us the opportunity to look back at the career of the honoree, both mathematical and otherwise. In the case of Joachim Schwermer, his career was very rich.

Schwermer received his Doctor rerum naturalium from the Rheinische-Friedrich-Wilhelms Universität, Bonn, in 1977, under the supervision of G. Harder and F. Hirzebruch, and his Habilitation there in 1982. After leaving Bonn in 1986, Schwermer held Professorships in Eichstätt (1986–1998) and Düsseldorf (1998–2000) before taking up residence as Professor at the University of Vienna in 2000. Simultaneously with his move to Vienna, he became Deputy Director of the Erwin Schrödinger Institute (ESI). He became Director in 2004 and in 2011 oversaw the transition of the ESI from an independent research institute supported by the Austrian Government to a research institute under the auspices of the University of Vienna. He retired from the ESI in 2016 and the University of Vienna in 2017.

There have been three separate but related focuses of Schwermer's career. The first is his mathematical career. His research interest has always been the cohomology of arithmetic groups and its relation with the theory of automorphic forms. This volume represents this side of his career, with many of the articles responses to the interests of Schwermer. The article of Grbac, in particular, describes his collaboration with Schwermer. But there are two more sides that have produced value in their scholarship and in the fostering of scholarship of others. The first of these is represented by Schwermer's interest in the history of mathematics, particularly that of the nineteenth and early twentieth centuries. This interest can be seen in his various articles and in particular in his books, one with Goldstein and Schappacher on Gauss' *Disquisitiones* and then with Dumbaugh on Emil Artin and Class Field Theory. Equally important is Schwermer's work in what might be termed the administration of mathematics. We would like to mention two aspects of this. One is the series of Oberwolfach meetings organized with S. Kudla, usually on cohomology of arithmetic groups, automorphic forms, representation theory, etc. These were very influential and well attended, and we owe them both a debt of gratitude for organizing these. More importantly is Schwermer's stint as the Director of the

ESI. Here too he hosted many workshops on similar topics, and many of us have enjoyed his hospitality and owe a debt of gratitude for the mathematics we have produced there. The ESI has been an important research center in Europe, and many of us recall Schwermer's heroic efforts to keep it open and move it under the umbrella of the University of Vienna. For a period it was unclear that the ESI would survive, and to a large part, its survival in its current form (which seems almost indistinguishable from its former form) is due to his efforts.

When we came together in June of 2016 to celebrate the mathematics of Joachim Schwermer, we did not forget his efforts in both history and administration. Knowing of them made his contributions to mathematics that much richer, both for the depth his historical interest provided and the appreciation of his work in light of his other concomitant administrative achievements. Not all of the speakers at the conference are represented in these proceedings, but we thank those that did contribute. We would also like to thank all of our colleagues that contributed to the volume by serving as anonymous referees for the contributions. We also thank Springer for seeing this volume through to fruition in spite our missing of various deadlines. And mostly, we thank Joachim himself for providing us with the opportunity to thank him for all he has done.

Columbus, USA James W. Cogdell
Bonn, Germany Günter Harder
Toronto, Canada Stephen Kudla
West Lafayette, USA Freydoon Shahidi

Contents

Globally Analytic p-adic Representations of the Pro–p Iwahori Subgroup of $GL(2)$ and Base Change, II: A Steinberg Tensor Product Theorem

Laurent Clozel

Abstract In this paper, which is a sequel to Clozel (Globally analytic p-adic representations of the pro-p Iwahori subgroup of GL(2) and base change, I: Iwasawa algebras and a base change map, to appear in Bull. Iran Math Soc, [4]), we exploit the base change map for globally analytic distributions constructed there, relating distributions on the pro-p Iwahori subgroup of $GL(2)$ over $\mathbb{Q}_p$ and those on the pro-p Iwahori subgroup of $GL(2, L)$ where L is an unramified extension of $\mathbb{Q}_p$. This is used to obtain a functor, the 'Steinberg tensor product', relating globally analytic p-adic representations of these two groups. We are led to extend the theory, sketched by Emerton (Locally analytic vectors in representations of locally p-adic analytic groups, [6]), of these globally analytic representations. In the last section we show that this functor exhibits, for principal series, Langlands' base change (at least for the restrictions of these representations to the pro-p Iwahori subgroups.)

Keywords 11R23 · 11F70 · 14G22

Introduction

This is part II of a paper, the first part of which is [4]. In that article we considered the Iwasawa algebra of the pro–p Iwahori subgroup of $GL(2, L)$ for an unramified extension L of degree r of $\mathbb{Q}_p$ and gave a presentation of it by generators and relations, imitating [3]. A natural base change map then appears that, however, is well–defined

---

Dedicated to Joachim schwermer on his 66th birthday

L. Clozel (✉)
Université de Paris-Sud, Orsay, France
e-mail: clozel@math.u-psud.fr

J. W. Cogdell et al. (eds.), *Cohomology of Arithmetic Groups*,
Springer Proceedings in Mathematics & Statistics 245,
https://doi.org/10.1007/978-3-319-95549-0_1

only for the *globally analytic distributions* on the groups, seen as rigid–analytic spaces.

In Sect. 1.1 of [4], we stated that this should be related to a construction of base change for representations of these groups, similar to Steinberg's tensor product theorem [13] for algebraic groups over finite fields.

In this paper we give such a construction, and we show that it is compatible with the (p–adic) Langlands correspondence in the case of the principal series for $GL(2)$.

By the previous remark, we have to limit ourselves to globally analytic representations. These representations have been considered by Emerton in his exhaustive introduction (unfortunately unpublished) to p–adic representation theory [6]. See in particular Sects. 3.3, 5.1 in his paper; the restriction of scalars, central to our constructions, is considered in his Sect. 2.3.

The first section of this paper contains preliminaries about rigid–analytic groups. The group associated to the pro–p Iwahori is (by Lazard's description) very simple, a product of copies of the rigid–analytic closed unit ball. In particular the algebras of functions we consider are all Tate algebras. We must, however, systematically consider restriction of scalars. Even for such simple spaces, this functor does not behave trivially, as was pointed out to me by Gaëtan Chenevier. See [1, 14]. However, this is the case for unramified extensions (Sect. 1.1.) It is then an easy matter to describe the natural functorial maps between Tate algebras (Defintion 1.4) and, dually, between (global) distribution algebras (Sect. 1.2). Nevertheless, the distribution algebra for a product is not a tensor product (even a completed tensor product.) This causes problems in the representation theory, which will be mentioned below; these "pathologies" are reviewed in the Appendix.

In Sect. 2 of this paper, we review the properties of these representations, adding some complements to Emerton's results. In particular, we study tensor products of representations (Theorem 2.3).

In contrast with the category of locally analytic representations, we can work here with (p–adic) Banach spaces rather than with Fréchet spaces, or spaces of compact type [6, 12]. Indeed, the spaces $\mathcal{A}$ and $\mathcal{D}$ of globally analytic functions (resp. distributions) are Banach spaces. The unfortunate consequence is that they are not reflexive. In particular we cannot systematically use duality as in the admissible Banach theory [11] or the locally analytic theory [12]. A related problem is that the spaces $\mathcal{D}$ of distributions are not Nœtherian. See the remarks in Sect. 2.3, as well as the Appendix.

In Sect. 3, we take up the construction of the base change functor, i.e., the Steinberg tensor product. Once the requisite property of the tensor product has been established in Sect. 2, this is totally natural. The main point is that a globally analytic representation will automatically extend, from the L–points of a rigid–analytic group G over L (we consider only very special groups, cf. Sect. 2) to the F–points for any finite extension F of L. Although this is not explicit in [6], it follows from his definitions. The construction is given in Sect. 3.2.

Of course this is meaningful only if it is compatible with the expected Langlands correspondence. The end of Sect. 3 is devoted to the proof of this fact for the principal series. We start with the pro–p Iwahori G of $GL(2, \mathbb{Q}_p)$. We must of course consider only the representations which have globally analytic vectors. This condition is specified in (3.4). In Proposition 3.6, we show that (under the same assumption as in [12]) the globally analytic representation of the pro–p Iwahori subgroup of $GL(2, \mathbb{Q}_p)$ is topologically irreducible.

In Sect. 3.4, we extend these results to the pro–p Iwahori subgroup of $GL(2, L)$ where $L/\mathbb{Q}_p$ is unramified. Here the similar irreducibility result is suggested by the work of of Orlik and Strauch [8]. We show that the formation of the Steinberg tensor product is compatible with Langlands functoriality (cf. Definition 3.2); the final result is Theorem 3.11 which exhibits base change in this context.

These results concern only the pro–p Iwahori subgroups, not the full groups $GL(2, \mathbb{Q}_p)$, $GL(2, L)$. In Sect. 3.5 we make some tentative remarks about the extension of base change to the full groups. Finally, the Appendix reviews some questions concerning the tensor products of distributions and the non-Noetherian character of these algebras.

While writing this paper I had the benefit of discussions or correspondence with Berthelot, Breuil, Chenevier, Raynaud and Schneider. Ariane Mézard corrected some mistakes in an early version. I am very grateful to them, and especially to Peter Schneider who explained to me the facts reviewed in the Appendix. I also thank the referee for useful comments.

1 Restriction of Scalars and Base Change Maps for Analytic Functions and Distributions on Rigid–Analytic Unit Ball Groups

1.1 .

We consider an unramified extension L/L_0, of degree r, of p–adic fields (finite extensions of $\mathbb{Q}_p$). Let $X = B^1/L$ be the closed unit ball over L, a rigid–analytic space whose affinoid algebra is

$$\mathcal{T}_L^1 = L < x > ,$$

the algebra of power series in x with coefficients tending to zero.

There is a functor of restriction of scalars, which to $X = X/L$ associates a rigid–analytic space $Y = \mathrm{Res}_{L/L_0} X/L_0$.

Lemma 1.1 (L/L$_0$ unramified)- *Y is isomorphic to the r–th power of B^1/L_0.*

This is a special case of the more general results of Bertapelle [1]. Let (e_i) be a basis of $\mathcal{O}_L$ over $\mathcal{O}_{L_0}$, and let B be an affinoid L_0–algebra. Consider $f \in Hom_L(L < x >, B \bigotimes_{L_0} L)$, thus

$$f(x) = \sum b_i e_i \qquad (b_i \in B) .$$

We want to define canonically

$$g \in Hom_{L_0}(L_0 < x_1, \dots x_r >, B), \text{ with } g(x_i) = b_i\,.$$

(Thus $Y \cong B^r/L_0$ is canonical, given the choice of the basis (e_i).) This is possible if, and only if, $\|b_i\|_{\text{Sup}} \le 1$ assuming $\|b\|_{\text{Sup}} \le 1$ where the sup norms are relative to the affinoid algebras B and $B \bigotimes_{L_0} L$.

Assume first that B is a finite field extension of L_0. Then $B \bigotimes_{L_0} L$ is a product of finite, unramified extensions B_α of L, and the integers $\mathcal{O}(B \bigotimes_{L_0} L) = \prod_\alpha \mathcal{O}(B_\alpha)$ satisfy, the extensions being unramified, $\mathcal{O}(B \bigotimes_{L_0} L) = \mathcal{O}_B \bigotimes_{\mathcal{O}_{L_0}} \mathcal{O}_L = \bigoplus \mathcal{O}_B e_i$. To say that $\|b\|_{\text{Sup}} \le 1$ for $b \in B \otimes L$ is to say that $b_\alpha \in \mathcal{O}(B_\alpha)$, or $b \in \mathcal{O}(B \bigotimes_{L_0} L)$.

This implies that $\|b_i\| \le 1$.

Now let B be a general affinoid algebra over L_0, and $B' = B \bigotimes_{L_0} L$. If $b = \sum b_i e_i \in B'$ $(b_i \in B)$, the computation in [1, p. 444] shows that

$$\|b\|_{\text{Sup}} = \operatorname*{Sup}_{y \in \text{Max}\, B} \operatorname*{Max}_{\substack{x \in \text{Max}\, B' \\ x|y}} \left\| \left(\sum b_i e_i \right)(x) \right\|_{\text{Sup}}.$$

However, y corresponds to a finite extension K_0 of L_0, x to a finite extension K of L contained in $L \bigotimes_{K_0} L_0$ so unramified over L. The previous result implies that $\| \sum b_i e_i(x)\|_{\text{Sup}} = \text{Sup}\|b_i(x)\|$. Thus $\|b_i\|_{\text{Sup}} \le 1$ if $\|b\|_{\le 1}$. We note that we have in fact:

Lemma 1.2 *The isomorphism $Y \overset{\cong}{\rightarrow} (B^1/L_0)^r$ is canonically defined by the choice of the basis (e_i).*

In fact, the function g (for instance if $B = K_0$ is a field extension of L_0) is defined by

$$g(x_1, \dots x_r) = f(\Sigma e_i x_i)\,. \tag{1.1}$$

$(|x_i| \le 1)$. The e_i being integral, it is easy to check that for $f \in \mathcal{T}_L^1$, the infinite series in the right is convergent.

Since restriction of scalars is compatible with direct products [1, Proposition 1.8] we have likewise

$$\text{Res}_{L/L_0}(B^1/L)^d = (B^1/L_0)^{dr}$$

the isomorphism being canonical once we have fixed the basis (e_i).

1.2 .

We now consider a rigid–analytic group G_L over L, isomorphic as a rigid–analytic space to $(B^1/L)^d$. (In particular $G_L(L)$ is dense in G_L for the Zariski topology.) Let $\mathcal{A}(G_L) \cong \mathcal{T}_d(L) = L < x_1, ..., x_d >$ [2, 5.1] be the space of analytic functions on G_L. The multiplication in G_L is associated to a (comultiplication) morphism

$$m^* : \mathcal{A}(G_L) \longrightarrow \mathcal{A}(G_L)\widehat{\otimes}\mathcal{A}(G_L)$$

(completed tensor product). In this case, the product is given, in co–ordinates, by integral functions, [2, Corollary 5.1.3.5] so

$$m^* : \mathcal{A}^0(G_L) \longrightarrow \mathcal{A}^0(G_L)\widehat{\otimes}\mathcal{A}^0(G_L)\,.$$

Then m^* defines naturally a map

$$\text{Res}\, m^* : \mathcal{A}^0(\text{Res}\, G_L) \longrightarrow \mathcal{A}^0(\text{Res}\, G_L)\widehat{\otimes}\mathcal{A}^0(\text{Res}\, G_L)\,,$$

Res being the restriction of scalars of G_L, a group over L_0.

Assume now that the group G_L is actually defined over L_0, i.e., is obtained by *extension* of scalars from L_0. Then $\mathcal{A}(G_L) = \mathcal{A}(G_{L_0}) \otimes L$. The map m^* is obtained by extension of scalars from

$$m_0^* : \mathcal{A}(G_{L_0}) \longrightarrow \mathcal{A}(G_{L_0})\widehat{\otimes}\mathcal{A}(G_{L_0})\,.$$

The integrality property for G_L and the property for G_{L_0} are equivalent.

Now the previous construction associates to $f \in \mathcal{A}(G_L)$ (with L–coefficients, i.e. in $\mathcal{T}_d(L)$) a function g in $\mathcal{A}(\text{Res}\, G_L) \otimes L$ (the function g defined by (1.1) will have coefficients in L). In particular we get a map $\mathcal{A}(G_{L_0}) \to \mathcal{A}(\text{Res}\, G_L) \otimes L$ by composition with the previous "tautological" map $\mathcal{A}(G_{L_0}) \to \mathcal{A}(G_L)$.

Definition 1.3 This map b_1: $\mathcal{A}(G_{L_0}) \to \mathcal{A}(\text{Res}\, G_L) \otimes L$ is the holomorphic base change map.

This map commutes with the comultiplications m_0^* and Res m^*: it is obvious if we consider m_0^* and m^*, and for m^* and Res m^* it follows from the formal properties of restriction of scalars. Furthermore b_1 sends $\mathcal{A}^0(G_{L_0})$ to $\mathcal{A}^0(\text{Res}\, G_L \otimes L)$.

The unramified extension L/L_0 is Galois. Thus the Galois group $\Sigma = \text{Gal}(L/L_0)$ acts naturally on G_L (by σ–linear automorphisms of the Tate algebra) and acts on Res G_L by L_0–automorphisms.

Definition 1.4 The map $b : \mathcal{A}(G_{L_0}) \to \mathcal{A}(\text{Res}\, G_L) \otimes L$ is defined by

$$b(f) = \prod_{\sigma \in \Sigma} b_1(f)^\sigma\,.$$

Since b_1 commutes with the comultiplication, the same is true for the product $\Pi\, b_1^\sigma$. We also note the following: Assume we *extend* scalars from L_0 to L for the L_0–groups. Then

$$\mathrm{Res}\; G_L \otimes_{L_0} L$$

is naturally isomorphic to $\prod\limits_\sigma G_L$. Indeed, if B is an L–algebra (in particular an affinoid algebra), $B \otimes_{L_0} L \cong \bigoplus\limits_\sigma B_\sigma$ where $B_\sigma = \{\beta \in B \otimes L : \lambda_1 \beta = \lambda_2^\sigma \beta$ where $\lambda \in L$ and λ_1 is the action of λ on $B \otimes L$ by the first component, λ_2 by the second component. Now, $\mathcal{A}$ denoting a Tate algebra:

$$\begin{aligned} &\mathrm{Hom}_L\; (\mathcal{A}(\mathrm{Res} G_L) \otimes_{L_0} L, B) \qquad (B/L) \\ &\quad = \mathrm{Hom}_{L_0}(\mathcal{A}(\mathrm{Res}\; G_L), B_0) \\ &\quad\ (B_0 \text{ being equal to } B/L_0) \\ &\quad = \mathrm{Hom}_L(\mathcal{A}(G_L), B_0 \otimes L) \\ &\quad = \bigoplus_\sigma \mathrm{Hom}_L(\mathcal{A}(G_L), B_\sigma)\,. \end{aligned}$$

In particular, after extension of scalars to L, $\mathcal{A}(\mathrm{Res}\; G_L) \otimes L \cong \widehat{\bigotimes\limits_\sigma} \mathcal{A}(G_L)$. The map b is then a tensor product: b_1 sends $\mathcal{A}(G_{L_0})$ to the functions on G_L that are L–holomorphic (given by power series $\Sigma a_m \underline{x}^m$, $\underline{x} = (x_1, \ldots, x_d)$ being the variable) while the component associated to σ sends a power series in $\mathcal{A}(G_0)$ to $\Sigma a_m \sigma(\underline{x})^m$.

We now agree to consider all Tate algebras as having coefficients in L, and we denote them by $\mathcal{A}_L$.

Summarizing, we now have the following result:

Proposition 1.5 *(i) There exists a natural map* b_1: $\mathcal{A}_L(G_{L_0}) \to \mathcal{A}_L(\mathrm{Res}\; G_L)$. *It commutes with the comultiplications.*

(ii) There exists a natural map $b = \prod\limits_{\sigma \in \Sigma} b_1^\sigma : \mathcal{A}_L(G_{L_0}) \to \mathcal{A}_L(\mathrm{Res}\; G_L)$. *It commutes with the comultiplications.*

(iii) In the isomorphism $\mathcal{A}_L(\mathrm{Res}\;\; G_L) \cong \widehat{\bigotimes\limits_\sigma} \mathcal{A}(G_L) \quad (\mathcal{A}(G_L) = \mathcal{A}_L(G_L))$, $b = \bigotimes\limits_\sigma b_1^\sigma$.

(iv) The maps b_1 *and* b *send the unit balls* $\mathcal{A}_L^0(G_{L_0})$ *to* $\mathcal{A}_L^0(\mathrm{Res}\; G_L)$. *(The norm being the sup norm of coefficients).*

We now consider the spaces of (L–valued) global distributions on G_{L_0} and $\mathrm{Res}\; G_L$. We denote them by $\mathcal{D}_L(G_{L_0})$, $\mathcal{D}_L(\mathrm{Res}\; G_L)$. These are the Banach spaces dual to the Banach spaces of analytic functions (for the sup norms). We obtain, dually, a map

$$b_1^* : \mathcal{D}_L(\mathrm{Res}\; G_L) \to \mathcal{D}_L(G_{L_0})$$

and also

$$b^* : \mathcal{D}_L(\text{Res } G_L) \to \mathcal{D}_L(G_{L_0}) .$$

These are homomorphisms, for the convolution of distributions. Using (iii) in the Proposition, we can write

$$\bigotimes_\sigma \mathcal{D}_L(G_L) \subset \mathcal{D}_L(\text{Res } G_L)$$

and b^* is then, on this subspace, given by

$$\bigotimes_\sigma T_\sigma \mapsto \mathop{*}_\sigma T_\sigma$$

(where $T_\sigma \in \mathcal{D}_L(G_{L_0})$ is σ–holomorphic). However, $\widehat{\bigotimes_\sigma} \mathcal{D}_L(G_L)$ is **not** equal to $\mathcal{D}_L(\text{Res } G_L)$. Since, after extension of scalars, our groups become products, this can be seen as follows.

We may forget for a moment the restriction of scalars, and consider two groups G, H isomorphic (as rigid–analytic spaces) to $(B^1)^d$, $(B^1)^{d'}$ over L. The spaces of analytic functions are $\mathcal{T}_d(L)$, $\mathcal{T}_{d'}(L)$, with the sup norm. The dual $\mathcal{D}_L(G)$ of the space of functions

$$f(x) = \sum_n a_n \, \underline{x}^n , \qquad a_n \to 0$$

($n \in \mathbb{N}^d$, $\underline{x} = (x_1, \dots x_d)$, $|x_i| \leq 1$) is the space of distributions

$$T = \sum_n c_n \, \delta_n \qquad (|c_n| \leq C)$$

where $\delta_n(f) = a_n = n ! \, \frac{\partial^n f}{\partial \underline{x}^n}(0)$. It is a Banach space, the norm being $\sup |c_n|$. The same description applies to a distribution S on H, and a distribution on $G \times H$. However, these Banach spaces are ℓ^∞ spaces in the indexes, and for three (countable) sets X, Y, $X \times Y$, it is not true that

$$\ell^\infty(X) \mathbin{\widehat{\otimes}} \ell^\infty(Y) = \ell^\infty(X \times Y) .$$

In order to form tensor products, we must consider the unit balls in $\mathcal{D}_L(G)$, $\mathcal{D}_L(H)$ (with their weak topology) and apply a result of Lazard. This was explained to me by Peter Schneider; we will return to it at the end.

2 Globally Analytic Representations

2.1 .

In this section we review some basic properties of globally analytic representations of a rigid–analytic group on a Banach space, mostly following Emerton [6]. We assume given L and G/L as in Sect. 1.2. We denote by $\mathcal{A} \cong \mathcal{T}_d(L)$ the space of globally analytic functions on G. We will often write G for $G(L)$ if this does not lead to confusion ; $G(L)$ is dense in G for the Zariski topology.

2.2 .

Let V be a Banach space over a field K containing L. We assume again K finite over $\mathbb{Q}_p$. If $g \mapsto \pi(g)$ is a continuous representation of G on V, we say that π (or V) is a globally analytic representation if the map

$$g \mapsto g \cdot v = \pi(g)v$$

is (globally) analytic on G for all $v \in V$. Thus, in coordinates $(x_1, \dots x_n)$:

$$g \cdot v = \sum_m \underline{x}^m v_m$$

where $v_m \in V$ and $\|v_m\| \to 0$.

Here $m = (m_1, \dots, m_d)$ and $\underline{x}^m = x_1^{m_1} \cdots x_d^{m_d}$, $m_i \in \mathbb{N}$. Such a representation is automatically differentiable. We will simply use the term “analytic” for “globally analytic”. Note that it is relative to the L–structure on V.

In this situation V is endowed with two natural norms, the given norm and

$$\|v\|_\omega = \operatorname{Sup}_m \|v_m\| .$$

The second norm is the norm of the map $g \mapsto gv$ in the Banach space $\mathcal{A}(G, V) = \mathcal{A}(G)\hat{\otimes}V$ (for this isomorphism cf. e.g. [6, Sect. 2.1]). The map $(V, \| \ \|_\omega) \to (V, \| \ \|)$ is bijective and obviously continuous. Since V, with the norm $\| \ \|_\omega$, is complete [6, 3.3.1, 3.3.3] it is bicontinuous by Banach’s isomorphism theorem [9, Corollary 8.7].

We recall the proof of the completeness of $(V, \| \ \|_\omega)$, as we will require similar arguments. Thus let $(v^\alpha)_\alpha$ be a Cauchy sequence in V for $\| \ \|_\omega$. For each α, $(v_m^\alpha)_{m\in M}$ is an element of $C^0(M, V)$ where $M = \mathbb{N}^d$ is the set of exponents. Since this space is complete, $(v_m^\alpha)_m \mapsto (v_m)$ in $C^0(M, V)$ for an element $(v_m) \in C^0(M, V)$. In particular $v^\alpha = v_0^\alpha \to v := v_0 \in V$. Now $gv = \lim_\alpha gv_0^\alpha$ $(g \in G)$, so $gv = \lim_\alpha (\sum_m \underline{x}^m v_m^\alpha)$.

Since

$$\Big\|\sum_m \underline{x}^m(v_m^\alpha - v_m)\Big\| \leq \operatorname{Sup}_m \Big\|v_m^\alpha - v_m\Big\| \longrightarrow 0 \qquad (\alpha \to \infty)$$

we see that $gv = \sum \underline{x}^m v_m$, which implies that $\|v - v^\alpha\|_\omega \to 0$.

Corollary 2.1 *There exists a constant C_V (depending on V) such that $\|v\|_\omega \leq C_V\|v\|$ $(v \in V)$.*

In particular $\|gv\| \leq C_V\|v\|$ $(g \in G)$.

In fact the original norm can be replaced by an equivalent norm such that $\|gv\| = \|v\|$: see Emerton [6, Sect. 6.5].

Lemma 2.2 *Let $(V, \|\ \|)$ be a continuous Banach representation of G, and let $W \subset V$ be a subspace comprised of analytic vectors. Assume that $\|w\|_\omega \leq C\|w\|$ $(C > 0)$ for $w \in W$. Then any vector of $\bar{W} \subset V$ (the closure for the topology of V) is analytic.*

Proof — Consider a sequence $(w^\alpha)_\alpha$ of vectors in W, such that $\|w^\alpha - v\| \to 0$ $(v \in V)$. Then w^α is a Cauchy sequence for $\|\ \|$, so also for $\|\ \|_\omega$. If

$$g \cdot w^\alpha = \sum_m \underline{x}^m w_m^\alpha\,,$$

the sequence $(w_m^\alpha)_{m\in M}$ has a limit (v_m) in $C^0(M, V)$. In particular $v_0 = v$. Again

$$g\,w^\alpha = \sum_m \underline{x}^m w_m^\alpha \longrightarrow gv \qquad (\alpha \to \infty)$$

and $\|\sum_m \underline{x}^m(w_m^\alpha - v_m)\| \leq \operatorname{Sup}_m\|w_m^\alpha - v_m\| \to 0$ $(\alpha \to \infty)$ which implies that $gv = \sum_m \underline{x}^m v_m$.

Consider now two rigid analytic groups G, H verifying our assumptions. Let V, W be analytic representation of G, H on Banach spaces. We assume the norms invariant, using Emerton's result. Then $G \times H$ acts on the algebraic tensor product $V \otimes W$. By [9, Proposition 2.1.7.5] this action extends to $V\hat{\otimes}W$, with $\|(g, h)u\| = \|u\|$ $(u \in V\hat{\otimes}W)$.

Now $V \otimes W$ is dense in $V\hat{\otimes}W$, and is comprised of analytic vectors : if $v \in V$, $w \in W$ and

$$gv = \sum_m \underline{x}^m v_m\,, \quad hw = \sum_p \underline{y}^p w_p$$

$(g \in G, h \in H)$ then

$$(g, h)(v \otimes w) = \sum_{m,p} \underline{x}^m \underline{y}^p v_m \otimes w_p\,.$$

Since $\|v_m \otimes w_p\| = \|v_m\| \, \|w_p\|$ (Schneider [9, Proposition 17.4]), this yields an analytic expansion.

Now endow $V \otimes W$ with its analytic norm $\| \ \|_\omega$, for the action of $G \times H$. We have

$$\begin{aligned}\|v \otimes w\|_\omega &= \operatorname*{Max}_{m,p} \|v_m \otimes w_p\| \\ &= \operatorname{Max}\|v_m\| \operatorname{Max}\|w_p\| \\ &= \|v\|_\omega \, \|w\|_\omega .\end{aligned}$$

Now consider any vector $u \in V \otimes W$. The tensor product norm is defined by

$$\|u\| = \inf \operatorname*{Max}_i \|v_i\| \, \|w_i\|$$

over the decompositions $u = \sum v_i \otimes w_i$. Choose $\varepsilon > 0$, and a decomposition such that

$$\begin{aligned}\|u\| &\geq \operatorname{Max}\|v_i\| \, \|w_i\| - \varepsilon. \\ \text{Then } \|u\|_\omega &\leq \operatorname*{Max}_i \|v_i \otimes w_i\|_\omega \\ &\leq C_V C_W \operatorname*{Max}_i \|v_i \otimes w_i\| \leq C_V C_W (\|u\| + \varepsilon).\end{aligned}$$

Thus $\|u\|_\omega \leq C_V C_W \|u\|$, and $V \otimes W \subset V \hat{\otimes} W$ verifies the assumption of the Lemma. This implies:

Theorem 2.3 *If V, W are (globally) analytic representations of G, H, $V \hat{\otimes} W$ is a globally analytic representation of $G \times H$.*

(For a similar result, but for locally analytic representations, see Emerton [6, 3.6.18]).

We also note the following property. Let $\mathfrak{g}$ be the Lie algebra of G (over $\mathbb{Q}_p$).

Proposition 2.4 *If V is a globally analytic representation of G and $W \subset V$ is a closed subspace, W is G–invariant if and only if W is invariant by the enveloping algebra $U(\mathfrak{g})$.*

(Recall from [12] that the Lie algebra, or $U(\mathfrak{g})$, acts on a space of analytic vectors). If W is G–invariant, it contains the derivatives $Xw = \lim_{t \to 0} \left(\frac{e^{tX}-1}{t}\right) w$ of its vectors by elements $X \in \mathfrak{g}$. Conversely, if

$$gw = \sum_m \underline{x}^m v_m$$

then $v_m = \frac{1}{m!} \frac{d^m}{d\underline{x}^m}\Big|_0 (gv)$, the derivative being computed with respect to the variables $\underline{x}$. However the enveloping algebra (acting via $uf = (u * f)(0)$ for $u \in U(\mathfrak{g})$, f an analytic function on G) also spans the space of derivatives at 0. If W is invariant by $U(\mathfrak{g})$, the coefficient v_m belong to W and therefore $gw \in W$.

By contrast with the case of complex unitary representations, we do not know if $V\hat{\otimes}W$ is (topologically) irreducible if V, W are topologically irreducible. The only, obvious, property is that $V\hat{\otimes}W$ is topologically cyclic (i.e., the closed subspace generated by a suitable vector is equal to $V\hat{\otimes}W$) if V and W are - in particular if they are irreducible. Indeed, if v spans V and w spans W, $v \otimes w$ spans $V\hat{\otimes}W$.

2.3 .

Finally, we also recall from Emerton's paper that there is a duality theory for globally analytic representations, similar to the duality for locally analytic (or Banach admissible) representations. If V is a globally analytic representation, the distribution algebra $\mathcal{D}_K(G)$ acts on the dual V'. There is a duality between closed submodules of $(\mathcal{A}(G) \otimes K)^n$ and quotients of $\mathcal{D}_K(G)^n$. See [6, Theorem 5.1.15]. We will not be able to use this, however. There are two obstacles: the algebra $\mathcal{D}_K(G)$ is not Nœtherian; furthermore, as noticed at the end of Sect. 1, it does not behave well with respect to the product of groups.

Let us define an *admissible globally analytic* representation as a globally analytic Banach representation that is a closed submodule of $(\mathcal{A}(G) \otimes K)^n$. Recall also from [6, 10] that there is a category of admissible (continuous) Banach representations and of admissible locally analytic Banach representations on spaces of compact type [11]. In general, an admissible globally analytic representation is not an admissible locally analytic representation (an infinite–dimensional Banach space is not of compact type) and is not an admissible Banach representation. Indeed, if E is such a representation and E^0 is its unit ball (for a given G–invariant norm), and if ϖ is a uniformising parameter of K, it is known that $E^0/\varpi E^0 = \bar{E}$ is a smooth admissible representation of G over the finite residue field k of K [10], [6, 6.5.7]. However, $\mathcal{A}(G)$ does not have that property.

For instance, if G is the additive unit ball, so $V = \mathcal{A}(G) \otimes K = \mathcal{T}_1(K)$, its unit ball is translation–invariant and the subgroup $\varpi_L \mathcal{O}_L$ of $G(L) = \mathcal{O}_L$ acts trivially on $\bar{V} = k[x]$, so this representation is not admissible.

Assume however that E is an admissible Banach representation. Then E is a closed subspace of $\mathcal{C}(G, K)^n$ for some n [10],[6, Sect. 6]. Let $V = E^{an}$ be the space of globally analytic vectors. Emerton's results (see the proof recalled before Corollary 2.1) show that V is complete for the norm $\| \ \|_\omega$. It is an analytic Banach representation [6, Corollary 3.3.6].

Assume $V = E^{an}$ is dense in E. Since $\mathcal{C}(G, K)^{an}$ is equal to $\mathcal{A}(G) \otimes K$, V is sent to $(\mathcal{A}(G) \otimes K)^n$. Let $j = (j_i)_{i=1,\dots,n}$ be the closed embedding $E \to \mathcal{C}(G, K)^n$. By Banach's theorem $\|v\| \geq C \operatorname{Sup}_i \|j_i(v)\|$ for $v \in E$, C being a > 0 constant. This implies, the embedding being equivariant, that $\|v\|_\omega \geq C \operatorname{Sup}_i \|j_i(v)\|_\omega$ for $v \in V$. The canonical norm $\| \ \|_\omega$ on $\mathcal{A}(G)$ is the usual norm – the sup norm on coefficients. (See Proposition 2.7 below.) Thus V is a closed subspace of $(\mathcal{A}(G) \otimes K)^n$. Conversely, if V is such a subspace, we can consider its closure $E \subset \mathcal{C}(G, K)^n$. It is an admissible Banach representation in which V is dense. Clearly $V \subset E^{an}$, but it does not seem to follow that V is equal to E^{an}. To summarise:

Proposition 2.5 *Any admissible globally analytic representation is a dense subspace of an admissible Banach representation. If E is an admissible Banach representation, E^{an} is an admissible globally analytic representation.*

The admissible analytic representations have further interesting properties. Recall that in general, if V is an analytic representation, there is an action of $\mathcal{D}(G) \otimes K$ on the continuous dual V' [6, 5.1.8]. If V is admissible, we can say more.

Assume $T \in \mathcal{D}(G)$ (we forget the extension of scalars for simplicity of notation.) If $f \in \mathcal{A}(G)$, we can define a function $T * f$ by

$$T * f(x) = \int T(z) f(z^{-1}x) dz \tag{2.1}$$

in functional notation, i.e. T applied to the function of z, $z \mapsto z^{-1}x$. Since $f(z^{-1}x)$ is in the Tate algebra of $G \times G$, this is well–defined and, moreover, defines a function in $\mathcal{A}(G)$. Thus $\mathcal{D}(G)$ acts by convolution on $\mathcal{A}(G)$, and this is compatible with the convolution product.

Assume now that $V \subset \mathcal{A}(G)$ is a closed invariant subspace. Then V is invariant by the differential operators $\frac{1}{m!}\frac{d^m}{dx^m}$. If $f \in \mathcal{A}(G)$ and

$$T = \sum_m c_m \frac{1}{m!}\frac{d^m}{dx^m}\Big|_0$$

$\in \mathcal{D}(G)$ (with c_m bounded), $T * f$ is the limit in $\mathcal{A}(G)$ of $T_X f$,

$$T_X f = \sum_{|m| \leq X} c_m \frac{1}{m!}\frac{d^m}{dx^m} f$$

as can be seen by expanding the function $f(z^{-1}x)$ in (2.1) in the Tate algebra of $G \times G$. Therefore V is invariant by $\mathcal{D}(G)$. The same extends to an embedding $V \to \mathcal{A}(G)^n$. Thus:

Proposition 2.6 *If V is an admissible globally analytic representation, the distribution algebra $\mathcal{D}(G)$ acts naturally on V. The action is continuous if $\mathcal{D}(G)$ is equipped with its weak dual topology.*

The continuity follows from the previous argument. It implies in particular that the action is intrinsic.

We recall that for locally analytic representations this construction is due to Schneider and Teitelbaum [11, Sect. 3]. However their proof relies on an isomorphism

$$\mathcal{L}(\mathcal{D}_{loc}(G), V) \cong \mathcal{A}_{loc}(G, V)$$

([11, Theorem 2.2]; here $\mathcal{A}_{loc}(G)$ is the space of locally analytic functions and $\mathcal{D}_{loc}(G)$ its dual space, and V is a suitable topological space. The analogue is not true in our context. Indeed

$$\mathcal{A}(G, V) = \mathcal{A}(G)\hat{\otimes}V \cong C_0(M, V)$$

where M is our set of exponents, while $\mathcal{D}(G) \cong \ell^\infty(M, L)$. Since $\ell^\infty(\mathbb{N})'$ is distinct from $C_0(\mathbb{N})$, we see a fortiori that these spaces are not isomorphic.

Because the comultiplication is given by integral series, we also have:

Proposition 2.7 *Consider the admissible representation $V = \mathcal{A}(G) \otimes K$ of G, with its usual norm (sup of the coefficients.) Then*

(i) V is a unitary representation.

(ii) On V, $\| \|_\omega = \| \|$.

(iii) For $T \in \mathcal{D}^0 = (V')^0$ and $f \in \mathcal{A}^0$, the function $g \mapsto < T, gf >$ is in $\mathcal{A}^0(G, K)$.

These facts easily follow from the property of the coproduct. Since an admissible analytic representation embeds as a closed subspace of $(\mathcal{A}(G) \otimes K)^n$, it follows that:

Corollary 2.8 *Properties (i)–(iii) of Proposition 2.7 are true for an admissible analytic representation.*

3 Unramified Base Change : The Pro–p Iwahori for $GL(2)$

3.1 .

The content of this section is twofold: we first describe a functor producing, for an unramified extension L/L_0 and a globally analytic representation of $G(L_0)$ (the assumptions are those of Sect. 1), a representation of $G(L)$ of the same kind. In fact, as in Sect. 1 for distribution algebras, there are two such functors. The first produces a "holomorphic" extension to $G(L)$. The second ("full base change") is the one that should be related to Langlands functoriality. It is the "Steinberg tensor product" described at the end of Sect. 1.1 of [4].

We then show that for $GL(2)$ and principal series representations of the pro–p Iwahori subgroup, this is compatible with base change for the principal series described by Schneider–Teitelbaum and Orlik–Strauch [8, 11]. In particular, we show that certain globally analytic representations are irreducible.

3.2 .

Let L/L_0 denote an unramified extension of p–adic fields and G a rigid analytic group over L_0 verifying the conditions of Sect. 1. We fix a p–adic field K (finite over $\mathbb{Q}_p$) and an injection $\iota : L \subset K$. If $\sigma \in \mathrm{Gal}(L/L_0)$, we then have the injection $\iota \circ \sigma : L \to K$.

Let V denote a (globally) analytic representation of $G(L_0)$ on a K–Banach space.

Proposition 3.1 *(i) V extends naturally to an analytic representation of $G(L)$.*
(ii) If V is admissible, the corresponding representation of $G(L)$ is admissible.

The group $G(L_0)$ acts on V by

$$g \cdot v = \sum_m \underline{x}^m v_m \tag{3.1}$$

with the notations of Sect. 2, and $v_m \to 0$. If $g \in G(L)$, the same expansion (with $\underline{x} = (x_1, \dots, x_d) \in \mathcal{O}_L^d$) is convergent, and we define $g \cdot v$ by (3.1).We must check that this defines a group representation of $G(L)$. The map

$$\begin{array}{c} (g, h) \mapsto gh.v = F(g, h) \\ G(L) \times G(L) \longrightarrow V \end{array}$$

is the composition of the map $(g, h) \mapsto gh$, analytic in the two variables, and of an analytic map $G(L) \to V$. It is analytic in the two variables.

On the other hand we have for $g, h \in G(L_0)$:

$$g(hv) = g\,F(1, h)\,. \tag{3.2}$$

Write $\underline{y}$ for the coordinates of g and $\underline{x}$ for the co-ordinates of h. Then

$$F(1, h) = hv = \sum_m \underline{x}^m v_m\,.$$

On the other hand, for any v_m,

$$g v_m = \sum_p \underline{y}^p v_{m,p}$$

with $v_{m,p} \to 0$ ($|p| \to \infty$).

Since $\|v_m\| \le C_V \|v\|$ for any m and $v \in V$,
$gF(1, h) = \sum_{m,p} \underline{x}^m \underline{y}^p v_{mp}$,
the double sum being convergent: if $|m| + |p| \to \infty$, either $m \to \infty$ and $\|v_{m,p}\| \le C_V \|v_m\| \to 0$ or m is bounded and, again, $\|v_{m,p}\| \to 0$. Thus the function $gF(1, h)$: $G(L_0) \times G(L_0) \to V$ is a Tate series (with coefficients in V) in the two variables, and

extends to an analytic function $G(L) \times G(L) \to V$. Since $F(g,h) = gF(1,h)$ for $g \in G(L_0)$, these two analytic functions coincide: indeed $G(L_0)$ is Zariski–dense in G, and the result follows (for instance, evaluate the two functions against a continuous linear form $\lambda \in V'$).

This proves (i). Assume now V is a closed subspace of $\mathcal{A}(G_{L_0}, K)$. Note that the same argument applies to $\mathcal{A}(G_{L_0}, K)$, an analytic representation of $G(L_0)$. But $\mathcal{A}(G_{L_0}, K) = \mathcal{A}(G_L, K)$ and now V, as a representation of $G(L)$, is a closed subspace of $\mathcal{A}(G_L, K)$.

We will call the extension of Proposition 3.1 the holomorphic base change of V. Its coefficients are L–analytic (for the given embedding $L \to K$): it is L–analytic in the sense of Emerton [6].

If $\sigma \in \mathrm{Gal}(L/L_0)$ we write V^σ for the representation of $G(L)$ associated to $\iota \circ \sigma$. It is L–analytic for $\iota \circ \sigma$.

Definition 3.2 The full base change of V is the globally analytic representation of $\mathrm{Res}_{L/L_0} G(L)$ on $W = \widehat{\bigotimes_\sigma} V^\sigma$.

It is analytic for $\mathrm{Res}_{L/L_0} G(L)$ by the results of Sect. 1. (Note that L/L_0 being unramified, $\mathrm{Res}_{L/L_0} G(L)$ is again a group of the same type.) The fact that the completed tensor product is globally analytic follows from Sect. 2.

When V is the restriction to $G(L)$ — the L–points of a rigid–analytic group deduced from a suitable integral structure on a reductive group $\mathcal{G}/L$ — of a representation (still denoted by V) of $\mathcal{G}(L)$, we conjecture that this will be compatible, in some sense, with Langlands base change (still conjectural) for p–adic Banach representations of $\mathcal{G}(L)$. Of course the relation between admissible Banach representations and globally analytic Banach representations (for $G(L)$) is not one–to–one, cf. Proposition 2.5. It would be interesting to determine which Banach spaces E give rise to a given V, for instance if V is irreducible. Furthermore, even in the case of irreducible principal series V for $\mathcal{G}(L)$, the restriction to $G(L)$ is not irreducible. The full base change of Definition 3.2 then describes only certain of its submodules. This will be clear for the principal series.

3.3 .

We now consider the case of the principal series for $GL(2)$. For simplicity we assume $L_0 = \mathbb{Q}_p$. We assume $p > 2$. Let G be the rigid-analytic group over $\mathbb{Q}_p$ defined by the pro-p Iwahori subgroup of $GL(2, \mathbb{Q}_p)$. It is checked in [4] that G is naturally a rigid-analytic group, with comultiplication given by integral series.[1] As a space G is a product of 1–dimensional balls. Thus $G(\mathbb{Q}_p) = \left\{ g \in GL(2, \mathbb{Z}_p) : g \equiv \begin{pmatrix} 1 & 0 \\ * & 1 \end{pmatrix} [p] \right\}$.

[1] In that paper p is assumed greater than 5, in order to apply Lazard's theory. However this particular computation only requires $p > 2$.

(Recall that we write G for $G(\mathbb{Q}_p)$ if the meaning is clear.) The principal series is then described by Schneider and Teitelbaum [11]. (They define the Iwahori subgroup by matrices that are lower triangular $\mod p$ while in [4] we consider upper–triangular matrices. We have followed their choice.) We fix K (and an embedding $L \hookrightarrow K$) as in Sect. 3.2.

Let $B = \left\{ g \in GL(2, \mathbb{Z}_p) : g \equiv \begin{pmatrix} * & 0 \\ * & * \end{pmatrix} [p] \right\}$, so our group $G = G(\mathbb{Q}_p)$ is a subgroup of B. Let $P_0 \supset T_0$ be the set of upper triangular (resp. diagonal) matrices in B. Let $\chi : T_0 \to K^\times$ be a locally analytic character, such that

$$\chi \begin{pmatrix} t^{-1} & \\ & t \end{pmatrix} = \exp(c(\chi) \log(t))$$

for $t \in T_0 = (\mathbb{Z}_p^\times)^2$ when t is sufficiently close to 1. Thus $c(\chi) \in K$.

We consider first, as they do, the locally analytic induced representation of B

$$J_{loc} = \mathrm{ind}_{P_0}^B(\chi) = \{ f \in \mathcal{A}_{loc}(B, K) : f(gb) = \chi(b^{-1}) f(g) \}$$

($b \in P_0$), where χ is naturally extended to P_0. We have

$$B = U P_0 , \quad U = \left\{ \begin{pmatrix} 1 & \\ z & 1 \end{pmatrix} , \quad z \in \mathbb{Z}_p \right\} . \tag{3.3}$$

Note that since χ is fixed, the restriction of the functions of J_{loc} to $G \subset B$ is injective. With

$$Q_0 = P_0 \cap G = \left\{ \begin{pmatrix} s & x \\ & t \end{pmatrix} : s, t \equiv 1 , \ x \equiv 0 \, [p] \right\}$$

we see that the space of J_{loc} is

$$I_{loc} = \{ f \in \mathcal{A}_{loc}(G, K) : f(gb) = \chi(b^{-1}) f(g) \}$$

($b \in Q_0$). With (3.3) replaced by $G = U Q_0$, we see that $I_{loc} \cong \mathcal{A}_{loc}(\mathbb{Z}_p, K)$ where $\mathbb{Z}_p$ is seen as the rigid analytic (additive) group $B^1(\mathbb{Z}_p)$. The group G acts by left translations, thus by $f(g) \mapsto f(h^{-1}g)$. We now have [11, Lemma 5.2]:

Lemma 3.3 *For $y \in \mathbb{Z}_p$, $x \in p\mathbb{Z}_p$, $s, t \equiv 1 \, [p]$:*

$$(i) \quad \begin{pmatrix} 1 & \\ y & 1 \end{pmatrix} f(z) = f(z - y)$$

$$(ii) \quad \begin{pmatrix} s & \\ & t \end{pmatrix} f(z) = f(st^{-1}z)\chi(s, t)$$

$$(iii) \quad \begin{pmatrix} 1 & x \\ & 1 \end{pmatrix} f(z) = f\left(\tfrac{z}{1-xz}\right)\chi((1 - xz)^{-1}, \ 1 - xz) .$$

We now seek conditions such that $\mathcal{A}(B^1, K)$, where B^1 is the unit ball in the z–variable, is a globally analytic representation. We simply denote this space by $\mathcal{A}$; we will similarly drop the subscript K in this section.

Lemma 3.4 *It suffices to check analyticity separately for the* 1*–parameter (rigid–analytic) subgroups of which G is the product.*

Changing notation, denote by x, y, u, w the variables in $\mathbb{Z}_p$ deduced from the natural variables. (So x is $p^{-1}x'$ where x' is the coordinate in (iii)). Assume for instance $yf = \sum_o^\infty y^m f_m$, $\|f_m\| \to 0$ for any f, where $\| \ \|$ is the natural norm on $\mathcal{A}$. Then, with obvious notation:

$$xyf = x \sum_o^\infty y^m f_m$$
$$= \sum_{m=0}^\infty y^m \sum_{p=0}^\infty x^p f_{m,p}$$

where, for each m, $f_{m,p} \to 0$ with p.

However, the norm on $\mathcal{A}$ is equivalent to the norm $\| \ \|_{\omega,x}$ deduced from the action of the (rigid–analytic) x–group. Thus we can assume that $\|f_{m,p}\| \le C\|f_m\|$. Then $\|f_{m,p}\| \to 0$ when $|m| + |p| \to \infty$. The same argument applies to any number of variables.

For $f \in \mathcal{A}$ and $z \in \mathbb{Z}_p$, $f \mapsto f(z)$ is a continuous linear form. For $s = t$, (ii) yields:

$$\begin{pmatrix} s & \\ & s \end{pmatrix} f(z) = f(z)\chi(s,s)\,.$$

If the action is analytic, we see that $\chi(s,s)$ must be an analytic function of s for $s \equiv 1\ [p]$. Now $\chi(s,t) = \chi(t,t)\chi(st^{-1}, 1)$. We may then consider

$$\begin{pmatrix} s & \\ & 1 \end{pmatrix} f(z) = f(sz)\chi(s,1)\,.$$

Taking $f = 1$, we see that $\chi(s, 1)$ must be analytic. Moreover, if

$$f(z) = \sum_{m\ge 0} a_m z^m$$

and $s = 1 + pu$, then

$$f(sz) = \sum_n (pu)^n \sum_{m\ge n} a_m \binom{m}{n} z^m = \sum_n u^n f_n(z)$$

yields an analytic expansion, in $\mathcal{A}$, of $f(sz)$.

The condition on the analyticity of $\chi(s,t)$ is as follows. Write $\chi = (\alpha, \beta)$ with

$$\alpha(1+pu) = e^{a\log(1+pu)}, \ \beta(1+pu) = e^{b\log(1+pu)}$$

$(a, b \in K)$ for $u \in \mathbb{Z}_p$ close to 0. The exponential is analytic (in K) in the domain $v_p(z) > \frac{e}{p-1}$ where $e = e(K)$; v_p is always the normalized valuation, $v_p(p) = 1$. Now

$$v_p(a\ \log(1+pu)) = v_p(a) + 1 + v_p(u)$$

since $p > 2$, so we must have $v_p(a) + 1 > \frac{e}{p-1}$, i.e.:

$$\begin{aligned} v_p(a), v_p(b) &> \frac{e}{p-1} - 1 \\ &= \frac{-p}{p-1} \quad \text{if } K \text{ is unramified.} \end{aligned} \tag{3.4}$$

Henceforth we assume that α, β verify these conditions (" α, β are analytic" for short.) Now the action of $\begin{pmatrix} s & \\ & t \end{pmatrix}$ is a twist of the action associated to $\chi = 1$ by an analytic character. Thus (i), (ii) yield analytic actions.

Now $\alpha(1+v)$ belongs to the Tate algebra on the ball $|v| \le p^{-1}$, so $\alpha(1-xz)$ belongs to the Tate algebra of two variables on $B^1 = B(1) \times B(p^{-1})$. In particular it has a convergent expression

$$\sum_{m\ge 0} x^m\ \alpha_m(z)\,, \qquad \alpha_m \in \mathcal{A}$$

on this domain, convergent (for $|x| \le p^{-1}$) as a series in $\mathcal{A}$. Now for $|v| < 1$

$$(1-v)^{-m} = \sum_{q=0}^{\infty} \binom{m+q-1}{q} v^q$$

so, for $f = \sum_0^\infty a_m z^m$,

$$\begin{aligned} f\Big(\frac{z}{1-xz}\Big) &= \sum_0^\infty a_m z^m \sum_{q=0}^\infty \binom{m+q-1}{q} x^q z^q \\ &= \sum_{q=0}^\infty x^q \sum_{m=0}^\infty \binom{m+q-1}{q} a_m z^{m+q}\,. \end{aligned}$$

We have to remember that $x \in p\mathbb{Z}_p$, so the analyticity of the action (iii) must be seen in the variable $\xi = \frac{x}{p} \in B^1$. The expression now becomes

$$\sum_{q=0}^{\infty} \xi^q p^q \sum_{m=0}^{\infty} \binom{m+q-1}{q} a_m z^{m+q}$$
$$= \sum_{q=0}^{\infty} \xi^q f_q(z)$$

with obviously $\|f_q\|_{\mathcal{A}} \le p^{-q}\|f\|_{\mathcal{A}}$. Thus the action (iii) is analytic.

Let $\mathcal{A}_{\text{loc}} \supset \mathcal{A}$ be the space of locally analytic functions. The representation of G on $\mathcal{A}_{\text{loc}}$ is studied by Schneider and Teitelbaum in [11]. Let $\mathcal{D}_{\text{loc}}$ be the space of distributions on $U = \left\{ \begin{pmatrix} 1 & \\ * & 1 \end{pmatrix} \right\}$ in their sense, i.e. the topological dual of $\mathcal{A}_{\text{loc}}$. We recall that $\mathcal{A}_{\text{loc}} = \varinjlim_n \mathcal{A}(n)$ where $\mathcal{A}(n)$ is the space of functions globally analytic on each ball of radius p^{-n}. Thus $\mathcal{A} = \mathcal{A}(0)$. The transition maps are injective and compact, with dense image. Dually we have $\mathcal{D}_{\text{loc}} = \varprojlim \mathcal{D}(n)$. This is a projective limit of Banach spaces, the projection maps being compact with dense image; $\mathcal{D} = \mathcal{D}(0)$.

Similarly for the rigid–analytic group G, we have $\mathcal{A}_{\text{loc}}(G)$, $\mathcal{D}_{\text{loc}}(G)$ with similar properties. Consider the maps

$$r : \mathcal{D}_{\text{loc}} \longrightarrow \mathcal{D} = \mathcal{A}' \qquad \text{(continuous dual of } \mathcal{A})$$
$$R : \mathcal{D}_{\text{loc}}(G) \longrightarrow \mathcal{D}(G)\,.$$

We have natural actions of $\mathcal{D}_{\text{loc}}(G)$ on $\mathcal{D}_{\text{loc}}$ and of $\mathcal{D}(G)$ on $\mathcal{D}$ (see 2.3), which we denote by the convolution sign.

Lemma 3.5 *For $T \in \mathcal{D}_{\text{loc}}(G)$, $F \in \mathcal{D}_{\text{loc}}$, $r(T * F) = R(T) * r(F)$.*

The maps r and R are continuous. The map $(t, f) \mapsto t * f$ ($t \in \mathcal{D}(G)$, $f \in \mathcal{D}$) is continuous in t; similarly $(T, F) \mapsto T * F$ ($T \in \mathcal{D}_{\text{loc}}(G)$, $F \in \mathcal{D}_{\text{loc}}$) is continuous [11]. Furthermore the finite group algebra $K[G]$ is dense in $\mathcal{D}_{\text{loc}}(G)$. It suffices then to check the formula for a single Dirac measure $T = \delta_g$, where it is obvious. The results of Schneider–Teitelbaum now easily imply:

Proposition 3.6 *If $b - a \notin \mathbb{N} = \{0, 1, ...\}$, the globally analytic representation of G on $\mathcal{A}$ is topologically cocyclic and admissible.*

Here 'topologically cocyclic' means that its dual is topologically cyclic.

Consider the G–map $r : \mathcal{D}_{\text{loc}} \to \mathcal{D}$, and let $X \subset \mathcal{D}$ be a closed submodule (for the action of G). Then $r^{-1}X \subset \mathcal{D}_{\text{loc}}$ is a closed submodule, invariant by $\mathcal{D}_{\text{loc}}(G)$. In [11], Schneider and Teitelbaum consider in fact the action of $\mathcal{D}_{\text{loc}}(B)$. By [11, Theorem 5.4], $\mathcal{D}_{\text{loc}}$ is (algebraically) irreducible under $\mathcal{D}_{\text{loc}}(B)$. However a glance at their proof shows that it remains irreducible under $\mathcal{D}_{\text{loc}}(G)$: the proof involves only the action of the Lie algebra, except for the argument at the bottom of p. 460. Here it must be checked that a submodule V of $\mathcal{D}_{\text{loc}}$, under the action of B, is generated by distributions, the Amice transform of which has only zeroes in the set of elements of the form $\zeta - 1$ where ζ is a root of unity (in $\mathbb{C}_p$) of p^n-order. The argument relies on the action of T_0; however it is easily seen that the action of the group of elements congruent to 1 mod p (contained in G) is sufficient.

Thus $r^{-1}X$ is null or equal to $\mathcal{D}_{\text{loc}}$. Since X is closed and $r : \mathcal{D}_{\text{loc}} \to \mathcal{D}$ has dense image, we deduce that X is equal to $\mathcal{D}$ in the second case. If $r^{-1}X$ is null, $X \cap Im(\mathcal{D}_{loc}) = \{0\}$. Since $Im(\mathcal{D}_{loc})$ is dense, choosing a suitable vector in $\mathcal{D}$ implies that $\mathcal{D}$ is cyclic. However $\mathcal{D}$ is the Banach dual of $\mathcal{A}$. Thus $\mathcal{A}$ is cocyclic.

Finally, the representation on $\mathcal{A}$ is admissible: indeed, $\mathcal{A}$ is the subspace of $\mathcal{A}(G)$ defined by the conditions $f(gb) \equiv \chi(b^{-1})f(g)$ (f is then analytic on G since χ is so) and this is a closed subspace.

This suggests the stronger result:

Theorem 3.7 *If $b - a \notin \mathbb{N}$, the globally analytic representation of G on $\mathcal{A}$ is topologically irreducible (and admissible).*

Assume that $\mathcal{X} \subset \mathcal{A}$ is a closed, G-invariant subspace. Then it is stable by the action of the enveloping algebra of $\mathfrak{g}$. Let X, Y be the usual infinitesimal generators of the upper and lower unipotent subgroups, and let H be an infinitesimal generator of the diagonal subgroup T with entries $(s, 1)$. Thus T is identified with $\{s \in \mathbb{Z}_p^\times, s \equiv 1[p]\}$. We deduce from Lemma 3.3 the action of these elements on a function $f \in \mathcal{A}$. For

$$f(z) = \sum_0^\infty a_n z^n$$

the image of f by $s \in T$ is

$$\alpha(s)f(sz) = \sum_0^\infty \alpha(s)s^n a_n z^n.$$

Therefore, since $\frac{d}{dt}|_0(s^n) = nH$ for $s = \exp(tH)$, $H \in \mathbb{Z}_p$ (for instance $H = 1$),

$$\sum_0^\infty n a_n z^n \in \mathcal{X}. \tag{3.5}$$

Moreover,

$$\begin{aligned} -Yf &= f' = \sum_0^\infty n a_n z^{n-1} \\ Xf &= -(d/dx)|_0(f\left(\tfrac{z}{1+xz}\right)\alpha((1+xz)^{-1})\beta(1+xz)) \\ &= z^2 f'(z) + (a-b)zf(z). \end{aligned}$$

Let τ_N be the natural truncation $\mathcal{A} \to K[z]_N (N \geq 0)$ where $K[z]_N$ is the space of polynomials of degree $\leq N$. Then τ_N is equivariant for the action of T. On the finite-dimensional space $K[z]_N$, the operators given by $s \in T$ are simultaneously diagonalizable, in the basis $\{z^n\}$. The associated characters of T are linearly independent. Thus $\tau_N(\mathcal{X})$ is a direct sum of the monomials for the exponents $m \in M_N \subset [0, N]$. If $N \leq N'$, the surjectivity of $K[z]_{N'} \to K[z]_N$ implies that $M_N \subset M_{N'}$ and in fact $M_N = M_{N'} \cap [0, N]$

Taking M equal to the union of the M_N for all N, we see that there exists $M \subset \mathbb{N}$ such that

$$(a) \quad f \in \mathcal{X} \Rightarrow a_n = 0 \quad (n \notin M)$$
$$(b) \quad n \in M \Rightarrow z^n \in \tau_N(\mathcal{X}) \ for\ any\ N \geq n$$

In particular, if $n \in M$, there exists

$$f = z^n + \sum_{m>N} a_m z^m \in \mathcal{X} \tag{3.6}$$

for any $N \geq n$. Furthermore, using the action of Y, we see that $n - 1 \in M$ if $n \in M$, so $M = [0, N](N \geq 0)$ or $M = \mathbb{N}$.

Lemma 3.8 *The constant function z^0 belongs to $\mathcal{X}$.*

The proof will rely on an analogue of the operator of 'ordinary projection' in Iwasawa theory. Start with

$$f = a_0 + \sum_{m>0} a_m z^m \in \mathcal{X} \quad (a_0 \neq 0).$$

Then the function deduced from $H^{p-1} f$,

$$\sum_{m>0} m^{p-1} a_m z^m \in \mathcal{X},$$

so

$$Af := a_0 + \sum_{m>0} (1 - m^{p-1}) a_m z^m \in \mathcal{X}.$$

If $p|m$, $(1 - m^{p-1})^n \to 1$ if $v_p(n) \to \infty$. If p does not divide m, this power tends to 0. Thus we see, with $E = \lim A^n$ (for such values of n) that

$$\begin{aligned} Ef &:= a_0 + \sum_{p|m} a_{pm} z^m \\ &= \sum_{m \geq 0} a_{pm} z^{pm} \in \mathcal{X}. \end{aligned}$$

Applying again the transformation given by formula (3.5), dividing by p, and iterating, we see that $\sum_{m>0} m^k a_{pm} z^{pm} \in \mathcal{X}$. Therefore

$$A_1 f := a_0 + \sum_{m>0} (1 - m^{p-1}) a_{pm} z^{pm} \in \mathcal{X}.$$

Defining E_1 as the limit of A_1^n for $v_p(n) \to \infty$, we see that $a_0 + \sum_{m>0} a_{p^2m} z^{p^2m} \in \mathcal{X}$. Iterating again, we see finally that the constant $a_0 z^0$ belongs to $\mathcal{X}$.

We can now finish the proof of Theorem 3.7. For $f = z^m$, the formula for the action of X yields $Xf = (m + a - b)z^{m+1}$. If $b - a \notin \mathbb{N}$, we see, starting with z^0, that all monomials are in $\mathcal{X}$. Since $\mathcal{X}$ is closed, it is equal to $\mathcal{A}$. On the other hand, if $b - a = N \in \mathbb{N}$, it is easy to see from Lemma 3.3, or from the derived representation, that the space of polynomials of degree $\leq N$ is stable by G: the corresponding representation is the irreducible representation of G of degree $N + 1$ and central character $\alpha\beta$.

3.4 .

Let now L be an unramified extension of $\mathbb{Q}_p$, of degree r. Denote by $I_{\mathbb{Q}_p}(\chi)$ the previous representation of G, on globally analytic functions. Let $I_L(\chi)$ be its extension to $G(L)$ by holomorphic base change (Proposition 3.1). It is an L-analytic representation , still given by the formulas of Lemma 3.3. Note that $L/\mathbb{Q}_p$ being unramified, the holomorphic extension of χ to $(\mathcal{O}_L[1])^2$, where $\mathcal{O}_L[1] = \{x \in \mathcal{O}_L^\times : x \equiv 1[p]\}$, verifies (3.4). The representation of $G(L)$ is realised on the L-analytic functions on B^1 , seen as an L-analytic space. More generally, if $\underline{\chi} = (\underline{\alpha}, \underline{\beta})$ is a pair of characters of $\mathcal{O}_L[1]$ verifying the condition extending (3.4) (see before Theorem 3.11), we can consider the globally $\mathbb{Q}_p$-analytic vectors in the induced representation of $G(L) = Res_{L/\mathbb{Q}_p} G(\mathbb{Q}_p)$ on $\mathcal{A}_{loc}(Res_{L/\mathbb{Q}_p} U(\mathbb{Q}_p), K)$. This representation will be denoted by $I(\underline{\chi})$.

Proposition 3.9 *If $b - a \notin \mathbb{N}$, the holomorphic base change $I_L(\chi)$ of $I_{\mathbb{Q}_p}(\chi)$ is topologically irreducible.*

(The irreducibility clearly follows from the irreducibility under $G(\mathbb{Q}_p)$.)

We now compare the base change functor we have constructed with the natural consequences of a (conjectural) Langlands functoriality for p–adic representations. We refer the reader to the Introduction to [3] for more motivation. The principal series representation of $G(\mathbb{Q}_p)$ is one of two summands (under $G(\mathbb{Q}_p)$) of an irreducible representation π of $\mathcal{G}(\mathbb{Q}_p) = GL(2, \mathbb{Q}_p)$ [11, Sect. 5], the principal series associated to the representation of the Galois group

$$\sigma \mapsto \alpha(\sigma) \oplus \beta(\sigma)$$

($\sigma \in \mathrm{Gal}(\bar{\mathbb{Q}}_p)$). Here we have assumed α, β extended to $\mathbb{Q}_p^\times$, thus giving characters of the Weil group $W_{\mathbb{Q}_p}$, and $\alpha(p)$, $\beta(p)$ units so the representation of $W_{\mathbb{Q}_p}$ actually extends to the Galois group.

In conformity with the general formalism, the base change π_L of π should be associated to the couple of characters $(\alpha \circ N_{L/\mathbb{Q}_p}, \beta \circ N_{L/\mathbb{Q}_p})$. Thus, instead of

$$\alpha(1+pu) = e^{a\log(1+pu)} \qquad (u \in \mathcal{O}_L)$$

we should consider

$$\begin{aligned}\underline{\alpha}(1+pu) &= \prod_{\sigma} \alpha(1+p\sigma(u)) \\ &= e^{a\Sigma\log(1+p\sigma(u))}\,.\end{aligned}$$

In other terms, we should induce the character given on $1+p\mathcal{O}_L$ by $\prod_{\sigma}\alpha^{\sigma}$, and similarly for $\underline{\beta}$. Furthermore, the representation should not be realized on a space of (L–holomorphic) functions on $U(L)$, but on a space of $\mathbb{Q}_p$–holomorphic functions on $\mathrm{Res}_{L/\mathbb{Q}_p}U(\mathbb{Q}_p)$.

Therefore we must consider the full base change of $I_{\mathbb{Q}_p}(\chi)$ to $\mathrm{Res}_{L/\mathbb{Q}_p}G(\mathbb{Q}_p)$.

Definition 3.10 The (Langlands) base change of $I_{\mathbb{Q}_p}(\chi)$ is the representation of $\mathrm{Res}_{L/\mathbb{Q}_p}G(\mathbb{Q}_p)$ on

$$\widehat{\bigotimes_{\sigma}}(I_L(\chi))^{\sigma} := I(\chi\circ N)\,.$$

Note that each factor is topologically irreducible for $b-a\notin\mathbb{N}$ and admits the same description as in Theorem 3.7; the embedding $\iota : L\to K$ must be replaced by $\iota\circ\sigma$. Each factor is holomorphic for this embedding.

Now the space of the representation $I(\chi\circ N)$ is $\widehat{\bigotimes_{\sigma}}\mathcal{A}(U,K)^{\sigma} = \mathcal{A}(\mathrm{Res}_{L/\mathbb{Q}_p}U,K)$. The representation is now a space of globally analytic vectors (by Theorem 3.7) in the locally analytic representation $I_{loc}(\chi\circ N)$ of $\mathrm{Res}_{L/\mathbb{Q}_p}G$. The representation of the full Iwahori subgroup on I_{loc} and its dual, the "locally analytic" distributions on $U(\mathcal{O}_L)$, have been studied by Orlik and Strauch. They prove the analogue of Theorem 5.4 in [10]: assume that we are given characters $(\underline{\alpha},\underline{\beta})$ of $L^{\times}$ with values in K. They give us a family of logarithmic parameters (a_σ,b_σ) for the embeddings $L\to K$. Then the space of distributions, dual to $I_{loc}(\chi\circ N)$, is irreducible under $\mathcal{D}_{loc}(\mathrm{Res}_{L/\mathbb{Q}_p}G,K)$ if the Verma module for $U(\mathfrak{g}\otimes L)\equiv\bigotimes_{\sigma}U(\mathfrak{g})$ induced from the *dual* of the linear form

$$(\mathbf{a},\mathbf{b}) : (X,Y)\in\mathfrak{t}\otimes L\mapsto\sum(a_\sigma X_\sigma + b_\sigma Y_\sigma)$$

is irreducible. Here $\mathfrak{t}$ is the Lie algera of the diagonal torus. See [8, Theorem 3.4.12]. Note that their result is again stated for the full Iwahori subgroup, but that, their argument being differential, it extends to G.

In our case the logarithmic parameters are parallel, given by (a,b). The highest weight to be considered is then $(-a,-b)$. The Verma module will be *reducible* if $(X,-X)\mapsto(b-a)X$, where $(X,-X)$ is in the Lie algebra of the diagonal torus in $SL(2)$, is the highest weight of a finite–dimensional representation: that is, if $b-a\in\mathbb{N}$. Otherwise, all the factors are irreducible, and the corresponding algebraic tensor product, under $U(\mathfrak{g}\otimes L)$, is irreducible. This implies as before that the tensor product is topologically cocyclic. It is also cyclic (see the end of Sect. 2.2.) We can in fact imitate the proof of Theorem 3.7 to prove the following, more general result, without assuming 'parallel weights'. Assume that $\underline{\alpha}(z) = \prod_0^{r-1} \exp(a_i \log(\sigma^i z))(z \in \mathcal{O}_L^\times)$ and similarly that $\underline{\beta}$ is associated with (b_i). We assume that the a's and b's verify the conditions (3.4). Then the representation induced from $(\underline{\alpha}, \underline{\beta})$ contains globally analytic vectors; their space is the space of analytic functions on $Res_{L/\mathbb{Q}_p}(U)$.

Theorem 3.11 *(i) Assume $b_i - a_i \notin \mathbb{N}$ for $i = 0, ..., r-1$. Then the globally analytic representation induced from $\underline{\chi} = (\underline{\alpha}, \underline{\beta})$ is topologically irreducible.*

(ii) In particular, if $\chi = (\alpha, \beta)$ is an analytic character of $(1 + p\mathbb{Z}_p)^2$, with $b - a \notin \mathbb{N}$, $\widehat{\bigotimes_\sigma} I_L(\chi)^\sigma$ is topologically irreducible, and is the representation of $G(L)$ on the space of globally analytic vectors in the representation of $G(L)$ induced from $\underline{\chi} = \chi \circ N$.

Denote by z_1 the natural variable on $U(L) = B^1(L)$ and by $z_i (i = 2, ..., r)$ the function $\sigma^{i-1}(z_1)$. The Tate algebra of $Res_{L/\mathbb{Q}_p}$ (with values in K) is then $\mathcal{A}_r := \mathcal{T}_r = K < z_1, ..., z_r >$. We write, for $f \in \mathcal{A}_r$, $f = \sum_n a_n z^n$ where $n = (n_1, ..., n_r)$ and $z^n = z_1^{n_1} ... z_r^{n_r}$; also, $|n| = n_1 + ... + n_r$. Let $\mathcal{X} \subset \mathcal{A}_r$ be a closed $G(L)$-submodule. Let $K[z]_N$ be the space of polynomials in the r variables of total degree $\leq N$. With $T_r = \{s \in \mathcal{O}_K^\times, s \equiv 1[p]\}$, the characters $\prod (s^{\sigma^i})^{n_i}$ are linearly independent for n ranging over $\mathbb{N}^r$. Thus there exists a subset $M \subset \mathbb{N}^r$ having properties (a), (b) as above.

Now $(Res_{L/\mathbb{Q}} G_L) \otimes L \cong G_L^r$, with its obvious action, componentwise, on $\mathcal{A}_r$. Using the representation of the ith component, we see that $z^q \in X$ if $q \leq n$, i.e., $q_i \leq n_i$. It follows that M is a product of intervals equal to $[0, N_i]$ or $\mathbb{N}$.

In order to prove the theorem, it will suffice to show, as in the case of Theorem 3.7, that a non-zero closed submodule $\mathcal{X}$ contains the constant function z^0. For then the action of X, separately on each component, and assuming $b_i - a_i \notin \mathbb{N}$, allows us to increase the partial degree of a monomial z^n for the ith variable. However, the proof of Lemma 3.8 extends to this case. Indeed, assume we have a series in $\mathcal{X}$

$$f(z) = \sum_0^\infty z_1^n f_n(z_2, ... z_r)$$

where we may assume that f_0 has a non-trivial constant term (since $z^0 \in M$). Applying the argument for Lemma 3.8 to the first factor, replacing the constants a_n by the functions f_n, we see that $z_1^0 f_0 \in M$. Proceeding inductively we see that $z^0 \in M$. (If $b_i - a_i \in \mathbb{N}$ for a subset of the indices i, $\mathcal{A}_r$ contains a tensor product of finite-dimensional representations (for these i) and of irreducible globally analytic principal series for the other indices.)

3.5 .

It is not our intention here to explore the situation if one considers the full groups $\mathcal{G}(\mathbb{Q}_p)$, $\mathcal{G}(L)$. The Bruhat decomposition yields a decomposition of the induced representation π of $\mathcal{G}(\mathbb{Q}_p)$ (say, on locally analytic functions), under the Iwahori subgroup, as a sum of two subspaces, one (the representations we have considered), π_w, associated to the big orbit, and another, say

$$\pi = \pi_w \oplus \pi_1 .$$

The corresponding irreducibility results for π_1, under the Iwahori subgroup, are proved by Schneider–Teitelbaum, and by Orlik and Strauch in the general case. We expect that the analogue of Proposition 3.9 remains true for the other summand (with now the condition $a - b \notin \mathbb{N}$). Now the full completed tensor product will be

$$\widehat{\bigotimes_{\sigma}}(I(w)^{\sigma} \oplus I(1)^{\sigma}) , \tag{3.7}$$

where I denotes the globally analytic representation. On the other hand, the representation of $G(L) = (\mathrm{Res}_{L/\mathbb{Q}_p} G)(\mathbb{Q}_p)$ is

$$\widehat{\bigotimes_{\sigma}} I(w)^{\sigma} \oplus \widehat{\bigotimes_{\sigma}} I(1)^{\sigma} . \tag{3.8}$$

(The Weyl group of $\mathcal{G}(L)$ still has two elements). Thus the mixed terms of (3.7), of the form

$$\widehat{\bigotimes_{\sigma\in\Sigma_1}}(I(w)^{\sigma} \otimes \widehat{\bigotimes_{\sigma\in\Sigma_2}} I(1)^{\sigma}) , \tag{3.9}$$

where $\Sigma = \Sigma_1 \cup \Sigma_2$ is a proper partition of Σ, should be deleted.

A Appendix: "Pathologies"

We review some properties of the distribution algebras on our groups, relative to tensor products or the Nœtherian property. They introduce some difficulties explained in the main text.

A.1 $\mathcal{D}(X \times Y) \not\equiv \mathcal{D}(X)\widehat{\otimes}\mathcal{D}(Y)$

Here X, Y are rigid–analytic spaces isomorphic to products of unit balls, over a p–adic field L. Then $\mathcal{A}(X \times Y)$ is naturally isomorphic to $\mathcal{A}(X)\widehat{\otimes}\mathcal{A}(Y)$. Moreover, each space of analytic functions is, as a Banach space, isomorphic to $C_0(M)$ where M is the set of exponents: $M = \mathbb{N}^d$, $d = \dim(X)$. Here $C_0(M) = C_0(M, L)$. We have, for two countable sets M, N.

$$C_0(M \times N) \cong C_0(M)\widehat{\otimes}C_0(N)$$

(Schneider [9, p. 112]) which yields the requisite (well–known) isomorphism for the Tate algebras. We can assume that our sets M, N are equal to $\mathbb{N}$. We now have the following result, certainly well–known:

Proposition A.1 *The natural map*

$$\ell^\infty(\mathbb{N})\widehat{\otimes}\ell^\infty(\mathbb{N}) \longrightarrow \ell^\infty(\mathbb{N} \times \mathbb{N})$$

is injective, but is not an isomorphism.

Here $\ell^\infty(\mathbb{N}) = \ell^\infty(\mathbb{N}, L)$ is the Banach space of bounded sequences, the dual of $C_0(\mathbb{N})$. We will denote by V' the dual of a Banach space V, with its strong topology (the topology as a Banach space.) We denote by $\mathcal{L}(V, W)$ the space of continuous linear maps $V \to W$, again with the topology of the norm. Now we have [2, 2.1.7.2]

$$\mathcal{L}(V\widehat{\otimes}W, X) \xrightarrow{\approx} \mathcal{L}(V, \mathcal{L}(W, X))$$

(isometric isomorphism) for three Banach spaces, the map being the natural one, so

$$(C_0\widehat{\otimes}C_0)' = \mathcal{L}(C_0\widehat{\otimes}C_0, L) \cong \mathcal{L}(C_0, \ell^\infty)\,.$$

Since $(C_0\widehat{\otimes}C_0)' = \ell^\infty(\mathbb{N} \times \mathbb{N})$, it suffices to check:

Proposition A.2 *The natural map*

$$\ell^\infty\widehat{\otimes}\ell^\infty \longrightarrow \mathcal{L}(C_0, \ell^\infty)$$

obtained by completion from

$$\ell^\infty\otimes\ell^\infty \longrightarrow \mathcal{L}(C_0, \ell^\infty)$$

is injective, but is not an isomorphism.

Here $\mathcal{L}(C_0, \ell^\infty)$ is provided with the strong ($=$ Banach) topology. Schneider [9, Proposition 18.2] shows that this map is an isomorphism onto its image, whence the injectivity in Proposition A.1. He also shows that its image is the space

$$\mathcal{CC}(C_0, \ell^\infty)$$

of completely continuous operators. Thus it suffices to show

$$\mathcal{CC}(C_0, \ell^\infty) \neq \mathcal{L}(C_0, \ell^\infty) . \tag{A.1}$$

Now $F : C_0 \to \ell^\infty$ is in $\mathcal{CC}$ if, and only if, $F(B)$ is compactoid for any bounded set $B \subset C_0(X)$. We can simply consider the unit ball.

Recall also that $\Omega \subset V$ (a Banach space) is compactoid [9, p. 71] if $\forall r > 0$ $\exists (v_i)_{i=1,\ldots N}$, $v_i \in V$, such that

$$\Omega \subset B_V(r) + \sum_1^N \mathcal{O}_L v_i .$$

We consider the identity map $F : C_0 \to \ell^\infty$ and show that it is not completely continuous. If it were, we would have for $f \in C_0$:

$$\|f\|_\infty \leq 1 \Longrightarrow \forall r \quad \|f - \sum_1^{N(r)} z_i f_i\|_\infty \leq r$$

for some functions $f_i \in \ell^\infty$, and integers z_i, depending on r (but not on f).

For simplicity assume $L = \mathbb{Q}_p$. Fix $r = p^{-n}, n \geq 0$. The f_i take values in $p^{-M}\mathbb{Z}_p$, $M \geq 0$. The function f defines $\bar{f} : \mathbb{N} \to \mathbb{Z}_p/p^n\mathbb{Z}_p$, and

$$\bar{f}(x) = \sum_1^{N(r)} \bar{z_i} \bar{f_i}(x) , \tag{A.2}$$

a linear combination of functions $\mathbb{N} \to p^{-M}\mathbb{Z}_p/p^n\mathbb{Z}_p$. Since f is arbitrary in $B(1) \subset C_0$, $\bar{f}$ can be an arbitrary function with finite support $\mathbb{N} \to \mathbb{Z}_p/p^n\mathbb{Z}_p$. However, when (z_i) varies in $\mathbb{Z}_p^{N(r)}$, the set of functions on the right–hand side of (A2) with values in $p^{-M}\mathbb{Z}_p/p^n\mathbb{Z}_p$ is finite, and this is a contradiction.

A.2 $\mathcal{D}(G)$ (As a Convolution Algebra) is not Nœtherian for Left- or Right- Ideals

Here the rigid–analytic group is assumed to verify the assumptions in Sect. 1. We start with $B(1)$ (rigid–analytic ball) seen as an additive group. Then

$$\mathcal{A}(G) = \Big\{ \sum_0^\infty a_n x^n,\ a_n \to 0 \Big\},$$
$$a_n = \frac{1}{n!}\Big(\frac{d^n}{dx^n} f\Big)(0).$$

The algebra $\mathcal{D}(G)$ is isomorphic to ℓ^∞ by taking the basis dual to the x^n; we can write a distribution $T \in \mathcal{D}(G)$ as

$$T = \sum_0^\infty c_n \frac{1}{n!}\Big(\frac{d^n}{dx^n}\Big)_0, \quad (c_n) \in \ell^\infty$$

and then $\mathcal{D}$ is naturally isomorphic to an algebra of divided power series:

$$T = \sum_0^\infty \frac{c_n}{n!} t^n, \quad t = \Big(\frac{d}{dx}\Big)_0, \quad (c_n) \in \ell^\infty.$$

As pointed out by Berthelot, this algebra is not Nœtherian.[2] Since it is a Banach algebra for convolution (which becomes here the product of the series), it suffices to check that there is an ideal which is not closed [2, 3.7.2.2]. In fact:

Lemma A.3 (Berthelot) *The ideal $(t) \in \mathcal{D}$ is not closed.*

Indeed if $T \in (t)$,

$$\begin{aligned} T = tS &= t\Big(\sum_0^\infty \frac{c_n}{n!} t^n\Big) \\ &= \sum_1^\infty t^n \frac{d_n}{n!} \end{aligned}$$

with $d_n = nc_{n-1}$, so

$$T \in (t) \Leftrightarrow d_0 = 0 \text{ and } \Big|\frac{d_n}{n}\Big| \le C.$$

But this subspace of $\mathcal{D}$ is clearly not closed. For instance, if $T = (d_n)$ has support on $\{n = p^r\}$ ($r \ge 0$ even) with

$$\begin{aligned} d_{p^r} &= p^{r/2} \text{ (so } T \in \ell^\infty) \\ \Big|\frac{d_{p^r}}{p^r}\Big| &= |p^{-r/2}| \longrightarrow \infty \text{ (so } T \notin (t)), \end{aligned}$$

T is the limit in $\mathcal{D}$ of the truncated series T^α with $d^\alpha_{p^r} = p^{r/2}$ ($r \le \alpha$), $d^\alpha_{p^r} = 0$ ($r > \alpha$) which obviously belong to the ideal.

Consider now a rigid–analytic group G, isomorphic to a product $B(1)^d$ over $\mathbb{Q}_p$ as a rigid–analytic space, the coproduct being then given by Tate series with integral

[2] Berthelot considered the subalgebra given by $(c_n) \in C_0$, but this makes no difference.

coefficients. We further assume (as is the case in this paper) that the factors are (additive) analytic subgroups, and their distribution algebras are therefore as in the previous proof.

Proposition A.4 *Under these assumptions $\mathcal{D}(G)$ is not Nœtherian (for left- or right-ideals).*

Indeed we have, as in the commutative case:

Lemma A.5 (Schneider–Teitelbaum) *If $\mathcal{D}(G)$ is (left) Nœtherian, any (left) ideal is closed.*

See Proposition 2.1 in [12]. For completeness we provide a proof. In the commutative case this is [2, 3.7.2.2]. A glance at their proof shows that it suffices to prove Nakayama's lemma for the ideal $A^\vee = \{a \in A : \|a\| < 1\}$ in A^0, where we have written $A = \mathcal{D}(G)$. We may assume that the (Banach) norm on A is submultiplicative [2, 1.2.1.2]. Since the argument in [2, 1.2.3.6] for Nakayama's lemma uses determinants, we rephrase it (using moreover $A^\vee$ rather than the set $\check{A}$ of topologically nilpotent elements):

Lemma A.6 *Let M be an A–module, and N a submodule of M such that there exist $x_1, \dots x_n \in M$ with $M \subset N + \sum_1^n A^\vee x_i$. Then $N = M$.*

As in [2], loc. cit., we can write

$$\underline{x} = \underline{y} + C\underline{x}$$

where $\underline{x}$ is a column vector in M^n with coordinates x_i, and $\underline{y}$ has coordinates in N, and the matrix $C \in M_n(A)$ has entries in $A^\vee$. Thus $\underline{y} = (1 - C)\underline{x}$. The matrix $1 - C$ is invertible: if $M_n(A)$ is endowed with the operator norm, this norm is submultiplicative, and $\|C\| < 1$. This implies that $\underline{x} = (1 - C)^{-1}\underline{y} \in N^n$. (It is not clear to us that the argument applies for $\check{A}$.)

We now return to the proof of Proposition A.4. Assume that (as a rigid analytic space)

$$G = G_1 \times \cdots \times G_d$$

where each G_i is a rigid–analytic group isomorphic to the additive unit ball over $\mathbb{Q}_p$.

In particular we have a bijection

$$\begin{aligned} \mathbb{Z}_p^d = G_1(\mathbb{Q}_p) \times \cdots \times G_d(\mathbb{Q}_p) &\longrightarrow G(\mathbb{Q}_p) \\ (g_1, g_2, \dots, g_d) &\longmapsto g_1 \cdots g_d \,. \end{aligned}$$

The Tate algebra $\mathcal{A}_G$ is isomorphic with $\widehat{\bigotimes_{i=1,\dots d}} \mathcal{A}_{G_i}$ where each $\mathcal{A}_{G_i}$ is the Tate algebra in one variable. Evaluated on the points of $G(\mathbb{Q}_p)$, this yields the map $f \mapsto f(g_1, \dots, g_d)$ $(f \in \mathcal{A}_G)$.

Each injection $j_i : G_i \longrightarrow G$ is an homomorphism, and the restriction $\mathcal{A}_G \longrightarrow \mathcal{A}_{G_i}$ is therefore compatible with the coproduct. Dually, we get

$$(j_i)_* : \mathcal{D}_{G_i} \longrightarrow \mathcal{D}_G ,$$

compatible with convolution. If we denote by x_i the local variable on G_i, a function $f \in \mathcal{A}_G$ being then in the Tate algebra in the x_i, an element of $\mathcal{D}_G$ can be written

$$T = \sum_n c_n \frac{1}{n!} \partial_1^{n_1} \cdots \partial_d^{n_d} := \sum_n c_n \delta_n \tag{A.3}$$

with $n = (n_1, \ldots n_d) \in \mathbb{N}^d$, $n! = \Pi(n_i)!$,

$$\partial_i f(x_1, \ldots, x_d) = \frac{d}{dx_i} f(x_1, \ldots, x_d)(0), \text{ and } (c_n) \in \ell^\infty(\mathbb{N}^d).$$

Let us write $\mathcal{D}_1$ for the subalgebra $(j_1)_*(\mathcal{D}_{G_1})$, given by $c_n = 0$ if $n_i > 0$ for some $i \geq 2$. This is clearly a closed subalgebra. The element $\partial_1 = \delta_1$ - abuse of notation for $\delta_{(1,0,\ldots,0)}$ - is equal to $(j_1)_*((\frac{d}{dx})_0)$. Moreover, for the convolution product in G, we have

$$\delta_1 * \delta_n = \frac{1}{n!} \frac{d}{dx_1} \left(\frac{d}{dx_1}\right)^{n_1} \cdots \left(\frac{d}{dx_d}\right)^{n_d}$$

(evaluated at $(0, \ldots, 0)) = (n_1 + 1) \frac{1}{(n_1+1)! n_2! \cdots n_d} (\frac{d}{dx_1})^{n_1+1} \cdots (\frac{d}{dx_d})^{n_d} = (n_1 + 1)\delta_{n'}$ where $n' = (n_1 + 1, \ldots, n_d)$.

We will show that the left ideal $\delta_1 \mathcal{D}(G)$ is not closed in $\mathcal{D}(G)$. Indeed, if it were, its intersection with the closed subalgebra $\mathcal{D}_1$ would be so. But the subalgebra is isomorphic (as an algebra) with the algebra

$$\mathcal{D}_{G_1} = \left\{ \sum_{m \geq 0} c_m \delta_m \right\} = \left\{ \sum_{m \geq 0} c_m \frac{1}{m!} \left(\frac{d}{dx}\right)^m \right\}$$

in one variable, and the previous computation shows that the intersection is the ideal $(\frac{d}{dx}) \subset \mathcal{D}_{G_1}$ considered in the first part of the proof. Therefore $\delta_1 \mathcal{D}(G)$ is not closed; this completes the proof of Lemma A.6.

A.3 .

Finally, we note that there is a possible substitute for the consideration of $\mathcal{D}(G)$, which could obviate the problems we encountered. (This was pointed out by Schneider.)[3]

[3] In this section we will skip details, as we only want to outline an alternative approach.

This algebra was already introduced by Lazard [7, III.3.3.3] who calls in $Ala(G)$. To use Lazard's results, we keep the assumptions of Sect. A.2 and assume moreover that the factors $G_i(\mathbb{Q}_p) \cong \mathbb{Z}_p$ form a Lazard basis (a "base ordonnée" in the sense of [7, III.2.2.4]. The Iwasawa algebra of G (with integral coefficients) is then given by the series

$$\sum_n a_n z_1^{n_1} \cdots z_d^{n_d} = \sum a_n z^n$$

with $a_n \in \mathbb{Z}_p$ and $z_i = \delta_1 - \delta_0$ in the Iwasawa algebra of $G_i(\mathbb{Q}_p) = \mathbb{Z}_p$. Lazard defines the algebra $Sat\ Al(G)$, where $Al(G) = \Lambda_G$ is the Iwasawa algebra, and $Sat\ Al(G)$ is given by $val(a_n) \geq -|n|$. (Thus $Sat Al(G)$ is contained in the completion of $\Lambda_G \otimes \mathbb{Q}_p$).

Recall that $\Lambda_G \subset \mathcal{D}_G$.

Lazard defines $Ala(G) \subset Sat\ Al(G)$ as the compact $\mathbb{Z}_p$–module generated by the $\frac{z^n}{n!}$, with the usual notation. If we recall that the basis elements z^n are dual to the Mahler basis $\binom{x}{n}$ of $\mathcal{C}(G)$, we see that this is the unit ball in the Banach dual of the functions $\sum_n c_n n! \binom{x}{n}$ $(c_n \to 0)$ in $\mathcal{C}(G)$, with the weak topology. By Amice's theorem [5, I.4] this space of functions is $\mathcal{A}(G)$. Thus $Ala(G) \subset \mathcal{D}_G$, and $Ala(G)$ is the unit ball in $\mathcal{D}_G$, with the weak topology. Note that both the $n!\binom{x}{n}$ and the x^n $(x = (x_1, \dots x_d))$ are orthonormal bases of $\mathcal{A}_G$: see Colmez [5, I.4.3]. Thus this coincides with the description of $\mathcal{D}_G$ as an ℓ^∞ space on the set of powers n.

Now if we consider these integral spaces (isomorphic to $\mathbb{Z}_p^N$, N the set of indices, with the product topology), we obtain indeed, for a product $G \times H$, an isomorphism $Ala(G \times H) \cong Ala(G) \widehat{\bigotimes}_{\mathbb{Z}_p} Ala(H)$. This is stated (but not proven) by Lazard [7, III.3.]. The completed tensor product (*over* $\mathbb{Z}_p$) is defined in [7, I.3.2.6, I.3.2.9].

Recall that we have assumed that the coproduct for G (and H, a group of the same type) was given by integral Tate expansions. Then $\mathcal{D}_G(\mathbb{Z}_p)$, the distributions with integral coefficients, is the unit ball in $\mathcal{D}_G$, and its weak topology is as we saw, the product topology. The tensor product $\mathcal{D}_G(\mathbb{Z}_p) \times \mathcal{D}_H(\mathbb{Z}_p) \to \mathcal{D}_{G\times H}(\mathbb{Z}_p)$ is simply given, M being the set of exponents for H, by

$$(z_n, w_m) \longmapsto (z_n w_m)_{n,m}$$

where $N \times M$ is the set of exponents for $G \times H$. It is easy to see that it yields, as asserted by Lazard, an isomorphism $\mathbb{Z}_p^N \widehat{\otimes} \mathbb{Z}_p^M \to \mathbb{Z}_p^{N\times M}$. In particular, under our standing assumptions on the groups (in particular, the integrality conditions), we have:

Proposition A.7 *Let* $\mathcal{D}_G^0$, $\mathcal{D}_H^0$ *denote the unit balls of* $\mathcal{D}_G$, $\mathcal{D}_H$ *with their weak topology. Then*

$$\mathcal{D}^0_{G\times H} \cong \mathcal{D}^0_G \widehat{\bigotimes_{\mathbb{Z}_p}} \mathcal{D}^0_H .$$

Now assume V, W are globally analytic Banach representations of G, H. The algebra $\mathcal{D}_G$ acts on V', by

$$< Tv', v >= \int < v', gv > dT(g)$$

where we have used the integral notation for T.

Now write V'_0 for the unit ball in V', and V_0 for the unit ball in V. We may assume, as in Sect. 2.1, that the action of G on V preserves the norm. Then if $v \in V_0$ and $v' \in V'_0$ $< v', gv >$ is a function f on G, given by a Tate series, and such that $\mathrm{Sup}_{g\in G}|f(g)| \leq 1$.

If moreover V is admissible, the Tate norm $\|f\|$ is ≤ 1 by Corollary 2.8. Thus Tv' belongs to V'_0 for $T \in \mathcal{D}^0_G$. We obtain an action of $\mathcal{D}^0_G$ on the unit ball of V', compatible with the weak topologies.

If now we consider another group H, an admissible globally analytic representation W of H, and the product $G \times H$, we see that $\mathcal{D}^0_G \widehat{\otimes} \mathcal{D}^0_H$ acts on $V'_0 \widehat{\otimes} W'_0$. Replacing the distribution algebras by their unit balls, we are very close to the original construction of Schneider–Teitelbaum [10] for Banach representations.

Since $\mathcal{D}_G$ is not Nœtherian, however, it easily follows that $\mathcal{D}^0_G$ is not Nœtherian.

References

1. Bertapelle, A.: Formal Néron models and Weil restriction. Math. Ann. **316**(3), 437–463 (2000)
2. Bosch, S., Güntzer, U., Remmert, R.: Non-Archimedean Analysis. A Systematic Approach to Rigid Analytic Geometry. Grundlehren der Mathematischen Wissenschaften, vol. 261. Springer, Berlin (1984)
3. Clozel, L.: Presentation of an Iwasawa algebra: the case of $\Gamma_1 SL(2, \mathbb{Z}_p)$. Doc. Math. **16**, 545–559 (2011)
4. Clozel, L.: Globally analytic p-adic representations of the pro-p Iwahori subgroup of GL(2) and base change, I : Iwasawa algebras and a base change map, to appear in Bull. Iran. Math. Soc
5. Colmez, P.: Fonctions d'une variable p-adique. Astérisque **330**, 13–59 (2010)
6. Emerton, M.: Locally analytic vectors in representations of locally p-adic analytic groups
7. Lazard, M.: Groupes analytiques p-adiques. Inst. Hautes Etudes Sci. Publ. Math. **26**, 389–603 (1965)
8. Orlik, S., Strauch, M.: On the irreducibility of locally analytic principal series representations. Represent. Theory **14**, 713–746 (2010)
9. Schneider, P.: Nonarchimedean Functional Analysis. Springer Monographs in Mathematics. Springer, Berlin (2002)
10. Schneider, P., Teitelbaum, J.: Banach space representations and Iwasawa theory. Isr. J. Math. **127**, 359–380 (2002)
11. Schneider, P., Teitelbaum, J.: Locally analytic distributions and p-adic representation theory, with applications to GL_2. J. Am. Math. Soc. **15**(2), 443–468 (2002)

12. Schneider, P., Teitelbaum, J.: Algebras of p-adic distributions and admissible representations. Invent. Math. **153**(1), 145–196 (2003)
13. Steinberg, R.: Representations of algebraic groups. Nagoya Math. J. **22**, 33–56 (1963)
14. Wahle, C.: Weil restriction of p-adic analytic spaces

Eisenstein Cohomology and Automorphic L-Functions

Neven Grbac

To Joachim Schwermer, with gratitude and admiration, for the occasion of his 66th birthday

Abstract During the past ten years of the most inspiring and very fruitful collaboration with Joachim Schwermer, we have carefully studied the non-vanishing conditions for certain summands in the decomposition along the cuspidal support of the (square-integrable) Eisenstein cohomology of a reductive group over a totally real number field. These conditions form a subtle combination of geometric conditions, arising from cohomological considerations, and arithmetic conditions, arising from the analytic properties of Eisenstein series and given in terms of automorphic L-functions. This paper is a survey of the most important results of our long-lasting collaboration.

Keywords Automorphic cohomology · Eisenstein cohomology
Square-integrable cohomology · Automorphic forms · Eisenstein series
Automorphic L-functions · Non-vanishing conditions · Franke filtration

2010 Mathematics Subject Classification Primary 11F75 · Secondary 11F70
22E40 · 22E55 · 11F67

Author's work has been supported by Croatian Science Foundation under the project 9364 and by University of Rijeka research grant 13.14.1.2.02.

N. Grbac (✉)
Department of Mathematics, University of Rijeka, Radmile Matejčić 2,
51000 Rijeka, Croatia
e-mail: neven.grbac@math.uniri.hr

J. W. Cogdell et al. (eds.), *Cohomology of Arithmetic Groups*,
Springer Proceedings in Mathematics & Statistics 245,
https://doi.org/10.1007/978-3-319-95549-0_2

1 Introduction

The cohomology of an arithmetic subgroup of a reductive group over a number field is closely related to automorphic forms with respect to that arithmetic subgroup. On the other hand, it has another interpretation in terms of geometric objects. In that way, it provides a direct link between arithmetic algebraic geometry and number theory. The flow of information may go in both directions. Using geometric constructions one may hope to show the existence of automorphic forms, satisfying certain properties, with respect to some arithmetic subgroups. In the other direction, one may use the decomposition and structure of the spaces of automorphic forms to get information about cohomology. The latter direction is exactly the main theme of my collaboration with Joachim Schwermer, and the subject of this survey paper. Schwermer has written several excellent overview papers about the cohomology of arithmetic groups, emphasizing both, the geometric and the number theoretic aspects [1–3]. Hence, we omit here the preliminaries and details regarding the setting, wider scope and applications of cohomology of arithmetic groups, and focus on the results obtained in our collaboration.

The main object of concern, when expressing the cohomology of arithmetic groups in terms of automorphic forms, is the so-called automorphic cohomology of a reductive group. It captures, in the adèlic setting, the information about cohomology of congruence arithmetic subgroups. It is defined as the relative Lie algebra cohomology of the space of automorphic forms on the adèlic group.

In the collaboration with Joachim Schwermer, we study the automorphic cohomology of a reductive group using information about automorphic forms on the adèlic points of that group. However, complete explicit information regarding the structure of spaces of automorphic forms is often not known. Hence, we are exploiting the interconnection of the two interpretations, geometric and automorphic, to show that only the automorphic forms with certain properties may possibly contribute to cohomology. These arguments exclude many difficult situations in the theory of automorphic forms from consideration. For the remaining possibilities, we use the information from the theory of automorphic forms to show the existence of certain cohomology classes represented by (non-cuspidal) square-integrable automorphic forms and study the internal structure of cohomology.

The paper is organized as follows. In Sect. 2 we define the main objects of concern and provide just the preliminaries required to state the results. Section 3 introduces the necessary conditions for non-vanishing of certain summands in cohomology and explains their consequences. In Sect. 4 the focus is on the subtle interplay of geometric and arithmetic necessary non-vanishing conditions for summands in square-integrable cohomology. Finally, Sect. 5 deals with internal structure of the full automorphic cohomology, in particular, with the existence of non-trivial cohomology classes in the summands of square-integrable cohomology.

* * *

When I first met Joachim Schwermer at the Erwin Schrödinger Institute in Vienna in 2006, I was a young mathematician, just finishing my PhD, and looking for new problems. My thesis was about L^2 spaces of automorphic forms and their spectral decomposition, the Eisenstein series and automorphic L-functions and their analytic behavior. My adviser was Goran Muić at the University of Zagreb. At that time, I did not know anything about cohomology. In our discussions back in 2006, I learned that the subject of my thesis is closely related to the cohomology of arithmetic groups, and we were soon working together on many problems in cohomology of arithmetic groups, looking at the possible applications of the structural information about spaces of automorphic forms.

I will always be grateful to Schwermer, who introduced me to the subject and taught me everything I know about cohomology. His kindness and patience, especially at the beginning of our collaboration, gave me the courage and confidence to continue the quest. I hope that he has enjoyed our collaboration as much as I have, and that it will continue for many years to come. *Happy birthday Joachim!*

2 Automorphic Cohomology and Cohomology of Arithmetic Groups

In this section we define the automorphic cohomology and its decomposition along the cuspidal support. The individual summands in that decomposition are the main objects of our concern in this paper. For simplicity of exposition and to avoid some technical issues, we work with a semi-simple group, instead of a reductive group, and over the field $\mathbb{Q}$ of rational numbers, instead of any totally real number field.

Hence, let G be a semi-simple connected linear algebraic group defined over the field of rational numbers $\mathbb{Q}$. For a finite prime p, let $\mathbb{Q}_p$ be the field of p-adic numbers. For $p = \infty$, we have $\mathbb{Q}_\infty = \mathbb{R}$. Let $\mathbb{A}$ denote the ring of adèles of $\mathbb{Q}$, and $\mathbb{A}_f$ the subring of finite adèles. Let $\mathfrak{g}_\infty$ be the real Lie algebra of the Lie group $G(\mathbb{R})$. We fix, once for all, a minimal parabolic $\mathbb{Q}$-subgroup P_0 of G, and a maximal compact subgroup K of $G(\mathbb{A})$. We may assume that $K = \prod_p K_p$, where K_p is a maximal compact subgroup of $G(\mathbb{Q}_p)$, hyperspecial for almost all p, and that K is in good position with respect to P_0 as in [4, Sect. I.1.4].

Let $\mathcal{A} = \mathcal{A}(G(\mathbb{Q})\backslash G(\mathbb{A}))$ be the space of automorphic forms on $G(\mathbb{A})$ as defined in [5]. It carries the structure of a $(\mathfrak{g}_\infty, K_\infty; G(\mathbb{A}_f))$-module arising from right translation. Let E be a finite-dimensional irreducible algebraic representation of G. We define the automorphic cohomology of G with respect to E as the relative Lie algebra cohomology of the $(\mathfrak{g}_\infty, K_\infty; G(\mathbb{A}_f))$-module $\mathcal{A}$ of automorphic forms on $G(\mathbb{A})$ with respect to E, that is,

$$H^*(G, E) = H^*(\mathfrak{g}_\infty, K_\infty; \mathcal{A} \otimes_\mathbb{C} E).$$

It carries the structure of a $G(\mathbb{A}_f)$-module. This object captures the information about the cohomology of congruence arithmetic subgroups of G. This fact is proved in [6–8].

More precisely, according to [7], the Eilenberg–MacLane cohomology of an arithmetic subgroup Γ of G is isomorphic to the de Rham cohomology of the corresponding locally symmetric space $\Gamma\backslash X$, where $X = G(\mathbb{R})/K_\infty$. This, in turn, is isomorphic to the relative Lie algebra cohomology of the $(\mathfrak{g}_\infty, K_\infty)$-module of smooth functions on $\Gamma\backslash G(\mathbb{R})$. The regularization theorem of Borel, proved in [6], shows that in this last cohomology space the same space is obtained if smooth functions are replaced with smooth functions of uniform moderate growth. For a congruence subgroup Γ, Franke proved in [8] that instead of smooth functions of moderate growth, the same cohomology space is obtained as the relative Lie algebra cohomology of the $(\mathfrak{g}_\infty, K_\infty)$-module of automorphic forms on $G(\mathbb{R})$ with respect to Γ. Writing this in the adèlic setting, and passing to the direct limit with respect to the inclusion of open compact subgroups of $G(\mathbb{A}_f)$, one obtains the automorphic cohomology $H^*(G, E)$ as defined above. Conversely, given a compact open subgroup C of $G(\mathbb{A}_f)$, one can recover the cohomology of the corresponding congruence arithmetic subgroup Γ by taking the C-invariants of the $G(\mathbb{A}_f)$ action on the automorphic cohomology $H^*(G, E)$.

The first step in the study of automorphic cohomology is to apply Wigner's lemma [7, Sect. I.4], which says that only automorphic forms whose infinitesimal character is compatible with E may possibly contribute to $H^*(G, E)$. Thus, we let $\mathcal{J}$ be the ideal of finite codimension in the center $\mathcal{Z}$ of the universal enveloping algebra of the complexification of $\mathfrak{g}_\infty$ which annihilates the conjugate dual of E. Then

$$H^*(G, E) \cong H^*(\mathfrak{g}_\infty, K_\infty; \mathcal{A}_\mathcal{J} \otimes_\mathbb{C} E),$$

where $\mathcal{A}_\mathcal{J}$ consists of automorphic forms that are annihilated by a power of $\mathcal{J}$.

The space $\mathcal{A}_\mathcal{J}$ admits a decomposition along the cuspidal support. Let $\{P\}$ be the associate class of parabolic $\mathbb{Q}$-subgroups of G, represented by a standard parabolic $\mathbb{Q}$-subgroup P with a Levi decomposition $P = M_P N_P$. Let ϕ_π be an associate class of cuspidal automorphic representations of the Levi factors of parabolic subgroups in $\{P\}$, represented by a cuspidal automorphic representation π of $M_P(\mathbb{A})$. More precisely, $\phi_\pi = (\phi_Q)_{Q\in\{P\}}$, where ϕ_Q is the finite set of cuspidal automorphic representations of the Levi factor $M_Q(\mathbb{A})$ which are $G(\mathbb{Q})$-conjugate to π. Note that we do not assume that π is unitary.

Let $\check{\mathfrak{a}}_P = X^*(P) \otimes_\mathbb{Z} \mathbb{R}$, where $X^*(P)$ is the $\mathbb{Z}$-module of $\mathbb{Q}$-rational characters of P, and let $\check{\mathfrak{a}}_{P,\mathbb{C}}$ be its complexification. It is well known, see for example [4, Sect. I], that the elements of $\check{\mathfrak{a}}_{P,\mathbb{C}}$ give rise to characters of $M_P(\mathbb{A})$. Abusing the notation, for $\nu \in \check{\mathfrak{a}}_{P,\mathbb{C}}$, we denote the corresponding character of $M_P(\mathbb{A})$ by the same letter ν. Since we may replace the representatives π and P with their conjugates, we will always assume, without loss of generality, that $\pi \cong \pi_0 \otimes \nu_0$, where π_0 is a unitary cuspidal automorphic representation of $M_P(\mathbb{A})$, and ν_0 is a character of $M_P(\mathbb{A})$ corresponding to an element $\nu_0 \in \check{\mathfrak{a}}_P$ which belongs to the closure of the positive Weyl chamber determined by P.

We now define, following [9, Sect. 1.3], the space $\mathcal{A}_{\{P\},\phi_\pi}$ of automorphic forms supported in the associate class ϕ_π of cuspidal automorphic representations of the Levi factors of parabolic subgroups in the associate class $\{P\}$. This definition is equivalent to the definition given in [4, Sect. III.2.6] according to [9, Theorem 1.4]. Let $\pi \cong \pi_0 \otimes \nu_0$ as above. For simplicity of exposition, we suppose that π_0 is of multiplicity one in the space of cuspidal automorphic forms on $M_P(\mathbb{A})$. Let W_{π_0} be the space of smooth right K-finite functions f on $N_P(\mathbb{A})M_P(\mathbb{Q})\backslash G(\mathbb{A})$ such that the function $f_g(m) = f(mg)$ on $M_P(\mathbb{A})$ belongs to the space of π_0 for all $g \in G(\mathbb{A})$. Given $f \in W_{\pi_0}$, we define the Eisenstein series, at least formally, by the series

$$E(f,\nu)(g) = \sum_{\gamma \in P(\mathbb{Q})\backslash G(\mathbb{Q})} (\nu + \rho_P)(\gamma g) f(\gamma g)$$

where $\nu \in \check{\mathfrak{a}}_{P,\mathbb{C}}$, $g \in G(\mathbb{A})$, and $\rho_P \in \check{\mathfrak{a}}_P$ is the half-sum of positive roots in N_P. We view $\nu + \rho_P$ as a character of $G(\mathbb{A})$ extended from a character of $M_P(\mathbb{A})$ via Iwasawa decomposition trivially on $N_P(\mathbb{A})$ and K. The defining series of $E(f,\nu)(g)$ converges absolutely and locally uniformly in a positive cone deep enough in the positive Weyl chamber of $\check{\mathfrak{a}}_{P,\mathbb{C}}$ determined by P. It has the analytic continuation to a meromorphic function of ν on the whole space $\check{\mathfrak{a}}_{P,\mathbb{C}}$. The poles in the closure of the positive Weyl chamber are along the singular hyperplanes which form a locally finite family. For these properties of Eisenstein series see [4, Sect. IV.1] or [10]. We refer to the Eisenstein series $E(f,\nu)$, with $f \in W_{\pi_0}$, as the Eisenstein series associated to π_0.

We are interested in the analytic behavior of the Eisenstein series $E(f,\nu)$ at $\nu = \nu_0$. Since ν_0 is in the closure of the positive Weyl chamber and the family of singular hyperplanes is locally finite around ν_0, there is a (possibly empty) finite set of singular hyperplanes passing through ν_0. Hence, there is a polynomial $F(\nu)$ such that $F(\nu)E(f,\nu)$ is holomorphic around $\nu = \nu_0$. Then, the space $\mathcal{A}_{\{P\},\phi_\pi}$ is defined as the span of all the coefficients in the Taylor expansions of $F(\nu)E(f,\nu)$ around $\nu = \nu_0$, with f ranging over W_{π_0}. Although $F(\nu)$ is not unique, this definition does not depend on that choice.

The automorphic forms in $\mathcal{A}_{\{P\},\phi_\pi}$ are compatible with E, that is, belong to the space $\mathcal{A}_{\mathcal{J}}$, if and only if the cuspidal support ϕ_π is compatible with E as in [9, Sect. 1.3]. We denote by $\Phi_{\mathcal{J},\{P\}}$ the set of all associate classes ϕ_π of cuspidal automorphic representations of the Levi factors of parabolic subgroups in $\{P\}$ which are compatible with E. If $\phi_\pi \in \Phi_{\mathcal{J},\{P\}}$ we write $\mathcal{A}_{\mathcal{J},\{P\},\phi_\pi}$ for $\mathcal{A}_{\{P\},\phi_\pi}$. Then, the space $\mathcal{A}_{\mathcal{J}}$ admits the decomposition along the cuspidal support

$$\mathcal{A}_{\mathcal{J}} = \bigoplus_{\{P\}} \bigoplus_{\phi_\pi \in \Phi_{\mathcal{J},\{P\}}} \mathcal{A}_{\mathcal{J},\{P\},\phi_\pi}$$

as a $(\mathfrak{g}_\infty, K_\infty; G(\mathbb{A}_f))$-module. The summands indexed by the full group $\{G\}$ consist of cuspidal automorphic forms.

The decomposition of $\mathcal{A}_{\mathcal{J}}$ gives rise to the corresponding decomposition in cohomology. Thus, the automorphic cohomology admits the decomposition along the cuspidal support

$$H^*(G, E) = \bigoplus_{\{P\}} \bigoplus_{\phi_\pi \in \Phi_{\mathcal{J},\{P\}}} H^*(\mathfrak{g}_\infty, K_\infty; \mathcal{A}_{\mathcal{J},\{P\},\phi_\pi} \otimes_{\mathbb{C}} E).$$

Since the summands indexed by the full group $\{G\}$ come from cuspidal automorphic forms, their sum is called cuspidal cohomology. The natural complement of cuspidal cohomology consists of summands indexed by $\{P\} \neq \{G\}$. Their sum is called Eisenstein cohomology, because the cohomology classes can be represented by derivatives of Eisenstein series and their residues.

The main object of our concern is the non-vanishing and structural description of the individual summands of Eisenstein cohomology in the decomposition along the cuspidal support. In other words, the problem is to determine for which cuspidal supports ϕ_π the summand

$$H^*(\mathfrak{g}_\infty, K_\infty; \mathcal{A}_{\mathcal{J},\{P\},\phi_\pi} \otimes_{\mathbb{C}} E), \quad \{P\} \neq \{G\} \tag{A}$$

is not trivial, and in that case determine the structure of that cohomology space.

Let $\mathcal{L}$ denote the $(\mathfrak{g}_\infty, K_\infty; G(\mathbb{A}_f))$-submodule of $\mathcal{A}$ consisting of square-integrable automorphic forms. The space $\mathcal{L}$ admits the decomposition along the cuspidal support. We denote by $\mathcal{L}_{\mathcal{J},\{P\},\phi_\pi}$ the summand in that decomposition supported in ϕ_π. It is the space of square-integrable automorphic forms in $\mathcal{A}_{\mathcal{J},\{P\},\phi_\pi}$. The Langlands spectral decomposition (cf. [4, 10]) implies that $\mathcal{L}_{\mathcal{J},\{P\},\phi_\pi}$ is spanned by all square-integrable iterated residues at $\nu = \nu_0$ of the Eisenstein series $E(f, \nu)$ associated to π_0. Recall that we write here $\pi \cong \pi_0 \otimes \nu_0$ as before.

The inclusion of $\mathcal{L}_{\mathcal{J},\{P\},\phi_\pi}$ into $\mathcal{A}_{\mathcal{J},\{P\},\phi_\pi}$ gives rise to a map in cohomology

$$H^*(\mathfrak{g}_\infty, K_\infty; \mathcal{L}_{\mathcal{J},\{P\},\phi_\pi} \otimes_{\mathbb{C}} E) \longrightarrow H^*(\mathfrak{g}_\infty, K_\infty; \mathcal{A}_{\mathcal{J},\{P\},\phi_\pi} \otimes_{\mathbb{C}} E).$$

However, this map is not necessarily injective. Its image is the summand supported in ϕ_π in the so-called square-integrable cohomology, and denoted by

$$H^*_{(\mathrm{sq})}(\mathfrak{g}_\infty, K_\infty; \mathcal{A}_{\mathcal{J},\{P\},\phi_\pi} \otimes_{\mathbb{C}} E).$$

The square-integrable cohomology consists of cohomology classes that can be represented by square-integrable automorphic forms. In the case $\{P\} = \{G\}$, the summand in square-integrable cohomology is the same as the corresponding summand in full cohomology, because the unitary cuspidal automorphic forms are square-integrable. Thus, the interesting part of square-integrable cohomology lies in Eisenstein cohomology. We are interested in describing the square-integrable cohomology

$$H^*_{(\mathrm{sq})}(\mathfrak{g}_\infty, K_\infty; \mathcal{A}_{\mathcal{J},\{P\},\phi_\pi} \otimes_{\mathbb{C}} E), \quad \{P\} \neq \{G\} \tag{L}$$

in the summand (A) in the decomposition of Eisenstein cohomology along the cuspidal support.

3 Non-Vanishing Conditions

In this section we review the necessary conditions for non-vanishing of the summand (A) in the decomposition of Eisenstein cohomology along the cuspidal support. These conditions arise from the representation theoretic consideration of the space $\mathcal{A}_{\mathcal{J},\{P\},\phi_\pi}$ of automorphic forms supported in π. We omit the details and refer to [11, 12].

The crucial point is that there is a finite filtration of the space $\mathcal{A}_{\mathcal{J},\{P\},\phi_\pi}$, defined through the analytic behavior at $\nu = \nu_0$ of the Eisenstein series $E(f, \nu)$ associated to π_0, such that the successive quotients of the filtration are parabolically induced representations. It is the so-called Franke filtration introduced in [8, Sect. 6]. If the Eisenstein series $E(f, \nu)$ associated to π_0 is holomorphic at $\nu = \nu_0$, then $E(f, \nu_0)$, together with its derivatives at $\nu = \nu_0$, defines an intertwining map of the representation parabolically induced from $\pi_0 \otimes \nu_0$ to $G(\mathbb{A})$, tensored by the symmetric algebra $S(\check{\mathfrak{a}}_{P,\mathbb{C}})$ of derivatives with respect to ν, into the space of automorphic forms. Otherwise, if the Eisenstein series has a pole at $\nu = \nu_0$, one should consider the residues of Eisenstein series, and use the degenerate Eisenstein series on $G(\mathbb{A})$, supported in π_0, to construct intertwining maps between certain induced representations and the filtration quotients of the Franke filtration.

Hence, roughly speaking, the relative Lie algebra cohomology of $\mathcal{A}_{\mathcal{J},\{P\},\phi_\pi}$ vanishes, if the relative Lie algebra cohomology of certain induced representation vanishes. Applying Frobenius reciprocity reduces this latter cohomology to the cohomology of $\pi_0 \otimes \nu_0$ with respect to the coefficient system given by the Lie algebra cohomology $H^*(\mathfrak{n}_P, E)$, where $\mathfrak{n}_P$ is the Lie algebra of the unipotent radical $N_P(\mathbb{R})$. Then, the non-vanishing conditions arise from the non-vanishing of the cuspidal cohomology at the level of Levi factors. To make these conditions precise we need some more notation.

Given a standard parabolic $\mathbb{Q}$-subgroup P of G, let A_P denote the maximal $\mathbb{Q}$-split torus in the center of the Levi factor M_P of P. The Lie algebra $\mathfrak{a}_P$ of $A_P(\mathbb{R})$ is isomorphic to $\mathfrak{a}_P \cong X_*(A_P) \otimes \mathbb{R}$, where $X_*(A_P)$ denotes the $\mathbb{Z}$-module of $\mathbb{Q}$-rational cocharacters of A_P. In the case of the minimal parabolic $\mathbb{Q}$-subgroup P_0, we write simply A_0 and $\mathfrak{a}_0$, instead of A_{P_0} and $\mathfrak{a}_{P_0}$, respectively. We retain the notation $\check{\mathfrak{a}}_P = X^*(P) \otimes_{\mathbb{Z}} \mathbb{R}$, and write $\check{\mathfrak{a}}_0$ in the case $P = P_0$. There is a natural pairing of $\mathfrak{a}_P$ and $\check{\mathfrak{a}}_P$. The inclusion of A_P into A_0 gives rise to a map of $\mathfrak{a}_P$ into $\mathfrak{a}_0$. On the other hand, the restriction of characters from P to P_0 gives rise to a map of $\check{\mathfrak{a}}_P$ to $\check{\mathfrak{a}}_0$. These two maps provide natural decompositions

$$\mathfrak{a}_0 = \mathfrak{a}_P \oplus \mathfrak{a}_0^P \quad \text{and} \quad \check{\mathfrak{a}}_0 = \check{\mathfrak{a}}_P \oplus \check{\mathfrak{a}}_0^P .$$

The space $\check{\mathfrak{a}}_0^P$ may be viewed as the space of infinitesimal characters of representations of $M_P(\mathbb{R})$. The projection of an element $\lambda \in \check{\mathfrak{a}}_0$ to $\check{\mathfrak{a}}_P$ and $\check{\mathfrak{a}}_0^P$ is obtained by restriction to $\mathfrak{a}_P$ and $\mathfrak{a}_0^P$ and is thus denoted by $\lambda|_{\mathfrak{a}_P}$ and $\lambda|_{\mathfrak{a}_0^P}$, respectively.

Let ρ_0 denote the half-sum of positive roots in the absolute root system for G, viewed as an element of $\check{\mathfrak{a}}_0$. Let Λ denote the highest weight of the finite-dimensional representation E of $G(\mathbb{C})$, viewed as an element of $\check{\mathfrak{a}}_0$.

Let W be the absolute Weyl group of G. For a standard parabolic $\mathbb{Q}$-subgroup P, let W_P be the absolute Weyl group of its Levi factor, viewed as a subgroup of W. We denote by W^P the set of coset representatives for $W_P \backslash W$, which are of minimal length in their coset.

According to [13], the Lie algebra cohomology of the unipotent radical is

$$H^q(\mathfrak{n}_P, E) = \bigoplus_{\substack{w \in W^P \\ \ell(w)=q}} F_{\mu_w},$$

where $\ell(w)$ is the length of w, and F_{μ_w} is the irreducible finite-dimensional representation of $M_P(\mathbb{C})$ of highest weight

$$\mu_w = w(\Lambda + \rho_0) - \rho_0.$$

Then, the non-vanishing conditions arise from the non-vanishing conditions for the cuspidal cohomology of the Levi factor with respect to coefficient systems F_{μ_w}. The results are summarized in the following theorem.

Theorem 3.1 (Necessary non-vanishing conditions) *Let $P = M_P N_P$ be a standard proper parabolic $\mathbb{Q}$-subgroup of G. Let $\pi = \pi_0 \otimes \nu_0$ be a cuspidal automorphic representation of $M_P(\mathbb{A})$, where π_0 is unitary, and ν_0 an element of the closure of the positive Weyl chamber in $\check{\mathfrak{a}}_P$ viewed as a character of $M_P(\mathbb{A})$.*

Then, the summand (A), *that is,*

$$H^*(\mathfrak{g}_\infty, K_\infty; \mathcal{A}_{\mathcal{J},\{P\},\phi_\pi} \otimes_{\mathbb{C}} E)$$

in the decomposition of Eisenstein cohomology is trivial except possibly if the following conditions

(1) $\nu_0 = -w(\Lambda + \rho_0)|_{\mathfrak{a}_P}$,
(2) $-w(\Lambda + \rho_0)|_{\mathfrak{a}_0^P}$ *is the infinitesimal character of* π_0,
(3) $-w_{l,P}\left(\mu_w|_{\mathfrak{a}_0^P}\right) = \mu_w|_{\mathfrak{a}_0^P}$, *where* $w_{l,P}$ *is the longest element of* W_P,
(4) $\pi_{0,\infty}$ *is cohomological,*

are all satisfied with the same minimal coset representative $w \in W^P$.

Regarding the proof of these facts, the first two conditions come from the compatibility with $\mathcal{J}$, and were proved in [12, Corollary 3.5] and [12, p. 55], respectively.

The third condition follows from [14]. The last condition is clear from the definition of cohomological representations. Recall that a representation of $G(\mathbb{R})$ is called cohomological, if it has non-trivial cohomology with respect to some coefficient system [15].

As an example of the strength of these necessary conditions for non-vanishing, we have the following theorem. We state here the theorem only for $\mathbb{Q}$-split classical groups, although an analogous result holds for the general linear group as well [16, 17]. The case of non-split classical groups can be handled in the same way, because the non-vanishing conditions are given in terms of absolute root systems and absolute Weyl groups. The precise statements for non-split classical groups can thus be deduced from the split case. For example, the case of unitary groups is considered in our work in progress [18].

Theorem 3.2 ([16, 17]) *Let G be one of the $\mathbb{Q}$-split classical groups Sp_n, SO_{2n+1}, SO_{2n} of $\mathbb{Q}$-rank n. Let $P = M_P N_P$ be the standard parabolic subgroup with the Levi factor $M_P \cong GL_{n_1} \times \cdots \times GL_{n_k} \times G'$, where G' is a (possibly trivial) smaller group of the same type. Let $\pi_0 \cong \tau_1 \otimes \cdots \otimes \tau_k \otimes \sigma$ be a unitary cuspidal automorphic representation of $M_P(\mathbb{A})$, where τ_i, resp. σ, is a unitary cuspidal automorphic representation of $GL_{n_i}(\mathbb{A})$, resp. $G'(\mathbb{A})$. Let $\nu_0 \in \check{\mathfrak{a}}_P$ correspond to the character $|\det|^{s_1} \otimes \cdots \otimes |\det|^{s_k}$ of $M_P(\mathbb{A})$, where $s_i \in \mathbb{R}$. Let*

$$\pi \cong \pi_0 \otimes \nu_0 \cong \tau_1 |\det|^{s_1} \otimes \cdots \otimes \tau_k |\det|^{s_k} \otimes \sigma.$$

Then, the summand (A) *in the decomposition of Eisenstein cohomology along the cuspidal support, that is,*

$$H^*(\mathfrak{g}_\infty, K_\infty; \mathcal{A}_{\mathcal{J},\{P\},\phi_\pi} \otimes_{\mathbb{C}} E),$$

is trivial except possibly if all $s_i \in \frac{1}{2}\mathbb{Z}$.

We call this property the half-integrality condition. Its importance lies in the fact that further study of the summand (A) requires understanding of the analytic properties of the Eisenstein series $E(f, \nu)$ at $\nu = \nu_0$, and this is simplified by the half-integrality restriction on the possible s_i. For example, if σ is globally generic in the theorem above, the Langlands–Shahidi method [4, 10, 19, 20], tells us that the poles of the Eisenstein series at $\nu = \nu_0$ are determined by the poles of the complete automorphic L-functions in its constant term. In the cases considered in the theorem, the automorphic L-functions in the constant term of the Eisenstein series are the Rankin–Selberg automorphic L-functions of pairs for the general linear group times the classical group, and the symmetric and exterior square automorphic L-functions attached to τ_i at the value $2s_i$ of their complex parameter. For the generic representations, by the global functorial lifting from classical groups to the general linear group [21, 22], the former are related to the Rankin–Selberg automorphic L-functions of pairs for the general linear group, which are well understood. For the latter, according to the above theorem, only $2s_i \in \mathbb{Z}$, that is, the analytic behavior at integral values of the complex parameter, matters for cohomology. This excludes the critical strip $0 < Re(s) < 1$ from the consideration. The holomorphy of the symmetric and exterior square (complete) automorphic L-functions for the values of its complex parameter in the critical strip is only recently proved [23, 24], using Arthur's endoscopic classification of automorphic representations in the discrete spectrum for G as in the theorem [25]. At the time of our study, Arthur's classification was still conjectural, and we used the above theorem to get unconditional results on automorphic cohomology.

4 Square-Integrable Cohomology

The first step towards complete understanding of the internal structure of the summand (A) in the decomposition of Eisenstein cohomology along the cuspidal support is understanding the summand (L) in the square-integrable cohomology, that is,

$$H^*_{(sq)}(\mathfrak{g}_\infty, K_\infty; \mathcal{A}_{\mathcal{J},\{P\},\phi_\pi} \otimes_{\mathbb{C}} E), \quad \{P\} \neq \{G\}.$$

This is due to the fact that the (possibly trivial) space $\mathcal{L}_{\mathcal{J},\{P\},\phi_\pi}$ is a $(\mathfrak{g}_\infty, K_\infty; G(\mathbb{A}_f))$-submodule of $\mathcal{A}_{\mathcal{J},\{P\},\phi_\pi}$, which always forms the lowest filtration step in the Franke filtration of the latter.

The summand (L) in square-integrable cohomology is, of course, trivial if the summand (A) in the full Eisenstein cohomology is trivial. Hence, all necessary conditions for non-vanishing stated in Theorem 3.1 should be satisfied in order that (L) is possibly non-trivial. Since these conditions are obtained from cohomological considerations, we refer to them as geometric conditions.

On the other hand, the summand (L) is certainly trivial if the space $\mathcal{L}_{\mathcal{J},\{P\},\phi_\pi}$ is trivial. The space $\mathcal{L}_{\mathcal{J},\{P\},\phi_\pi}$ is spanned by the square-integrable iterated residues at $\nu = \nu_0$ of the Eisenstein series associated to π_0. The existence of (non-zero) such residues is determined by the analytic behavior of the Eisenstein series at $\nu = \nu_0$, which is closely related to the analytic behavior of the automorphic L-functions in its constant term, according to the Langlands–Shahidi method [4, 10, 19, 20]. Hence, the necessary conditions for non-vanishing of the summand (L) arising from non-triviality of the space $\mathcal{L}_{\mathcal{J},\{P\},\phi_\pi}$ can be written in terms of the analytic properties of automorphic L-functions. We refer to them as arithmetic non-vanishing conditions.

The geometric and arithmetic necessary conditions for non-vanishing of the summand (L) of square-integrable cohomology form a subtle combination, which provides a strong restriction on possible contributions to square-integrable cohomology. As an example, we state here a theorem for summands (L) supported in the Siegel maximal proper parabolic $\mathbb{Q}$-subgroup of the symplectic group over $\mathbb{Q}$. This is a special case of the results obtained in [26], which are dealing with arbitrary maximal proper parabolic $\mathbb{Q}$-subgroup of the symplectic group. The analogous results for the case of odd special orthogonal group are obtained by Gotsbacher and Grobner in [27].

Theorem 4.1 ([26]) *Let $G = Sp_n$ be the symplectic group over $\mathbb{Q}$ of $\mathbb{Q}$-rank n. Let P be the Siegel standard maximal proper parabolic $\mathbb{Q}$-subgroup, that is, the Levi factor is $M_P \cong GL_n$. Let π_0 be a unitary cuspidal automorphic representation of $M_P(\mathbb{A}) \cong GL_n(\mathbb{A})$. Let $\nu_0 \in \check{\mathfrak{a}}_P$ correspond to the character $|\det|^{s_0}$, with $s_0 \geq 0$, of $M_P(\mathbb{A}) \cong GL_n(\mathbb{A})$. Let $\pi \cong \pi_0 \otimes |\det|^{s_0}$. Let $\Lambda = \sum_{i=1}^{n} \lambda_i e_i$, with $\lambda_i \in \mathbb{Z}$ and $\lambda_1 \geq \cdots \geq \lambda_n \geq 0$, be the highest weight of the finite-dimensional representation E of G, where e_i is the projection of a fixed maximal split torus of G onto its ith component.*

Then, the summand (L) in square-integrable cohomology supported in π, that is,

$$H^*_{(\mathrm{sq})}(\mathfrak{g}_\infty, K_\infty; \mathcal{A}_{\mathcal{J},\{P\},\phi_\pi} \otimes_{\mathbb{C}} E),$$

is trivial, except possibly if the following conditions are satisfied.

(1) $s_0 = 1/2$,
(2) *the exterior square automorphic L-function $L(s, \pi_0, \wedge^2)$ has a pole at $s = 1$,*
(3) *the principal automorphic L-function $L(s, \pi_0)$ is non-zero at $s = 1/2$,*
(4) *the rank n is even,*
(5) *the highest weight Λ satisfies $\lambda_{2j-1} = \lambda_{2j}$ for $j = 1, \ldots, n/2$,*
(6) *the infinite component $\pi_{0,\infty}$ of π_0 is the tempered representation of $GL_n(\mathbb{R})$ which is isomorphic to the fully induced representation*

$$\pi_{0,\infty} \cong \mathrm{Ind}_{Q(\mathbb{R})}^{GL_n(\mathbb{R})} \left(\bigotimes_{j=1}^{n/2} D(2\mu_j + 2n - 4j + 4) \right),$$

where $\mu_j = \lambda_{2j-1} = \lambda_{2j}$, $D(k)$ with $k \geq 2$ is the discrete series representation of $GL_2(\mathbb{R})$ of lowest $O(2)$-type k, and Q is the standard parabolic subgroup of GL_n with the Levi factor isomorphic to a direct product of $n/2$ copies of GL_2.

From the above theorem one may get the idea how difficult it is to determine the very existence of a cuspidal support which may possibly contribute non-trivially to square-integrable cohomology. In the special case treated in the theorem, the unitary part π_0 of the cuspidal support should have a precisely determined infinite component, while at the same time should satisfy both arithmetic conditions: the exterior square L-function attached to π_0 should have a pole at $s = 1$ and the principal L-function attached to π_0 should be non-zero at $s = 1/2$.

On the other hand, the theorem excludes from consideration many possibilities. It shows, for example, that if the rank of the symplectic group is odd, then there is no contribution to square-integrable cohomology from automorphic forms supported in the Siegel maximal proper parabolic subgroup. It also shows that for some coefficient systems, namely those not satisfying condition (5) of the theorem, there is no contribution to square-integrable cohomology from automorphic forms supported in the Siegel maximal proper parabolic subgroup.

To further illustrate the subtle combination of geometric and arithmetic necessary conditions for non-vanishing, we consider a simple low-rank example in which some of the more technical conditions of the above theorem are avoided.

Corollary 4.2 *Let $G = Sp_2$ be the symplectic group over $\mathbb{Q}$ of $\mathbb{Q}$-rank two. Let P be the Siegel standard maximal proper parabolic $\mathbb{Q}$-subgroup of G, that is, the Levi factor $M_P \cong GL_2$. Let π_0 be a unitary cuspidal automorphic representation of $M_P(\mathbb{A}) \cong GL_2(\mathbb{A})$. Let $\nu_0 \in \check{\mathfrak{a}}_P$ correspond to the character $|\det|^{s_0}$, with $s_0 \geq 0$, of $M_P(\mathbb{A}) \cong GL_2(\mathbb{A})$. Let $\pi \cong \pi_0 \otimes |\det|^{s_0}$. Let the coefficient system E be the trivial representation of G.*

Then, the summand (L) *in square-integrable cohomology supported in π, that is,*

$$H^*_{(\mathrm{sq})}(\mathfrak{g}_\infty, K_\infty; \mathcal{A}_{\mathcal{J},\{P\},\phi_\pi} \otimes_{\mathbb{C}} E),$$

is trivial, except possibly if the following conditions are satisfied.

(1) $s_0 = 1/2$,
(2) the central character of π_0 is trivial,
(3) the principal automorphic L-function $L(s, \pi_0)$ is non-zero at $s = 1/2$,
(4) the infinite component $\pi_{0,\infty}$ of π_0 is the discrete series representation of $GL_2(\mathbb{R})$ of lowest $O(2)$-type 4.

Hence, to show the existence of a cuspidal support π in the Siegel parabolic subgroup that could possibly contribute non-trivially to square-integrable cohomology of Sp_2, with respect to the trivial coefficient system, one should show the existence of a unitary cuspidal automorphic representation π_0 of $GL_2(\mathbb{A})$ with trivial central character, discrete series of lowest $O(2)$-type 4 as the infinite component, and such that $L(1/2, \pi_0) \neq 0$. In terms of classical automorphic forms, this may be rephrased as the existence problem for a holomorphic modular form of weight 4, trivial Nebentypus, arbitrary level, and such that $L(1/2, \pi_0) \neq 0$. This problem in wider generality was studied by Trotabas in [28]. A consequence of his work is the existence of a Hilbert modular form of any given even weight, trivial Nebentypus, arbitrary level, and such that $L(1/2, \pi_0) \neq 0$. In particular, it shows the existence of π_0 with the required properties.

In a recent preprint [29], we study the existence of non-trivial cohomology classes in square-integrable cohomology for the split symplectic and special orthogonal groups of rank two, as well as the exceptional group G_2, over a totally real number field. In that work, we encounter various arithmetic conditions, and to show the existence of the cuspidal support satisfying the necessary non-vanishing conditions, we use not only the result of Trotabas, but also a construction of monomial representations of $GL_2(\mathbb{A})$ via automorphic induction from the appropriate Hecke characters of the group of idèles.

However, the existence of the cuspidal support satisfying the necessary non-vanishing conditions is still not sufficient to imply the non-vanishing of the summand (L) in the decomposition of square-integrable cohomology along the cuspidal support. This requires further study of the internal structure of the summand (A) in the decomposition of full Eisenstein cohomology, which is the subject of the following section.

5 Internal Structure of Cohomology

As already mentioned above, once we establish the existence of a cuspidal support ϕ_π, satisfying all the necessary non-vanishing conditions for the summand (L) in the decomposition of square-integrable cohomology along the cuspidal support, that is,

$$H^*_{(\mathrm{sq})}(\mathfrak{g}_\infty, K_\infty; \mathcal{A}_{\mathcal{J},\{P\},\phi_\pi} \otimes_{\mathbb{C}} E), \quad \{P\} \neq \{G\},$$

the final step is to show that this summand is indeed non-trivial.

There exist two approaches to settle this problem. The first is due to Rohlfs–Speh [30, Theorem I.1 = III.1]. They show that, given a cuspidal support ϕ_π satisfying the necessary non-vanishing conditions, the summand (L) of square-integrable cohomology is non-trivial in the lowest degree in which the relative Lie algebra cohomology

$$H^*(\mathfrak{g}_\infty, K_\infty; \mathcal{L}_{\mathcal{J},\{P\},\phi_\pi} \otimes_{\mathbb{C}} E)$$

is non-trivial. In other words, they show that the map in cohomology induced by the inclusion of $\mathcal{L}_{\mathcal{J},\{P\},\phi_\pi}$ into $\mathcal{A}_{\mathcal{J},\{P\},\phi_\pi}$ is non-zero in that lowest degree. Their method of proving this fact is using the geometric expression of the automorphic cohomology in terms of differential forms.

We use the result of Rohlfs–Speh in the preprint [29], already mentioned above, to show the actual non-vanishing of certain summands in the decomposition of square-integrable cohomology for the split symplectic and special orthogonal groups of rank two, and the exceptional group G_2, defined over a totally real number field. As an example, we present here the special case already considered in Corollary 4.2.

Theorem 5.1 ([29]) *Let $G = Sp_2$ be the symplectic group over $\mathbb{Q}$ of $\mathbb{Q}$-rank two. Let P be the Siegel standard maximal proper parabolic $\mathbb{Q}$-subgroup of G, that is, the Levi factor $M_P \cong GL_2$. Let the coefficient system E be the trivial representation of G.*

Then, there exists a unitary cuspidal automorphic representation π_0 of $M_P(\mathbb{A}) \cong GL_2(\mathbb{A})$ such that, for the cuspidal support $\pi \cong \pi_0 \otimes |\det|^{1/2}$, the summand (L) *in square-integrable cohomology supported in π, that is,*

$$H^*_{(\mathrm{sq})}(\mathfrak{g}_\infty, K_\infty; \mathcal{A}_{\mathcal{J},\{P\},\phi_\pi} \otimes_{\mathbb{C}} E),$$

does not vanish in degree $q = 2$. These classes, represented by square-integrable residues of Eisenstein series, contribute non-trivially in degree $q = 2$ to the full Eisenstein cohomology space $H^*_{\mathrm{Eis}}(Sp_2, \mathbb{C})$.

The second approach to the non-vanishing of summands (L) is using the Franke filtration. The Franke filtration, originally defined in [8, Sect. 6], and its refinement introduced in [9, Theorem 1.4], is a finite descending filtration of the $(\mathfrak{g}_\infty, K_\infty; G(\mathbb{A}_f))$-module $\mathcal{A}_{\mathcal{J},\{P\},\phi_\pi}$. The main advantage of using this filtration is that the successive quotients of the filtration may be described as parabolically induced representations from certain Levi factors of G. This allows not only showing the non-vanishing of the summands (L) in square-integrable cohomology, but also the explicit calculation of the internal structure of the summands (A) in the full Eisenstein cohomology. The idea is to calculate first the cohomology of filtration quotients, and then use, step-by-step, the long exact sequences in cohomology.

The main disadvantage of the Franke filtration approach is that it may be very difficult to write the filtration in a form feasible for explicit calculation. The reason is that the definition of the filtration depends on the analytic behavior, not only of the Eisenstein series associated to π_0, but also all possible degenerate Eisenstein series with same cuspidal support, on the Levi factors of parabolic subgroups containing an element of the associate class $\{P\}$. The Franke filtration for the case of the cuspidal support in the associate class $\{P\}$ of a maximal proper parabolic subgroup P is described in [31].

The Franke filtration approach was pursued in [32] for the case of the symplectic group of rank two defined over a totally real number field. As an example of the Franke filtration approach to the calculation of cohomology, we present here again the same example as in Corollary 4.2 and Theorem 5.1. Observe that using the Franke filtration provides more information about the structure of cohomology. This result over $\mathbb{Q}$ was earlier obtained by Schwermer in [33] using a completely different method.

Theorem 5.2 ([32, 33]) *Let $G = Sp_2$ be the symplectic group over $\mathbb{Q}$ of $\mathbb{Q}$-rank two. Let P be the Siegel standard maximal proper parabolic $\mathbb{Q}$-subgroup of G, that is, the Levi factor $M_P \cong GL_2$. Let the coefficient system E be the trivial representation of G. Assume that π_0 is a unitary cuspidal automorphic representation of $GL_2(\mathbb{A})$, such that the cuspidal support ϕ_π, represented by $\pi \cong \pi_0 \otimes |\det|^{1/2}$, satisfies the necessary conditions for non-vanishing of the summand* (L)*.*

Then, the summand (A) *in the decomposition of full Eisenstein cohomology along the cuspidal support is isomorphic to*

$$H^q(\mathfrak{g}_\infty, K_\infty; \mathcal{A}_{\mathcal{J},\{P\},\phi_\pi}) \cong$$

$$\begin{cases} H^2_{(\mathrm{sq})}(\mathfrak{g}_\infty, K_\infty; \mathcal{A}_{\mathcal{J},\{P\},\phi_\pi}) = H^2(\mathfrak{g}_\infty, K_\infty; \mathcal{L}_{\mathcal{J},\{P\},\phi_\pi}) \neq 0, & \text{for } q = 2, \\ \text{a submodule of } H^3(\mathfrak{g}_\infty, K_\infty; \mathcal{A}_{\mathcal{J},\{P\},\phi_\pi}/\mathcal{L}_{\mathcal{J},\{P\},\phi_\pi}), & \text{for } q = 3, \\ H^4_{(\mathrm{sq})}(\mathfrak{g}_\infty, K_\infty; \mathcal{A}_{\mathcal{J},\{P\},\phi_\pi}), \text{ possibly trivial}, & \text{for } q = 4, \\ 0, & \text{otherwise.} \end{cases}$$

In particular, under the assumption that π_0 with the required properties exists, the square-integrable cohomology is non-trivial in degree $q = 2$.

The non-vanishing of the map in cohomology induced by $\mathcal{L}_{\mathcal{J},\{P\},\phi_\pi} \hookrightarrow \mathcal{A}_{\mathcal{J},\{P\},\phi_\pi}$ in general was studied by Grobner in [34] using the Franke filtration. In that paper he shows that this map is injective in a certain range of cohomology degrees (all low degrees up to a certain bound). This, as a consequence, reproves the result of Rohlfs–Speh. However, in order to show actual non-vanishing, it is still necessary to check the existence of the cuspidal support ϕ_π such that the cohomology space

$$H^*(\mathfrak{g}_\infty, K_\infty; \mathcal{L}_{\mathcal{J},\{P\},\phi_\pi} \otimes_\mathbb{C} E)$$

is non-trivial in some of the degrees in which the map in cohomology is injective. This again boils down to the subtle combination of geometric and arithmetic conditions, as in Sects. 3 and 4, which were studied in [29] for aforementioned rank two cases. In particular, the existence of representations π_0 as in the theorem is obtained in [29].

References

1. Li, J.-S., Schwermer, J.: Automorphic representations and cohomology of arithmetic groups. In: Challenges for the 21st century (Singapore, 2000), pp. 102–137. World Scientific Publishing, River Edge, NJ (2001)
2. Schwermer, J.: The cohomological approach to cuspidal automorphic representations. Automorphic Forms and L-Functions I. Global Aspects. Contemporary Mathematics, vol. 488, pp. 257–285. American Mathematical Society, Providence (2009)
3. Schwermer, J.: Geometric cycles, arithmetic groups and their cohomology. Bull. Am. Math. Soc. (N.S.) **47**(2), 187–279 (2010)
4. Mœglin, C., Waldspurger, J.-L.: Spectral Decomposition and Eisenstein Series, Cambridge Tracts in Mathematics, vol. 113. Cambridge University Press, Cambridge (1995)
5. Borel, A., Jacquet, H.: Automorphic Forms and Automorphic Representations. In: Automorphic forms, representations and L-functions, Proceedings of Symposia Pure Mathematics, XXXIII, Part 1, pp. 189–202. American Mathematical Society, Providence (1979)
6. Borel, A.: Regularization theorems in Lie algebra cohomology. Applications. Duke Math. J. **50**(3), 605–623 (1983)
7. Borel, A., Wallach, N.: Continuous Cohomology, Discrete Subgroups, and Representations of Reductive Groups. Mathematical Surveys and Monographs, vol. 67, 2nd edn. American Mathematical Society, Providence (2000)
8. Franke, J.: Harmonic analysis in weighted L_2-spaces. Ann. Sci. École Norm. Super. (4) **31**(2), 181–279 (1998)
9. Franke, J., Schwermer, J.: A decomposition of spaces of automorphic forms, and the Eisenstein cohomology of arithmetic groups. Math. Ann. **311**(4), 765–790 (1998)
10. Langlands, R.P.: On the Functional Equations Satisfied by Eisenstein Series. Lecture Notes in Mathematics, vol. 544. Springer, Berlin (1976)
11. Li, J.-S., Schwermer, J.: On the Eisenstein cohomology of arithmetic groups. Duke Math. J. **123**(1), 141–169 (2004)
12. Schwermer, J.: Kohomologie Arithmetisch Definierter Gruppen und Eisensteinreihen. Lecture Notes in Mathematics, vol. 988. Springer, Berlin (1983)
13. Kostant, B.: Lie algebra cohomology and the generalized Borel-Weil theorem. Ann. Math. **74**(2), 329–387 (1961)
14. Borel, A., Casselman, W.: L_2-cohomology of locally symmetric manifolds of finite volume. Duke Math. J. **50**(3), 625–647 (1983)
15. Vogan Jr., D.A., Zuckerman, G.J.: Unitary representations with nonzero cohomology. Compos. Math. **53**(1), 51–90 (1984)

16. Grbac, N., Schwermer, J.: On Eisenstein series and the cohomology of arithmetic groups. C. R. Math. Acad. Sci. Paris **348**(11–12), 597–600 (2010)
17. Grbac, N., Schwermer, J.: Eisenstein series, cohomology of arithmetic groups, and automorphic L-functions at half integral arguments. Forum Math. **26**(6), 1635–1662 (2014)
18. Grbac, N., Schwermer, J.: Eisenstein cohomology for unitary groups I. The case of relative rank one, in preparation
19. Shahidi, F.: A proof of Langlands' conjecture on Plancherel measures; complementary series for p-adic groups. Ann. Math. (2) **132**(2), 273–330 (1990)
20. Shahidi, F.: Eisenstein Series and Automorphic L-Functions. American Mathematical Society Colloquium Publications, vol. 58. American Mathematical Society, Providence (2010)
21. Cogdell, J.W., Kim, H.H., Piatetski-Shapiro, I.I., Shahidi, F.: On lifting from classical groups to GL_N. Publ. Math. Inst. Hautes Études Sci. **93**, 5–30 (2001)
22. Cogdell, J.W., Kim, H.H., Piatetski-Shapiro, I.I., Shahidi, F.: Functoriality for the classical groups. Publ. Math. Inst. Hautes Études Sci. **99**, 163–233 (2004)
23. Grbac, N.: On the residual spectrum of split classical groups supported in the Siegel maximal parabolic subgroup. Monatsh. Math. **163**(3), 301–314 (2011)
24. Grbac, N., Shahidi, F.: Endoscopic transfer for unitary groups and holomorphy of Asai L-functions. Pac. J. Math. **276**(1), 185–211 (2015)
25. Arthur, J.: The Endoscopic Classification of Representations. American Mathematical Society Colloquium Publications, vol. 61. American Mathematical Society, Providence (2013)
26. Grbac, N., Schwermer, J.: On residual cohomology classes attached to relative rank one Eisenstein series for the symplectic group. Int. Math. Res. Not. IMRN (7), 1654–1705 (2011)
27. Gotsbacher, G., Grobner, H.: On the Eisenstein cohomology of odd orthogonal groups. Forum Math. **25**(2), 283–311 (2013)
28. Trotabas, D.: Non annulation des fonctions L des formes modulaires de Hilbert au point central. Ann. Inst. Fourier (Grenoble) **61**(1), 187–259 (2011)
29. Grbac, N., Schwermer, J.: A non-vanishing result for the residual Eisenstein cohomology of arithmetic groups of low rank (preprint)
30. Rohlfs, J., Speh, B.: Pseudo Eisenstein forms and the cohomology of arithmetic groups III: residual cohomology classes. On Certain L-Functions. Clay Mathematics Proceedings, vol. 13, pp. 501–523. American Mathematical Society, Providence (2011)
31. Grbac, N.: The Franke filtration of the spaces of automorphic forms supported in a maximal proper parabolic subgroup. Glas. Mat. Ser. III. **47**(67)(2), 351–372 (2012)
32. Grbac, N., Grobner, H.: The residual Eisenstein cohomology of Sp_4 over a totally real number field. Trans. Am. Math. Soc. **365**(10), 5199–5235 (2013)
33. Schwermer, J.: On arithmetic quotients of the Siegel upper half space of degree two. Compos. Math. **58**(2), 233–258 (1986)
34. Grobner, H.: Residues of Eisenstein series and the automorphic cohomology of reductive groups. Compos. Math. **149**, 1061–1090 (2013)

Eisenstein Cohomology for $\mathrm{Sl}_2(\mathbb{Z}[i])$ and Special Values of L-Functions

Günter Harder

Abstract In this note we discuss a special case of a general theme: How do special values of L-functions influence the structure of the cohomology of arithmetic groups?

Keywords Cohomology of arithmetic groups · Eisenstein cohomology
Algebraic Hecke characters · Special values

Introduction: This paper is a continuation of (or appendix to) my paper [3]. In that paper I investigate arithmetic properties of cohomology classes, which are represented by Eisenstein-differential forms. As an application I could prove certain rationality results for special values of L-function attached to algebraic Hecke characters. (See [3], Sects. 2.2 and 4). The study of the integrality properties of these Eisenstein-cohomology classes was postponed to a later moment in time ([3], bottom p. 106).

These integrality properties of Eisenstein cohomology classes are governed by the arithmetic properties of special values of L-functions. We consider rational Eisenstein classes, which become integral if, we restrict them to the boundary, but which are not integral themselves, they have a denominator. The general philosophy predicts that this denominator is related to a certain expression in special L-values.

In [6], Volume-III.pdf, Chaps. 3 and 5. I discuss the analogous problem for the group $\Gamma = \mathrm{Sl}_2(\mathbb{Z})$ and the cohomology $H^1(\Gamma\backslash\mathbb{H}, \tilde{\mathcal{M}}_n)$. In this case it turns out that the denominator is equal to the numerator of $\zeta(-1-n)$ (Theorem 5.5.1).

G. Harder (✉)
Max Planck Institut für Mathematik, Vivatsgasse 7, 53111
Bonn, Germany
e-mail: harder@mpim-bonn.mpg.de

G. Harder
Universität Bonn Mathematisches Institut Endenicher, Allee 60,
53115 Bonn, Germany

J. W. Cogdell et al. (eds.), *Cohomology of Arithmetic Groups*,
Springer Proceedings in Mathematics & Statistics 245,
https://doi.org/10.1007/978-3-319-95549-0_3

In the present paper I consider a special class of congruence subgroups of $\mathrm{Sl}_2(\mathbb{Z}[i])$ and cohomology with trivial coefficients. As in the above case the integrality of the Eisenstein classes can be tested by evaluating theses classes on certain specific cycles, the capped modular symbols. We show that the Eisenstein classes are not necessarily integral, their denominators can be computed in terms of special values of L-functions. It turns out that we get much smaller denominators than in the above case (Theorem 1). These denominators "create" non trivial cohomology classes.

The big question, which I have not been able to decide, is whether these classes are torsion classes or non torsion classes in the inner cohomology. (Section 5, dichotomy A /B).

In this paper I concentrated on a rather special class of groups because for these groups D. Yasaki provides some experimental data (See Sect. 5), so we could decide the issue for the the primes $p < 20,000$. So far we only found torsion classes. This is in contrast to what happens in the case $\mathrm{Sl}_2(\mathbb{Z})$ and what seems to happen in the case of $\mathrm{Sp}_2(\mathbb{Z})$ (See [5, 6]).

In his talk of at this conference A. Reid mentioned the works of Grunewald, Schwermer and Rohlfs on the cohomology of Bianchi groups. This was a subject of intense discussions, when all these people were together at Bonn.

A. Reid's talk inspired me to resume the theme.

1 A Special Family of Bianchi Groups

We only consider a very special case, in principle this is an example that could have been discussed at the end of [3]. Our starting point is the semi-simple group scheme $G_0/\mathbb{Z}[i] = \mathrm{Sl}_2/\mathbb{Z}[i]$. Let $T_0/\mathbb{Z}[i]$, $B_0/\mathbb{Z}[i]$ and $U_0/\mathbb{Z}[i]$ be the standard split diagonal torus, the standard Borel subgroup of upper diagonal matrices and its unipotent radical. We identify $\mathbb{G}_m/\mathbb{Z}[i] = T_0/\mathbb{Z}[i]$ via the isomorphism $t \mapsto \begin{pmatrix} t & 0 \\ 0 & t^{-1} \end{pmatrix}$. Let $G/\mathbb{Z} = R_{\mathbb{Z}[i]/\mathbb{Z}}(G_0)$, $T/\mathbb{Z}$, $B/\mathbb{Z}$, $U/\mathbb{Z}$ be the respective Weil restriction to $\mathbb{Z}$. Throughout this paper we put $w = \begin{pmatrix} 0 & -1 \\ 1 & 0 \end{pmatrix}$, it is the standard representative for the Weyl group of T_0. In the following we put $F = \mathbb{Q}(i)$.

The group $\Gamma = \mathrm{Sl}_2(\mathbb{Z}[i]) = G(\mathbb{Z})$ is an arithmetic subgroup of $G(\mathbb{R})$. We choose a prime $p \equiv 5 \bmod 8$, we choose a prime ideal $\mathfrak{p} \subset \mathbb{Z}[i]$ which divides p, let $\varpi_\mathfrak{p}$ be a generator of this ideal. We define the arithmetic group

$$\Gamma_0(\mathfrak{p}) = \{\gamma = \begin{pmatrix} a & b \\ c & d \end{pmatrix} \in \Gamma | c \equiv 0 \bmod \mathfrak{p}\}, \tag{1}$$

it is the inverse image of the Borel subgroup $\bar{B} = B(\mathbb{Z}[i]/\mathfrak{p}) \subset \bar{G} = \mathrm{Sl}_2(\mathbb{Z}[i]/\mathfrak{p})$ under the reduction map. The map $r_\mathfrak{p} : \gamma \mapsto a \bmod \mathfrak{p}$ defines a surjective homomorphism from $\Gamma_0(\mathfrak{p})$ to the multiplicative group $(\mathbb{Z}[i]/\mathfrak{p})^\times = \mathbb{F}_p^\times$, we put

$$\Gamma_1(\mathfrak{p}) = \ker(r_\mathfrak{p}) = \{\gamma \in \Gamma | a \equiv 1, c \equiv 0 \bmod \mathfrak{p}\}.$$

Finally let $\Gamma^{(1)}(\mathfrak{p})$ be the full congruence subgroup mod $\mathfrak{p}$.

The group $\mathbb{F}_p^\times$ contains a primitive fourth root of unity i_p more precisely we have $\mathbb{F}_p^\times = \{i_p^\nu\}_{\{\nu=0,\dots,3\}} \times \mathbb{F}_p^{\times,\mathrm{odd}}$. Hence we have a character

$$\chi : \Gamma_0(\mathfrak{p}) \to \mu_4 = \{i^\nu\}_{\{\nu=0,1,2,3\}} \subset \mathbb{Z}[i]^\times,$$

which sends i_p to -1. We define still another congruence subgroup, namely $\Gamma_0'(\mathfrak{p}) = \ker(\chi)$.

The group $G(\mathbb{R}) = \mathrm{Sl}_2(\mathbb{C})$ contains $SU(2) = K_\infty$ as maximal compact subgroup and $\mathbb{H}_3 = G(\mathbb{R})/K_\infty$ is the three dimensional hyperbolic space, which can be identified to $\mathbb{H}_3 = \mathbb{C} \times \mathbb{R}_{>0}$. The maximal compact subgroup K_∞ is the stabilizer of the point (0, 1) (See [6], Chap. 1, Sect. 1.2.1). The group Γ acts properly discontinuously on $\mathbb{H}_3$ we consider quotient spaces $\Gamma'\backslash\mathbb{H}_3$, where Γ' is a congruence subgroup of Γ. These spaces are not compact, we can embed them into their Borel–Serre compactification $\Gamma'\backslash\bar{\mathbb{H}}_3$, this embedding is a homotopy equivalence.

For later purpose we briefly recall this construction (See [6]). For any point $\alpha \in \mathbb{P}^1(F)$ we add a copy $X_\alpha = \mathbb{C}$ to $\mathbb{H}_3$, i.e. form the space

$$\bar{\mathbb{H}}_3 = \mathbb{H}_3 \cup \bigcup_\alpha X_\alpha$$

endowed with a suitable topology. The group Γ' acts on $\bar{\mathbb{H}}_3$, it has finitely many orbits on $\mathbb{P}^1(F)$ and the quotient $\Gamma'\backslash\bar{\mathbb{H}}_3$ is compact. If $\mathcal{C} = \{\alpha_1, \dots, \alpha_h\}$ is a set of representatives for the orbits of Γ' on $\mathbb{P}^1(F)$, then the quotient

$$\Gamma'\backslash \bigcup_\alpha X_\alpha = \bigcup_{\alpha\in\mathcal{C}} \Gamma'_\alpha\backslash X_\alpha = \partial(\Gamma\backslash\mathbb{H}_3)$$

is the boundary. Here of course Γ'_α is the stabilizer of α.

In this note we are interested in the cohomology with compact supports $H_c^q(\Gamma_0'(\mathfrak{p})\backslash\mathbb{H}_3, \mathbb{Z}[i])$, in the cohomology without supports $H^q(\Gamma_0'(\mathfrak{p})\backslash\mathbb{H}_3, \mathbb{Z}[i])$ and in the cohomology of the boundary $H^q(\partial(\Gamma_0'(\mathfrak{p})\backslash\mathbb{H}_3), \mathbb{Z}[i])$. These three versions of cohomology are related by the fundamental exact sequence below (See (3)).

On these cohomology groups we have an action of the cyclic group $\mu_4 = \{i^\nu\}$ of order 4. We put $R = \mathbb{Z}[i, 1/2]$ then we can decompose these cohomology groups into eigenspaces under this action. In this paper we are mainly interested in the χ eigenspaces.

For the following see for instance [6], Chap. 2. We introduce the induced $\bar{G} = \mathrm{Sl}_2(\mathbb{Z}[i]/\mathfrak{p})-$ module

$$I_{\bar{B}}^{\bar{G}}\chi = \mathrm{Ind}_{\Gamma_0(\mathfrak{p})}^{\Gamma}\chi := \{h : \mathrm{Sl}_2(\mathbb{F}_p) \to R \mid h(bx) = \chi(b)h(x)\ \forall\, b \in B(\mathbb{F}_p), x \in \mathrm{Sl}_2(\mathbb{F}_p)\}, \quad (2)$$

This is of course also a $\Gamma-$ module and therefore it provides a sheaf $\widetilde{I_{\bar{B}}^{\bar{G}}\chi}$ on $\Gamma\backslash\mathbb{H}_3$. Then the lemma of Shapiro asserts

$$H^1_?(\Gamma\backslash\mathbb{H}_3, \widetilde{I_{\bar{B}}^{\bar{G}}\chi}) = H^1_?(\Gamma'_0(\mathfrak{p})\backslash\mathbb{H}_3, R)[\chi].$$

Of course the corresponding statement also holds for the cohomology of the boundary. We have the fundamental exact sequence which relates the different cohomology groups

$$\to H^1_c(\Gamma\backslash\mathbb{H}_3, \widetilde{I_{\bar{B}}^{\bar{G}}\chi}) \to H^1(\Gamma\backslash\mathbb{H}_3, \widetilde{I_{\bar{B}}^{\bar{G}}\chi}) \xrightarrow{r} H^1(\partial(\Gamma\backslash\mathbb{H}_3), \widetilde{I_{\bar{B}}^{\bar{G}}\chi}) \to H^2_c(\Gamma\backslash\mathbb{H}_3, \widetilde{I_{\bar{B}}^{\bar{G}}\chi}) \to . \tag{3}$$

On all these modules we have an action of the Hecke algebra, this action commutes with all the arrows in the sequence. For any prime $\mathfrak{q} \neq \mathfrak{p}$ there exists a Hecke operator $T_{\mathfrak{q}}$ which induces an endomorphism on any of the modules in the sequence. Moreover these induced endomorphisms commute with each other.

As usual we also introduce the "inner cohomology" (See [6], Chap. 2, Sect. 2.1.4)

$$H^1_!(\Gamma\backslash\mathbb{H}_3, \widetilde{I_{\bar{B}}^{\bar{G}}\chi}) = \ker(r). \tag{4}$$

2 The Cohomology of the Boundary and the Eisenstein Section

The term $H^1(\partial(\Gamma\backslash\mathbb{H}_3), \widetilde{I_{\bar{B}}^{\bar{G}}\chi})$ can be computed explicitly (See [3] p. 66, [4]II, Sect. 2.3). The set of Borel subgroups of $G/\mathbb{Q}$ which are defined over $\mathbb{Q}$ is equal to $\mathbb{P}^1(F)$. Since Γ acts transitively on $\mathbb{P}^1(F)$ we get that the quotient $\Gamma\backslash\mathbb{H}_3$ has only one cusp ([6], Sect. 2.2.2), this cusp can represented by $\infty \in \mathbb{P}^1(F)$. The stabilizer of this cusp is

$$B(\mathbb{Z}) = \Gamma_\infty = \left\{\begin{pmatrix} i^\nu & u \\ 0 & i^{-\nu}\end{pmatrix} | u \in \mathbb{Z}[i]\right\} \supset U(\mathbb{Z}) = \Gamma'_\infty = \left\{\begin{pmatrix} 1 & u \\ 0 & 1\end{pmatrix} | u \in \mathbb{Z}[i]\right\}.$$

Under the action of $\bar{B} = B(\mathbb{Z}/(p))$ the module $I_{\bar{B}}^{\bar{G}}\chi$ decomposes into the two summands of those functions h which are supported on $\bar{B}$ and those functions which are supported on $\bar{B}w\bar{U}$. This means that the $\bar{B}$-module $\bar{G}$ on $I_{\bar{B}}^{\bar{G}}\chi$ decomposes into two summands

$$I_{\bar{B}}^{\bar{G}}\chi = R \otimes \chi \oplus R[R/\mathfrak{p}] \otimes \chi^{-1},$$

where $R[R/\mathfrak{p}]$ is of course the group ring of the finite group $\mathbb{F}_p = R/\mathfrak{p}$. On this group ring the element $\begin{pmatrix} i & 0 \\ 0 & i^{-1}\end{pmatrix}$ acts by multiplication by -1. We have the obvious identification $\Gamma'_\infty = R$, the cohomology $H^1(\partial(\Gamma\backslash\mathbb{H}_3), \widetilde{I_{\bar{B}}^{\bar{G}}\chi})$ is the module of coinvariants under the action of Γ'_∞ and hence

$$H^1(\partial(\Gamma\backslash\mathbb{H}_3), \widetilde{I_{\bar{B}}^{\bar{G}}\chi}) = \mathrm{Hom}(R, R) \oplus \mathrm{Hom}(\mathfrak{p}, R).$$

For the second term we again use the lemma of Shapiro. Now we have two special elements Id (the identity) and $\mathbf{c}$ (the complex conjugation) in $\mathrm{Hom}(R, R)$, and since we inverted 2, we get

$$\mathrm{Hom}(\Gamma'_\infty, I_{\bar{B}}^{\bar{G}}\chi) = \mathbb{Z}\left[\frac{1}{2}\right] \mathrm{Id} \oplus \mathbb{Z}\left[\frac{1}{2}\right] \mathbf{c} \oplus \mathbb{Z}\left[\frac{1}{2}\right] \mathrm{Id}\ \varpi_{\mathfrak{p}}^{-1} \oplus \mathbb{Z}\left[\frac{1}{2}\right] \mathbf{c}\ \bar{\varpi}_{\mathfrak{p}}^{-1}, \tag{5}$$

this is a torsion free module. We also know that the Hecke operators $T_{\mathfrak{q}}$ act by multiplication by scalars on the on the cohomology of the boundary. More precisely:

$$\begin{cases} T_{\mathfrak{q}} = \phi(\mathfrak{q}) + \phi(\bar{\mathfrak{q}}) & \text{if } \mathfrak{q} \text{ is split and not} = \bar{\mathfrak{p}} \\ 0 & \text{if } \mathfrak{q} \text{ is inert} \end{cases} \tag{6}$$

The goal of this paper is to determine the image of the restriction map r, we will not be completely successful. In Sect. 5 we formulate a partial result, we end up with a dichotomy between two cases.

2.1 *The Rational Image*

For this we start with the identification of

$$\mathrm{Im}(r) \otimes \mathbb{Q} \subset H^1(\partial(\Gamma\backslash\mathbb{H}_3), \widetilde{I_{\bar{B}}^{\bar{G}}\chi} \otimes \mathbb{Q}).$$

We have to recall the description of the boundary cohomology from [4]. Let $J(\mathfrak{p})$ be the group of fractional ideals in F which are prime to $\mathfrak{p}$. Then we have a unique Hecke character $\phi_f : J(\mathfrak{p}) \to \mathbb{Q}[i]^\times$ which is defined by the rule:

$$\text{If } \mathfrak{a} \in J(\mathfrak{p}) \text{ and if } \mathfrak{a} = (\alpha) \text{ with } \alpha \bmod \mathfrak{p} \in \mathbb{F}_p^{\times,\mathrm{odd}} \text{ then } \phi_f(\mathfrak{a}) = \alpha^2. \tag{7}$$

Let $\mathbb{A} = \mathbb{A}_{\mathbb{Q}} = \mathbb{R} \times \mathbb{A}_f$ be the ring of adeles. Elements in the adele ring or in groups with values in adele rings will be denoted by underlined letters, and we write $\underline{x} = (x_\infty, \underline{x}_f)$, where x_∞(resp. $\underline{x}_f$) are the component at infinity (resp. finite component.)

The above character ϕ_f is the finite part of a uniquely defined algebraic Hecke character

$$\phi = \phi_\infty \times \phi_f : \mathbb{G}_m(F)\backslash\mathbb{G}_m(\mathbb{A}_F) \to \mathbb{C}^\times$$

where $\phi_\infty(z) = z^{-2}$.

We relate ϕ to χ. Using the above identification $\mathbb{G}_m = T_0$ we may also view ϕ as a Hecke character $\phi : T(\mathbb{Q})\backslash T(\mathbb{A}) \to \mathbb{C}^\times$. Let $\mathfrak{U}_f = \prod_{\mathfrak{q}} \mathcal{O}_{\mathfrak{q}}^\times \subset T(\mathbb{A}_f)$ be the maximal compact subgroup of units. Then $\phi_{\mathfrak{q}}$ is unramified for all primes $\mathfrak{q} \neq \mathfrak{p}$. The local component $\phi_{\mathfrak{p}}$ factors over $(\mathcal{O}_{\mathfrak{p}}/\mathfrak{p})^\times = \mathbb{F}_p^\times$ and this factorization is our character χ.

2.1.1 Interlude on Algebraic Hecke Characters and Their L-Functions

An algebraic Hecke character $\psi : T(\mathbb{Q})\backslash T(\mathbb{A}) \to \mathbb{C}^\times$ is of type (a, b) if $\psi_\infty(z) = z^{-a}\bar{z}^{-b}$, hence our above character ϕ of type $(2, 0)$. Such a Hecke character ψ is ramified in a finite set of places $S(\psi)$ and we define its L-function

$$L(\psi, s) = \prod_{\mathfrak{q}\notin S(\psi)} \frac{1}{1 - \psi(\mathfrak{q})N\mathfrak{q}^{-s}}.$$

The completed L-function is defined by

$$\Lambda(\psi, s) = \frac{\Gamma(s)}{(2\pi)^s} L(\psi, s)$$

We may twist our Hecke character by an integral power of the norm, i.e. we may consider $\psi \circ N_{F/\mathbb{Q}}^n = \psi'$. Then ψ' is of type $(a', b') = (a - n, b - n)$. There is a unique $n_0 \in \mathbb{Z}$ such that $a', b' \geq 0$ and such that one of these two numbers is zero, i.e. $a'b' = 0$. Let us call the resulting character $\psi^* = \psi \circ N_{F/\mathbb{Q}}^{n_0}$. We define the *cohomological L -function* by

$$L^{\text{coh}}(\psi, s) = L(\psi^*, s) = L(\psi, s + n_0), \tag{8}$$

and the *completed cohomological $L-$function* (see also [6] Chap. 7)

$$\Lambda^{\text{coh}}(\psi, s) = \Lambda(\psi^*, s). \tag{9}$$

This cohomological $L-$function $L^{\text{coh}}(\psi, s)$ has the virtue that it does not change if we twist ψ by a power of the norm $N_{F/\mathbb{Q}}^n$. If we define the weight of ψ to be $k(\psi) = |a - b|$ then this weight also remains unchanged under twists. The completed cohomological $L-$function satisfies a functional equation

$$\Lambda^{\text{coh}}(\psi, s) = W(\psi)(N\mathfrak{f})^{\frac{k(\psi)+1}{2} - s} \Lambda^{\text{coh}}(\bar{\psi}, k(\psi) + 1 - s). \tag{10}$$

We apply this to our character ϕ. Then the completed $L-$function is a holomorphic function in the entire s plane and satisfies the functional equation

$$\Lambda^{\mathrm{coh}}(\phi, s) = W_{\mathfrak{p}}(\phi)(4p)^{3/2-s}\Lambda^{\mathrm{coh}}(\bar{\phi}, 3-s), \tag{11}$$

where $W_{\mathfrak{p}}(\phi) = \pm\frac{\varpi_{\mathfrak{p}}^2}{p}$. We have an explicit rule to determine the sign.

2.1.2 Back to the Boundary Cohomology

We return to the description of the boundary cohomology. We introduce the induced representation

$$\mathrm{Ind}_{B(\mathbb{A})}^{G(\mathbb{A})}\phi = \{f : G(\mathbb{A}) \to \mathbb{C} \mid f(\underline{b}\underline{x}) = \phi(\underline{b})h(\underline{x})\ \forall\ \underline{b} \in B(\mathbb{A}), x \in G\mathbb{A})\} = I_{\phi_\infty} \otimes I_{\phi_f}. \tag{12}$$

The group $G(\mathbb{A})$ is acting by right translations: $(R(\underline{x})f)(\underline{g}) = f(\underline{g}\underline{x})$.

The finite part I_{ϕ_f} is a $G(\mathbb{A}_f)$ -module. Let $K_f^{(0)} = \prod \mathrm{Sl}_2(\mathcal{O}_{F,\mathfrak{q}}) \subset G(\mathbb{A}_f)$ be the standard maximal compact open subgroup, let $K_f^{(1)}(\mathfrak{p})$ be the open compact subgroup where the factor at $\mathfrak{p}$ is replaced by the full congruence subgroup $\bmod \mathfrak{p}$. Then $K_f^{(0)}/K_f^{(1)}(\mathfrak{p}) = \mathrm{Sl}_2(\mathbb{F}_p)$. Then we have the following well known equality

$$I_{\phi_f}^{K_f^{(1)}(\mathfrak{p})} = I_{\bar{B}}^{\bar{G}}\chi \otimes \mathbb{C}. \tag{13}$$

If $K_{0,f}(\mathfrak{p}) \subset K_f^{(0)}$ is the inverse image of the standard Borel subgroup $B(\mathbb{F}_p)$, then

$$I_{\phi_f}^{K_{0,f}(\mathfrak{p})} = (I_{\bar{B}}^{\bar{G}}\chi)^{\bar{U}} \otimes \mathbb{C} = \mathbb{C}h_0 \oplus \mathbb{C}h_\infty \tag{14}$$

where h_0 has support $\bar{B}$ and takes value 1 at the identity e and h_∞ has support $\bar{B}w\bar{B}$ and takes value 1 at w. Then it is shown in [3, 4] that

$$H^1(\partial(\Gamma_0'(\mathfrak{p})\backslash\mathbb{H}_3), \mathbb{C})[\chi] = H^1(\mathfrak{g}, K_\infty, I_{\phi_\infty}) \otimes I_{\phi_f}^{K_{0,f}(\mathfrak{p})} \oplus H^1(\mathfrak{g}, K_\infty, I_{\bar{\phi}_\infty}) \otimes I_{\phi_f}^{K_{0,f}(\mathfrak{p})}. \tag{15}$$

The Lie-algebra cohomology $H^1(\mathfrak{g}, K_\infty, I_{\phi_\infty})$ is easy to compute, it is a one dimensional $\mathbb{C}$ vector space with a canonical generator. By definition it is the first cohomology of the complex

$$\mathrm{Hom}_{K_\infty}(\Lambda^\bullet(\mathfrak{g}/\mathfrak{k}), I_{\phi_\infty}) \xrightarrow{\sim} \mathrm{Hom}_{K_\infty^T}(\Lambda^\bullet(\mathfrak{t}/\mathfrak{t}^K), \mathrm{Hom}(\Lambda^\bullet(\mathfrak{u}), \mathbb{C} \otimes \phi_\infty)),$$

the isomorphism is Frobenius reciprocity. We are interested in degree $\bullet = 1$ and here in the subspace

$$\mathrm{Hom}_{K_\infty^T}(\Lambda^0(\mathfrak{t}/\mathfrak{t}^K), \mathrm{Hom}(\Lambda^1(\mathfrak{u}), \mathbb{C}\otimes\phi_\infty)) = \mathrm{Hom}_{K_\infty^T}(\Lambda^1(\mathfrak{u}), \mathbb{C}\otimes\phi_\infty)$$

The Lie-algebra $\mathfrak{u} = \mathbb{Q}\begin{pmatrix}0 & 1\\ 0 & 0\end{pmatrix} \oplus \mathbb{Q}\begin{pmatrix}0 & i\\ 0 & 0\end{pmatrix}$ and the following two elements yield a basis of $\mathfrak{u}\otimes F$:

$$U_\alpha = \frac{1}{2}\left(\begin{pmatrix}0 & 1\\ 0 & 0\end{pmatrix}\otimes 1 - \begin{pmatrix}0 & i\\ 0 & 0\end{pmatrix}\otimes i\right);\ U_{\bar\alpha} = \frac{1}{2}\left(\begin{pmatrix}0 & 1\\ 0 & 0\end{pmatrix}\otimes 1 + \begin{pmatrix}0 & i\\ 0 & 0\end{pmatrix}\otimes i\right). \quad (16)$$

Now $e(\varphi) = \begin{pmatrix}e^{i\varphi} & 0\\ 0 & e^{-i\varphi}\end{pmatrix} \in K_\infty^T$ acts by conjugation on U_α(resp. $U_{\bar\alpha}$) and this action is by multiplication by $e^{i2\varphi}$(resp. $e^{-i2\varphi}$). Hence it is clear that the elements of the dual basis

$$U_\alpha^\vee \in \mathrm{Hom}_{K_\infty^T}(\Lambda^1(\mathfrak{u}), \mathbb{C}\otimes\phi_\infty) \text{ resp. } U_{\bar\alpha}^\vee \in \mathrm{Hom}_{K_\infty^T}(\Lambda^1(\mathfrak{u}), \mathbb{C}\otimes\bar\phi_\infty)$$

are generators of these two respective one dimensional vector spaces. Then (15) becomes

$$H^1(\partial(\Gamma_0'(\mathfrak{p})\backslash\mathbb{H}_3), \mathbb{C})[\chi] = \mathbb{C}U_\alpha^\vee \otimes I_{\phi_f}^{K_{0,f}(\mathfrak{p})} \oplus \mathbb{C}U_{\bar\alpha}^\vee \otimes I_{\phi_f}^{K_{0,f}(\mathfrak{p})}. \quad (17)$$

Comparing this to our formula (5) we have that $U_\alpha^\vee = \mathrm{Id}$, $U_{\bar\alpha}^\vee = \mathbf{c}$, and if $h = ah_0 + bh_\infty$ then

$$U_\alpha^\vee \otimes h = (a, 0, \frac{b}{\varpi_\mathfrak{p}}, 0),\ U_{\bar\alpha}^\vee \otimes h = \left(0, \bar a, 0, \frac{\bar b}{\bar\varpi_\mathfrak{p}}\right).$$

We are still working on the image of r. We try to write down a section, we start by writing down an embedding

$$\mathrm{Eis}_\infty : \mathrm{Ind}_{B(\mathbb{R})}^{G(\mathbb{R})} I_{\phi_\infty} \otimes I_{\bar B}^{\bar G}\chi \to \mathcal{A}(\Gamma\backslash(G(\mathbb{R})\times\Gamma/\Gamma^{(1)}(\mathfrak{p})) \quad (18)$$

which is defined by the infinite summation

$$\mathrm{Eis}_\infty(f_\infty\otimes h)(g_\infty)(a) = \sum_{\gamma\in B(\mathbb{Z})\backslash G(\mathbb{Z})} f_\infty(\gamma g_\infty)h(\gamma a^{-1}) \quad (19)$$

where $g_\infty \in G(\mathbb{R})$, $a \in \Gamma/\Gamma^{(1)}$ and f_∞ is a $\mathcal{C}_\infty$ function on $G(\mathbb{R})$.

Here a comment is in order. In this paper we do not care about questions of convergence. The theory of Eisenstein series by Langlands provides a well known method to overcome this kind of difficulties with divergent series, they are defined by analytic continuation. These issues are discussed in [3, 4, 6].

The notation suggests that $\mathrm{Eis}_\infty(f_\infty \otimes h)(g_\infty)$ is a function on $\Gamma\backslash G(\mathbb{R})$ with values in the group ring $\mathbb{C}[\Gamma/\Gamma^{(1)}(\mathfrak{p})]$ and therefore we can evaluate at $a \in \mathbb{C}[\Gamma/\Gamma^{(1)}(\mathfrak{p})]$). For $b \in \bar{B}$ we get $\mathrm{Eis}_\infty(f_\infty \otimes h)(g_\infty)(ba) = \mathrm{Eis}_\infty(f_\infty \otimes b^{-1}h)(g_\infty)(a)$. Recall $h = uh_0 + vh_\infty$. We have $b^{-1}h_0 = \chi^{-1}(b)h_0$, $bh_\infty = \chi(b)h_\infty$. Since χ is a quadratic character we conclude

$$\mathrm{Eis}_\infty(U_\alpha^\vee \otimes h)(g_\infty)(ba) = \chi(b)\mathrm{Eis}_\infty(U_\alpha^\vee \otimes h)(g_\infty)(a), \tag{20}$$

and hence we can interpret $\mathrm{Eis}_\infty(U_\alpha^\vee \otimes h)(g_\infty)(a)$ as a 1-form on $\Gamma\backslash\mathbb{H}_3$ with coefficients in $I_{\bar{B}}^{\bar{G}}\chi \otimes \mathbb{C}$.

This yields an inclusion of cohomology classes

$$\mathrm{Eis}^{(1)} : \mathbb{C}U_\alpha^\vee \otimes I_{\phi_f}^{K_{0,f}(\mathfrak{p})} = \mathbb{C}U_\alpha^\vee \otimes (I_{\bar{B}}^{\bar{G}}\chi)^{\bar{U}} \to H^1(\Gamma_0'(\mathfrak{p})\backslash\mathbb{H}_3, \mathbb{C})[\chi]).$$

We restrict these cohomology classes to the boundary, for this we have to compute the constant Fourier coefficient. We have the well known formula

$$\mathcal{F} \circ \mathrm{Eis}^{(1)}(U_\alpha^\vee \otimes h) = U_\alpha^\vee \otimes h + \frac{\pi}{2}\frac{L(\phi,-1)}{L(\phi,0)}\, U_{\bar{\alpha}}^\vee \otimes T_{\mathfrak{p}}^{\mathrm{loc}}(h), \tag{21}$$

where $T_{\mathfrak{p}}^{\mathrm{loc}} : I_{\phi_f}^{K_{0,f}(\mathfrak{p})} \to I_{\phi_f}^{K_{0,f}(\mathfrak{p})}$ is the local intertwining operator at $\mathfrak{p}$. It is given by the following formula

$$T_{\mathfrak{p}}^{\mathrm{loc}}(h_0) = \frac{1}{p}h_\infty \,;\, T_{\mathfrak{p}}^{\mathrm{loc}}(h_\infty) = h_0 + \left(1 - \frac{1}{p}\right)h_\infty. \tag{22}$$

The ratio of L-values can be written in terms of the completed cohomological $L-$functions. We invoke the functional equation, and the formula for the constant term becomes

$$\mathcal{F} \circ \mathrm{Eis}^{(1)}(U_\alpha^\vee \otimes h) = U_\alpha^\vee \otimes h + 2\sqrt{p}\, W_{\mathfrak{p}}(\phi)\frac{\Lambda^{\mathrm{coh}}(\bar{\phi},2)}{\Lambda^{\mathrm{coh}}(\phi,2)}\, U_{\bar{\alpha}}^\vee \otimes T_{\mathfrak{p}}^{\mathrm{loc}}(h).$$

All this up to here is already-in a slightly different language- in [3]. Theorem 2.1. in [3] says: If

$$c(\phi) = 2\sqrt{p}\, W_{\mathfrak{p}}(\phi)\frac{\Lambda^{\mathrm{coh}}(\bar{\phi},2)}{\Lambda^{\mathrm{coh}}(\phi,2)} = \frac{\pi}{2}\frac{L^{\mathrm{coh}}(\phi,1)}{L^{\mathrm{coh}}(\phi,2)}, \tag{23}$$

then the image is described by

$$\mathrm{Im}(r) \otimes \mathbb{Q} = \{U_\alpha^\vee \otimes h + c(\phi)U_{\bar{\alpha}}^\vee \otimes T_{\mathfrak{p}}^{\mathrm{loc}}(h)\} \mid h \in I_{\phi_f}^{K_f^{(1)}(\mathfrak{p})}\}. \tag{24}$$

Of course this implies $c(\phi) \in F$.

2.2 Integrality of Eisenstein Classes

Now we come to the question which was raised at the end of [3]. For any $\mathbb{Z}$ -module A we denote by A_{int} the quotient of A by its torsion submodule. This quotient is canonically isomorphic to the image of A under the canonical homomorphism $A \to A \otimes \mathbb{Q}$.

We consider the intersection

$$\mathrm{Im}(r)_{\text{int}} = \mathrm{Im}(r) \otimes \mathbb{Q} \cap H^1(\partial(\Gamma_0'(\mathfrak{p})\backslash\mathbb{H}_3), R)[\chi]. \tag{25}$$

These are those classes $U_\alpha^\vee \otimes h + c(\phi) U_{\bar{\alpha}}^\vee \otimes T_{\mathfrak{p}}^{\text{loc}}(h)$ which are integral.

We write

$$c(\phi) = \frac{a(\phi)}{b(\phi)} \text{ with } a(\phi), b(\phi) \in \mathbb{Z}[i] \text{ and coprime.} \tag{26}$$

Hence if $h = ah_0 + bh_\infty$ then $U_\alpha^\vee \otimes (ah_0 + bh_\infty) + U_{\bar{\alpha}}^\vee \otimes c(\phi)(bh_0 + (b + \frac{a-b}{p})h_\infty)$ is integral if and only if both summands are integral. This means that

$$b \in b(\phi)R \text{ and } \frac{a-b}{\bar{\varpi}_{\mathfrak{p}}} \in b(\phi)R.$$

For simplicity we also invert p, i.e. we replace R by $R\left[\frac{1}{p}\right]$, then we get

$$\mathrm{Im}(r)_{\text{int}} = \{b(\phi) U_\alpha^\vee \otimes h + a(\phi) U_{\bar{\alpha}}^\vee \otimes T_{\mathfrak{p}}^{\text{loc}}(h) \mid h \in I_{\phi_f}^{K_f^{(1)}(\mathfrak{p})}\}. \tag{27}$$

We consider the image $\mathrm{Im}(r)$ in the fundamental exact sequence. Clearly $\mathrm{Im}(r) \subset \mathrm{Im}(r)_{\text{int}}$ and the quotient is a finite torsion module which injects under the boundary map

$$\delta : \mathrm{Im}(r)_{\text{int}}/\mathrm{Im}(r) \to H_c^2(\Gamma\backslash\mathbb{H}_3, \widetilde{I_{\bar{B}}^{\bar{G}}\chi}).$$

On the finitely generated $R-$ module $H^1(\Gamma\backslash\mathbb{H}_3, \widetilde{I_{\bar{B}}^{\bar{G}}\chi})$ we have an action of the Hecke algebra. If we tensorize it by $\mathbb{Q}$, then we get a splitting (Manin-Drinfeld principle)

$$H^1(\Gamma\backslash\mathbb{H}_3, \widetilde{I_{\bar{B}}^{\bar{G}}\chi}) \otimes \mathbb{Q} = H_!^1(\Gamma\backslash\mathbb{H}_3, \widetilde{I_{\bar{B}}^{\bar{G}}\chi}) \otimes \mathbb{Q} \oplus H_{\text{Eis}}^1(\Gamma\backslash\mathbb{H}_3, \widetilde{I_{\bar{B}}^{\bar{G}}\chi}) \otimes \mathbb{Q}, \tag{28}$$

where the first summand is the kernel of r and the second summand is spanned by the Eisenstein classes $\mathrm{Eis}^{(1)}(U_\alpha^\vee \otimes h), h \in I_{\phi_f}^{K_f^{(1)}(\mathfrak{p})}$. It is the unique submodule for which $H^1_{\mathrm{Eis}}(\Gamma\backslash\mathbb{H}_3, \widetilde{I_{\bar{B}}^{\bar{G}}\chi}) \otimes \mathbb{Q} \to \mathrm{Im}(r)$ is an isomorphism of modules for the Hecke algebra.

This induces a decomposition up to isogeny

$$H^1(\Gamma\backslash\mathbb{H}_3, \widetilde{I_{\bar{B}}^{\bar{G}}\chi}) \supset H^1_!(\Gamma\backslash\mathbb{H}_3, \widetilde{I_{\bar{B}}^{\bar{G}}\chi}) \oplus H^1_{\mathrm{Eis}}(\Gamma\backslash\mathbb{H}_3, \widetilde{I_{\bar{B}}^{\bar{G}}\chi}),$$

and we can define $\mathrm{Im}_{\mathrm{Eis}}(r)$ as the image of $H^1_{\mathrm{Eis}}(\Gamma\backslash\mathbb{H}_3, \widetilde{I_{\bar{B}}^{\bar{G}}\chi})$ under r. Clearly

$$\mathrm{Im}(r)_{\mathrm{int}} \supset \mathrm{Im}(r) \supset \mathrm{Im}_{\mathrm{Eis}}(r). \tag{29}$$

It is the main result of this paper that we have criteria for $\mathrm{Im}(r)_{\mathrm{int}}/\mathrm{Im}_{\mathrm{Eis}}(r) \neq 0$ in terms of special values of L-functions attached to algebraic Hecke characters (see Theorem 1 in Sect. 5 and its consequence (67)).

If we want to understand the quotient $\mathrm{Im}(r)_{\mathrm{int}}/\mathrm{Im}_{\mathrm{Eis}}(r)$ we start from a class $\xi = b(\phi)U_\alpha^\vee \otimes h + a(\phi)U_{\bar{\alpha}}^\vee \otimes T_{\mathfrak{p}}^{\mathrm{loc}}(h) \in \mathrm{Im}(r)_{\mathrm{int}}$. We consider the Eisenstein cohomology class

$$[\mathrm{Eis}^{(1)}(U_\alpha^\vee \otimes b(\phi)h)] \in H^1_{\mathrm{Eis}}(\Gamma\backslash\mathbb{H}_3, \widetilde{I_{\bar{B}}^{\bar{G}}\chi})) \otimes \mathbb{Q},$$

we know that the restriction of this class to the boundary cohomology is our class ξ. Now the class $[\mathrm{Eis}^{(1)}(U_\alpha^\vee \otimes b(\phi)h)]$ is not necessarily integral. We look for the "smallest" $\Delta(h) \in R$ such that

$$\Delta(h)[\mathrm{Eis}^{(1)}(U_\alpha^\vee \otimes b(\phi)h)] \in H^1_{\mathrm{Eis}}(\Gamma\backslash\mathbb{H}_3, \widetilde{I_{\bar{B}}^{\bar{G}}\chi})), \tag{30}$$

this number is the *denominator* of the Eisenstein class.

In the next section we determine this denominator by evaluating the Eisenstein class on certain homology classes, which will be provided by "capped modular symbols".

3 Homology and Modular Symbols

To our module $I_{\bar{B}}^{\bar{G}}\chi$ we attached the sheaf $\widetilde{I_{\bar{B}}^{\bar{G}}\chi}$, we can also attach a dual object to it namely the cosheaf $\underline{I_{\bar{B}}^{\bar{G}}\chi}$. (See [6], Chap. 2, Sect. 2.1.3) We can define the homology groups $H_i(\Gamma\backslash\bar{\mathbb{H}}_3, \underline{I_{\bar{B}}^{\bar{G}}\chi})$, ... and again we have a fundamental long exact sequence

$$H_1(\partial(\Gamma\backslash\mathbb{H}_3), \underline{I_{\bar{B}}^{\bar{G}}\chi}) \to H_1(\Gamma\backslash\bar{\mathbb{H}}_3, \underline{I_{\bar{B}}^{\bar{G}}\chi}) \to H_1(\Gamma\backslash\mathbb{H}_3, \partial(\Gamma\backslash\bar{\mathbb{H}}_3), \underline{I_{\bar{B}}^{\bar{G}}\chi}) \to H_0(\partial(\Gamma\backslash\bar{\mathbb{H}}_3), \underline{I_{\bar{B}}^{\bar{G}}\chi}) \to \ldots. \tag{31}$$

These homology groups are the homology of the chain complexes

$$\begin{array}{ccccccc}
 & 0 & & 0 & & 0 & \\
 & \downarrow & & \downarrow & & \downarrow & \\
\to & C_{i+1}(\partial(\Gamma\backslash\mathbb{H}_3), \underline{I_{\bar{B}}^{\bar{G}}\chi}) & \to & C_i(\partial(\Gamma\backslash\mathbb{H}_3), \underline{I_{\bar{B}}^{\bar{G}}\chi}) & \to & C_{i-1}(\partial(\Gamma\backslash\mathbb{H}_3), \underline{I_{\bar{B}}^{\bar{G}}\chi}) & \to \\
 & \downarrow & & \downarrow & & \downarrow & \\
\to & C_{i+1}(\Gamma\backslash\bar{\mathbb{H}}_3, \underline{I_{\bar{B}}^{\bar{G}}\chi}) & \to & C_i(\Gamma\backslash\bar{\mathbb{H}}_3, \underline{I_{\bar{B}}^{\bar{G}}\chi}) & \to & C_{i-1}(\Gamma\backslash\bar{\mathbb{H}}_3, \underline{I_{\bar{B}}^{\bar{G}}\chi}) & \to, \\
 & \downarrow & & \downarrow & & \downarrow & \\
\to & \bar{C}_{i+1}(\Gamma\backslash\bar{\mathbb{H}}_3, \underline{I_{\bar{B}}^{\bar{G}}\chi}) & \to & \bar{C}_i(\Gamma\backslash\bar{\mathbb{H}}_3, \underline{I_{\bar{B}}^{\bar{G}}\chi}) & \to & \bar{C}_{i-1}(\Gamma\backslash\bar{\mathbb{H}}_3, \underline{I_{\bar{B}}^{\bar{G}}\chi}) & \to \\
 & \downarrow & & \downarrow & & \downarrow & \\
 & 0 & & 0 & & 0 &
\end{array} \tag{32}$$

where the complex at the bottom line is simply the complex in the second line divided by the complex in the first line.

We recall the definition the of module $C_i(\Gamma\backslash\bar{\mathbb{H}}_3, \underline{I_{\bar{B}}^{\bar{G}}\chi})$: We start from the tensor product

$$C_i(\mathbb{H}_3) \otimes I_{\bar{B}}^{\bar{G}}\chi = \left\{\sum \sigma \otimes m_\sigma | \text{ finite sums and } \sigma : \Delta^i \to \mathbb{H}_3 \text{ is an } i \text{ simplex}\right\}.$$

On this module we have an action of Γ and

$$C_i(\Gamma\backslash\bar{\mathbb{H}}_3, \underline{I_{\bar{B}}^{\bar{G}}\chi}) := (C_i(\mathbb{H}_3) \otimes I_{\bar{B}}^{\bar{G}}\chi)_\Gamma \tag{33}$$

is the module of coinvariants.

Inside $C_1(\Gamma\backslash\bar{\mathbb{H}}_3, \underline{I_{\bar{B}}^{\bar{G}}\chi})$ we have some special elements, the modular symbols. We pick two (different) points $r, s \in \mathbb{P}^1(F)$, the stabilizers of these points are Borel subgroups $B_r, B_s \subset G/\mathbb{Q}$. The intersection of these two Borel subgroups is a torus $T = T_{r,s}/\mathbb{Q}$. The group of real points of this torus is $T(\mathbb{R}) = \mathbb{R}^\times_{>0} K^T_\infty$, where $K^T_\infty \xrightarrow{\sim} S^1$. The set of fixed points of K^T_∞ is an orbit under $T(\mathbb{R})$, hence, if $x_0 \in \mathbb{H}_3$ is any fixed point for K^T_∞, then $\sigma_\infty : t_\infty \mapsto t_\infty x_0$ yields a diffeomorphism

$$\sigma_\infty : T(\mathbb{R})/K^T_\infty = \mathbb{R}^\times_{>0} \xrightarrow{\sim} T(\mathbb{R})x_0 = \mathbb{H}_3^{K^T_\infty}. \tag{34}$$

We can extend σ_∞ to $\bar{\sigma}_\infty : [0, \infty] \to \bar{\mathbb{H}}_3$ i.e. we have limits $\bar{\sigma}_\infty(0)$, and $\bar{\sigma}_\infty(\infty) \in \partial(\Gamma\backslash\mathbb{H}_3)$. Then $T(\mathbb{R})x_0$ is the "geodesic line" joining the two points $\bar{\sigma}_\infty(0), \bar{\sigma}_\infty(\infty) \in \partial\bar{\mathbb{H}}_3$. If now $f \in I_{\bar{B}}^{\bar{G}}\chi$ then $\sigma_\infty \otimes f$ is a 1-chain, its boundary lies in the boundary, so it is a relative 1-cycle. Since it depends only- up to orientation -on the torus we denote it by $\mathfrak{z}_T \otimes f$.

On the "open" part of these cycles we have the translation invariant vector field $D_\sigma(x\frac{\partial}{\partial x})$. Therefore these cycles together with these vector fields define currents, we can integrate them against one-forms with compact support.

If the class $[\mathfrak{Z}_T \otimes f]$ maps to zero in $H_0(\partial(\Gamma\backslash\mathbb{H}_3), \underline{I_{\bar{B}}^{\bar{G}}\chi})$ under the boundary map, then we can lift this class to a class $[\widetilde{\mathfrak{Z}_T \otimes f}] \in H_1(\Gamma\backslash\mathbb{H}_3, \underline{I_{\bar{B}}^{\bar{G}}\chi})]$, this gives a way to construct possibly non trivial homology classes. We will show that we can find explicit generators for $H_1(\Gamma\backslash\mathbb{H}_3, \underline{I_{\bar{B}}^{\bar{G}}\chi})$ consisting of such classes. We even give explicit cycles which represent these classes $[\widetilde{\mathfrak{Z}_T \otimes f}] \in H_1(\Gamma\backslash\mathbb{H}_3, \underline{I_{\bar{B}}^{\bar{G}}\chi})]$: We write

$$\partial(\mathfrak{Z}_T \otimes f) = \partial c_f \ , \ \text{with } c_f \in C_1(\partial(\Gamma\backslash\mathbb{H}_3), \underline{I_{\bar{B}}^{\bar{G}}\chi}),$$

and then

$$\widetilde{\mathfrak{Z}_T \otimes f} = \mathfrak{Z}_T \otimes f - c_f \in C_1(\Gamma\backslash\bar{\mathbb{H}}_3, \underline{I_{\bar{B}}^{\bar{G}}\chi}) \tag{35}$$

is a 1-cycle which represents $[\widetilde{\mathfrak{Z}_T \otimes f}]$. These cycles $\widetilde{\mathfrak{Z}_T \otimes f}$ are the *capped* modular symbols. (This terminology was suggested to me by G. Stevens)

The whole point of this construction is that we can give a formula for the evaluation of the Eisenstein class $[\mathrm{Eis}^{(1)}(U_\alpha^\vee \otimes b(\phi)h)]$ on this homology class $[\widetilde{\mathfrak{Z}_T \otimes f}]$.

We discuss this evaluation in the special case of the standard torus $T/\mathbb{Q} = R_{F/\mathbb{Q}}(T_0)$, where T_0 is the standard diagonal torus in Sl_2/F. This torus corresponds to the choice of points $r = (0, 0), s = (0, \infty) \in \mathbb{C} \times (\mathbb{R}_{>0} \cup \{0, \infty\})$ and our 1-chain is the image of the geodesic joining these two points, i.e. the line $\{(0, t)\}_{t\in[0,\infty]} \in \bar{\mathbb{H}}_3$.

The modular symbols are of the form $\mathfrak{Z}_T \otimes f$ where $f \in I_{\bar{B}}^{\bar{G}}\chi$. Here we keep in mind that $\mu_4 - T(\mathbb{Z}) \subset \Gamma_{(0,t)}$. Since the costalk is $(I_{\bar{B}}^{\bar{G}}\chi)_{\Gamma_{(0,t)}}$ we find that the 1-chain $\mathfrak{Z}_T \otimes f = 0$ if $f = (1-\zeta)g$, with $\zeta \in \mu_4, g \in I_{\bar{B}}^{\bar{G}}\chi$. For later reference we formulate

Proposition 1 *To represent the modular symbol $\mathfrak{Z}_T \otimes f$ we can choose $f \in (I_{\bar{B}}^{\bar{G}}\chi)^{\mu_4}$.*

This obvious because we inverted 2.

We compute the boundary $\partial(\mathfrak{Z}_T \otimes f)$. Our modular symbol can be written as a sum of two 1-chains

$$\mathfrak{Z}_T \otimes f = [(0, 0), (0, 1)] \otimes f + [(0, 1), (0, \infty)] \otimes f.$$

The element $w = \begin{pmatrix} 0 & -1 \\ 1 & 0 \end{pmatrix}$ exchanges $(0, 0)$ and $(0, \infty)$ and fixes $(0, 1)$. Hence we see that in $C_1(\Gamma\backslash\bar{\mathbb{H}}_3, \underline{I_{\bar{B}}^{\bar{G}}\chi})$ we have the equality

$$\mathfrak{z}_T \otimes f = [(0,\infty),(0,1)] \otimes wf + [(0,1),(0,\infty)] \otimes f,$$

and the boundary of this chain is

$$\partial(\mathfrak{z}_T \otimes f) = (0,\infty) \otimes wf - (0,1) \otimes wf + (0,1) \otimes f - (0,\infty) \otimes f.$$

Since the 0-chain $-(0,1) \otimes wf + (0,1) \otimes f$ is zero we get

$$\partial(\mathfrak{z}_T \otimes f) = (0,\infty) \otimes (wf - f).$$

We claim: The class

$$[(0,\infty) \otimes (wf - f)] = 0 \iff \exists f_1 \in I_{\bar{B}}^{\bar{G}}\chi \text{ such that } wf - f = f_1 - \begin{pmatrix} 1 & 1 \\ 0 & 1 \end{pmatrix} f_1 \tag{36}$$

If $wf - f = f_1 - \begin{pmatrix} 1 & 1 \\ 0 & 1 \end{pmatrix} f_1$, then we consider the map $\tau_\infty : [0,1] \to X_\infty$, which is defined by $\tau_\infty : y \mapsto \begin{pmatrix} 1 & y \\ 0 & 1 \end{pmatrix} (0,\infty) = (y,\infty)$ and consider the 1-chain $c_{f_1} = \tau_\infty \otimes f_1$. Clearly we have $\partial(\tau_\infty \otimes f_1) = (0,\infty) \otimes (f_1 - \begin{pmatrix} 1 & 1 \\ 0 & 1 \end{pmatrix} f_1)$, and this proves the claim.

It is easy to decide for which choices of $f \in I_{\bar{B}}^{\bar{G}}\chi$ we can solve $wf - f = f_1 - \begin{pmatrix} 1 & 1 \\ 0 & 1 \end{pmatrix} f_1$. We start from the Bruhat decomposition

$$\mathrm{Sl}_2(\mathbb{F}_p) = B(\mathbb{F}_p) \cup \left(\cup_{u \in U(\mathbb{F}_p)} Bw \begin{pmatrix} 1 & u \\ 0 & 1 \end{pmatrix} \right).$$

We define a basis for $I_{\bar{B}}^{\bar{G}}\chi$. By construction the elements in $I_{\bar{B}}^{\bar{G}}\chi$ are R valued functions f on $\mathrm{Sl}_2(\mathbb{F}_p)$ which satisfy $f(bg) = \chi(b) f(g)$. The first basis element is $f_0^{(0)}$, it is supported on $B(\mathbb{F}_p)$ and assumes value one at the identity element. The other elements are $f_u^{(\infty)}$, $u \in \mathbb{F}_p$, they are supported on $B(\mathbb{F}_p) w \begin{pmatrix} 1 & u \\ 0 & 1 \end{pmatrix}$ and take value one at w. Then we have

$$\begin{pmatrix} 1 & 1 \\ 0 & 1 \end{pmatrix} f_0^{(0)} = f_0^{(0)};\ \begin{pmatrix} 1 & 1 \\ 0 & 1 \end{pmatrix} f_u^{(\infty)} = f_{u+1}^{(\infty)}$$

Hence we see that an element $f = a_0 f_0^{(0)} + \sum b_u f_u^{(\infty)} \in I_{\bar{B}}^{\bar{G}}\chi$ can be written as $f - \begin{pmatrix} 1 & 1 \\ 0 & 1 \end{pmatrix} f$ if and only if $a_0 = 0, \sum b_u = 0$.

Now we look at the elements $f - wf$. We have $wf_0^{(0)} = f_0^{(\infty)}$, $wf_0^{(\infty)} = wf_0^{(0)}$ and for $u \in \mathbb{F}_p, u \neq 0$ we have $wf_u^{(\infty)} = \chi(u) f_{-u^{-1}}^{(\infty)}$. Hence it is clear that for $f = a_0 h_0^{(0)} + \sum b_u f_u^{(\infty)}$ we can solve $f - wf = f_1 - \begin{pmatrix} 1 & 1 \\ 0 & 1 \end{pmatrix} f_1$ if and only if $a_0 = b_0 = 0$. We say that $f \in I_{\bar{B}}^{\bar{G}} \chi(0)$.

Proposition 2 *Let T be still our standard torus. Then the cycles $\mathfrak{z}_T \otimes f$ generate the homology $H_1(\Gamma\backslash\mathbb{H}_3, \partial(\Gamma\backslash\mathbb{H}_3), \underline{I_{\bar{B}}^{\bar{G}}\chi})$. If f runs through $I_{\bar{B}}^{\bar{G}}\chi(0)$ the classes*

$$\widetilde{\mathfrak{z}_T \otimes f} = \mathfrak{z}_T \otimes f - c_{f_1}$$

generate $H_1(\Gamma\backslash\mathbb{H}_3, \underline{I_{\bar{B}}^{\bar{G}}\chi})$ modulo the image of $H_1(\partial(\Gamma\backslash\mathbb{H}_3), \underline{I_{\bar{B}}^{\bar{G}}\chi})$.

See for instance [11], Proposition 5.2.3.

4 The Evaluation of the Eisenstein Class on Capped Modular Symbols

We have to recall that the non degenerate pairing

$$H_1(\Gamma\backslash\mathbb{H}_3, \underline{I_{\bar{B}}^{\bar{G}}\chi}) \times H^1(\Gamma\backslash\mathbb{H}_3, \widetilde{I_{\bar{B}}^{\bar{G}}\chi}) \to R$$

is induced by the pairing between the coefficient systems

$$<,>_\chi : I_{\bar{B}}^{\bar{G}}\chi \times I_{\bar{B}}^{\bar{G}}\chi \to R; (f_1, f_2) \mapsto \sum_{x \in B(\mathbb{F}_p)\backslash G(\mathbb{F}_p)} f_1(x) f_2(x). \quad (37)$$

For $f_\mathfrak{p} \in I_{\bar{B}}^{\bar{G}}\chi(0)$ for this $f_\mathfrak{p}$, we solve the Eq. (36), we choose $h_\mathfrak{p} \in (I_{\bar{B}}^{\bar{G}}\chi)^{\bar{U}}$. Then we evaluate $\mathrm{Eis}_\infty(U_\alpha^\vee \otimes b(\phi) h_\mathfrak{p})$ on the homology class $\widetilde{\mathfrak{z}_T \otimes f_\mathfrak{p}}$. By definition this evaluate is given by the the integral

$$< \widetilde{\mathfrak{z}_T \otimes f_\mathfrak{p}}, \mathrm{Eis}_\infty(U_\alpha^\vee \otimes b(\phi) h_\mathfrak{p} >= \int_{\mathfrak{z}_T \otimes f_\mathfrak{p} - c_{f_1}} \mathrm{Eis}_\infty(U_\alpha^\vee \otimes b(\phi) h_\mathfrak{p}). \quad (38)$$

(For further details I refer to standard textbooks in topology, but this will also be treated in [6]. There we will also discuss the issues concerning convergence, which are swept under the carpet in this paper.)

This integral is the sum of two integrals, an integral over the modular symbol, and an integral is the integral against the cap.

At first we compute the integral against the cap. From the definition of c_{f_1} (see (36)) we get

$$\int_{\tau_\infty} < \frac{\partial}{\partial y} \otimes f_1, U_\alpha^\vee \otimes b(\phi)h_\mathfrak{p} + c(\phi)U_{\bar{\alpha}}^\vee \otimes T_\mathfrak{p}^{\mathrm{loc}}(b(\phi)h)_\mathfrak{p} > . \tag{39}$$

Since $< \frac{\partial}{\partial y}, U_\alpha^\vee > = < \frac{\partial}{\partial y}, U_{\bar{\alpha}}^\vee > = 1$ we find for the contribution of the cap

$$< f_\mathfrak{p}, b(\phi)h_\mathfrak{p} + c(\phi)T_\mathfrak{p}^{\mathrm{loc}} b(\phi)h_\mathfrak{p} > \in R. \tag{40}$$

Now we compute the integral over the modular symbol. For $f_\mathfrak{p} \in I_{\bar{B}}^{\bar{G}}\chi$ the expression $x\frac{\partial}{\partial x} \otimes f_\mathfrak{p}$ is a vector field on $T(\mathbb{R})/K_\infty^T$ with values in $I_{\bar{B}}^{\bar{G}}\chi$ and $\sigma^*(\mathrm{Eis}_\infty(U_\alpha^\vee \otimes b(\phi)h_\mathfrak{p}))$ is a one-form with values in $I_{\bar{B}}^{\bar{G}}\chi$. We can evaluate the one-form on the vector field, the result is a function

$$t_\infty \mapsto < x\frac{\partial}{\partial x} \otimes f_\mathfrak{p}, \sigma^*(\mathrm{Eis}_\infty(U_\alpha^\vee \otimes h_\mathfrak{p})) >_\infty (t_\infty)$$

on $T(\mathbb{R})/K_\infty^T$ with values in $I_{\bar{B}}^{\bar{G}}\chi \otimes I_{\bar{B}}^{\bar{G}}\chi \otimes \mathbb{C}$. To these values we can apply the pairing $< , >_\chi$ and then it is clear from the definition that the integral against the modular symbol is given by

$$b(\phi) \int_{T(\mathbb{R})/K_\infty^T} << x\frac{\partial}{\partial x} \otimes f_\mathfrak{p}, \sigma^*(\mathrm{Eis}_\infty(U_\alpha^\vee \otimes h_\mathfrak{p})) >_\infty (t_\infty) >_\chi \frac{dt_\infty}{t_\infty}.$$

We recall the definition of the pairing $< , >_\chi$: We have the following system of representatives for $B(\mathbb{F}_p)\backslash G(\mathbb{F}_p)$:

$$\mathcal{Y}_\mathfrak{p} = \left\{ \begin{pmatrix} 1 & 0 \\ 0 & 1 \end{pmatrix}, \left\{ w \begin{pmatrix} 1 & u \\ 0 & 1 \end{pmatrix} \right\}_{u \in \mathbb{F}_p, u \neq 0}, \begin{pmatrix} 0 & -1 \\ 1 & 0 \end{pmatrix} \right\} \tag{41}$$

In the following we drop the factor $b(\phi)$ and remember that at the end we have to multiply by this factor. We get for the above expression

$$\sum_{y_\mathfrak{p} \in \mathcal{Y}_\mathfrak{p}} \int_{T(\mathbb{R})/K_\infty^T} < x\frac{\partial}{\partial x}(t_\infty) \otimes f_\mathfrak{p}(y_\mathfrak{p}), \sigma^*(\mathrm{Eis}_\infty(U_\alpha^\vee \otimes h_\mathfrak{p}))(t_\infty)(y_\mathfrak{p}) >_\infty \frac{dt_\infty}{t_\infty}.$$

Now we observe that the condition $f_\mathfrak{p} \in I_{\bar{B}}^{\bar{G}}\chi(0)$ means that for the first and the last term in this summation we have

$$f_\mathfrak{p}\left(\begin{pmatrix} 1 & 0 \\ 0 & 1 \end{pmatrix}\right) = f_\mathfrak{p}(w) = 0$$

and hence we only have to sum over the $y_\mathfrak{p}(u) = w \begin{pmatrix} 1 & u \\ 0 & 1 \end{pmatrix}$ with $u \in \mathbb{F}_p^\times$. Hence we have to compute

$$\sum_{u\in\mathbb{F}_p^\times}\int_{T(\mathbb{R})/K_\infty^T} < x\frac{\partial}{\partial x}(t_\infty)\otimes f_{\mathfrak{p}}(y_{\mathfrak{p}}(u)), \sigma^*(\mathrm{Eis}_\infty(U_\alpha^\vee\otimes h_{\mathfrak{p}})(t_\infty)(y_{\mathfrak{p}}(u)) >_\infty \frac{dt_\infty}{t_\infty}. \tag{42}$$

We enlarge our group slightly, we pass to $\tilde{G}_0 = \mathrm{Gl}_2/\mathbb{Z}[i]$, the other groups which we define in Sect. 1 will be enlarged accordingly and we denote them by also putting a $\tilde{}$ atop of them. The group $\tilde{\Gamma}_0(\mathfrak{p})/\tilde{\Gamma}_1(\mathfrak{p}) = \mathbb{F}_p^\times\times\mathbb{F}_p^\times$. This group contains $\Gamma_0(\mathfrak{p})/\Gamma_1(\mathfrak{p}) = \{x, x^{-1}\}$, and our character χ , which is defined on this subgroup, extends to a character $\eta : \mathbb{F}_p^\times\times\mathbb{F}_p^\times \to \mu_4$ which in addition is trivial on the center $\{x, x\}$. This extension is not unique, we choose one, the other choice is given by $\bar{\eta} = \eta^{-1}$.

We define the $\tilde{\Gamma}$-modules $I_{\bar{B}}^{\bar{G}}\eta, I_{\bar{B}}^{\bar{G}}\eta^{-1}$ in analogy to (2). If we restrict these modules to Γ they become $I_{\bar{B}}^{\bar{G}}\chi$. The symmetric space remains unchanged, we replace $K_\infty\subset \mathrm{Sl}_2(\mathbb{C})$ by $\tilde{K}_\infty = K_\infty\times\mathbb{C}^\times$ where of course $\mathbb{C}^\times$ is the center of $\mathrm{Gl}_2(\mathbb{C})$. Then $\mathbb{H}_3 = \mathrm{Gl}_2(\mathbb{C})/\tilde{K}_\infty$. Our group $\tilde{\Gamma} = \mathrm{Gl}_2(\mathbb{Z}[i])$ acts upon $\mathbb{H}_3$ and we consider the cohomology groups

$$H^1(\tilde{\Gamma}\backslash\mathbb{H}_3, \widetilde{I_{\bar{B}}^{\bar{G}}\eta}), H^1(\tilde{\Gamma}\backslash\mathbb{H}_3, \widetilde{I_{\bar{B}}^{\bar{G}}\eta^{-1}}).$$

Of course we again have the cohomology with compact support, the cohomology of the boundary and the long exact sequence. Actually these new cohomology groups for $\tilde{\Gamma}$ are very close to the previous ones: We have the exact sequence

$$1\to\Gamma\to\tilde{\Gamma}\xrightarrow{\det}\mu_4\to 1$$

and since $H^0(\Gamma, I_{\bar{B}}^{\bar{G}}\chi) = (I_{\bar{B}}^{\bar{G}}\chi)^\Gamma = 0$ the Hochschild Serre spectral sequence yields

$$H^1(\tilde{\Gamma}\backslash\mathbb{H}_3, \widetilde{I_{\bar{B}}^{\bar{G}}\eta}) = H^1(\Gamma\backslash\mathbb{H}_3, \widetilde{I_{\bar{B}}^{\bar{G}}\eta})^{\mu_4} \tag{43}$$

Let $Z/\mathbb{Q}$ be the center of $\tilde{G}/\mathbb{Q}$, then $\tilde{T}/Z = T^{\mathrm{ad}} \xrightarrow{\sim} R_{\mathbb{Z}[i])/\mathbb{Z}}(\mathbb{G}_m)$. The last identification is given by $x\mapsto\begin{pmatrix}x&0\\0&1\end{pmatrix}$. Our Hecke character ϕ is a homomorphism $\phi : T(\mathbb{Q})\backslash T(\mathbb{A})\to\mathbb{C}^\times$. It is easy to see that it extends to a Hecke character $\psi : \tilde{T}(\mathbb{Q})\backslash\tilde{T}(\mathbb{A})\to\mathbb{C}^\times$ which is trivial on $Z(\mathbb{A})$, hence it yields a Hecke character $\psi : T^{\mathrm{ad}}(\mathbb{Q})\backslash T^{\mathrm{ad}}(\mathbb{A})\to\mathbb{C}^\times$. This character is unramified at all finite places except $\mathfrak{p}$ and at $\mathfrak{p}$ it is our character η. We can do the same with $\bar{\phi}$ and then $\bar{\psi}$ gives us η^{-1} at $\mathfrak{p}$. We transfer our considerations in Sects. 2.1 and 2.1.2 to this case, there are some minor changes. We always have to observe that $B\backslash G = \tilde{B}\backslash\tilde{G}$.

Again we consider the induced representation

$$I_\psi = I_{\psi_\infty}\otimes I_{\psi_f} = \mathrm{Ind}_{\tilde{B}(\mathbb{A})}^{\tilde{G}(\mathbb{A})}\psi,$$

it it easy to see that $I_\psi = I_\phi$ as vector spaces, but on I_ψ we have extended the representation of $\mathrm{Sl}_2(\mathbb{A})$ to a representation of $\tilde{G}(\mathbb{A})$. This representation has a trivial central character hence we can consider it as a representation of $G^{\mathrm{ad}}(\mathbb{A}) = \tilde{G}(\mathbb{A})/Z(\mathbb{A})$. We have an extension of the equality (13))

$$I_{\bar{B}}^{\bar{G}}\eta = I_{\psi_f}^{\tilde{K}_f^{(1)}(\mathrm{p})} = I_{\phi_f}^{K_f^{(1)}(\mathrm{p})}. \tag{44}$$

For $h \in (I_{\bar{B}}^{\bar{G}}\eta)^{\bar{U}}$ we define as in (19)

$$\widetilde{\mathrm{Eis}}_\infty(f_\infty \otimes h)(g_\infty)(a) = \sum_{\gamma \in B(\mathbb{Z})\backslash G(\mathbb{Z})} f_\infty(\gamma g_\infty) h(\gamma a^{-1}), \tag{45}$$

where now $g_\infty \in \tilde{G}(\mathbb{R})$, $a \in \tilde{G}(\mathbb{F}_p)$. There is now a minor difference: We have the two special elements h_0, h_∞, then the Eq. (20) becomes

$$\begin{aligned} \widetilde{\mathrm{Eis}}_\infty(U_\alpha^\vee \otimes h_0)(g_\infty)(ba) &= \eta(b)^{-1}\widetilde{\mathrm{Eis}}_\infty(U_\alpha^\vee \otimes h_0)(g_\infty)(a) \\ \widetilde{\mathrm{Eis}}_\infty(U_\alpha^\vee \otimes h_\infty)(g_\infty)(ba) &= \eta(b)\widetilde{\mathrm{Eis}}_\infty(U_\alpha^\vee \otimes h_\infty)(g_\infty)(a), \end{aligned} \tag{46}$$

hence these expressions are 1-forms with coefficients in $\widetilde{I_{\bar{B}}^{\bar{G}}\eta}$ and $\widetilde{I_{\bar{B}}^{\bar{G}}\eta^{-1}}$ respectively. Our element f_p can be viewed as an element in $I_{\bar{B}}^{\bar{G}}\eta^{-1}$ or $I_{\bar{B}}^{\bar{G}}\eta$ we can evaluate these one forms on the modular symbol $\mathfrak{z}_T \otimes f_\mathrm{p}$. We choose $h_\mathrm{p} = h_0$ or $h_\mathrm{p} = h_\infty$, then we have to to compute the expression

$$\sum_{u \in \mathbb{F}_p^\times} \int_{\tilde{T}(\mathbb{R})/\tilde{K}_\infty^T} < x\frac{\partial}{\partial x}(t_\infty) \otimes f_\mathrm{p}(y_\mathrm{p}(u)), \sigma^*(\widetilde{\mathrm{Eis}}_\infty(U_\alpha^\vee \otimes h_\mathrm{p})(t_\infty)(y_\mathrm{p}(u)) >_\infty \frac{dt_\infty}{t_\infty}. \tag{47}$$

Here we have to take care that the expression should not depend on the choice of the representatives y_p and hence we have to choose f_p in the appropriate space $I_{\bar{B}}^{\bar{G}}\eta^{-1}$ or $I_{\bar{B}}^{\bar{G}}\eta$, we will say that f_p and h_p are matching.

We gained something from passing to Gl_2 : We can write

$$y_\mathrm{p}(u) = w\begin{pmatrix} u & 0 \\ 0 & 1 \end{pmatrix}\begin{pmatrix} 1 & 1 \\ 0 & 1 \end{pmatrix}\begin{pmatrix} u^{-1} & 0 \\ 0 & 1 \end{pmatrix} \tag{48}$$

We insert this into the expression above. We get

$$f_\mathrm{p}\left(w\begin{pmatrix} u & 0 \\ 0 & 1 \end{pmatrix}\begin{pmatrix} 1 & 1 \\ 0 & 1 \end{pmatrix}\begin{pmatrix} u^{-1} & 0 \\ 0 & 1 \end{pmatrix}\right) = \eta^\epsilon(u) f_\mathrm{p}\left(\begin{pmatrix} 1 & -1 \\ 1 & 0 \end{pmatrix}\begin{pmatrix} u^{-1} & 0 \\ 0 & 1 \end{pmatrix}\right) \tag{49}$$

where $\epsilon = \pm 1$. For the second factor we get

$$\widetilde{\mathrm{Eis}}_\infty(U_\alpha^\vee \otimes h_\mathfrak{p})(t_\infty)\left(w\begin{pmatrix}u&0\\0&1\end{pmatrix}\begin{pmatrix}1&1\\0&1\end{pmatrix}\begin{pmatrix}u^{-1}&0\\0&1\end{pmatrix}\right)\right) = \\ \widetilde{\mathrm{Eis}}_\infty(U_\alpha^\vee \otimes \begin{pmatrix}1&0\\0&u\end{pmatrix}h)(t_\infty)\left(\begin{pmatrix}1&-1\\1&0\end{pmatrix}\begin{pmatrix}u^{-1}&0\\0&1\end{pmatrix}\right) \tag{50}$$

If now $h_\mathfrak{p} = h_0$ or $h = h_\infty$ then $\begin{pmatrix}1&0\\0&u\end{pmatrix} h_\mathfrak{p}) = \eta(u)^{\epsilon'} h$ and in any case we can choose $f_\mathfrak{p}$ in $I_{\tilde{B}}^{\tilde{G}}\eta^{-1}$ or $I_{\tilde{B}}^{\tilde{G}}\eta$ so that $f_\mathfrak{p}$ and $h_\mathfrak{p}$ match. Then our expression in (47) becomes

$$\sum_{u\in\mathbb{F}_p^\times}\int_{\tilde{T}(\mathbb{R})/\tilde{K}_\infty^T}$$

$$< x\tfrac{\partial}{\partial x}(t_\infty)\otimes f_\mathfrak{p}\left(\begin{pmatrix}1&-1\\1&0\end{pmatrix}\begin{pmatrix}u^{-1}&0\\0&1\end{pmatrix}\right), \sigma^*\left(\widetilde{\mathrm{Eis}}_\infty(U_\alpha^\vee\otimes h_\mathfrak{p})(t_\infty)\right)\left(\begin{pmatrix}1&-1\\1&0\end{pmatrix}\begin{pmatrix}u^{-1}&0\\0&1\end{pmatrix}\right) >_\infty \frac{dt_\infty}{t_\infty}. \tag{51}$$

We pass to the language of adeles. Let $\mathfrak{U}_f \subset \mathbb{A}_F^\times$ be the maximal compact subgroup of units. For any open subgroup $K_f \subset K_f^{(0)}$ for which the map det $K_f \to \mathfrak{U}_f$ is surjective, we have

$$\tilde{G}(\mathbb{Q}) \cap K_f\backslash\tilde{G}(\mathbb{R}) = \tilde{G}(\mathbb{Q})\backslash\tilde{G}(\mathbb{R}) \times \tilde{G}(\mathbb{A}_f)/K_f,$$

for this we need strong approximation for Sl_2 and we need that the class number of $\mathbb{Z}[i]$ is one. We have the Eisenstein intertwining operator

$$\widetilde{\mathrm{Eis}} : \mathrm{Ind}_{\tilde{B}(\mathbb{A})}^{\tilde{G}(\mathbb{A})}\psi \to \mathcal{A}(\tilde{G}(\mathbb{Q})\backslash\tilde{G}(\mathbb{A})) \tag{52}$$

and for $K_f = \tilde{K}_f^{(0)}$ or $K_f = \tilde{K}_{0,f}^{(1)}(\mathfrak{p})$ we get

$$\widetilde{\mathrm{Eis}} : (\mathrm{Ind}_{\tilde{B}(\mathbb{A})}^{\tilde{G}(\mathbb{A})}\psi)^{K_f} \to \mathcal{A}(\tilde{G}(\mathbb{Q})\backslash\tilde{G}(\mathbb{A})/K_f) = \mathcal{A}(\tilde{G}(\mathbb{Q}) \cap K_f\backslash\tilde{G}(\mathbb{R})).$$

Under this identification we get

$$\widetilde{\mathrm{Eis}}_\infty(f_\infty \otimes h_\mathfrak{p})(g_\infty)(a) = \widetilde{\mathrm{Eis}}(f_\infty \otimes h_\mathfrak{p})(g_\infty, \underline{a}_f^{-1}) \tag{53}$$

where $\underline{a}_f \in \tilde{K}_f^{(0)}$ maps to $a \in \tilde{\Gamma}/\tilde{\Gamma}^{(1)}(\mathfrak{p})$.

We define the chain complexes and especially our modular symbols in (32) in terms of adeles. Instead of looking at the projection $\mathbb{H}_3 \to \Gamma\backslash\mathbb{H}_3$ we consider the projection

$$\pi_G : \mathbb{H}_3 \times \tilde{G}(\mathbb{A}_f) \to \tilde{G}(\mathbb{Q})\backslash\mathbb{H}_3 \times \tilde{G}(\mathbb{A}_f)/\tilde{K}_f^{(0)},$$

and we define the adelic version of the sheaf $\widetilde{I_{\bar{B}}^{\bar{G}}\eta}$ and the cosheaf $\underline{I_{\bar{B}}^{\bar{G}}\eta}$ attached to the $K_f^{(0)}$-module $I_{\bar{B}}^{\bar{G}}\chi$. For any point $\underline{\bar{x}} \in \tilde{G}(\mathbb{Q})\backslash\mathbb{H}_3 \times \tilde{G}(\mathbb{A}_f)/\tilde{K}_f^{(0)}$ we introduce the module

$$\mathcal{C}(\underline{\bar{x}}, I_{\bar{B}}^{\bar{G}}\eta) := \{\underline{f} : \pi_G^{-1}(\underline{\bar{x}}) \to I_{\bar{B}}^{\bar{G}}\eta \mid \underline{f}((x_\infty, \underline{g}_f)) = \underline{f}(\gamma(x_\infty, \underline{g}_f)) \text{ for all } \gamma \in \tilde{G}(\mathbb{Q})\}. \tag{54}$$

We require a continuity condition and a finiteness condition: We consider subgroups $K_f^* \subset \tilde{K}_f^{(0)}$ which are of the form $K_f^* = \left\{\begin{pmatrix} \underline{z}_f & 0 \\ 0 & \underline{z}_f \end{pmatrix} K'_f\right\}$ where K'_f is a deep congruence subgroup in $K_f^{(0)}$ and where $\underline{z}_f = (\dots, z_{\mathfrak{q}}, \dots)$ is an idele in $\mathfrak{U}_f$ whose $\mathfrak{p}$ component satisfies $z_{\mathfrak{p}} \bmod \mathfrak{p} \in \mathbb{F}_p^{\times,\mathrm{odd}}$.

Then the required continuity condition says that for any $\underline{f}$ we can find a group K_f^* such that $\underline{f}((x_\infty, \underline{g}_f\underline{k}_f)) = \underline{f}((x_\infty, \underline{g}_f))$ for all $\underline{k}_f \in \underline{K'_f}$.

The finiteness condition says that $\underline{g}_f \to \underline{f}((x_\infty, \underline{g}_f))$ has compact support.

Now strong approximation gives us that $\underline{f} \in \mathcal{C}(\underline{\bar{x}}, I_{\bar{B}}^{\bar{G}}\eta)$ is determined by its values $\underline{f}((\gamma x_\infty, e))$. We have an operation of $\tilde{K}_f^{(0)}$ on $\mathcal{C}(\underline{\bar{x}}, I_{\bar{B}}^{\bar{G}}\eta)$ which is defined by

$$(\underline{k}_f\underline{f})((x_\infty, \underline{g}_f)) = \underline{k}_f(\underline{f}((x_\infty, \underline{g}_f\underline{k}_f))). \tag{55}$$

We recall that by definition the elements in $I_{\bar{B}}^{\bar{G}}\eta$ are functions on $\tilde{\Gamma}/\tilde{\Gamma}^{(1)}(\mathfrak{p}) = \tilde{K}_f^{(0)}/\tilde{K}_f^{(1)}(\mathfrak{p})$ and then the above operation is defined by

$$(\underline{k}_f\underline{f})((x_\infty, \underline{g}_f))(\underline{u}_f) = \underline{f}((x_\infty, \underline{g}_f\underline{k}_f))(\underline{u}_f\underline{k}_f).$$

Now we can define the costalk $\underline{I_{\bar{B}}^{\bar{G}}\eta_{\underline{\bar{x}}}}$ of $\underline{I_{\bar{B}}^{\bar{G}}\eta}$ at the point $\underline{\bar{x}} \in \tilde{G}(\mathbb{Q})\backslash\mathbb{H}_3 \times \tilde{G}(\mathbb{A}_f)/\tilde{K}_f^{(0)}$ as the module of coinvariants of $\mathcal{C}(\underline{\bar{x}}, I_{\bar{B}}^{\bar{G}}\eta)$ under the action of $\tilde{K}_f^{(0)}$.

Previously we defined these cosheaves on $\tilde{\Gamma}\backslash\mathbb{H}_3$ by essentially the same procedure, for a point $\bar{x}_\infty \in \tilde{\Gamma}\backslash\mathbb{H}_3$ we considered

$$\mathcal{C}_\infty(\bar{x}_\infty, I_{\bar{B}}^{\bar{G}}\eta) = \{f : \pi^{-1}(\bar{x}_\infty) \to I_{\bar{B}}^{\bar{G}}\eta \mid f \text{ has finite support}\}$$

together with the action by $\tilde{\Gamma}$

$$\gamma(f)(x_\infty) = \gamma(f(\gamma^{-1}x_\infty)).$$

We defined the costalk of the cosheaf by taking the coinvariants. The identification $\tilde{\Gamma}\backslash\mathbb{H}_3 = \tilde{G}(\mathbb{Q})\backslash\mathbb{H}_3 \times \tilde{G}(\mathbb{A}_f)/\tilde{K}_f^{(0)}$ is induced by

$$x_\infty \mapsto (x_\infty, \underline{e}_f)$$

where of course $\underline{e}_f$ is the identity element in $\tilde{G}(\mathbb{A}_f)$. This of course provides the identification between the two different kinds of (co-) sheaves.

In the adelic language our modular symbol is a 1-chain $\underline{\sigma} \otimes \underline{f}^*_{\mathfrak{p}}$ where

$$\bar{\underline{\sigma}} : T(\mathbb{R})/K^T_\infty \to \tilde{G}(\mathbb{Q})\backslash \mathbb{H}_3 \times \tilde{G}(\mathbb{A}_f)/\tilde{K}^{(0)}_f$$

is the composition of $t_\infty \mapsto t_\infty x_0 \times \underline{e}_f$ and the projection $\mathbb{H}_3 \times \tilde{G}(\mathbb{A}_f) \to \tilde{G}(\mathbb{Q})\backslash \mathbb{H}_3 \times \tilde{G}(\mathbb{A}_f)/\tilde{K}^{(0)}_f$. The value $\underline{f}^*_{\mathfrak{p},\bar{\underline{\sigma}}(t_\infty)}$ of the cosection $\underline{f}^*_{\mathfrak{p}}$ is represented by a map from $\pi^{-1}_G(\bar{\underline{\sigma}}(t_\infty))$ to $I^{\bar{G}}_{\bar{B}}\eta$. To define such a map we choose a map which is supported on $\bar{\underline{\sigma}}(t_\infty) \times \tilde{K}^{(0)}_f \subset \pi^{-1}_G(\bar{\underline{\sigma}}(t_\infty))$. The value of this map at $(t_\infty x_0, \underline{k}_f)$ must be an element in $I^{\bar{G}}_{\bar{B}}\eta$. We identify $I^{\bar{G}}_{\bar{B}}\eta$ with the right induced module

$$R\mathrm{Ind}^{\tilde{K}^{(0)}_f}_{\tilde{K}_{0,f}(\mathfrak{p}))}\eta = \{f : \tilde{K}^{(0)} \to R \mid f(\underline{x}_f \underline{b}_f^{-1}) = \eta(\underline{b}_f)h(\underline{x}_f)\ \forall\, \underline{x}_f \in \tilde{K}^{(0)}_f, \underline{b}_f \in \tilde{K}_{0,f}(\mathfrak{p}))\}.$$

The identification is given by $f \mapsto \iota(f)$ where $\iota(f)(\underline{k}_f) = f(\underline{k}_f^{-1})$. We put $\underline{f}_{\mathfrak{p}} := \iota(f_{\mathfrak{p}})$, then the value of $\underline{f}^*_{\mathfrak{p},\bar{\underline{\sigma}}(t_\infty)}$ at $(t_\infty, \underline{k}_f)$ given by

$$\underline{f}^*_{\mathfrak{p},\bar{\underline{\sigma}}(t_\infty)}(\underline{k}_f)(\underline{x}_f) := \underline{f}_{\mathfrak{p}}(\underline{k}_f \underline{x}_f). \tag{56}$$

We define the element $\underline{v}_{\mathfrak{p}} \in \tilde{G}(\mathbb{A})$: At all places $\mathfrak{q} \neq \mathfrak{p}$ it is the identity and at the place $\mathfrak{p}$ we have $v_{\mathfrak{p}} = \begin{pmatrix} -1 & 1 \\ -1 & 0 \end{pmatrix}$. Then $\underline{h}_{\mathfrak{p}}$ will be the same object as $h_{\mathfrak{p}}$ but now viewed as an element in $I^{\tilde{K}_{0,f}(\mathfrak{p})}_{\phi_f}$ (See (14)). The expression under the integral in (51) becomes

$$< x\frac{\partial}{\partial x} \otimes \underline{f}_{\mathfrak{p}}\left(\left(t_\infty, \begin{pmatrix} \underline{x}_f & 0 \\ 0 & 1\end{pmatrix} \underline{v}_{\mathfrak{p}}\right)\right), \sigma^*(\widetilde{\mathrm{Eis}}(U^\vee_\alpha \otimes \underline{h}_{\mathfrak{p}})\left(\left(t_\infty, \begin{pmatrix} \underline{x}_f & 0 \\ 0 & 1\end{pmatrix}\right) \underline{v}_{\mathfrak{p}}\right) >_\infty \frac{dt_\infty}{t_\infty},$$

where $\underline{x}_f$ is an element that represent $u \in \mathbb{F}^*_p$. Of course the value of this expression does not depend on the choice of this representative. Hence the summation over $u \in \mathbb{F}^\times_p$ written as an integral

$$\int_{T^{\mathrm{ad}}(\mathbb{R})/K^T_\infty \times \mathfrak{U}_f} < x\tfrac{\partial}{\partial x} \otimes \underline{f}_{\mathfrak{p}}\left(\left(t_\infty, \begin{pmatrix} \underline{u}_f & 0 \\ 0 & 1\end{pmatrix}\right) \underline{v}_{\mathfrak{p}}\right), \sigma^*(\widetilde{\mathrm{Eis}}(U^\vee_\alpha \otimes \underline{h}_{\mathfrak{p}})\left(\left(t_\infty, \begin{pmatrix} \underline{u}_f & 0 \\ 0 & 1\end{pmatrix}\right) \underline{v}_{\mathfrak{p}}\right) >_\infty \tfrac{dt_\infty}{t_\infty} \times d\underline{u}_f \tag{57}$$

where $d\underline{u}_f$ gives volume one to $\mathfrak{U}^{(1)}_f(\mathfrak{p})$.

Since F has class number 1 we know that $T^{\mathrm{ad}}(\mathbb{Q})\backslash T^{\mathrm{ad}}(\mathbb{A}) = \mu_4\backslash T^{\mathrm{ad}}(\mathbb{R}) \times \mathfrak{U}_f$. We can choose $\underline{f}_{\mathfrak{p}} \in (I_{\bar{B}}^{\bar{G}}\eta)^{\mu_4}$ (see Proposition 1).

$$\int_{T^{\mathrm{ad}}(\mathbb{Q})\backslash T^{\mathrm{ad}}(\mathbb{A})} < x\frac{\partial}{\partial x} \otimes \underline{f}_{\mathfrak{p}}\left(\left(\begin{pmatrix} \underline{t} & 0 \\ 0 & 1 \end{pmatrix}\underline{v}_{\mathfrak{p}}\right)\right), \sigma^*(\widetilde{\mathrm{Eis}}(U_\alpha^\vee \otimes \underline{h}_{\mathfrak{p}})\left(\left(\begin{pmatrix} \underline{t} & 0 \\ 0 & 1 \end{pmatrix}\underline{v}_{\mathfrak{p}}\right)\right) >_\infty d^\times \underline{t}, \tag{58}$$

where $d^\times \underline{t} = \frac{dt_\infty}{t_\infty} \times d^\times \underline{t}_f$ and where $\mathrm{vol}(d^\times \underline{t}_f(\mathfrak{U}_f^{(1)}(\mathfrak{p}))) = 1$

This is now a standard computation, for the case of compact modular symbols it is already in [3]. We write down the definition of the Eisenstein series (we are a little bit sloppy and do not care about convergence. This will be (has been) discussed in [3, 6]). We have to compute

$$\int_{T^{\mathrm{ad}}(\mathbb{Q})\backslash T^{\mathrm{ad}}(\mathbb{A})/K_\infty^{T^{\mathrm{ad}}}} < \left(x\frac{\partial}{\partial x} \otimes \underline{f}_{\mathfrak{p}}\right)\left(\begin{pmatrix} \underline{t} & 0 \\ 0 & 1 \end{pmatrix}\underline{v}_{\mathfrak{p}}\right), \sum_{a \in B(\mathbb{Q})\backslash G(\mathbb{Q})} \left(U_\alpha^\vee \otimes h)(a\begin{pmatrix} \underline{t} & 0 \\ 0 & 1 \end{pmatrix}\underline{v}_{\mathfrak{p}}\right) >_\infty d^\times \underline{t}.$$

The torus $T^{\mathrm{ad}}(\mathbb{Q})$ acts on $B(\mathbb{Q})\backslash G(\mathbb{Q})$. It has the two fixed points $B(\mathbb{Q})e$, $B(\mathbb{Q})w$, and for $a = e$, $a = w$ clearly

$$\int_{T^{\mathrm{ad}}(\mathbb{Q})\backslash T^{\mathrm{ad}}(\mathbb{A})/K_\infty^{T^{\mathrm{ad}}}} < \left(x\frac{\partial}{\partial x} \otimes \underline{f}_{\mathfrak{p}}\right)\left(\begin{pmatrix} \underline{t} & 0 \\ 0 & 1 \end{pmatrix}\underline{v}_{\mathfrak{p}}\right), (U_\alpha^\vee \otimes h)\left(a\begin{pmatrix} \underline{t} & 0 \\ 0 & 1 \end{pmatrix}\underline{v}_{\mathfrak{p}}\right) >_\infty d^\times \underline{t} = 0.$$

(The tangent vector $x\frac{\partial}{\partial x}$ is orthogonal to $U_{\pm\alpha}^\vee$). If $a \neq e, w$ then we can write

$$a = w\begin{pmatrix} 1 & b \\ 0 & 1 \end{pmatrix} \text{ where } b \in F^\times.$$

From this we get

$$\mathrm{Eis}(U_\alpha^\vee \otimes h \otimes \underline{h}_{\mathfrak{p}})\left(\begin{pmatrix} \underline{t} & 0 \\ 0 & 1 \end{pmatrix}\underline{v}_{\mathfrak{p}}\right) = \sum_{b \in T^{\mathrm{ad}}(\mathbb{Q})} (U_\alpha^\vee \otimes \underline{h}_{\mathfrak{p}})\left(w\begin{pmatrix} b & 0 \\ 0 & 1 \end{pmatrix}\begin{pmatrix} 1 & 1 \\ 0 & 1 \end{pmatrix}\begin{pmatrix} b^{-1} & 0 \\ 0 & 1 \end{pmatrix}\begin{pmatrix} \underline{t} & 0 \\ 0 & 1 \end{pmatrix}\underline{v}_{\mathfrak{p}}\right).$$

The first factor $\begin{pmatrix} b & 0 \\ 0 & 1 \end{pmatrix}$ drops out. We have to evaluate

$$\int_{T(\mathbb{Q})\backslash T^{\mathrm{ad}}(\mathbb{A})/K_\infty^{T^{\mathrm{ad}}}} \sum_{b \in T^{\mathrm{ad}}(\mathbb{Q})}$$

$$< \left(x\frac{\partial}{\partial x} \otimes \underline{f}_{\mathfrak{p}}\right)\left(\begin{pmatrix} \underline{t} & 0 \\ 0 & 1 \end{pmatrix}\underline{v}_{\mathfrak{p}}\right), \sum_{b \in T^{\mathrm{ad}}(\mathbb{Q})} \left(U_\alpha^\vee \otimes \underline{h}_{\mathfrak{p}}\right)\left(\begin{pmatrix} 0 & -1 \\ 1 & 1 \end{pmatrix}\begin{pmatrix} b^{-1}\underline{t} & 0 \\ 0 & 1 \end{pmatrix}\underline{v}_{\mathfrak{p}}\right) >_\infty d^\times \underline{t}.$$

Since $x\frac{\partial}{\partial x}$ is invariant under $t_\infty \to bt_\infty$ we get from (54)

$$\left(x\frac{\partial}{\partial x} \otimes \underline{f}_{\mathfrak{p}}\right)\left(\begin{pmatrix} \underline{t} & 0 \\ 0 & 1 \end{pmatrix}\underline{v}_{\mathfrak{p}}\right) = (x\frac{\partial}{\partial x} \otimes \underline{f}_{\mathfrak{p}})\left(\begin{pmatrix} b^{-1}\underline{t} & 0 \\ 0 & 1 \end{pmatrix}\underline{v}_{\mathfrak{p}}\right)$$

and we have to compute

$$\int_{T^{\mathrm{ad}}(\mathbb{A})} < \left(x\frac{\partial}{\partial x}\otimes \underline{f}_{\mathfrak{p}}\right)\left(\left(\begin{pmatrix} \underline{t} & 0\\ 0 & 1\end{pmatrix}\underline{v}_{\mathfrak{p}}\right), (U_\alpha^\vee\otimes \underline{h}_{\mathfrak{p}})\left(\begin{pmatrix} 0 & -1\\ 1 & 1\end{pmatrix}\begin{pmatrix} \underline{t} & 0\\ 0 & 1\end{pmatrix}\underline{v}_{\mathfrak{p}}\right)\right) >_\infty d^\times \underline{t}.$$

Our integral becomes a product of local integrals over all places,, i.e. it is equal to

$$\begin{gathered}\int_{T^{\mathrm{ad}}(\mathbb{R})} < x\tfrac{\partial}{\partial x}(t_\infty), (U_\alpha^\vee)\left(\begin{pmatrix} 0 & -1\\ t_\infty & 1\end{pmatrix}\right) >_\infty d^\times t_\infty\\ \times\\ \times\prod_{\mathfrak{q}}\int_{T^{\mathrm{ad}}(F_{\mathfrak{q}})} f_{\mathfrak{q}}\left(\begin{pmatrix} t_{\mathfrak{q}} & 0\\ 0 & 1\end{pmatrix}v_{\mathfrak{q}}\right) h_{\mathfrak{q}}\left(\begin{pmatrix} 0 & -1\\ t_{\mathfrak{q}} & 1\end{pmatrix}v_{\mathfrak{q}}\right) d^\times t_{\mathfrak{q}}.\end{gathered}\tag{59}$$

We look at the case $\mathfrak{q}\neq\mathfrak{p}$ first. Then $v_{\mathfrak{q}}$ is the identity, $f_{\mathfrak{p}}(t_{\mathfrak{q}})=1$ and $h_{\mathfrak{q}}$ is the normalized spherical function. We have to evaluate

$$\int_{T^{\mathrm{ad}}(F_{\mathfrak{q}})} h_{\mathfrak{q}}\left(\begin{pmatrix} 0 & -1\\ 1 & 1\end{pmatrix}\begin{pmatrix} t_{\mathfrak{q}} & 0\\ 0 & 1\end{pmatrix}\right) d^\times t_{\mathfrak{q}} = \int_{T^{\mathrm{ad}}(F_{\mathfrak{q}})} h_{\mathfrak{q}}\left(\begin{pmatrix} 0 & -1\\ t_{\mathfrak{q}} & 1\end{pmatrix}\right) d^\times t_{\mathfrak{q}}$$

We write $t_{\mathfrak{q}} = u\varpi_{\mathfrak{q}}^{\nu}$, with $\nu\in\mathbb{Z}$ and $u\in\mathcal{O}_{\mathfrak{p}}^\times$. Then we have the Iwasawa decompositions

$$\begin{pmatrix} 0 & -1\\ u\varpi_{\mathfrak{q}}^{\nu} & 1\end{pmatrix} = \begin{cases}\begin{pmatrix} u\varpi_{\mathfrak{q}}^{\nu} & -1\\ 0 & 1\end{pmatrix}\begin{pmatrix} 1 & 0\\ u\varpi_{\mathfrak{q}}^{\nu} & 1\end{pmatrix} & \text{if } \nu\ge 0\\ \begin{pmatrix} 1 & 0\\ 0 & u\varpi_{\mathfrak{q}}^{\nu}\end{pmatrix}\begin{pmatrix} 0 & -1\\ 1 & u\varpi_{\mathfrak{q}}^{-\nu}\end{pmatrix} & \text{if } \nu < 0.\end{cases}\tag{60}$$

We get the infinite sum

$$\sum_{\nu\in\mathbb{Z}} h_{\mathfrak{q}}\left(\begin{pmatrix} 0 & -1\\ u\varpi_{\mathfrak{q}}^{\nu} & 1\end{pmatrix}\right).$$

and the Iwasawa decomposition gives us

$$h_{\mathfrak{q}}\left(\begin{pmatrix} 0 & -1\\ u\varpi_{\mathfrak{q}}^{\nu} & 1\end{pmatrix}\right) = \begin{cases}\psi(\mathfrak{q})^{\nu} & \text{if } \nu\ge 0\\ \psi(\mathfrak{q})^{-\nu} & \text{if } \nu\le 0\end{cases},$$

and hence we finally get for our local integral

$$\int_{T^{\mathrm{ad}}(F_{\mathfrak{q}})} h_{\mathfrak{q}}\left(\begin{pmatrix} 0 & -1\\ 1 & 1\end{pmatrix}\begin{pmatrix} t_{\mathfrak{q}} & 0\\ 0 & 1\end{pmatrix}\right) d^\times t_{\mathfrak{q}} = \frac{1+\psi(\mathfrak{q})}{1-\psi(\mathfrak{q})} = \frac{1-\phi(\mathfrak{q})}{(1-\psi(\mathfrak{q}))^2}.\tag{61}$$

Here we used $\psi(\mathfrak{q})^2 = \phi(\mathfrak{q})$.

The next case is $\mathfrak{q} = \mathfrak{p}$. We have to evaluate

$$\int_{T^{\mathrm{ad}}(F_\mathfrak{p})} f_\mathfrak{p}\left(\begin{pmatrix} t_\mathfrak{p} & 0 \\ 0 & 1 \end{pmatrix} v_\mathfrak{p}\right) h\left(\begin{pmatrix} 0 & -1 \\ t_\mathfrak{p} & 1 \end{pmatrix} v_\mathfrak{p}\right) d^\times t_\mathfrak{p} =$$

$$\sum_{\nu\in\mathbb{Z}} \sum_{u\in\mathbb{F}_p^\times} f_\mathfrak{p}\left(\begin{pmatrix} u\varpi_\mathfrak{p}^\nu & 0 \\ 0 & 1 \end{pmatrix} v_\mathfrak{p}\right) h_\mathfrak{p}\left(\begin{pmatrix} 0 & -1 \\ u\varpi_\mathfrak{p}^\nu & 1 \end{pmatrix} v_\mathfrak{p}\right).$$

We claim that that the terms with $\nu \neq 0$ give zero. To see this we look at the Iwasawa decomposition (60). We observe that the second factor in the decomposition is congruent to a constant matrix mod $\mathfrak{p}$ this means that the value $h_\mathfrak{p}(\begin{pmatrix} 0 & -1 \\ u\varpi_\mathfrak{p}^\nu & 1 \end{pmatrix})$ does not depend on the second factor. We have the element $i_p \in \mathbb{F}_p^\times$ and then it follows that

$$h_\mathfrak{p}\left(\begin{pmatrix} 0 & -1 \\ i_p u\varpi_\mathfrak{p}^\nu & 1 \end{pmatrix}\right) = \eta^{\pm 1}(i_p) h_\mathfrak{p}\left(\begin{pmatrix} 0 & -1 \\ u\varpi_\mathfrak{p}^\nu & 1 \end{pmatrix}\right).$$

On the other hand we assumed that $f_\mathfrak{p}$ is invariant under the action of $\mu_4 = \{i_p^\mu\}$ and hence we see that the summation over $u \in \mathbb{F}_p^\times$ gives zero.

For $\nu = 0$ we have to look at the sum

$$\sum_{u\in\mathbb{F}_p^\times} f_\mathfrak{p}\left(\begin{pmatrix} u & 0 \\ 0 & 1 \end{pmatrix} v_\mathfrak{p}\right) h_\mathfrak{p}\left(\begin{pmatrix} 0 & -1 \\ u & 1 \end{pmatrix} v_\mathfrak{p}\right) = \sum_{u\in\mathbb{F}_p^\times} f_\mathfrak{p}\left(\begin{pmatrix} u & 0 \\ 0 & 1 \end{pmatrix} v_\mathfrak{p}\right) h_\mathfrak{p}\left(\begin{pmatrix} 1 & -1 \\ -1-u & 1 \end{pmatrix}\right)$$

Recall that our function $h_\mathfrak{p}$ was either h_0 or h_∞. Then it is clear that

$$h_\mathfrak{p}\left(\begin{pmatrix} 1 & -1 \\ -1-u & 1 \end{pmatrix}\right) = \begin{cases} \eta^{\pm 1}(u) h_\mathfrak{p}(w) & \text{if } u \neq -1 \\ h_\mathfrak{p}(1) & \text{if } u \neq -1 \end{cases}$$

If we group the summation over $u \in \mathbb{F}_p^\times$ into a sum over the orbits under μ_4 then the previous argument shows that all orbits contribute by zero except the orbit which contains the element $-1 \in \mathbb{F}_p^\times$. So we see that the above sum is equal to

$$< f_\mathfrak{p}, h_\mathfrak{p} >_\mathfrak{p} := -f_\mathfrak{p}(-1) h_\mathfrak{p}(w) \text{ or } f_\mathfrak{p}(-1) h_\mathfrak{p}(e) \tag{62}$$

depending on in which case we are.

So we are left with the place $v = \infty$. We have to evaluate the integral

$$\int_0^\infty < H, U_\alpha^\vee \otimes f_\infty > \begin{pmatrix} 0 & -1 \\ x & 1 \end{pmatrix} \frac{dx}{x}.$$

The Iwasawa decomposition yields

$$\begin{pmatrix} 0 & -1 \\ x & 1 \end{pmatrix} = \begin{pmatrix} \frac{x}{\sqrt{1+x^2}} & * \\ 0 & \sqrt{1+x^2} \end{pmatrix} \begin{pmatrix} \frac{1}{\sqrt{1+x^2}} & \frac{-x}{\sqrt{1+x^2}} \\ \frac{x}{\sqrt{1+x^2}} & \frac{1}{\sqrt{1+x^2}} \end{pmatrix},$$

and hence

$$< H, U_\alpha^\vee \otimes f_\infty > \begin{pmatrix} 0 & -1 \\ x & 1 \end{pmatrix} = \frac{x}{1+x^2} < H, \begin{pmatrix} \frac{1}{\sqrt{1+x^2}} & \frac{x}{\sqrt{1+x^2}} \\ \frac{-x}{\sqrt{1+x^2}} & \frac{1}{\sqrt{1+x^2}} \end{pmatrix} U_\alpha^\vee > .$$

The second factor is a matrix coefficient for the action of K_∞ on $\mathfrak{g}/\mathfrak{k}$, and Sect. 1.4.1 in [3] says that this coefficient is equal to $\frac{x}{1+x^2}$. Hence we have to evaluate

$$\int_0^\infty \frac{x}{(1+x^2)^2} dx = \frac{1}{2}.$$

We can collect these results: If we take the definition of the cohomological $L-$function into account then we find

$$\begin{aligned} &< \widetilde{\mathfrak{z}_T \otimes f_\mathfrak{p}}), \mathrm{Eis}(U_\alpha^\vee \otimes b(\phi) h_\mathfrak{p}) >= \int_{\widetilde{\mathfrak{z}_T \otimes f}} \mathrm{Eis}(U_\alpha^\vee \otimes b(\phi) h_\mathfrak{p}) = \\ &\frac{1}{2} b(\phi) < f_\mathfrak{p}, h_\mathfrak{p} > \frac{L^{\mathrm{coh}}(\psi, 1)^2}{L^{\mathrm{coh}}(\phi, 2)} + < f_1, h_\mathfrak{p} + c(\phi) T_\mathfrak{p}^{\mathrm{loc}} h_\mathfrak{p} > . \end{aligned} \tag{63}$$

Of course we have to take into account that the class $[\widetilde{\mathfrak{z}_T \otimes f_\mathfrak{p}})]$ is not well defined, it depends on the choice of f_1. We can modify f_1 to $f_1 + g$ where $\sigma_\infty \otimes g$ is a cycle and hence defines a homology class $[\sigma_\infty \otimes g]$ in $H_1(\partial(\Gamma\backslash\mathbb{H}_3), \underline{I_{\bar{B}}^{\bar{G}} \chi})$ and the integral against the cap changes by the value of

$$< [\sigma_\infty \otimes g], \mathrm{Eis}(U_\alpha^\vee \otimes h)|_{\partial(\Gamma\backslash\mathbb{H}_3)} =< [\sigma_\infty \otimes g], U_\alpha^\vee \otimes h + c(\phi) U_{\bar{\alpha}}^\vee \otimes T_\mathfrak{p}^{\mathrm{loc}}(h) > .$$

If we are in the case that $U_\alpha^\vee \otimes h + c(\phi) U_{\bar{\alpha}}^\vee \otimes T_\mathfrak{p}^{\mathrm{loc}}(h)$ is integral, then we get that

$$< [\widetilde{\mathfrak{z}_T \otimes f}], \mathrm{Eis}(U_\alpha^\vee \otimes h) >$$

is well defined modulo R.

5 Integrality of Eisenstein Classes

We already discussed the question whether the restriction of an Eisenstein class to the boundary can be integral. (See (27)), For $h_{\mathfrak{p}} \in I_{\phi_f}^{K_{0,f}(\mathfrak{p})}$ the classes $[\mathrm{Eis}^{(1)}(U_\alpha^\vee \otimes b(\phi)h_{\mathfrak{p}})]$, are the classes which become integral when we restrict them to the boundary. If we want to know whether they are integral themselves we have to evaluate them on our special class of modular symbols. Our computation above implies (remember that we have to multiply by $b(\phi)$): The class $[\mathrm{Eis}^{(1)}(U_\alpha^\vee \otimes b(\phi)h_{\mathfrak{p}})]$ is integral if for all choices of $f_{\mathfrak{p}} \in I_{\bar{B}}^{\bar{G}}\chi(0)$ the number

$$b(\phi) < f_{\mathfrak{p}}, h_{\mathfrak{p}} >_{\mathfrak{p}} \frac{L^{\mathrm{coh}}(\psi, 1)^2}{L^{\mathrm{coh}}(\phi, 2)} \in R = \mathbb{Z}[i, \frac{1}{2\varpi_{\mathfrak{p}}}] \tag{64}$$

We define

$$\Delta(\psi) = \text{ denominator of } b(\phi)\frac{L^{\mathrm{coh}}(\psi, 1)^2}{L^{\mathrm{coh}}(\phi, 2)}. \tag{65}$$

Then our formula (62) implies

Theorem 1 *For $h = ah_0 + bh_\infty$ the class $[Eis^{(1)}(U_\alpha^\vee \otimes h]$ is integral if and only if*

$$\Delta(\psi)|(a-b).$$

The reader may ask, whether we have cases where $\Delta(\psi) \neq 1$. In fact this does happen, but it is a rare event. We looked at the first 200 primes $p \equiv 5 \bmod 8$ and we found the following list of pairs $\{\Delta(\psi), p)\}$ with $\Delta(\psi) \neq 1$:

$$\begin{gathered}\{5, 173\}, \{3, 229\}, \{3, 653\}, \{3, 709\}, \{3, 797\}, \{23, 1021\}, \{3, 1061\}, \{5, 1109\} \\ \{7, 1117\}, \{7, 1429\}, \{11, 1933\}, \{3, 2069\}, \{47, 2381\}, \{5, 3229\}, \{5, 3469\} \\ \{5, 3541\}, \{5, 3733\}, \{27, 4133\}, \{5, 4373\}, \{3, 4397\}, \{51, 4637\}, \{5, 4813\} \\ \{3, 4861\}, \{5, 5501\}, \{23, 5653\}, \{7, 5669\}, \{5, 5701\}, \{3, 5749\}, \{11, 6029\}.\end{gathered} \tag{66}$$

For this see also the next section.

We go back to the considerations in Sect. 2.2. Our above theorem says that in (29)

$$\mathrm{Im}(r)_{\mathrm{int}}/\mathrm{Im}_{\mathrm{Eis}}(r) = R/\Delta(\psi). \tag{67}$$

Clearly the boundary operator maps

$$\delta_\psi : \mathrm{Im}(r)_{\mathrm{int}}/\mathrm{Im}_{\mathrm{Eis}}(r) \to H_c^2(\Gamma\backslash\mathbb{H}_3, \widetilde{I_{\bar{B}}^{\bar{G}}\chi}). \tag{68}$$

and by definition the kernel of δ_ψ is $\mathrm{Im}(r)$.

So our question whether it happens that $\mathrm{Im}(r) \neq \mathrm{Im}_{\mathrm{Eis}}(r)$ amounts to the question whether δ_ψ is injective. Then we have the following dichotomy

A) If δ_ψ is injective then $H^2_c(\Gamma\backslash\mathbb{H}_3, \widetilde{I_{\bar{B}}^{\bar{G}}\chi})$ contains a torsion element of order $\Delta(\psi)$.

B) If δ_ψ is not injective, then it follows that $H^1_!(\Gamma\backslash\mathbb{H}_3, \widetilde{I_{\bar{B}}^{\bar{G}}\chi}) \neq 0$.

If we are in the case A) we know from (6) how the Hecke operators act on this torsion subgroup.

I was hoping that - at least under certain further conditions - we could find cases where B) happens. This would tell us that the divisibility of certain special values of $L-$functions would "create" non trivial inner cohomology classes. In the analogue situation for the case of $\mathrm{Sl}_2(\mathbb{Z})$ with a non trivial coefficient system (See [6], Chap. 2, Sect. 4.5 this is true and for the case of Siegel modular forms [5] this has been verified experimentally in many cases. (See also the article of van der Geer in that volume)

But a computer program written by D. Yasaki seems to destroy this hope. D. Yasaki computed the cohomology for all primes $p \leq 20000$ and he found that

$$H^1_!(\Gamma\backslash\mathbb{H}_3, \widetilde{I_{\bar{B}}^{\bar{G}}\chi}) = 0$$

in all the cases.

On the other hand a simple argument using Poincare duality and some standard exact sequences shows that in case A) and $\Delta(\psi) \neq 1$ we also find non trivial classes in $H^2(\Gamma\backslash\mathbb{H}_3, \widetilde{I_{\bar{B}}^{\bar{G}}\chi})$ which are annihilated by $\Delta(\psi)$. Yasakis code also computes this cohomology and he finds these torsion classes predicted by the theorem.

It seems that the hunt for torsion classes supplies a very meager prey. In the paper [13] Sengun raises the question 5.4 whether

$$\log(\# H^2(\Gamma_0'(\mathfrak{p})\backslash\mathbb{H}_3, R)[\chi]_{\mathrm{tors}}) \sim Cp.$$

Sengun considers the case $\Gamma_0(\mathfrak{p})$, but of course we can ask the same question in our case. If we look at Yasaki's tables, then we find some agreement if we try

$$C = \frac{1}{12\pi} V_1 \text{ where } V_1 = \mathrm{vol}(\Gamma\backslash\mathbb{H}_3) = 0.305321\ldots .$$

(See [13]). Hence we expect huge torsion groups. For instance for $p = 16661$ Yasaki gives

$$\#(H^2(\Gamma_0'(\mathfrak{p})\backslash\mathbb{H}_3, R)[\chi]_{\text{tors}}) =$$

$$281919015079139215995994222726262701807272659578941724 69 =$$

$$19 * 5521 * 1195578482743137877 * 224788935704970988378771028475403 \tag{69}$$

The number $\#(H^2(\Gamma_0'(\mathfrak{p})\backslash\mathbb{H}_3, R)[\chi]_{\text{tors}})$ has only four prime factors. This is the number of prime factors of a "random" integer of this size. Looking at Yasaki's tables we find, that this seems to be a general phenomenon:

The number of prime factors in $\#(H^2(\Gamma_0'(\mathfrak{p})\backslash\mathbb{H}_3, R)[\chi]_{\text{tors}})$ *is close to the number* $\log(\log(\#(H^2(\Gamma_0'(\mathfrak{p})\backslash\mathbb{H}_3, R)[\chi]_{\text{tors}}))$. *This is roughly the number of prime factors - predicted by analytic number theory - of a huge random integer.*

In other words, the torsion is huge in the archimedian sense, but the number of primes which occur in the torsion seems to be small.

In this respect it may be interesting to notice that in the above case $\Delta(\psi) = 19$, in other words the 19 is predicted by our theorem.

6 The Periods

We want to say something about the number $b(\phi)$. We recall some results, concerning the nature of the special values $\Lambda^{\text{coh}}(\phi, \nu)$ and $\Lambda^{\text{coh}}(\psi, 1)$ which essentially go back to Hurwitz (See [2, 8, 11]).

We consider the period

$$\Omega = \frac{1}{2\pi}\int_0^1 \frac{1}{\sqrt{x - x^3}}dx = 0.417313..$$

Our prime ideal $\mathfrak{p}$ is principal we write $\mathfrak{p} = (\varpi_{\mathfrak{p}})$. We recall that we have the functional equation

$$\Lambda^{\text{coh}}(\phi, s) = (-1)^{e(\mathfrak{p})}\frac{\varpi_{\mathfrak{p}}^2}{p}(4p)^{3/2-s}\Lambda^{\text{coh}}(\bar{\phi}, 3 - s) \tag{70}$$

where $e(\mathfrak{p}) \in \mathbb{Z}$. We define modified periods

$$\Omega(\phi) = e^{\frac{2pi}{8}e(\mathfrak{p})}\Omega^2 \cdot \bar{\varpi}_{\mathfrak{p}}^{-3/2} \tag{71}$$

and

$$\Omega(\psi) = e^{\frac{\pi i}{8}d(\mathfrak{p})}\begin{cases}\Omega \cdot \bar{\varpi}_{\mathfrak{p}}^{-3/4} & \text{if } p \equiv 13 \bmod 16\\ \Omega \cdot \bar{\varpi}_{\mathfrak{p}}^{-1/4} & \text{if } p \equiv 5 \bmod 16\end{cases} \tag{72}$$

where $d(\mathfrak{p}) \in \mathbb{Z}$, we do not give the rule how to determine this 16-th root of unity.

The choice of these modified periods is somewhat delicate, it depends on the arithmetic of the elliptic curve below, we have to determine a Neron differential. The most delicate point is the choice of the 16-th root of unity.

I think that the arguments [1, 2, 8] applied to the elliptic curve

$$E_{\mathfrak{p}} := \varpi_{\mathfrak{p}} y^2 = x - x^3.$$

should show that there are some "small" computable integers $n(\phi), n(\psi) > 0$ such that

$$\Lambda^{\mathrm{coh}}_{\mathrm{alg}}(\psi, 1) := \frac{\Lambda^{\mathrm{coh}}(\psi, 1)}{\Omega(\psi)} \in \frac{1}{n(\psi)}\mathbb{Z}[i]$$
$$\Lambda^{\mathrm{coh}}_{\mathrm{alg}}(\phi, \nu) := \frac{\Lambda^{\mathrm{coh}}(\phi, \nu)}{\Omega(\phi)} \in \frac{1}{n(\phi)}\mathbb{Z}[i] \text{ for } \nu = 1, 2\,. \tag{73}$$

For the first 700 primes we found numerically that we can choose $n(\phi) = 1, n(\psi) = 1$. We did not try to prove that this is always the case.

To our elliptic curve we attach an $L-$function, and it is well known that

$$L(E_{\mathfrak{p}}, s) = L^{\mathrm{coh}}(\psi, s)L^{\mathrm{coh}}(\bar{\psi}, s). \tag{74}$$

For this Lfunction we have the Birch and Swinnerton–Dyer conjecture which says that under the assumption $L(E_{\mathfrak{p}}, 1) \neq 0$

$$\frac{1}{\Omega_{\mathfrak{p}}} L(E_{\mathfrak{p}}, 1) = \frac{\#\text{Ш}(E_{\mathfrak{p}})}{\#\mathrm{Tors}(E_{\mathfrak{p}}(F))^2}, \tag{75}$$

here $\text{Ш}(E_{\mathfrak{p}})$ is of course the Tate-Shafarevic group, and $\Omega_{\mathfrak{p}}$ is a period, which is computed from the arithmetic data of our elliptic curve.

For the first 700 primes we never found a p for which $L(E_{\mathfrak{p}}, 1) = 0$ and hence for these primes $E_{\mathfrak{p}}/F$ never has a point of infinite order [12].

We are certainly not very far away from the truth if we say that

$$\Omega_{\mathfrak{p}} = \Omega(\psi)^2 c_{\mathfrak{p}}$$

where $c_{\mathfrak{p}}$ is a very small non zero "fudge" factor. Hence we see- if we believe in the Birch and Swinnerton–Dyer conjectures- that the denominator $n(\psi)$ must be very small, because we can bound the torsion of $E_{\mathfrak{p}}(F)$.

6.1 Back to $\Delta(\psi)$

If we now write $\Lambda^{\text{coh}}_{\text{alg}}(\phi, 2) = a + bi$ we assume $a, b \in \mathbb{Z}$. Then

$$c(\phi) = 2\bar{\varpi}_{\mathrm{p}} \frac{a - bi}{a + bi}. \tag{76}$$

If we want $\Delta(\psi)$ to be large, then $b(\phi)$ should be small. Hence we have to look at the greatest common divisor $d(\phi) = \gcd(a, b)$ and we put

$$a' = a/d(\phi), b' = b/d(\phi).$$

Then we get

$$b(\phi) = a' + b'i$$

and finally

$$\Delta(\psi) = \text{ denominator } ((a' + b'i)\frac{\Lambda^{\text{coh}}_{\text{alg}}(\psi, 1)^2}{d(\phi)}). \tag{77}$$

This means that we may expect that $\Delta(\psi)$ becomes large only if $d(\phi) = \gcd(a, b)$ becomes large. This does not happen too often.

Finally we would like to remark that the above rationality results (see (73)) are not essential for our paper. Since we have a cancellation of the periods these results imply

$$\frac{\pi}{2} \frac{L^{\text{coh}}(\phi, 1)}{L^{\text{coh}}(\phi, 2)} \in F\ ; \quad \frac{\Lambda^{\text{coh}}(\psi, 1)^2}{\Lambda^{\text{coh}}(\phi, 2)} \in F \tag{78}$$

but this is also clear from the cohomological interpretation. But this rationality result does not involve any period, hence we can speak of the prime factorization of these numbers.

Our results above say that we can read off $\Delta(\psi)$ from this prime factorization of these numbers. Returning to my habit in [3] I want to illustrate this by an explicit example. We take $p = 229$ the second entry in (66). We get

$$\frac{\pi}{2} \frac{L^{\text{coh}}(\phi, 1)}{L^{\text{coh}}(\phi, 2)} = -\frac{(1 + i)^2(2 + i)(15 + 2i)}{1 + 2i} \text{ and } \frac{L^{\text{coh}}(\psi, 1)^2}{L^{\text{coh}}(\phi, 2)} = -\frac{15 + 2i}{(1 + 2i)3}$$

Hence $b(\phi) = 1 + 2i$ and hence we see that

$$\Delta(\psi) = \text{ denominator } \left(b(\phi) \frac{L(\psi, 1)^2}{L(\phi, 2)} \right) = 3.$$

Here we used some very effective algorithms which compute the values of the L-function with a very high precision. Then the LLL-algorithm gives us their probably correct exact values. Of course this never proves that the values are the correct ones. These computations are the basis for the table (73). If we want to be sure that the values a really correct then we need to control the denominators $n(\phi), n(\psi)$.

Final remarks: (A) This note discusses a very special example in a much wider domain of interest. Of course we could consider arbitrary imaginary extensions of $\mathbb{Q}$ and arbitrary congruence subgroups. Of course we could also go back to [3]. In that paper we evaluate the Eisenstein classes on compact cycles which are obtained from anisotropic tori $T_1/F \subset \mathrm{Sl}_2/F$. They correspond to quadratic extensions E/F, and we can consider the Hecke character $\psi \circ N_{E/F}$. Then it is clear that we can sharpen our Corollary 4.1 in [3] and prove some boundedness for the denominators of the values $\Lambda^{\mathrm{alg}} L(\psi \circ N_{E/F}, 1)$.

But it is also conceivable that similar results can be proved for the cohomology of $\mathrm{Gl}_n(\mathbb{Z})$. In our paper [7] we prove a vast generalization of the results of the result in Sect. 2.1. We prove that a ratio of values of an $L-$function at two consecutive critical arguments, divided by a "relative" period, is an algebraic number. We also indicate that we can pin down the relative period, which is essentially defined up to a unit, and hence we can speak of the prime factorization of these values. Is there a generalization of Theorem 1, do we get some information about the existence of certain torsion classes, whose order is visible in this prime factorization?

We see some modular symbols (see [9, 10]), here the capping of the modular symbols may cause some serious problems.

(B) But also in the special case, treated in this paper, various questions remain open. Will we ever see a case where $H^1_!(\Gamma\backslash\mathbb{H}_3, \widetilde{I_{\bar{B}}^{\bar{G}}\chi}) \neq 0$? Will we ever see a case where $E_{\mathfrak{p}}(F)$ has a point of infinite order?

References

1. Damerell, R.M.: L-functions of elliptic curves with complex multiplication I. Acta Arith. **17**, 287–301 (1970)
2. Damerell, R.M.: L-functions of elliptic curves with complex multiplication II. Acta Arith. **19**, 311–317 (1971)
3. Harder, G.: Period integrals of cohomology classes which are represented by Eisenstein series. In: Proceedings of the Bombay Colloquium. Springer, Berlin (1979)
4. Harder, G.: Eisenstein cohomology of arithmetic groups. The case GL_2. Invent. Math. **89**(1), 37–118 (1987)
5. Harder, G.: A congruence between a Siegel and an elliptic modular form. The 1-2-3 of Modular Forms. Springer, Berlin (Unitext)
6. Harder, G.: Cohomology of arithmetic groups. Book in preparation. Preliminary version available at http://www.math.uni-bonn.de/people/harder/Manuscripts/buch/
7. Harder, G.- Raghuram, A.: Eisenstein cohomology for GL_N and ratios of critical values of Rankin–Selberg L -functions (submitted)
8. Hurwitz, A.: Ueber die Entwickelungs coefficienten der lemniscatischen Functionen. Math. Ann. **51**(2), 196–226 (1898)

9. Kasten, H., Schmidt, C.-G.: The critical values of Rankin–Selberg convolutions. Int. J. Number Theory **9**(1), 205–256 (2013)
10. Kazhdan, D., Mazur, B., Schmidt, C.-G.: Relative modular symbols and Rankin–Selberg convolutions. J. Reine Angew. Math. **2000**(519), 97–141 (2000)
11. König, H.: Eisenstein-Kohomologie von${}_2([i])$. Dissertation, Bonner Mathematische Schriften Nr. 222 (1991)
12. Rubin, K.: Tate-Shafarevich groups and L-functions of elliptic curves with complex multiplication. Invent. Math. **89**(3), 527–559 (1987)
13. Sengün, M.-H.: On the Integral cohomology of bianchi groups. Exp. Math. **20**(4), 487–505 (2011). https://doi.org/10.1080/10586458.2011.594671

Nearby Cycles of Automorphic Étale Sheaves, II

Kai-Wen Lan and Benoît Stroh

Dedicated to Joachim Schwermer on the occasion of his 66th birthday

Abstract We review some recent results of ours on the nearby cycles of automorphic étale sheaves, and record some improvements of the arguments.

Keywords Integral models of Shimura varieties · Compactifications
Automorphic étale sheaves · Nearby cycles

2010 Mathematics Subject Classification Primary 11G18 · Secondary 11G15, 11F75

K.-W. Lan—partially supported by the National Science Foundation under agreement No. DMS-1352216, by an Alfred P. Sloan Research Fellowship, and by the Université Paris 13.
B. Stroh—partially supported by the A.N.R. (Agence Nationale de la Recherche) under the program ANR-14-CE25-0002.

K.-W. Lan (✉)
University of Minnesota, Minneapolis, MN 55455, USA
e-mail: kwlan@math.umn.edu

B. Stroh
C.N.R.S. and Institut de Mathématiques de Jussieu–Paris Rive Gauche, 75252 Paris Cedex 05, France
e-mail: benoit.stroh@imj-prg.fr

J. W. Cogdell et al. (eds.), *Cohomology of Arithmetic Groups*, Springer Proceedings in Mathematics & Statistics 245,
https://doi.org/10.1007/978-3-319-95549-0_4

1 Introduction

In the context of the Langlands program, the étale cohomology of Shimura varieties serves as an important source of Galois representations associated with automorphic representations. Concretely, let X be a model of a Shimura variety defined over some number field contained in $\mathbb{C}$, which we temporarily assume to be $\mathbb{Q}$, for simplicity of exposition; let $\bar{\mathbb{Q}}$ denote the algebraic closure of $\mathbb{Q}$ in $\mathbb{C}$; and let $X_{\bar{\mathbb{Q}}}$ denote the base change of X to $\bar{\mathbb{Q}}$. Let ℓ be any prime number. Then the étale cohomology $H^i_{\text{ét}}(X_{\bar{\mathbb{Q}}}, \mathbb{Q}_\ell)$ is canonically a representation of the absolute Galois group $\mathrm{Gal}(\bar{\mathbb{Q}}/\mathbb{Q})$, and it is desirable to understand such a representation.

For this purpose, it is important to also understand the restrictions of such a representation to the decomposition groups $\mathrm{Gal}(\bar{\mathbb{Q}}_p/\mathbb{Q}_p)$, for all prime numbers p. This can be achieved by considering the canonical action of $\mathrm{Gal}(\bar{\mathbb{Q}}_p/\mathbb{Q}_p)$ on $H^i_{\text{ét}}(X_{\bar{\mathbb{Q}}_p}, \mathbb{Q}_\ell) \cong H^i_{\text{ét}}(X_{\bar{\mathbb{Q}}}, \mathbb{Q}_\ell)$ (see [11, Arcata, V, 3.3]), where $\bar{\mathbb{Q}}_p$ is any algebraic closure of $\mathbb{Q}_p$ containing $\bar{\mathbb{Q}}$. Then we can ask, for example, when $p \neq \ell$, whether $H^i_{\text{ét}}(X_{\bar{\mathbb{Q}}_p}, \mathbb{Q}_\ell)$ is an *unramified* representation of $\mathrm{Gal}(\bar{\mathbb{Q}}_p/\mathbb{Q}_p)$.

To answer such a question, a general method is to reduce it to the case where X has some model over $\mathbb{Z}_p$, and consider the so-called *nearby cycles* over the geometric special fiber of such a model. Let us explain this in more detail. (From now on, the symbol X will no longer denote a model of Shimura variety over $\mathbb{Q}$.)

Let us consider some more general base rings. Let R be a Henselian discrete valuation ring of residue characteristic $p > 0$, with fraction field $K = \mathrm{Frac}(R)$. Let $\bar{K}$ be an algebraic closure of K, and let $\bar{R}$ be the integral closure of R in $\bar{K}$. Let k denote the residue field of R, and let $\bar{k}$ denote the residue field of $\bar{R}$. Then we have the following commutative diagram

$$\begin{array}{ccccc}
\bar{s} := \mathrm{Spec}(\bar{k}) & \xrightarrow{\bar{i}} & \bar{S} := \mathrm{Spec}(\bar{R}) & \xleftarrow{\bar{j}} & \bar{\eta} := \mathrm{Spec}(\bar{K}) \\
\downarrow & & \downarrow & & \downarrow \\
s := \mathrm{Spec}(k) & \xrightarrow[i]{} & S := \mathrm{Spec}(R) & \xleftarrow[j]{} & \eta := \mathrm{Spec}(K)
\end{array}$$

of canonical morphisms. We shall denote pullbacks with subscripts $\bar{s}$ etc. as usual.

Suppose ℓ is a prime number different from p, and suppose Λ is a coefficient ring that is either $\mathbb{Z}/\ell^m\mathbb{Z}$ (for some integer $m \geq 1$), $\mathbb{Z}_\ell$, $\mathbb{Q}_\ell$, $\bar{\mathbb{Q}}_\ell$, or a finite extension of any of these. (These are the coefficient rings accepted in, for example, [25, 3.1].) For each scheme X separated and of finite type over S, we denote by $D^b_c(X_\eta, \Lambda)$ the bounded derived category of Λ-étale constructible sheaves over X_η, and by $D^b_c(X_{\bar{s}} \times \bar{\eta}, \Lambda)$ the bounded derived category of Λ-étale constructible sheaves over $X_{\bar{s}}$ with compatible continuous $\mathrm{Gal}(\bar{K}/K)$-actions. (See [12, 1.1] and [14] when Λ is not torsion.) Then we have the functor of *nearby cycles*:

$$R\Psi_X : D^b_c(X_\eta, \Lambda) \to D^b_c(X_{\bar{s}} \times \bar{\eta}, \Lambda) : \mathcal{F} \mapsto \bar{i}^* R\bar{j}_*(\mathcal{F}_{\bar{\eta}}), \tag{1.1}$$

where $\mathcal{F}_{\bar{\eta}}$ denotes the pullback of $\mathcal{F}$ to $X_{\bar{\eta}}$. (See [13, XIII], [11, Th. finitude, Sect. 3], and [25, Sect. 4] for more details.)

Suppose we have a morphism $\varphi : X \to Y$ of schemes of finite type over S. Then, on one hand, we have the adjunction morphisms

$$R\Psi_Y\, R\varphi_{\eta,*}(\mathcal{F}) \to R\varphi_{\bar{s},*}\, R\Psi_X(\mathcal{F}) \tag{1.2}$$

and

$$R\varphi_{\bar{s},!}\, R\Psi_X(\mathcal{F}) \to R\Psi_Y\, R\varphi_{\eta,!}(\mathcal{F}) \tag{1.3}$$

for pushforwards, which are isomorphisms when φ is *proper*, by the proper base change theorem (cf. [2, XII, 5.1] and [13, XIII, (2.1.7.1) and (2.1.7.3)].) On the other hand, we have the adjunction morphism

$$\varphi_{\bar{s}}^*\, R\Psi_Y(\mathcal{F}) \to R\Psi_X\, \varphi_{\eta}^*(\mathcal{F}) \tag{1.4}$$

for pullbacks, which is an isomorphism when φ is *smooth*, by the smooth base change theorem (see [2, XVI, 1.2] and [13, XIII, (2.1.7.2)].)

To see why these are useful, consider the special case where the structural morphism $X \to S$ is both *proper* and *smooth*, let $Y = S$, and let φ be the above structural morphism. Then we obtain from (1.4) and (1.2) the canonical isomorphisms

$$\Lambda \xrightarrow{\sim} R\Psi_X(\Lambda) \tag{1.5}$$

and

$$H^i(X_{\bar{\eta}}, \Lambda) \xrightarrow{\sim} H^i(X_{\bar{s}}, R\Psi_X(\Lambda)), \tag{1.6}$$

respectively, and hence their combination

$$H^i(X_{\bar{\eta}}, \Lambda) \xrightarrow{\sim} H^i(X_{\bar{s}}, \Lambda), \tag{1.7}$$

which are compatible with actions of $\mathrm{Gal}(\bar{K}/K)$. In particular, the action of $\mathrm{Gal}(\bar{K}/K)$ on the left-hand side $H^i(X_{\bar{\eta}}, \Lambda)$ of (1.7) is *unramified*, because the action of $\mathrm{Gal}(\bar{K}/K)$ on the right-hand side $H^i(X_{\bar{s}}, \Lambda)$ of (1.7) factors through $\mathrm{Gal}(\bar{k}/k)$.

In this article, we shall consider the more general situation where $\varphi : X \to S$ is some integral model of Shimura varieties that is neither proper nor smooth, and we shall also allow the trivial coefficient Λ to be replaced with certain automorphic étale sheaves valued in Λ-modules. (We will make these more precise in Sects. 2 and 3.) Although we cannot expect an isomorphism as in (1.7) in such a generality, we will show that, for most integral models we know, we still have an isomorphism as in (1.6), despite the lack of the properness assumption. Intuitively speaking, this means, at least for studying étale cohomology, the special fibers of these integral models have as many points as there should be—there are no *missing points*.

We will review some results and ideas in our previous work [42], but with some improvements of the statements and proofs. This is partly motivated by some recent developments (such as [31]) after [42] was written. Compared with [42], the main innovations of this article are the following: (i) a different argument in the proof of the key Theorem 4.1 using torsion automorphic coefficients instead of using Kuga families and their compactifications; and (ii) the inclusion of abelian-type cases in Corollaries 4.6 and 4.10, and in Theorems 4.13, 4.19, and 4.23.

We shall follow [35, Notation and Conventions] unless otherwise specified. We will sometimes use the terminologies introduced in [42] without repeating their definitions in detail, when their meanings can be understood from the context. (Nevertheless, we will still provide references to such definitions.)

2 Integral Models We Consider

Let us start by emphasizing that we cannot expect isomorphisms as in (1.6) to be true for all kinds of nonproper integral models. For an extreme example, in the context of Sect. 1, if we consider the *trivially wrong* model obtained by replacing X with X_η, in which case the whole special fiber X_s is removed, then the right-hand side of (1.6) is always zero, and this cannot be what we want. Thus, if we are to have meaningful generalizations of (1.6) without the properness assumption, we need to be precise about our choices of integral models.

Let us retain the notation K, R, etc. in Sect. 1. Suppose K is of characteristic zero, and suppose $\mathsf{X}_{\mathcal{H}}$ is the pullback to $S = \mathrm{Spec}(R)$ of one of the following models of Shimura varieties (or related moduli problems), where $\mathcal{H}$ is an open compact subgroup of $\mathrm{G}(\mathbb{A}^\infty)$ for some group scheme G over $\mathbb{Z}$ associated with the various constructions, and where R is now also an $\mathcal{O}_{F_0,(p)}$-algebra, with F_0 denoting the reflex field: (For more detailed references in the first four cases, see [42, Sect. 2.1].)

(Sm) A good reduction integral model defined by a smooth PEL moduli problem, as in [32, Sect. 5] and [35, Sects. 1.4.1–1.4.2].

(Nm) A flat integral model defined by taking normalization, as in [36, Sect. 6], of a characteristic zero PEL moduli problem over a product of good reduction integral models in Case (Sm) above. (This includes all normalizations of the PEL moduli defined by multichains of isogenies as in [56, Chaps. 3 and 6].)

(Spl) A flat integral model defined by taking normalization, as in [39, Sect. 2.4], of the pullback of a characteristic zero PEL moduli problem over the so-called *splitting models* defined as in [53, Sect. 15].

(Hdg) A flat integral model defined by taking normalization, as in [44, Introduction], of a Hodge-type Shimura variety over some Siegel moduli scheme.

(Ab) A flat integral model defined as in [29, 30], or [31], of an abelian-type Shimura variety.

We shall say that we are in Case (Sm), (Nm), (Spl), (Hdg), or (Ab) depending on the case above from where $\mathsf{X}_{\mathcal{H}}$ is pulled back.

Remark 2.1 Let us be more precise about the levels (i.e., the open compact subgroups $\mathcal{H}$ of $\mathrm{G}(\mathbb{A}^\infty)$) allowed in these cases. For simplicity, we shall assume $\mathcal{H}$ to be of the form $\mathcal{H} = \mathcal{H}^p\mathcal{H}_p$ for some neat open compact subgroup $\mathcal{H}^p$ of $\mathrm{G}(\mathbb{A}^{\infty,p})$ and for some open compact subgroup $\mathcal{H}_p$ of $\mathrm{G}(\mathbb{Q}_p)$, and we say $\mathcal{H}^p$ and $\mathcal{H}_p$ are the levels *away from p* and *at p*, respectively. In Case (Sm), the level at p is hyperspecial. (But $p = 2$ is excluded if simple factors of type D are involved, as in [35, Definition 1.2.1.15].) In Cases (Nm) and (Spl), we emphasize that we allow not only the parahoric levels at p as in [56, Chaps. 3 and 6] and [53, Sect. 15] defined by certain multichains of isogenies, but also arbitrarily higher levels and also arbitrary collections of isogenies. (Also, $p =$ 2 is allowed.) In Case (Hdg), the level at p is exactly the pullback of a hyperspecial level at p of a symplectic similitude group, which can be the hyperspecial levels at p as in [29, 30], by composing any Siegel embedding as in [29, Sect. 2.3] and [30, Sect. 4] with the embedding given by "Zarhin's trick" as in [36, Lemma 4.9] or [38, Lemma 2.1.1.9]. (In fact, this was explained in an earlier version of [44], but not in the 2015 version we cited.) However, there is some subtlety for Hodge-type Shimura varieties at parahoric levels, in addition to the requirements in [31] that $p > 2$ and that the group $\mathrm{G}_{\mathbb{Q}_p}$ is split over a tamely ramified extension of $\mathbb{Q}_p$, when $\mathcal{H}_p = K_p^\circ \neq K_p$ (in the notation of [31, Sect. 4.3]), in which case we have to defer them to Case (Ab) below. (That is, we have to treat some integral models of Hodge-type Shimura varieties with parahoric levels at p only as integral models of abelian-type Shimura varieties.) In Case (Ab), the level at p has to be either hyperspecial or parahoric at p, and in the latter case there are the above-mentioned requirements in [31] that $p > 2$ and that the group $\mathrm{G}_{\mathbb{Q}_p}$ is split over a tamely ramified extension of $\mathbb{Q}_p$.

Remark 2.2 None of the three PEL-type cases we consider is completely subsumed by the Hodge-type case, and the Hodge-type case is not subsumed by the abelian-type case either. We emphasize again that this is about the actual choices of integral models, but not about the classification in characteristic zero. As we consider more and more general Shimura varieties in characteristic zero, the integral models that are available to us also become more and more restrictive, and there are some subtleties due to the fact that not everything available in the literature has been written in the most generality or flexibility its arguments allowed.

In Cases (Sm), (Nm), (Spl), and (Hdg), we have a commutative diagram

$$\begin{array}{ccc} \mathsf{X}_{\mathcal{H}} & \xhookrightarrow{\ \jmath^{\mathrm{tor}}\ } & \mathsf{X}_{\mathcal{H}}^{\mathrm{tor}} \\ & \searrow^{\jmath^{\mathrm{min}}} & \downarrow \\ & & \mathsf{X}_{\mathcal{H}}^{\mathrm{min}} \end{array} \tag{2.3}$$

of canonical morphisms between noetherian normal schemes over $S = \mathrm{Spec}(R)$, where $\mathsf{X}_{\mathcal{H}}^{\mathrm{tor}}$ and $\mathsf{X}_{\mathcal{H}}^{\mathrm{min}}$ denote some projective toroidal and minimal compactifications. In Case (Sm), we use [35, Theorems 6.4.1.1, 7.2.4.1, and 7.3.3.4]. In Case (Nm), we use [40, Theorem 6.1] and [36, Proposition 6.4, and Theorems 12.1 and

12.16]. In Case (Spl), we use [39, Theorems 3.4.1 and 4.3.1]. In Case (Hdg), we use [44, Theorems 4.1.5 and 5.2.11]. (See [42, Proposition 2.2] for a detailed qualitative description of these compactifications, and see the proof there for further references to the literature.)

In Case (Ab), we expect similar results, but they are not yet available in the literature. Nevertheless, we still have the following crude constructions:

Proposition 2.4 *Given any* $\mathsf{X}_{\mathcal{H}}$ *in Case* (Ab), *there exists an open immersion*

$$J^{min} : \mathsf{X}_{\mathcal{H}} \hookrightarrow \mathsf{X}_{\mathcal{H}}^{min} \tag{2.5}$$

from a quasi-projective scheme to a normal projective scheme over $S = \mathrm{Spec}(R)$, *which we consider the minimal compactification of* $\mathsf{X}_{\mathcal{H}}$, *with the following properties:*

(1) *There exists a Galois finite étale extension* $R \to R^+$ *of discrete valuation rings of mixed characteristics* $(0, p)$ *and an integral model* $\widetilde{\mathsf{X}}_{\widetilde{\mathcal{H}}}$ *in Case* (Hdg) *defined over a subring* $\widetilde{R}$ *of* R^+ *such that* R^+ *is unramified over* $\widetilde{R}$ *and such that each connected component* $\mathsf{X}_{\mathcal{H},R^+}^+$ *of the base change* $\mathsf{X}_{\mathcal{H},R^+} := \mathsf{X}_{\mathcal{H}} \otimes_R R^+$ *is noetherian normal and has geometrically connected fiber over* K^+, *where* $K^+ := \mathrm{Frac}(R^+)$, *and is isomorphic to the quotient by the free action of a finite group* Δ^+ *of some (noetherian normal) connected component* $\widetilde{\mathsf{X}}_{\widetilde{\mathcal{H}},R^+}^+$ *of the base change* $\widetilde{\mathsf{X}}_{\widetilde{\mathcal{H}},R^+} := \widetilde{\mathsf{X}}_{\widetilde{\mathcal{H}}} \otimes_{\widetilde{R}} R^+$. *(The group* Δ^+ *depends not only on the levels* $\mathcal{H}$ *and* $\widetilde{\mathcal{H}}$, *but also on the actual connected components* $\mathsf{X}_{\mathcal{H},R^+}^+$ *and* $\widetilde{\mathsf{X}}_{\widetilde{\mathcal{H}},R^+}^+$.*)*

(2) *Let* $\widetilde{\mathsf{X}}_{\widetilde{\mathcal{H}}} \hookrightarrow \widetilde{\mathsf{X}}_{\widetilde{\mathcal{H}}}^{min}$ *denote the minimal compactification of* $\widetilde{\mathsf{X}}_{\widetilde{\mathcal{H}}}$ *as in* (2.3). *Then the action of* Δ^+ *on* $\widetilde{\mathsf{X}}_{\widetilde{\mathcal{H}},R^+}^+$ *extends to a (possibly non-free) action on the schematic closure* $\widetilde{\mathsf{X}}_{\widetilde{\mathcal{H}},R^+}^{min,+}$ *of* $\widetilde{\mathsf{X}}_{\widetilde{\mathcal{H}},R^+}^+$ *in* $\widetilde{\mathsf{X}}_{\widetilde{\mathcal{H}},R^+}^{min} := \widetilde{\mathsf{X}}_{\widetilde{\mathcal{H}}}^{min} \otimes_{\widetilde{R}} R^+$, *and the quotient of* $\widetilde{\mathsf{X}}_{\widetilde{\mathcal{H}},R^+}^{min,+}$ *by* Δ^+ *is isomorphic to the schematic closure* $\mathsf{X}_{\mathcal{H},R^+}^{min,+}$ *of* $\mathsf{X}_{\mathcal{H},R^+}^+$ *in* $\mathsf{X}_{\mathcal{H},R^+}^{min} := \mathsf{X}_{\mathcal{H}}^{min} \otimes_R R^+$.

Remark 2.6 In Proposition 2.4, the quotients of quasi-projective schemes by finite groups are defined by the same argument as in [50, Sect. 7, Theorem and Remark] (see also [20, V, 1.8]). Moreover, in (1) of Proposition 2.4, when the action of Δ^+ on $\widetilde{\mathsf{X}}_{\widetilde{\mathcal{H}},R^+}^+$ is free, the canonically induced morphism $\widetilde{\mathsf{X}}_{\widetilde{\mathcal{H}},R^+}^+ \to \widetilde{\mathsf{X}}_{\widetilde{\mathcal{H}},R^+}^+/\Delta^+ \cong \mathsf{X}_{\mathcal{H},R^+}^+$ is a *Galois finite étale cover* with Galois group Δ^+ by the same argument as in the proof of the last assertion of [50, Sect. 7, Theorem] (if we use completions of strict local rings instead of completions of local rings; see also [20, V, 2.4 and 2.6]).

Proof of Proposition 2.4 Note that (1) follows (up to slight reformulation) from the constructions in [29, Sect. 3.4], [30, Theorem 3.10], and [31, Sect. 4.6]. We note that we need [31, Proposition 4.3.7] (when $\mathcal{H}_p = K_p^\circ \neq K_p$ in the notation there; cf. Remark 2.1) and "Zarhin's trick" (as in [36, Lemma 4.9] or [38, Lemma 2.1.1.9]) to ensure that, when working with connected components over a finite étale base ring extension, we can indeed reduce to Case (Hdg) (where the level at p is exactly the pullback of a hyperspecial level at p of a symplectic similitude group, as explained above).

Our main task is to prove (2). By [55, 12.3], we have a canonical open immersion

$$\mathsf{X}_{\mathcal{H},K} := \mathsf{X}_{\mathcal{H}} \underset{R}{\otimes} K \hookrightarrow \mathsf{X}^{\min}_{\mathcal{H},K}, \tag{2.7}$$

where $\mathsf{X}^{\min}_{\mathcal{H},K}$ denotes the pullback to K of the canonical model of the minimal compactification over the reflex field (which is a subfield of K by assumption), and we would like to extend (2.7) to its analogue (2.5) over R (with the desired properties).

Let us temporarily assume that G^{ad} is simple as an algebraic group over $\mathbb{Q}$. By [31, the proof of Proposition 4.6.28], and by [37, the proof of Theorem 3.8], there exists an ample invertible sheaf $\mathcal{L}_0$ over $\mathsf{X}_{\mathcal{H}}$ whose pullback to $\mathsf{X}_{\mathcal{H},K}$ is isomorphic to a positive tensor power of the canonical bundle $\Omega^d_{\mathsf{X}_{\mathcal{H},K}/K}$, where $d := \dim(\mathsf{X}_{\mathcal{H},K})$. Note that [31, Proposition 4.6.28] assumed that G^{ad} is absolutely simple (not just simple), and that the level at p is very special (as in [54, Sect. 10.3.2]). Let us explain in the next two paragraphs why we can borrow their arguments without these assumptions.

As for the assumption that the level at p is very special, it was made because the goal of [31, Proposition 4.6.28] was to show that the integral models constructed there are canonical, and detailed properties of local models were used to ensure that the special fiber of each connected component is reduced and irreducible, so that $\mathcal{L}_0$ is the unique extension of its pullback to $\mathsf{X}_{\mathcal{H},K}$. But since we have a rather different goal here, we can drop this assumption.

As for the assumption that G^{ad} is absolutely simple, it was made to ensure that, in the Hodge-type case, in the notation of [31, Proposition 4.6.28], the pullback of the Hodge invertible sheaf ω_{GSp} under the Siegel embedding is a positive tensor power of ω_{G}. But we can drop this assumption because we know that a positive tensor power of this pullback is isomorphic to a positive tensor power of the canonical bundle—let us explain why. This pullback is an automorphic line bundle with a canonical model over the reflex field (see [48]), and for our purpose (up to replacing the invertible sheaf with a positive tensor power) we just have to identify its pullback from the reflex field to $\mathbb{C}$, which is associated with some one-dimensional representation of the Levi of a parabolic subgroup P of $\mathrm{G}_{\mathbb{C}}$ defined by the Hodge cocharacter determined by the Shimura datum. As explained in [10, 2.3.7] and [49, Sect. 10], on each simple factor of $\mathrm{G}_{\mathbb{C}}$, the pullback of the standard representation of the symplectic group under the Siegel embedding is a multiple of the fundamental weight representation associated with some underlined node in [10, Table 1.3.9] or starred node in [49, Sect. 10, pp. 528–531] (the latter reference containing one more case missing from the former), and these multiplicities are the same on all the simple factors of $\mathrm{G}_{\mathbb{C}}$. Each such fundamental weight representations has a two-step filtration whose stabilizer is the corresponding factor of P, which corresponds to a direct summand of the pullback of the Hodge filtration on the relative de Rham homology of the universal abelian scheme over the Siegel moduli, and the dual of the top exterior power of the top graded piece corresponds to a tensor factor of the pullback of the Hodge invertible sheaf (as in [44, Definition 5.1.2]). It can be easily checked (using explicit realizations in example-rich texts such as [17] or [19]) that, as a representation of the Levi of this factor of P, the weight of this tensor factor is exactly the pullback of the corresponding fundamental weight. Hence, it follows from [37, the proof of Theorem 3.8] that $h^\vee$ times this weight, where $h^\vee$ is the dual Coxeter number of the root system of this (and every other) simple factor of $\mathrm{G}_\mathbb{C}$, is the weight of the corresponding factor of the canonical bundle. Then the above assertion follows, because the exponents of the tensor factors from the simple factors of $\mathrm{G}_\mathbb{C}$ are all the same.

By [51, Proposition 3.4(b)] (whose assertion can be formulated in terms of the pushforward of the log canonical bundle of any toroidal compactification, and therefore is compatible with descent), $\mathcal{L}_0 \otimes_R K$ extends to an ample invertible sheaf $\mathcal{L}_1$ over $\mathsf{X}^{\min}_{\mathcal{H},K}$. Let $\mathsf{X}_{\mathcal{H},R^+} = \mathsf{X}_\mathcal{H} \otimes_R R^+$ and $\mathsf{X}^+_{\mathcal{H},R^+}$ be as in (1) of Proposition 2.4. Let $\mathsf{X}_{\mathcal{H},K^+} := \mathsf{X}_{\mathcal{H},K} \otimes_K K^+ \cong \mathsf{X}_\mathcal{H} \otimes_R K^+$ and $\mathsf{X}^{\min}_{\mathcal{H},K^+} := \mathsf{X}^{\min}_{\mathcal{H},K} \otimes_K K^+$. Let us denote by U the union of $\mathsf{X}_\mathcal{H}$ and $\mathsf{X}^{\min}_{\mathcal{H},K}$ (glued over the open subscheme $\mathsf{X}_{\mathcal{H},K}$), and by $\mathcal{L}$ the invertible sheaf over U whose restrictions to $\mathsf{X}_\mathcal{H}$ and $\mathsf{X}^{\min}_{\mathcal{H},K}$ are $\mathcal{L}_0$ and $\mathcal{L}_1$, respectively. Let $\mathsf{X}^+_{\mathcal{H},K^+} := \mathsf{X}^+_{\mathcal{H},R^+} \otimes_{R^+} K^+$, and let $\mathsf{X}^{\min,+}_{\mathcal{H},K^+}$ denote the schematic closure of $\mathsf{X}^+_{\mathcal{H},K^+}$ in $\mathsf{X}^{\min}_{\mathcal{H},K^+}$. Let $\mathsf{U}^+_{R^+}$ denote the union of $\mathsf{X}^+_{\mathcal{H},R^+}$ and $\mathsf{X}^{\min,+}_{\mathcal{H},K^+}$ (glued over the open subscheme $\mathsf{X}^+_{\mathcal{H},K^+}$), which is noetherian normal and is the connected component of $\mathsf{U}_{R^+} := \mathsf{U} \otimes_R R^+$ containing $\mathsf{X}^+_{\mathcal{H},R^+}$. Let $\mathcal{L}_{R^+}$ and $\mathcal{L}^+_{R^+}$ denote the pullbacks of $\mathcal{L}$ to U_{R^+} and $\mathsf{U}^+_{R^+}$, respectively.

Let $\widetilde{\mathsf{X}}^+_{\widetilde{\mathcal{H}},R^+}$ and $\widetilde{\mathsf{X}}^{\min,+}_{\widetilde{\mathcal{H}},R^+}$ be as in (1) and (2) of Proposition 2.4, respectively. Let $\widetilde{\mathsf{U}}^+_{R^+}$ denote the union of $\widetilde{\mathsf{X}}^+_{\widetilde{\mathcal{H}},R^+}$ and $\widetilde{\mathsf{X}}^{\min,+}_{\widetilde{\mathcal{H}},K^+} := \widetilde{\mathsf{X}}^{\min,+}_{\widetilde{\mathcal{H}},R^+} \otimes_{R^+} K^+$ in $\widetilde{\mathsf{X}}^{\min,+}_{\widetilde{\mathcal{H}},R^+}$, whose complement has codimension at least two because the morphism $\widetilde{\mathsf{X}}_{\widetilde{\mathcal{H}}} \hookrightarrow \widetilde{\mathsf{X}}^{\min}_{\widetilde{\mathcal{H}}}$ is fiberwise dense over $\widetilde{R}$. As explained in [31, the proof of Proposition 4.6.28] (with adjustments as explained above), up to replacing $\mathcal{L}$ with a positive tensor power, we may and we shall assume that the pullback of $\mathcal{L}^+_{R^+}$ to $\widetilde{\mathsf{U}}^+_{R^+}$, which we denote by $\widetilde{\mathcal{L}}^+_{R^+}$, extends to an ample invertible sheaf over the whole $\widetilde{\mathsf{X}}^{\min,+}_{\widetilde{\mathcal{H}},R^+}$, which we denote by $\widetilde{\mathcal{L}}^{\min,+}_{R^+}$. Since $\widetilde{\mathsf{X}}^{\min,+}_{\widetilde{\mathcal{H}},R^+}$ is noetherian normal, and since $\widetilde{\mathcal{L}}^{\min,+}_{R^+}$ is an invertible sheaf, we have $\widetilde{\mathcal{L}}^{\min,+}_{R^+} \cong (\widetilde{\mathsf{U}}^+_{R^+} \hookrightarrow \widetilde{\mathsf{X}}^{\min,+}_{\widetilde{\mathcal{H}},R^+})_* \widetilde{\mathcal{L}}^+_{R^+}$. Consequently, we have

$$\widetilde{\mathsf{X}}^{\min,+}_{\widetilde{\mathcal{H}},R^+} \cong \operatorname{Proj}\Big(\underset{k\geq 0}{\oplus} \Gamma(\widetilde{\mathsf{X}}^{\min,+}_{\widetilde{\mathcal{H}},R^+}, (\widetilde{\mathcal{L}}^{\min,+}_{R^+})^{\otimes k})\Big) \cong \operatorname{Proj}\Big(\underset{k\geq 0}{\oplus} \Gamma(\widetilde{\mathsf{U}}^+_{R^+}, (\widetilde{\mathcal{L}}^+_{R^+})^{\otimes k})\Big), \tag{2.8}$$

and the same holds up to replacing $\mathcal{L}$ and $\widetilde{\mathcal{L}}^+_{R^+}$ with the same positive tensor power.

At this point, we can explain how to remove the assumption that G^{ad} is simple over $\mathbb{Q}$, and drop this assumption. Since all we need is the ampleness and extensibility of certain invertible sheaves up to replacing them with the same positive tensor powers, we may work at a higher level defined by a finite index subgroup of $\mathcal{H}$, and pullback the desired sheaves from products of integral models of smaller Shimura varieties and their minimal compactifications—we already have the desired sheaves over the factors, as explained above. (The upshot is that we can use possibly different positive tensor powers of the Hodge invertible sheaves over different factors. See [43, Sect. 7] for a prototypical example of such an argument in PEL-type cases.)

Note that the action of Δ^+ on $\widetilde{\mathsf{X}}^+_{\widetilde{\mathcal{H}},K^+}$ extends to an action on $\widetilde{\mathsf{X}}^{\min,+}_{\widetilde{\mathcal{H}},K^+}$, which induces an isomorphism $\widetilde{\mathsf{X}}^{\min,+}_{\widetilde{\mathcal{H}},K^+}/\Delta^+ \xrightarrow{\sim} \mathsf{X}^{\min,+}_{\mathcal{H},K^+}$, because $\widetilde{\mathsf{X}}^+_{\widetilde{\mathcal{H}},K^+}$ and $\mathsf{X}^+_{\mathcal{H},K^+}$ have models over some number field whose pullbacks to $\mathbb{C}$ are arithmetic quotients of the same Hermitian symmetric domain, and Δ^+ is just the quotient of one such arithmetic subgroup by another, whose action extends to the minimal compactifications in [3] by their very constructions. (The adelic construction in [55, 6.1–6.2] has to be formulated in terms of finite disjoint unions of Hermitian symmetric domains, but the connected components of the minimal compactifications thus obtained are still the same projective normal varieties in [3]. We need the theory in [10] to relate the constructions in [29, Sect. 3.4], [30, Theorem 3.10], and [31, Sect. 4.6] to the complex analytic construction mentioned above, but we do not need to generalize the theory in [10] to an analogous theory for minimal compactifications.) Hence, the action of Δ^+ on $\widetilde{\mathsf{X}}^+_{\widetilde{\mathcal{H}},R^+}$ extends to an action on $\widetilde{\mathsf{U}}^+_{R^+}$, which induces an isomorphism $\widetilde{\mathsf{U}}^+_{R^+}/\Delta^+ \xrightarrow{\sim} \mathsf{U}^+_{R^+}$ extending the isomorphism $\widetilde{\mathsf{X}}^+_{\widetilde{\mathcal{H}},R^+}/\Delta^+ \xrightarrow{\sim} \mathsf{X}^+_{\mathcal{H},R^+}$. Consequently, $\widetilde{\mathcal{L}}^+_{R^+}$ is canonically and compatibly isomorphic to its pullbacks under the actions of the elements of Δ^+, and it follows from (2.8) that the action of Δ^+ on $\widetilde{\mathsf{U}}^+_{R^+}$ extends to the whole $\widetilde{\mathsf{X}}^{\min,+}_{\widetilde{\mathcal{H}},R^+}$. Thus, $\widetilde{\mathcal{L}}^{\min,+}_{R^+} \cong (\widetilde{\mathsf{U}}^+_{R^+} \hookrightarrow \widetilde{\mathsf{X}}^{\min,+}_{\widetilde{\mathcal{H}},R^+})_* \widetilde{\mathcal{L}}^+_{R^+}$ is also canonically and compatibly isomorphic to its pullbacks under the actions of the elements of Δ^+.

Let us form the quotient $\mathsf{X}^{\min,+}_{\mathcal{H},R^+} := \widetilde{\mathsf{X}}^{\min,+}_{\widetilde{\mathcal{H}},R^+}/\Delta^+$, which is a noetherian normal scheme over R^+ containing $\mathsf{U}^+_{R^+}$ as an open dense subscheme whose complement has codimension at least two. Let $\{\widetilde{V}_\alpha\}_{\alpha\in A}$ be an affine open covering of $\widetilde{\mathsf{X}}^{\min,+}_{\widetilde{\mathcal{H}},R^+}$ such that, for each $\alpha \in A$, the action of Δ^+ stabilizes $\widetilde{V}_\alpha$ and there exists some trivialization $\widetilde{f}_\alpha$: $\mathscr{O}_{\widetilde{V}_\alpha} \xrightarrow{\sim} \widetilde{\mathcal{L}}^{\min,+}_{R^+}|_{\widetilde{V}_\alpha}$. Given any $\alpha, \beta \in A$, we have a section $\widetilde{f}_{\alpha\beta} := \widetilde{f}_\beta^{-1} \circ \widetilde{f}_\alpha$ of $\mathscr{O}^\times_{\widetilde{V}_{\alpha\beta}}$, where $\widetilde{V}_{\alpha\beta} := \widetilde{V}_\alpha \cap \widetilde{V}_\beta$ is affine because $\widetilde{\mathsf{X}}^{\min,+}_{\widetilde{\mathcal{H}},R^+}$ is projective and hence separated. Then the collection $\{\widetilde{f}_{\alpha\beta}\}_{\alpha,\beta\in A}$ (with respect to the open covering $\{\widetilde{V}_\alpha\}_{\alpha\in A}$) defines the class of $\widetilde{\mathcal{L}}^{\min,+}_{R^+}$ in $H^1(\widetilde{\mathsf{X}}^{\min,+}_{\widetilde{\mathcal{H}},R^+}, \mathscr{O}^\times_{\widetilde{\mathsf{X}}^{\min,+}_{\widetilde{\mathcal{H}},R^+}}) \cong \operatorname{Pic}(\widetilde{\mathsf{X}}^{\min,+}_{\widetilde{\mathcal{H}},R^+})$ (cf. [21, 0_{I}, 5.4.7]). Since $\delta^*(\widetilde{\mathcal{L}}^{\min,+}_{R^+}) \cong \widetilde{\mathcal{L}}^{\min,+}_{R^+}$, for all $\delta \in \Delta^+$, the collection $\{\prod_{\delta\in\Delta^+} \delta^*(\widetilde{f}_{\alpha\beta})\}_{\alpha,\beta\in A}$ defines the class of $(\widetilde{\mathcal{L}}^{\min,+}_{R^+})^{\otimes|\Delta^+|}$. Given any $\alpha, \beta \in A$, let $V_\alpha := \widetilde{V}_\alpha/\Delta^+$, and let $V_{\alpha\beta} := V_\alpha \cap V_\beta \cong \widetilde{V}_{\alpha\beta}/\Delta^+$, which are affine open subschemes of $\mathsf{X}^{\min,+}_{\mathcal{H},R^+}$. Since the section $\prod_{\delta\in\Delta^+} \delta^*(\widetilde{f}_{\alpha\beta})$ of $\mathscr{O}^\times_{\widetilde{V}_{\alpha\beta}}$ is invariant under the action of Δ^+, it defines a section $f_{\alpha\beta}$ of $\mathscr{O}^\times_{V_{\alpha\beta}}$. Then the collection $\{f_{\alpha\beta}\}_{\alpha,\beta\in A}$ (with respect to the open covering $\{V_\alpha\}_{\alpha\in A}$) defines the class of an invertible sheaf in $\operatorname{Pic}(\mathsf{X}^{\min,+}_{\mathcal{H},R^+}) \cong H^1(\mathsf{X}^{\min,+}_{\mathcal{H},R^+}, \mathscr{O}^\times_{\mathsf{X}^{\min,+}_{\mathcal{H},R^+}})$, whose pullbacks to $\widetilde{\mathsf{X}}^{\min,+}_{\widetilde{\mathcal{H}},R^+}$ and U_{R^+} are isomorphic to $(\widetilde{\mathcal{L}}^{\min,+}_{R^+})^{\otimes|\Delta^+|}$ and $(\mathcal{L}^+_{R^+})^{\otimes|\Delta^+|}$, respectively. For simplicity, let us replace $\mathcal{L}^+_{R^+}$, $\widetilde{\mathcal{L}}^+_{R^+}$, and $\widetilde{\mathcal{L}}^{\min,+}_{R^+}$ with their respective $|\Delta^+|$th tensor powers, and denote the above descended invertible sheaf over $\mathsf{X}^{\min,+}_{\mathcal{H},R^+}$ by $\mathcal{L}^{\min,+}_{R^+}$, whose pullback to $\widetilde{\mathsf{X}}^{\min,+}_{\widetilde{\mathcal{H}},R^+}$ is isomorphic to $\widetilde{\mathcal{L}}^{\min,+}_{R^+}$ by construction. As before, since $\mathsf{X}^{\min,+}_{\mathcal{H},R^+}$ is noetherian normal, we have $\mathcal{L}^{\min,+}_{R^+} \cong (\mathsf{U}^+_{R^+} \hookrightarrow \mathsf{X}^{\min,+}_{\mathcal{H},R^+})_* \mathcal{L}^+_{R^+}$. Also, we have $((\widetilde{\mathsf{X}}^{\min,+}_{\widetilde{\mathcal{H}},R^+} \to \mathsf{X}^{\min,+}_{\mathcal{H},R^+})_*(\widetilde{\mathcal{L}}^{\min,+}_{R^+})^{\otimes k})^{\Delta^+} \cong (\mathcal{L}^{\min,+}_{R^+})^{\otimes k}$ over $\mathsf{X}^{\min,+}_{\mathcal{H},R^+}$, for all k. Hence, $\oplus_{k\geq 0} \Gamma(\mathsf{X}^{\min,+}_{\mathcal{H},R^+}, (\mathcal{L}^{\min,+}_{R^+})^{\otimes k}) \cong \left(\oplus_{k\geq 0} \Gamma(\widetilde{\mathsf{X}}^{\min,+}_{\widetilde{\mathcal{H}},R^+}, (\widetilde{\mathcal{L}}^{\min,+}_{R^+})^{\otimes k})\right)^{\Delta^+}$ is finitely generated, and therefore there exists some integer $m \geq 1$ such that $\Gamma(\mathsf{X}^{\min,+}_{\mathcal{H},R^+}, (\mathcal{L}^{\min,+}_{R^+})^{\otimes m})$ generates $\oplus_{k\geq 0} \Gamma(\mathsf{X}^{\min,+}_{\mathcal{H},R^+}, (\mathcal{L}^{\min,+}_{R^+})^{\otimes mk})$, as algebras over R^+. Consequently,

$$\mathsf{X}^{\min,+}_{\mathcal{H},R^+} \cong \operatorname{Proj}\Big(\underset{k\geq 0}{\oplus} \Gamma(\mathsf{X}^{\min,+}_{\mathcal{H},R^+}, (\mathcal{L}^{\min,+}_{R^+})^{\otimes k})\Big) \cong \operatorname{Proj}\Big(\underset{k\geq 0}{\oplus} \Gamma(\mathsf{U}^+_{R^+}, (\mathcal{L}^+_{R^+})^{\otimes k})\Big) \quad (2.9)$$

is projective over R^+, and $\mathcal{L}^{\min,+}_{R^+}$ is ample because $(\mathcal{L}^{\min,+}_{R^+})^{\otimes m}$ is very ample. (The arguments in this paragraph are well known and already implicit in several references given above, but we still decided to spell them out for the sake of clarity.)

The disjoint union of quotients $\mathsf{X}^{\min,+}_{\mathcal{H},R^+} = \widetilde{\mathsf{X}}^{\min,+}_{\widetilde{\mathcal{H}},R^+}/\Delta^+$ extending $\mathsf{U}^+_{R^+} \cong \widetilde{\mathsf{U}}^+_{R^+}/\Delta^+$ as above, which we abusively denote by $\mathsf{X}^{\min}_{\mathcal{H},R^+}$, is a noetherian normal projective scheme over R^+ containing U_{R^+} and carrying an ample invertible sheaf $\mathcal{L}^{\min}_{R^+}$ extending $\mathcal{L}_{R^+}$, whose pullback to each $\widetilde{\mathsf{X}}^{\min,+}_{\widetilde{\mathcal{H}},R^+}$ as above is isomorphic to $\widetilde{\mathcal{L}}^{\min,+}_{R^+}$ (up to replacement with a positive tensor power, to account for the replacements of invertible sheaves with positive tensor powers on the connected components thus far—this is feasible because there are only finitely many connected components). By combining isomorphisms as in (2.9), we obtain

$$\mathsf{X}^{\min}_{\mathcal{H},R^+} \cong \operatorname{Proj}\Big(\underset{k\geq 0}{\oplus} \Gamma(\mathsf{X}^{\min}_{\mathcal{H},R^+}, (\mathcal{L}^{\min}_{R^+})^{\otimes k})\Big) \cong \operatorname{Proj}\Big(\underset{k\geq 0}{\oplus} \Gamma(\mathsf{U}_{R^+}, (\mathcal{L}_{R^+})^{\otimes k})\Big). \quad (2.10)$$

Since the pair $(\mathsf{U}_{R^+}, \mathcal{L}_{R^+})$ of a quasi-projective scheme and an ample invertible sheaf has a model $(\mathsf{U}, \mathcal{L})$ over R, it carries descent data with respect to the finite étale morphism $R \to R^+$. Hence, it follows from (2.10) that $\mathsf{X}^{\min}_{\mathcal{H},R^+}$ carries induced descent data and (by the theory of descent—see [20, VIII, 7.8]) has a model $\mathsf{X}^{\min}_{\mathcal{H}}$ over R, together with the open immersion (2.5) extending (2.7), as desired. □

Remark 2.11 There is also some crude construction of (integral models of) toroidal compactifications in Case (Ab), if we combine the constructions in [40, 44] in Case (Hdg) and pursue a similar strategy as in the proof of Proposition 2.4. But we have decided to omit it at the expense of excluding Case (Ab) in Theorem 4.1 below, not just because the details are more tedious to write up, but also because it would be more desirable to have a construction of toroidal compactifications which we can describe in as much detail as in other cases, and because we are still able to include Case (Ab) in Corollary 4.6 below. On the other hand, while it would also be more desirable to have a construction of minimal compactifications which we can better describe, we do need some crude construction as in Proposition 2.4 to at least define the intersection complexes of nearby cycles as in Theorem 4.13 and Corollary 4.15 below. This is admittedly a compromise, but it still allows the applications to the intersection cohomology in Theorems 4.19 and 4.23 below.

3 Automorphic Étale Sheaves

Let $\ell > 0$ be a rational prime number. Let us fix the choice of an algebraic closure $\bar{\mathbb{Q}}_\ell$ of $\mathbb{Q}_\ell$. For simplicity of notation, let us assume the following:

(1) In Case (Sm), we have $\ell \notin \Box$ and $\mathcal{H} = \mathcal{H}^\ell \mathcal{H}_\ell$ in $\mathrm{G}(\hat{\mathbb{Z}}^\Box)$ for some open compact subgroups $\mathcal{H}^\ell \subset \mathrm{G}(\hat{\mathbb{Z}}^{\Box \cup \{\ell\}})$ and $\mathcal{H}_\ell \subset \mathrm{G}(\mathbb{Z}_\ell)$. (Here $\Box$ is a set of rational primes as in [35, Notation and Conventions], which are the places where we have level structures in the definition of PEL moduli problems.)
(2) In Cases (Nm), (Spl), (Hdg), and (Ab), we have $\mathcal{H} = \mathcal{H}^\ell \mathcal{H}_\ell$ in $\mathrm{G}(\hat{\mathbb{Z}})$ for some open compact subgroups $\mathcal{H}^\ell \subset \mathrm{G}(\hat{\mathbb{Z}}^\ell)$ and $\mathcal{H}_\ell \subset \mathrm{G}(\mathbb{Z}_\ell)$.

As in [48, Chap. III], let us denote by G^c the quotient of $\mathrm{G}_\mathbb{Q}$ by the maximal $\mathbb{Q}$-anisotropic $\mathbb{R}$-split subtorus of the center of $\mathrm{G}_\mathbb{Q}$ (as algebraic groups over $\mathbb{Q}$). (In Cases (Sm), (Nm), (Spl), and (Hdg), we have $\mathrm{G}_\mathbb{Q} \cong \mathrm{G}^c$, but in Case (Ab) this is not true in general.) For any subgroup of $\mathrm{G}(\mathbb{A})$ (including those of $\mathrm{G}(\mathbb{Q})$, $\mathrm{G}(\mathbb{A}^\infty)$, $\mathrm{G}(\hat{\mathbb{Z}})$, $\mathrm{G}(\mathbb{Z}_\ell)$, etc.), we shall denote its image in $\mathrm{G}^c(\mathbb{A})$ with an additional superscript c. (We are not introducing a model of G^c over $\mathbb{Z}$.) Therefore, for example, we have an open compact subgroup $\mathcal{H}^c$ of $\mathrm{G}^c(\mathbb{A}^\infty)$, which is of the form $\mathcal{H}^c = \mathcal{H}^{\ell,c}\mathcal{H}^c_\ell$.

For each integer $r > 0$, let $\mathcal{U}_\ell(\ell^r) := \ker(\mathrm{G}(\mathbb{Z}_\ell) \to \mathrm{G}(\mathbb{Z}/\ell^r\mathbb{Z}))$, and consider $\mathcal{H}(\ell^r) := \mathcal{H}^\ell \mathcal{U}_\ell(\ell^r)$, which is contained in $\mathcal{H}$ when r is sufficiently large. For such sufficiently large r, in all cases considered above, we have a finite cover

$$\mathsf{X}_{\mathcal{H}(\ell^r)} \to \mathsf{X}_\mathcal{H}, \tag{3.1}$$

which induces a Galois finite étale cover

$$\mathsf{X}_{\mathcal{H}(\ell^r)} \underset{\mathbb{Z}}{\otimes} \mathbb{Q} \to \mathsf{X}_\mathcal{H} \underset{\mathbb{Z}}{\otimes} \mathbb{Q} \tag{3.2}$$

with Galois group $\mathcal{H}^c_\ell / \mathcal{U}_\ell(\ell^r)^c$ (cf. [48, Chap. III, Sect. 6, Remark 6.1]), where $\mathsf{X}_{\mathcal{H}(\ell^r)}$ is defined as in the case of $\mathsf{X}_\mathcal{H}$ but with $\mathcal{H}$ replaced with its normal subgroup $\mathcal{H}(\ell^r)$. If $\ell \neq p$, then the finite cover (3.1) is étale over all of S. (In Cases (Sm), (Nm), (Spl), and (Hdg), this is because (3.1) relatively represents a functor of level structures at $\ell \neq p$. In Case (Ab), this is because (3.1) can be étale locally identified with analogous morphisms in Case (Hdg); cf. (1) of Proposition 2.4 and Remark 2.6.)

For any finite-dimensional algebraic representation V of G^c over $\Lambda = \bar{\mathbb{Q}}_\ell$, which we also view as an algebraic representation of G by pullback (whose restriction to the maximal $\mathbb{Q}$-anisotropic $\mathbb{R}$-split subtorus of the center of $\mathrm{G}_\mathbb{Q}$ is trivial), there exists a canonical étale sheaf $\mathcal{V}$ over $\mathsf{X}_\mathcal{H}$ (with stalks isomorphic to V). Let us briefly review the construction, because we will need it in our later argument.

As explained in [23, Sect. III.2], by the Baire category theorem (see, for example, the proof of [6, 2.2.1.1] or the beginning of [58, Sect. 2]), there exists a finite extension E of $\mathbb{Q}_\ell$ in $\bar{\mathbb{Q}}_\ell$, and an $\mathcal{O}_E$-lattice V_0 with a continuous action of $\mathrm{G}(\mathbb{Z}_\ell)$ (with respect to the ℓ-adic topology), such that $V \cong V_0 \otimes_{\mathcal{O}_E} \bar{\mathbb{Q}}_\ell$ as continuous representations of $\mathrm{G}(\mathbb{Q}_\ell)$. For each $m > 0$, by the continuity of the action of $\mathrm{G}(\mathbb{Z}_\ell)$ on V_0, there exists an integer $r(m) > 0$ such that $\mathcal{H}(\ell^{r(m)}) \subset \mathcal{H}$ and $\mathcal{U}_\ell(\ell^{r(m)})$ acts trivially on the finite quotient $V_{0,\ell^m} := V_0 \otimes_{\mathbb{Z}_\ell} (\mathbb{Z}/\ell^m\mathbb{Z})$. By abuse of notation, let us also denote by $\underline{V}_{0,\ell^m}$ the constant group scheme over $\mathbb{Z}$, which then carries an action of $\mathcal{H}_\ell^c/\mathcal{U}_\ell(\ell^{r(m)})^c$ induced by that of $\mathcal{H}_\ell/\mathcal{U}_\ell(\ell^{r(m)})$. Let us define $\mathcal{V}_{0,\ell^m}$ to be the torsion étale sheaf of sections over $\mathsf{X}_\mathcal{H} \otimes_\mathbb{Z} \mathbb{Q}$ of the contraction product

$$(\mathsf{X}_{\mathcal{H}(\ell^{r(m)})} \underset{\mathbb{Z}}{\otimes} \mathbb{Q}) \overset{\mathcal{H}_\ell^c/\mathcal{U}_\ell(\ell^{r(m)})^c}{\times} \underline{V}_{0,\ell^m}, \tag{3.3}$$

whose pullback to $\mathsf{X}_{\mathcal{H}(\ell^{r(m)})} \otimes_\mathbb{Z} \mathbb{Q}$ is isomorphic to $\underline{V}_{0,\ell^m}$ because (3.2) is a Galois finite étale cover with Galois group $\mathcal{H}_\ell^c/\mathcal{U}_\ell(\ell^r)^c$; and define the étale sheaves

$$\mathcal{V}_0 := \varprojlim_m \mathcal{V}_{0,\ell^m} \tag{3.4}$$

and

$$\mathcal{V} := \mathcal{V}_0 \underset{\mathcal{O}_E}{\otimes} \bar{\mathbb{Q}}_\ell \tag{3.5}$$

over $\mathsf{X}_\mathcal{H} \otimes_\mathbb{Z} \mathbb{Q}$ as usual. Then it is elementary (though tedious) to verify that such a construction is independent of the various choices, is functorial in various natural senses, and allows us to define the Hecke actions on the cohomology groups if we take the limit over $\mathcal{H}$. (See [32, Sect. 6] and [23, Sect. III.2] for more details.)

When $\ell \neq p$, the same construction defines the étale sheaf extensions of $\mathcal{V}_0$ and $\mathcal{V}$ to all of $\mathsf{X}_\mathcal{H}$. This construction also works if we replace V with a continuous representation of $\mathrm{G}(\mathbb{Z}_\ell)^c$ on a (possibly torsion) finite $\mathbb{Z}_\ell$-module (or a $\bar{\mathbb{Z}}_\ell$-module if one prefers), without referring to any representation over $\bar{\mathbb{Q}}_\ell$.

For simplicity, we shall often denote the pullbacks of $\mathcal{V}$ by the same symbol.

Let us record the following observation, based on the above construction:

Lemma 3.6 *For any finite-dimensional algebraic representation V of G^c as above over $\Lambda = \bar{\mathbb{Q}}_\ell$, there exists a finite extension E of $\mathbb{Q}_\ell$ in $\bar{\mathbb{Q}}_\ell$, together with a projective system of torsion étale sheaves $\mathcal{V}_{0,m}$ of $\mathcal{O}_E$-modules indexed by integers $m > 0$ such that $\mathcal{V} \cong \mathcal{V}_0 \otimes_{\mathcal{O}_E} \bar{\mathbb{Q}}_\ell$ for $\mathcal{V}_0 := \varprojlim_m \mathcal{V}_{0,m}$ over $\mathsf{X}_\mathcal{H} \otimes_\mathbb{Z} \mathbb{Q}$, and such that, for each $m > 0$, there exist some integer $r(m) > 0$ such that $\mathcal{H}(\ell^{r(m)}) \subset \mathcal{H}$ and such that the pullback of $\mathcal{V}_{0,m}$ from $\mathsf{X}_\mathcal{H} \otimes_\mathbb{Z} \mathbb{Q}$ to $\mathsf{X}_{\mathcal{H}(\ell^{r(m)})} \otimes_\mathbb{Z} \mathbb{Q}$ is a constant étale sheaf. When $\ell \neq p$, the same assertions are true with all sheaves $\mathcal{V}$, $\mathcal{V}_0$, and $\mathcal{V}_{0,m}$, for all $m > 0$, defined over all of $\mathsf{X}_\mathcal{H}$. If we replace V with a continuous representations of $\mathrm{G}(\mathbb{Z}_\ell)^c$ on a (possibly torsion) finite $\mathbb{Z}_\ell$-module, then there exist a projective system of torsion étale sheaves $\mathcal{V}_{0,m}$ of $\mathbb{Z}_\ell$-modules indexed by integers $m > 0$ such that $\mathcal{V} \cong \varprojlim_m \mathcal{V}_{0,m}$ and such that the remaining assertions as above concerning $\mathcal{V}_{0,m}$ hold verbatim.*

4 Main Results

Let $\mathsf{X}_{\mathcal{H}}$ be as in Sect. 2, with the toroidal and minimal compactifications $J^{\text{tor}} : \mathsf{X}_{\mathcal{H}} \hookrightarrow \mathsf{X}_{\mathcal{H}}^{\text{tor}}$ and $J^{\text{min}} : \mathsf{X}_{\mathcal{H}} \hookrightarrow \mathsf{X}_{\mathcal{H}}^{\text{min}}$ as in (2.3), in Cases (Sm), (Nm), (Spl), and (Hdg); or with the compactification $J^{\text{min}} : \mathsf{X}_{\mathcal{H}} \hookrightarrow \mathsf{X}_{\mathcal{H}}^{\text{min}}$ as in Proposition 2.4, in Case (Ab). Let $\mathcal{V}$ be as in Sect. 3, which is associated with either a finite-dimensional algebraic representation V of G^c over $\Lambda = \bar{\mathbb{Q}}_\ell$, or with a continuous representations of $\mathrm{G}(\mathbb{Z}_\ell)^c$ on a (possibly torsion) finite $\mathbb{Z}_\ell$-module with $\Lambda = \mathbb{Z}_\ell$ (or $\bar{\mathbb{Z}}_\ell$, if one prefers). We will also consider the case with $\Lambda = \mathbb{F}_\ell$ (resp. $\bar{\mathbb{F}}_\ell$), which we view as a subcase of the case with $\Lambda = \mathbb{Z}_\ell$ (resp. $\bar{\mathbb{Z}}_\ell$). By abuse of language, we shall indicate the type of $\mathcal{V}$ we consider by simply specifying the type of the corresponding coefficient ring Λ. Let $d := \dim((\mathsf{X}_{\mathcal{H}})_\eta)$.

Then we have the following results:

Theorem 4.1 (cf. [42, Theorem 5.15]) *In Cases* (Sm), (Nm), (Spl), *and* (Hdg), *for all $\mathcal{V}$ over $\Lambda = \bar{\mathbb{Q}}_\ell$ or $\mathbb{Z}_\ell$ (or $\bar{\mathbb{Z}}_\ell$), the adjunction morphisms*

$$R\Psi_{\mathsf{X}_{\mathcal{H}}^{tor}} RJ_{\eta,*}^{tor}(\mathcal{V}) \to RJ_{\bar{s},*}^{tor} R\Psi_{\mathsf{X}_{\mathcal{H}}}(\mathcal{V}) \tag{4.2}$$

and

$$J_{\bar{s},!}^{tor} R\Psi_{\mathsf{X}_{\mathcal{H}}}(\mathcal{V}) \to R\Psi_{\mathsf{X}_{\mathcal{H}}^{tor}} J_{\eta,!}^{tor}(\mathcal{V}) \tag{4.3}$$

(see (2.3) *for the notation) are isomorphisms in $D_c^b((\mathsf{X}_{\mathcal{H}}^{tor})_{\bar{s}} \times \bar{\eta}, \Lambda)$.*

Remark 4.4 Compared with [42, Theorem 5.15], we note that, in the case of integral or torsion coefficients, we no longer require $\mathcal{V}$ to be of the form $\mathcal{W}_{0,M}$ over $\Lambda = \mathbb{Z}_\ell$ as in [42, Proposition 3.4] for some $\ell > c_W$. That is, roughly speaking, we no longer require the integral or torsion coefficients to be defined by representations of ℓ-small weights.

Remark 4.5 Theorem 4.1 is the logical foundation of all our other results in this article. We will explain the proof of this key theorem in Sect. 5. As we shall see, the main reason to consider the four kinds of integral models in Cases (Sm), (Nm), (Spl), and (Hdg) as in Sect. 2 is that they are known to have good toroidal compactifications, and Case (Ab) is not included in this theorem exactly because good integral models of toroidal compactifications of abelian-type Shimura varieties (with properties we need) are not yet available in the literature (cf. Remark 2.11).

Corollary 4.6 (cf. [42, Corollary 5.20]) *In all Cases* (Sm), (Nm), (Spl), (Hdg), *and* (Ab), *for all $\mathcal{V}$ over $\Lambda = \bar{\mathbb{Q}}_\ell$ or $\mathbb{Z}_\ell$ (or $\bar{\mathbb{Z}}_\ell$), the canonical adjunction morphisms*

$$H_{\acute{e}t}^i((\mathsf{X}_{\mathcal{H}})_{\bar{\eta}}, \mathcal{V}) \to H^i((\mathsf{X}_{\mathcal{H}})_{\bar{s}}, R\Psi_{\mathsf{X}_{\mathcal{H}}}(\mathcal{V})) \tag{4.7}$$

and

$$H_{\acute{e}t,c}^i((\mathsf{X}_{\mathcal{H}})_{\bar{s}}, R\Psi_{\mathsf{X}_{\mathcal{H}}}(\mathcal{V})) \to H_{\acute{e}t,c}^i((\mathsf{X}_{\mathcal{H}})_{\bar{\eta}}, \mathcal{V}) \tag{4.8}$$

of $\mathrm{Gal}(\bar{K}/K)$*-modules are isomorphisms, for all i.*

Proof In Cases (Sm), (Nm), (Spl), and (Hdg), this follows from Theorem 4.1 and the proper base change theorem, as in the proof of [42, Corollary 5.20].

In Case (Ab), the question is whether the morphisms (4.7) and (4.8) are isomorphisms, and we can ignore the $\mathrm{Gal}(\bar{K}/K)$-module structures when answering such a question. Also, by first reducing to the torsion case by Lemma 3.6, and by duality (see [25, 4.2]), it suffices to answer the question for the morphism (4.7). Hence, we are free to replace K with a finite extension inside $\bar{K}$ (and accordingly R with its integral closure in this finite extension), and to replace $\mathsf{X}_{\mathcal{H}}$ with its connected components. Because the construction of integral models in Case (Ab) (see (1) of Proposition 2.4 and Remark 2.6) was achieved by descent and by taking quotients of connected components of some integral models of Hodge-type Shimura varieties by the free actions of some finite groups (which is compatible with the construction of $\mathcal{V}$ in Sect. 3 when restricted to subgroups stabilizing the connected components), by using the Hochschild–Serre spectral sequence for étale cohomology (see, for example, [47, Chap. III, Sect. 2, Theorem 2.20] or [16, Chap. 9, Sect. 9.1, p. 501]), the isomorphism assertion is reduced to the known one in Case (Hdg), as desired. □

Remark 4.9 (cf. [42, Remark 5.42]) There are closely related results in [26, Theorem 4.2] and [27, Corollary 7.3] for the supercuspidal parts of cohomology. Concretely, although they have not shown that (4.7) and (4.8) (in the PEL-type cases they consider) are isomorphisms, they showed that the kernels and cokernels do not contain supercuspidal representations. Their method based on the consideration of adic spaces is quite flexible and of some independent interest.

Corollary 4.10 (cf. [42, Corollary 5.23]) *In all Cases* (Sm), (Nm), (Spl), (Hdg), *and* (Ab), *for all* $\mathcal{V}$ *over* $\Lambda = \bar{\mathbb{Q}}_\ell$ *or* $\mathbb{Z}_\ell$ *(or* $\bar{\mathbb{Z}}_\ell$*), the adjunction morphisms*

$$R\Psi_{\mathsf{X}_{\mathcal{H}}^{\mathrm{min}}}\, RJ_{\eta,*}^{\mathrm{min}}(\mathcal{V}) \to RJ_{\bar{s},*}^{\mathrm{min}}\, R\Psi_{\mathsf{X}_{\mathcal{H}}}(\mathcal{V}) \tag{4.11}$$

and

$$J_{\bar{s},!}^{\mathrm{min}}\, R\Psi_{\mathsf{X}_{\mathcal{H}}}(\mathcal{V}) \to R\Psi_{\mathsf{X}_{\mathcal{H}}^{\mathrm{min}}}\, J_{\eta,!}^{\mathrm{min}}(\mathcal{V}) \tag{4.12}$$

are isomorphisms in $D_c^b((\mathsf{X}_{\mathcal{H}}^{\mathrm{min}})_{\bar{s}} \times \bar{\eta}, \Lambda)$.

Proof In Cases (Sm), (Nm), (Spl), and (Hdg), this follows from Theorem 4.1 and the proper base change theorem, as in the proof of [42, Corollary 5.23].

In Case (Ab), by similar reduction steps as in the proof of Corollary 4.6, by Proposition 2.4 and Remark 2.6, and by using the Hochschild–Serre spectral sequence (see, for example, [16, Chap. 9, Sect. 9.1, p. 501]) for the (derived) direct images towards $\mathsf{X}_{\mathcal{H},R^+}^{\mathrm{min},+}$ (which is now viewed as a base scheme for all other schemes), the desired isomorphism assertion is reduced to the known one in Case (Hdg). □

Theorem 4.13 (cf. [42, Theorem 5.26]) *In all Cases* (Sm), (Nm), (Spl), (Hdg), *and* (Ab), *for all* $\mathcal{V}$ *over* $\Lambda = \bar{\mathbb{Q}}_\ell$ *or* $\mathbb{F}_\ell$ *(or* $\bar{\mathbb{F}}_\ell$*), we also have a canonically induced isomorphism*

$$R\Psi_{\mathsf{X}^{min}_{\mathcal{H}}} J^{min}_{\bar{\eta},!*}(\mathcal{V}[d]) \xrightarrow{\sim} J^{min}_{\bar{s},!*} R\Psi_{\mathsf{X}_{\mathcal{H}}}(\mathcal{V}[d]) \tag{4.14}$$

in the category of perverse sheaves over $(\mathsf{X}^{min}_{\mathcal{H}})_{\bar{s}}$ *with continuous* $\mathrm{Gal}(\bar{K}/K)$*-actions.*

Proof This follows from Corollary 4.10 by the same argument as in the proof of [42, Theorem 5.26], using the t-exactness of nearby cycle functors (as in [25, 4.5]). (Although [42, Theorem 5.26] was stated only in the case with $\Lambda = \bar{\mathbb{Q}}_\ell$, the argument in its proof also works in the case with $\Lambda = \mathbb{F}_\ell$.) □

Corollary 4.15 (cf. [42, Corollary 5.31]) *In all Cases* (Sm), (Nm), (Spl), (Hdg), *and* (Ab), *for all* $\mathcal{V}$ *over* $\Lambda = \bar{\mathbb{Q}}_\ell$ *or* $\mathbb{F}_\ell$ *(or* $\bar{\mathbb{F}}_\ell$*), we have a canonical isomorphism*

$$H^i_{\acute{e}t}((\mathsf{X}^{min}_{\mathcal{H}})_{\bar{\eta}}, (J^{min}_{\bar{\eta},!*}(\mathcal{V}[d]))[-d]) \xrightarrow{\sim} H^i((\mathsf{X}^{min}_{\mathcal{H}})_{\bar{s}}, (J^{min}_{\bar{s},!*}(R\Psi_{\mathsf{X}_{\mathcal{H}}}(\mathcal{V}[d])))[-d]) \tag{4.16}$$

for the ***intersection cohomology****, for all* i.

Proof This follows from Theorem 4.13 and the proper base change theorem, by the same argument as in the proof of [42, Corollary 5.31]. □

Remark 4.17 For the compatibility of the isomorphisms in Theorems 4.1 and 4.19 and Corollaries 4.6, 4.10 and 4.15 with Hecke actions (when they are defined), we refer the readers to [42, Remarks 5.35 and 5.41].

Remark 4.18 In Cases (Sm), (Nm), (Spl), and (Hdg), we also have analogues of Theorem 4.13 and Corollary 4.15 for $J^{\mathrm{tor}} : \mathsf{X}_{\mathcal{H}} \to \mathsf{X}^{\mathrm{tor}}_{\mathcal{H}}$ instead of $J^{\mathrm{min}} : \mathsf{X}_{\mathcal{H}} \to \mathsf{X}^{\mathrm{min}}_{\mathcal{H}}$, by essentially the same arguments, with Theorem 4.1 instead of Corollary 4.10 as an input. The same would be true in Case (Ab) as soon as we have good toroidal compactifications $J^{\mathrm{tor}} : \mathsf{X}_{\mathcal{H}} \to \mathsf{X}^{\mathrm{tor}}_{\mathcal{H}}$ for which Theorem 4.1 holds (cf. Remark 4.5).

As an application, let us answer the question we raised in Sect. 1 about cases where the étale cohomology of (possibly nonproper) Shimura varieties is unramified:

Theorem 4.19 (cf. [42, Theorems 6.1 and 6.7]) *Suppose we are in Case* (Sm), *or in Cases* (Hdg) *and* (Ab) *when the level at p is hyperspecial. Then* $\mathcal{V} \xrightarrow{\sim} R\Psi_{\mathsf{X}_{\mathcal{H}}}(\mathcal{V})$, *for all* $\mathcal{V}$ *over* $\Lambda = \bar{\mathbb{Q}}_\ell$ *or* $\mathbb{Z}_\ell$ *(or* $\bar{\mathbb{Z}}_\ell$*), because* $\mathsf{X}_{\mathcal{H}} \to S$ *is a smooth morphism. Moreover, we have the following canonical isomorphisms of* $\mathrm{Gal}(\bar{K}/K)$*-modules, for each* i*: for the usual cohomology,*

$$H^i_{\acute{e}t}((\mathsf{X}_{\mathcal{H}})_{\bar{\eta}}, \mathcal{V}) \xrightarrow{\sim} H^i_{\acute{e}t}((\mathsf{X}_{\mathcal{H}})_{\bar{s}}, \mathcal{V}); \tag{4.20}$$

for the compactly supported cohomology,

$$H^i_{\acute{e}t,c}((\mathsf{X}_{\mathcal{H}})_{\bar{s}}, \mathcal{V}) \xrightarrow{\sim} H^i_{\acute{e}t,c}((\mathsf{X}_{\mathcal{H}})_{\bar{\eta}}, \mathcal{V}); \tag{4.21}$$

and, when $\Lambda = \bar{\mathbb{Q}}_\ell$ *or* $\mathbb{F}_\ell$ *(or* $\bar{\mathbb{F}}_\ell$*), for the intersection cohomology*

$$H^i_{\acute{e}t}((\mathsf{X}^{min}_{\mathcal{H}})_{\bar{\eta}}, (J^{min}_{\bar{\eta},!*}(\mathcal{V}[d]))[-d]) \xrightarrow{\sim} H^i_{\acute{e}t}((\mathsf{X}^{min}_{\mathcal{H}})_{\bar{s}}, (J^{min}_{\bar{s},!*}(\mathcal{V}[d]))[-d]). \tag{4.22}$$

In particular, these $\mathrm{Gal}(\bar{K}/K)$*-modules are* ***unramified*** *(i.e., the inertia subgroup* $I_K := \ker(\mathrm{Gal}(\bar{K}/K) \to \mathrm{Gal}(\bar{k}/k))$ *acts trivially on them). If* $\Lambda = \bar{\mathbb{Q}}_\ell$*, and if* $\mathcal{V}$ *is pointwise pure of weight* m *(which is known, for example, in the cases considered in* [42, Proposition 3.2]*), then both sides of* (4.20) *(resp.* (4.21)*) are mixed of weights* $\geq i + m$ *(resp.* $\leq i + m$*), and both sides of* (4.22) *are pure of weight* $i + m$*.*

Proof As in the proofs of [42, Theorems 6.1 and 6.7], these follow from the smooth base change theorem, from Corollaries 4.6 and 4.15, and from [12, 3.3.4, 3.3.5, and 6.2.6] and [4, 5.3.2]. □

Theorem 4.23 (cf. [42, Theorems 6.8 and 6.13]) *Suppose we are in Cases* (Nm), (Spl), (Hdg), *and* (Ab) *where the levels at* p *are* ***parahoric****, and suppose* $p > 2$*. (Note that, as explained in Remark* 2.1*, we have to treat some integral models of Hodge-type Shimura varieties with parahoric levels at* p *only as abelian-type ones in Case* (Ab)*, although this is harmless for our purpose.) In Cases* (Nm) *and* (Spl)*, we consider the same cases as in* [42, Theorems 6.8 and 6.13]*, with a field extension* $\widetilde{K}$ *of* K *in* $\bar{K}$ *defined there. In Cases* (Hdg) *and* (Ab)*, we assume that* $\widetilde{K}$ *is a tamely ramified extension of* K *in* $\bar{K}$ *over which* $\mathrm{G}_{\mathbb{Q}_p}$ *is split. Then, for each* i*, we have the following canonical isomorphisms of* $\mathrm{Gal}(\bar{K}/K)$*-modules: for the usual cohomology,*

$$H^i_{\acute{e}t}((\mathsf{X}_\mathcal{H})_{\bar{\eta}}, \mathcal{V}) \xrightarrow{\sim} H^i_{\acute{e}t}((\mathsf{X}_\mathcal{H})_{\bar{s}}, R\Psi_{\mathsf{X}_\mathcal{H}}(\mathcal{V})); \tag{4.24}$$

for the compactly supported cohomology,

$$H^i_{\acute{e}t,c}((\mathsf{X}_\mathcal{H})_{\bar{s}}, R\Psi_{\mathsf{X}_\mathcal{H}}(\mathcal{V})) \xrightarrow{\sim} H^i_{\acute{e}t,c}((\mathsf{X}_\mathcal{H})_{\bar{\eta}}, \mathcal{V}); \tag{4.25}$$

and, when $\Lambda = \bar{\mathbb{Q}}_\ell$ *or* $\mathbb{F}_\ell$ *(or* $\bar{\mathbb{F}}_\ell$*), for the intersection cohomology,*

$$H^i_{\acute{e}t}((\mathsf{X}^{min}_\mathcal{H})_{\bar{\eta}}, (J^{min}_{\bar{\eta},!*}(\mathcal{V}[d]))[-d]) \xrightarrow{\sim} H^i_{\acute{e}t}((\mathsf{X}^{min}_\mathcal{H})_{\bar{s}}, (J^{min}_{\bar{s},!*} R\Psi_{\mathsf{X}_\mathcal{H}}(\mathcal{V}[d]))[-d]). \tag{4.26}$$

Moreover, the restrictions of the $\mathrm{Gal}(\bar{K}/K)$*-actions of these modules to the subgroup* $I_{\widetilde{K}} := \ker(\mathrm{Gal}(\bar{K}/\widetilde{K}) \to \mathrm{Gal}(\bar{k}/k))$ *(but not* I_K*) of* $\mathrm{Gal}(\bar{K}/K)$ *are all* ***unipotent****, and even* ***trivial*** *when the level at* p *is* ***very special*** *(as in* [54, Sect. 10.3.2]*).*

Proof As in the proofs of [42, Theorems 6.8 and 6.13], these follow from Corollaries 4.6 and 4.15, and from the results of local models in [54, Theorem 1.4, and more detailed results in Sect. 10.3] and [53, Theorem 13.1 and Remark 13.2] (see also [52, Remark 7.4]) in Cases (Nm) and (Spl); and in [31, Corollaries 0.5 and 4.7.3] in Cases (Hdg) and (Ab). □

Remark 4.27 (cf. [42, Remark 6.15]) The isomorphism (4.25) in Theorem 4.23 established [22, Conjecture 10.3] for all integral models of PEL-type Shimura varieties (with parahoric levels at p) considered in [53, 54]. Also, we have established its analogue for all integral models of Hodge-type and abelian-type Shimura varieties (with parahoric levels at p) considered in [31].

Remark 4.28 Theorems 4.19 and 4.23 might convey the impression that our main results are only useful at hyperspecial and parahoric levels, in which case we have good theories of local models, but this is not true. In fact, our initial motivation was to generalize Mantovan's formula (describing the cohomology of Shimura varieties in terms of the cohomology of Igusa varieties and Rapoport–Zink spaces) in [45, 46], and also Scholze's formula (in the context of Langlands–Kottwitz method, describing the cohomology of Shimura varieties in terms of twisted orbital integrals) in [57], to the noncompact case (i.e., removing from their works the assumption that the relevant Shimura varieties are compact). In these works, as explained in the introduction of [42], the analyses of the cohomology of nearby cycles were carried out without the compactness assumption. It is only in their initial steps—or final steps, depending on one's viewpoint—that they used the compactness assumption and the proper base change theorem to relate their results to the étale cohomology in characteristic zero, and our main results above removed the need of the compactness assumption. We emphasize that these generalizations do require our main results in Case (Nm) with arbitrarily high levels at p. See [42, Sect. 6.3] (for the compactly supported cohomology) and [41, Sect. 4.4] (for the usual cohomology, with boundary terms) for more details concerning the generalizations of Mantovan's formula, and see [42, Sect. 6.4] for more details concerning the generalization of Scholze's formula. Moreover, a combination of the generalizations of Mantovan's formula in [41, Sect. 4.4] and of Morel's formula in [41, Sect. 4.5] provides a formula for certain intersection complexes over the partial minimal compactifications of Newton strata, which is potentially useful for generalizing Caraiani and Scholze's results in [7] to the noncompact case.

5 Proof of the Key Theorem

In this section, we explain how to prove Theorem 4.1. It suffices to show that the globally defined morphisms (4.2) and (4.3) are isomorphisms *étale locally*.

The rough idea is that any toroidal compactification $\mathsf{X}_{\mathcal{H}} \hookrightarrow \mathsf{X}_{\mathcal{H}}^{\mathrm{tor}}$ we consider in Cases (Sm), (Nm), (Spl), and (Hdg) is étale locally at each point a product of some affine toroidal embedding $E \hookrightarrow E(\sigma)$, which we know everything about, and the identity morphism Id_C of some scheme C, which we do not need to know anything about. Then we can try to reduce the problem to its analogues over the individual factors $E \hookrightarrow E(\sigma)$ and $\mathrm{Id}_C : C \to C$, by using the Künneth isomorphisms as in [2, XVII, 5.4.3] and [4, 4.2.7], and by using Gabber's theorem (see [25, 4.7]) on nearby cycles over products of schemes of finite type over S. (We will explain these in more details. This idea can be traced back to the lemma [18, 7.1.4] due to Laumon. See also [42, Remark 5.33]. Now we can do more because we have a much better understanding of integral models of Shimura varieties and their toroidal compactifications.) To carry out this idea, we need to have a better control on the étale sheaf $\mathcal{V}$. (What we are about to do is different from what we did in [42, Sects. 4–5]. We will comment on the difference in Remark 5.14 below.)

By Lemma 3.6, we may assume that $\mathcal{V}$ is torsion and associated with some finite $\mathbb{Z}_\ell$-module (with $\Lambda = \mathbb{Z}_\ell$), and that there exists some $r > 0$ such that $\mathcal{H}(\ell^r) \subset \mathcal{H}$ and such that the pullback of $\mathcal{V}$ from $\mathsf{X}_\mathcal{H}$ to $\mathsf{X}_{\mathcal{H}(\ell^r)}$ is a constant étale sheaf. Let $\mathsf{X}^{\mathrm{tor}}_{\mathcal{H}(\ell^r)}$ be defined using the collection of cone decompositions induced by that for $\mathsf{X}^{\mathrm{tor}}_{\mathcal{H}}$, so that $\mathsf{X}^{\mathrm{tor}}_{\mathcal{H}(\ell^r)} \to \mathsf{X}^{\mathrm{tor}}_{\mathcal{H}}$ is *finite*. (In Case (Sm), this might move us from the context of [35] into that of [36], because the former assumed that the cone decompositions are always smooth, but this is harmless in practice.)

To understand this finite morphism better, we have the following proposition, in which we shall freely use the notation in [42, Proposition 2.2]: (First-time readers might assume that $\mathcal{V} = \Lambda$ is trivial, and skip all materials in Propositions 5.1 and 5.3 below involving objects at level $\mathsf{X}^{\mathrm{tor}}_{\mathcal{H}(\ell^r)}$, which are denoted with a prime.)

Proposition 5.1 *Let x be a point of a stratum $\mathsf{Z}_{[\sigma]}$ of $\mathsf{X}^{tor}_{\mathcal{H}}$. Then there exists an étale neighborhood $\overline{U} \to \mathsf{X}^{tor}_{\mathcal{H}}$ of x such that, if we denote by $\overline{U}'$ the pullback of $\overline{U}$ to $\mathsf{X}^{tor}_{\mathcal{H}(\ell^r)}$, then there is a commutative diagram*

$$\begin{array}{ccccc} \mathsf{X}^{tor}_{\mathcal{H}(\ell^r)} & \longleftarrow & \overline{U}' & \longrightarrow & \coprod\limits_{\mathsf{Z}'_{[\sigma']} \text{ lying above } \mathsf{Z}_{[\sigma]}} \big(E'(\sigma') \underset{S}{\times} C'\big) \\ \downarrow & & \downarrow & & \downarrow \\ \mathsf{X}^{tor}_{\mathcal{H}} & \longleftarrow & \overline{U} & \longrightarrow & E(\sigma) \times_S C \end{array} \tag{5.2}$$

with the following properties:

(1) All objects denoted with a prime are at level $\mathcal{H}(\ell^r)$.
(2) The vertical morphisms are equipped with compatible actions of the finite group $\mathcal{H}/\mathcal{H}(\ell^r) \cong \mathcal{H}_\ell/\mathcal{U}_\ell(\ell^r)$ (see Sect. 3—note that $\mathrm{G}_\mathbb{Q} \cong \mathrm{G}^c$ now) that are trivial on the targets and induce isomorphisms from quotients of the sources to the targets.
(3) The horizontal morphisms are étale, and the squares are Cartesian.
(4) The morphism $C' \to C$ is finite étale.
(5) The $E(\sigma)$ is defined by an affine toroidal embedding $E \hookrightarrow E(\sigma)$ for a split torus E, and the $E'(\sigma')$'s are all defined by similar affine toroidal embeddings $E' \hookrightarrow E'(\sigma')$ for a split tori E'. (For simplicity, we only define the tori E and E' over S here.) The morphisms $E'(\sigma') \to E(\sigma)$ are finite flat and tamely ramified, extending isogenies $E' \to E$ of tori of ℓ-power degrees.
(6) The preimages of $\mathsf{X}_\mathcal{H}$ and $E \times_S C$ in $\overline{U}$ are the same open subscheme U, and the preimages of U, $\mathsf{X}_{\mathcal{H}(\ell^r)}$, and $\coprod\limits_{\mathsf{Z}'_{[\sigma']} \text{ lying above } \mathsf{Z}_{[\sigma]}} \big(E' \underset{S}{\times} C'\big)$ in $\overline{U}'$ are the same subscheme U'.

Consequently, the pullback of $\mathcal{V}$ to U descends to an étale sheaf over $E \times_S C$, which we abusively still denote by $\mathcal{V}$, and the pullback of $\mathcal{V}$ to $\coprod\limits_{\mathsf{Z}'_{[\sigma']} \text{ lying above } \mathsf{Z}_{[\sigma]}} \big(E' \underset{S}{\times} C'\big)$ is constant, whose further pullback to U' is the same pullback of $\mathcal{V}$ from $\mathsf{X}_\mathcal{H}$.

Proof First note that the morphisms $C' \to C$, $E' \to E$, and $E'(\sigma') \to E(\sigma)$ are defined regardless of the étale neighborhood $\overline{U}$. The morphisms $C' \to C$ are finite étale because we can define them alternatively as morphisms relatively representing functors defining certain level structures at $\ell \neq p$, which are finite étale and hence must coincide with the construction by normalizations, by Zariski's main theorem (see [21, III-1, 4.4.3, 4.4.11]), since they already agree in characteristic zero (see [35, Sect. 6.2.4; see also the errata] and the reinterpretations in [38, Sect. 1.3.2] in Case (Sm); see [36, Sect. 8] in Case (Nm); see [39, Sect. 3.2] in Case (Spl); and see [44, Sects. 2.1.7 and 4.1–4.2] in Case (Hdg)). The morphisms $E' \to E$ are isogenies of tori of ℓ-power degrees, which are dual to homomorphisms $\mathbf{S} \to \mathbf{S}'$ of character groups that are in turn dual to inclusions of lattices of ℓ-power indices induced by $\mathcal{H}(\ell^r) \hookrightarrow \mathcal{H}$ (see [35, Lemma 6.2.4.4] and [33, Corollary 3.6.10] in Cases (Sm), (Nm), and (Spl); and see [44, Sect. 2.1.11] in Case (Hdg)). Consequently, because the collection of cone decompositions for $\mathsf{X}^{\mathrm{tor}}_{\mathcal{H}(\ell^r)}$ is induced by that for $\mathsf{X}^{\mathrm{tor}}_{\mathcal{H}}$, the induced morphisms $E'(\sigma') \to E(\sigma)$ are finite flat and tamely ramified. The rest of the proposition then follows from an analogue of the approximation argument in the proofs of [42, Proposition 2.2 and Corollary 2.4], by also approximating the objects at level $\mathcal{H}(\ell^r)$ and the action of the finite group $\mathcal{H}/\mathcal{H}(\ell^r) \cong \mathcal{H}_\ell/\mathcal{U}_\ell(\ell^r)$ on these objects. □

Proposition 5.3 *In the context of Proposition 5.1, we have the following commutative diagram*

$$\begin{array}{ccc}
\coprod_{\mathsf{Z}'_{[\sigma']} \text{ lying above } \mathsf{Z}_{[\sigma]}} (E' \underset{S}{\times} C') & \hookrightarrow & \coprod_{\mathsf{Z}'_{[\sigma']} \text{ lying above } \mathsf{Z}_{[\sigma]}} (E'(\sigma') \underset{S}{\times} C') \\
\downarrow & & \downarrow \\
\coprod_{\mathsf{Z}'_{[\sigma']} \text{ lying above } \mathsf{Z}_{[\sigma]}} (E \underset{S}{\times} C') & \hookrightarrow & \coprod_{\mathsf{Z}'_{[\sigma']} \text{ lying above } \mathsf{Z}_{[\sigma]}} (E(\sigma) \underset{S}{\times} C') \\
\downarrow & & \downarrow \\
E \times_S C & \longrightarrow & E(\sigma) \times_S C
\end{array} \tag{5.4}$$

of canonical morphisms, with the following properties:

(1) *All squares are Cartesian, and the composition of the vertical morphisms at the right-hand side is the canonical morphism in (5.2).*

(2) *The bottom two vertical morphisms are pullbacks of the finite étale morphism $\coprod_{\mathsf{Z}'_{[\sigma']} \text{ lying above } \mathsf{Z}_{[\sigma]}} C' \to C$, and the top two morphisms are pullbacks of the morphisms $\coprod_{\mathsf{Z}'_{[\sigma']} \text{ lying above } \mathsf{Z}_{[\sigma]}} E'(\sigma') \to \coprod_{\mathsf{Z}'_{[\sigma']} \text{ lying above } \mathsf{Z}_{[\sigma]}} E(\sigma)$.*

(3) *For each particular $\mathsf{Z}'_{[\sigma']}$ lying above $\mathsf{Z}_{[\sigma]}$, there exist finite groups*

$$1 \subset H_{-2} \subset H_{-1} \subset H := \mathcal{H}_\ell / \mathcal{U}_\ell(\ell^r) \tag{5.5}$$

such that $C' \to C$ is a Galois finite étale cover with Galois group H_{-1}/H_{-2}, and such that $E' \to E$ is a Galois finite étale cover with Galois group H_{-2}. Consequently, the pullback of the étale sheaf $\mathcal{V}$ under the finite étale morphism $E \times_S C' \to E \times_S C$, which we abusively denote by the same symbol $\mathcal{V}$, descends to an étale sheaf $\mathcal{V}_E$ over E (defined by the action of H_{-2} only). That is, $\mathcal{V} \cong \mathcal{V}_E \boxtimes_S \Lambda$ over $E \times_S C'$, and the pullback of $\mathcal{V}_E$ to E' is constant.

Proof These are because of the very constructions of these boundary objects, which can be compatibly realized as quotients of objects of higher principal levels. (See the same references given in the proof of Proposition 5.1.) □

Thus, in order to prove Theorem 4.1, it suffices to show that, for any point x of $\mathsf{X}^{\mathrm{tor}}_{\mathcal{H}}$ as in Proposition 5.1, together with the commutative diagrams (5.2) and (5.4) of morphisms (with properties as in Propositions 5.1 and 5.3), if we denote by

$$J_{E(\sigma)\times_S C'} = J_{E(\sigma)} \times \mathrm{Id}_{C'} : E \underset{S}{\times} C' \hookrightarrow E(\sigma) \underset{S}{\times} C' \tag{5.6}$$

the canonical open immersion, where $J_{E(\sigma)} : E \hookrightarrow E(\sigma)$ is the affine toroidal embedding and $\mathrm{Id}_{C'}$ is the identity morphism on C', then the adjunction morphisms

$$R\Psi_{E(\sigma)\times_S C'}\, RJ_{E(\sigma)\times_S C',\eta,*}(\mathcal{V}) \to RJ_{E(\sigma)\times_S C',\bar{s},*}\, R\Psi_{E\times_S C'}(\mathcal{V}) \tag{5.7}$$

and

$$J_{E(\sigma)\times_S C',\bar{s},!}\, R\Psi_{E\times_S C'}(\mathcal{V}) \to R\Psi_{E(\sigma)\times_S C'}\, J_{E(\sigma)\times_S C',\eta,!}(\mathcal{V}) \tag{5.8}$$

are isomorphisms.

Since $\mathcal{V} \cong \mathcal{V}_E \boxtimes_S \Lambda$ over $E \times_S C'$ for some étale sheaf $\mathcal{V}_E$ over E whose pullback to E' is constant (see Proposition 5.3), by the Künneth isomorphisms (see [2, XVII, 5.4.3] and [4, 4.2.7]), and by Gabber's theorem (see [25, 4.7]) on nearby cycles over products of schemes of finite type over S, we have canonical isomorphisms

$$\begin{aligned}(R\Psi_{E(\sigma)}\, RJ_{E(\sigma),\eta,*}(\mathcal{V}_E)) \overset{L}{\underset{\bar{s}}{\boxtimes}} (R\Psi_{C'}(\Lambda)) &\xrightarrow{\sim} R\Psi_{E(\sigma)\times_S C'}((RJ_{E(\sigma),\eta,*}(\mathcal{V}_E)) \overset{L}{\underset{\eta}{\boxtimes}} \Lambda) \\ &\xrightarrow{\sim} R\Psi_{E(\sigma)\times_S C'}\, RJ_{E(\sigma)\times_S C',\eta,*}(\mathcal{V}_E \underset{\eta}{\boxtimes} \Lambda),\end{aligned}$$

$$\begin{aligned}(RJ_{E(\sigma),\bar{s},*}\, R\Psi_E(\mathcal{V}_E)) \overset{L}{\underset{\bar{s}}{\boxtimes}} (R\Psi_{C'}(\Lambda)) &\xrightarrow{\sim} RJ_{E(\sigma)\times_S C',\bar{s},*}((R\Psi_E(\mathcal{V}_E)) \overset{L}{\underset{\bar{s}}{\boxtimes}} (R\Psi_{C'}(\Lambda))) \\ &\xrightarrow{\sim} RJ_{E(\sigma)\times_S C',\bar{s},*}\, R\Psi_{E\times_S C'}(\mathcal{V}_E \underset{\eta}{\boxtimes} \Lambda),\end{aligned}$$

$$\begin{aligned}(J_{E(\sigma),\bar{s},!}\, R\Psi_E(\mathcal{V}_E)) \overset{L}{\underset{\bar{s}}{\boxtimes}} (R\Psi_{C'}(\Lambda)) &\xrightarrow{\sim} J_{E(\sigma)\times_S C',\bar{s},!}((R\Psi_E(\mathcal{V}_E)) \overset{L}{\underset{\bar{s}}{\boxtimes}} (R\Psi_{C'}(\Lambda))) \\ &\xrightarrow{\sim} J_{E(\sigma)\times_S C',\bar{s},!}\, R\Psi_{E\times_S C'}(\mathcal{V}_E \underset{\eta}{\boxtimes} \Lambda),\end{aligned}$$

and

$$(R\Psi_{E(\sigma)}\, J_{E(\sigma),\eta,!}(\mathcal{V}_E)) \overset{L}{\underset{\bar{s}}{\boxtimes}} (R\Psi_{C'}(\Lambda)) \xrightarrow{\sim} R\Psi_{E(\sigma)\times_S C'}((J_{E(\sigma),\eta,!}(\mathcal{V}_E)) \overset{L}{\underset{\eta}{\boxtimes}} \Lambda)$$
$$\xrightarrow{\sim} R\Psi_{E(\sigma)\times_S C'}\, J_{E(\sigma)\times_S C',\eta,!}(\mathcal{V}_E \underset{\eta}{\boxtimes} \Lambda),$$

which are compatible with each other under the adjunction morphisms

$$R\Psi_{E(\sigma)}\, RJ_{E(\sigma),\eta,*}(\mathcal{V}_E) \to RJ_{E(\sigma),\bar{s},*}\, R\Psi_E(\mathcal{V}_E), \tag{5.9}$$

$$J_{E(\sigma),\bar{s},!}\, R\Psi_E(\mathcal{V}_E) \to R\Psi_{E(\sigma)}\, J_{E(\sigma),\eta,!}(\mathcal{V}_E), \tag{5.10}$$

and (5.7) and (5.8) (and the identity morphism on $R\Psi_{C'}(\Lambda)$). Hence, in order to show that the adjunction morphisms (5.7) and (5.8) are isomorphisms, it suffices to show that the simpler adjunction morphisms (5.9) and (5.10) are isomorphisms.

We can complete the collection consisting of σ and its faces into a cone decomposition Σ of $\mathbf{S}_{\mathbb{R}}^{\vee} = \mathrm{Hom}_{\mathbb{Z}}(\mathbf{S}, \mathbb{R})$, which defines a toroidal embedding

$$J_{\overline{E}} : E \hookrightarrow \overline{E} \tag{5.11}$$

over S, such that $\overline{E}$ is *proper* over S (see [28, Chap. I, Sect. 2, Theorem 8]), which contains $E(\sigma)$ as an open subscheme. Then it suffices to show that the corresponding adjunction morphisms

$$R\Psi_{\overline{E}}\, RJ_{\overline{E},\eta,*}(\mathcal{V}_E) \to RJ_{\overline{E},\bar{s},*}\, R\Psi_E(\mathcal{V}_E) \tag{5.12}$$

and

$$J_{\overline{E},\bar{s},!}\, R\Psi_E(\mathcal{V}_E) \to R\Psi_{\overline{E}}\, J_{\overline{E},\eta,!}(\mathcal{V}_E) \tag{5.13}$$

are isomorphisms. Since smooth refinements of cone decompositions induce *proper morphisms* between the corresponding toroidal embeddings (again, see [28, Chap. I, Sect. 2, Theorem 8]), by the proper base change theorem, as in the proof of [42, Lemma 5.5], we may and we shall assume in the above that $\overline{E}$ is proper and *smooth* over S, and that the boundary $\overline{E} - E$ with its reduced subscheme structure is a simple normal crossings divisor on $\overline{E}$.

Finally, let us consider the étale sheaf $\mathcal{V}_E$ over E, whose pullback under the isogeny $E' \to E$ of ℓ-power degree is constant, which is then *tamely ramified* along the boundary $\overline{E} - E$. (See Propositions 5.1 and 5.3 again for these assertions.) Therefore, the adjunction morphisms (5.12) and (5.13) are isomorphisms, by [13, XIII, 2.1.9]. The proof of Theorem 4.1 is now complete. □

Remark 5.14 Compared with the proof of [42, Theorem 5.15], we have not used the Kuga families and their toroidal compactifications as in [42, Sect. 4] (which were based on ideas in [34]). As a result, we do not have any restriction in the case of integral or torsion coefficients. (In [42, Theorem 5.15], we assumed that $\mathcal{V}$ is either over $\Lambda = \bar{\mathbb{Q}}_\ell$, or of the form $\mathcal{W}_{0,M}$ over $\Lambda = \mathbb{Z}_\ell$ as in [42, Proposition 3.4] for some $\ell > c_M$.) Our argument here can be considered a fulfilment of the strategy in [42, Remark 5.35], although its actual execution here is subtler than suggested there.

Remark 5.15 We remarked in the introduction of [42] that the argument in [24, Sect. 7], for the usual and compactly supported cohomology in Case (Sm), is unfortunately incomplete, because the first step in the proof of [24, Lemma 7.1] should require some tameness assumption as in [13, XIII, 2.1.9]. Nevertheless, our proof in this section shows that the tameness assumption can indeed be verified.

Acknowledgements It is our great honor and pleasure to dedicate this article to Joachim Schwermer on the occasion of his 66th birthday. His many works on the cohomology of noncompact locally symmetric spaces have been great sources of information and inspiration for us. We would also like to thank the anonymous referee for a careful reading and helpful suggestions.

References

1. Arthur, J., Ellwood, D., Kottwitz, R. (eds.), Harmonic analysis, the trace formula, and Shimura varieties. In: Clay Mathematics Proceedings, vol. 4, Proceedings of the Clay Mathematics Institute 2003 Summer School, The Fields Institute, Toronto, Canada, 2–27 June 2003. American Mathematical Society, Providence, Rhode Island, Clay Mathematics Institute, Cambridge, Massachusetts (2005)
2. Artin, M., Grothendieck, A., Verdier, J.-L. (eds.): Théorie des topos et cohomologie étale des schémas SGA 4, Tome 3. Lecture Notes in Mathematics, vol. 305. Springer, Berlin (1973)
3. Baily Jr., W.L., Borel, A.: Compactification of arithmetic quotients of bounded symmetric domains. Ann. Math. 2 **84**(3), 442–528 (1966)
4. Beilinson, A.A., Bernstein, J., Deligne, P.: Faisceaux pervers, Analyse et topologie sur les espaces singuliers (I) CIRM, 6–10 juillet 1981, Astérisque, vol. 100. Société Mathématique de France, Paris (1982)
5. Borel, A., Casselman, W. (eds.) Automorphic forms, representations and L-functions. In: Proceedings of Symposia in Pure Mathematics, vol. 33, Part 2, held at Oregon State University, Corvallis, Oregon, July 11–August 5 1977. American Mathematical Society, Providence, Rhode Island (1979)
6. Breuil, C., Mézard, A.: Multiplicités modulaires et représentations de $\mathrm{GL}_2(\mathbf{Z}_p)$ et de $\mathrm{Gal}(\overline{\mathbf{Q}}_p/\mathbf{Q}_p)$ en $\ell = p$. Duke Math. J. **115**(2), 205–310 (2002)
7. Caraiani, A., Scholze, P.: On the generic part of the cohomology of compact unitary Shimura varieties. Ann. Math. (2) **186**(3), 649–766 (2017)
8. Cartier, P., Illusie, L., Katz, N.M., Laumon, G., Manin, Y., Ribet, K.A. (eds.): The Grothendieck festschrift: A collection of articles written in honer of the 60th birthday of Alexander Grothendieck, vol. 2. Birkhäuser, Boston (1990)
9. Clozel, L., Milne, J.S. (eds.), Automorphic forms, Shimura varieties, and L-functions. Volume I In: Perspectives in Mathematics, vol. 10, Proceedings of a Conference held at the University of Michigan, Ann Arbor, 6–16 July 1988. Academic Press Inc., Boston (1990)
10. Deligne, P.: Variétés de Shimura: Interprétation modulaire, et techniques de construction de modèles canoniques, in Borel and Casselman [5], pp. 247–290
11. Deligne, P. (ed.): Cohomologie étale SGA $4\frac{1}{2}$. Lecture Notes in Mathematics, vol. 569. Springer, Berlin (1977)
12. Deligne, P.: La conjecture de Weil. II. Publ. Math. Inst. Hautes Étud. Sci. **52**, 137–252 (1980)

13. Deligne, P., Katz, N. (eds.): Groupes de monodromie en géométrie algébriqueSGA 7 II. Lecture Notes in Mathematics, vol. 340. Springer, Berlin (1973)
14. Ekedahl, T.: On the adic formalism, in Cartier et al. [8], pp. 197–218
15. Farkas, G., Morrison, I. (eds.): Handbook of Moduli: Volume II. Advanced Lectures in Mathematics, vol. 25. International Press, Somerville; Higher Education Press, Beijing (2013)
16. Fu, L.: Etale Cohomology Theory. Nankai Tracts in Mathematics, vol. 13. World Scientific, Singapore (2011)
17. Fulton, W., Harris, J.: Representation Theory: A First Course. Graduate Texts in Mathematics, vol. 129. Springer, Berlin (1991)
18. Genestier, A., Tilouine, J.: Systèmes de Taylor-Wiles pour GSp_4, in Tilouine et al. [59], pp. 177–290
19. Goodman, R., Wallach, N.R.: Symmetry, Representations, and Invariants. Graduate Texts in Mathematics, vol. 255. Springer, Berlin (2009)
20. Grothendieck, A. (ed.): Revêtements étales et groupe fondamental (SGA 1). Lecture Notes in Mathematics, vol. 224. Springer, Berlin (1971)
21. Grothendieck, A., Dieudonné, J.: Eléments de géométrie algébrique. Publications mathématiques de l'I.H.E.S, vol. 4, 8, 11, 17, 20, 24, 28, 32. Institut des Hautes Etudes Scientifiques, Paris (1960, 1961, 1961, 1963, 1964, 1965, 1966, 1967)
22. Haines, T.J.: Introduction to Shimura varieties with bad reductions of parahoric type, in Arthur et al. [1], pp. 583–658
23. Harris, M., Taylor, R.: The Geometry and Cohomology of Some Simple Shimura Varieties. Annals of Mathematics Studies, vol. 151. Princeton University Press, Princeton (2001)
24. Helm, D.: Towards a geometric Jacquet–Langlands correspondence for unitary Shimura varieties. Duke Math. J. **155**, 483–518 (2010)
25. Illusie, L.: Autour du théorème de monodromie locale, Périodes p-adiques (Bures-sur-Yvette, 1988), Astérisque, vol. 223. Société Mathématique de France, Paris, 9–57 (1994)
26. Imai, N., Mieda, Y.: Compactly supported cohomology and nearby cycles of open Shimura varieties of PEL type, preprint (2011)
27. Imai, N., Mieda, Y.: Potentially good reduction loci of Shimura varieties, preprint (2016)
28. Kempf, G., Knudsen, F., Mumford, D., Saint-Donat, B.: Toroidal Embeddings I. Lecture Notes in Mathematics, vol. 339. Springer, Berlin (1973)
29. Kisin, M.: Integral models for Shimura varieties of abelian type. J. Am. Math. Soc. **23**(4), 967–1012 (2010)
30. Kim, W., Madapusi Pera, K.: 2-adic integral canonical models. Forum Math. Sigma **4**, e28, 34 pp. (2016)
31. Kisin, M., Pappas, G.: Integral models of Shimura varieties with parahoric level structure, preprint (2015)
32. Kottwitz, R.E.: Points on some Shimura varieties over finite fields. J. Am. Math. Soc. **5**(2), 373–444 (1992)
33. Lan, K.-W.: Comparison between analytic and algebraic constructions of toroidal compactifications of PEL-type Shimura varieties. J. Reine Angew. Math. **664**, 163–228 (2012)
34. Lan, K.-W.: Toroidal compactifications of PEL-type Kuga families. Algebra Number Theory **6**(5), 885–966 (2012)
35. Lan, K.-W.: Arithmetic Compactification of PEL-type Shimura Varieties. London Mathematical Society Monographs, vol. 36. Princeton University Press, Princeton (2013). (errata and revision available online at the author's website)
36. Lan, K.-W.: Compactifications of PEL-type Shimura varieties in ramified characteristics. Forum Math. Sigma **4**, e1, 98 pp. (2016)
37. Lan, K.-W.: Vanishing theorems for coherent automorphic cohomology. Res. Math. Sci. **3**, 43 pp. (2016). (article no. 39)
38. Lan, K.-W.: Compactifications of PEL-Type Shimura Varieties and Kuga Families with Ordinary Loci. World Scientific, Singapore (2018)
39. Lan, K.-W.: Compactifications of splitting models of PEL-type Shimura varieties. Trans. Am. Math. Soc. **370**(4), 2463–2515 (2018)

40. Lan, K.-W.: Integral models of toroidal compactifications with projective cone decompositions. Int. Math. Res. Not. IMRN **2017**(11), 3237–3280 (2017)
41. Lan, K.-W., Stroh, B.: Compactifications of subschemes of integral models of Shimura varieties, preprint (2015)
42. Lan, K.-W., Stroh, B.: Nearby cycles of automorphic étale sheaves. Compos. Math. **154**(1), 80–119 (2018)
43. Lan, K.-W., Suh, J.: Vanishing theorems for torsion automorphic sheaves on general PEL-type Shimura varieties. Adv. Math. **242**, 228–286 (2013)
44. Madapusi Pera, K.: Toroidal compactifications of integral models of Shimura varieties of Hodge type, preprint (2015)
45. Mantovan, E.: On the cohomology of certain PEL-type Shimura varieties. Duke Math. J. **129**(3), 573–610 (2005)
46. Mantovan, E.: ℓ-adic étale cohomology of PEL type Shimura varieties with non-trivial coefficients. In: WIN-Woman in Numbers, Fields Institute Communications, vol. 60, pp. 61–83. American Mathematical Society, Providence (2011)
47. Milne, J.S.: Étale Cohomology. Princeton Mathematical Series, vol. 33. Princeton University Press, Princeton (1980)
48. Milne, J.S.: Canonical models of (mixed) Shimura varieties and automorphic vector bundles, in Clozel and Milne [9], pp. 283–414
49. Milne, J.S.: Shimura varieties and moduli, in Farkas and Morrison [15], pp. 467–548
50. Mumford, D.: Abelian Varieties. Tata Institute of Fundamental Research Studies in Mathematics, vol. 5. Oxford University Press, Oxford (1970). (with appendices by C. P. Ramanujam and Yuri Manin)
51. Mumford, D.: Hirzebruch's proportionality theorem in the non-compact case. Invent. Math. **42**, 239–272 (1977)
52. Pappas, G., Rapoport, M.: Local models in the ramified case, I. The EL-case. J. Algebr. Geom. **12**(1), 107–145 (2003)
53. Pappas, G., Rapoport, M.: Local models in the ramified case, II. Splitting models. Duke Math. J. **127**(2), 193–250 (2005)
54. Pappas, G., Zhu, X.: Local models of Shimura varieties and a conjecture of Kottwitz. Invent. Math. **194**, 147–254 (2013)
55. Pink, R.: Arithmetic compactification of mixed Shimura varieties, Ph.D. thesis, Rheinischen Friedrich-Wilhelms-Universität, Bonn (1989)
56. Rapoport, M., Zink, T.: Period Spaces for p-Divisible Groups. Annals of Mathematics Studies, vol. 141. Princeton University Press, Princeton (1996)
57. Scholze, P.: The Langlands–Kottwitz method and deformation spaces of p-divisible groups. J. Am. Math. Soc. **26**, 227–259 (2013)
58. Skinner, C.: A note on the p-adic Galois representations attached to Hilbert modular forms. Doc. Math. **14**, 241–258 (2009)
59. Tilouine, J., Carayol, H., Harris, M., Vignéras, M.-F. (eds.): Formes automorphes (II): Le cas du groupe GSp(4). Astérisque, vol. 302. Société Mathématique de France, Paris (2005)

On Slope Subspaces of Cohomology of p-adic Verma Modules

J. Mahnkopf

Abstract We determine bounds for the dimension of the slope subspaces of cohomology groups of arithmetic subgroups of semi simple algebraic groups $\mathbf{G}$ with coefficients in p-adic Verma modules.

Keywords Cohomology of arithmetic groups · Gouvea-Mazur conjecture
Verma module

1 Introduction

In the 1980s H. Hida began to develop his theory of p-adic families of ordinary modular forms (cf. e.g. [12, 13]). As main results of this theory he proved that (i) the dimension of the space $\mathcal{M}_k(\Gamma_0(pN))^{\text{ord}}$ of ordinary modular forms of weight k and level $\Gamma_0(pN)$ where $N \in \mathbb{N}$ is not divisible by the (fixed) prime $p \in \mathbb{N}$ depends on the weight $k \geq 2$ only modulo $(p-1)$ (mod 2 if $p = 2$) and (ii) any ordinary eigenform $f \in \mathcal{M}_k(\Gamma_0(pN))^{\text{ord}}$ fits in a p-adic family of eigenforms of varying weight k.

The space of ordinary modular forms is the same as the space of modular forms of slope 0 and later Gouvea and Mazur conjectured that Hida's results hold for all finite slope subspaces of $\mathcal{M}_k(\Gamma_0(pN))$ (cf. [11]). In particular, they conjectured in a precise form that for any $\beta \in \mathbb{Q}_{\geq 0}$ the dimension of the slope-β subspace $\mathcal{M}_k(\Gamma_0(pN))^{\beta}$ is locally constant in the p-adic topology as a function of the weight k. A weakened form of the conjecture which still implies local constancy of dimension of slope subspaces was proven by Wan (cf. [21]) and later Buzzard and Calegari found counter-examples showing that the original conjecture of Gouvea–Mazur does not hold (cf. [8]). In his proof Wan made use of work of Coleman, Katz, Gouvea–Mazur and used methods from rigid analytic geometry, p-adic Banach space theory and p-adic (overconvergent) modular forms. In particular, he deduced the Gouvea–Mazur Conjecture

J. Mahnkopf (✉)
Institut fur Mathematik, Universitat Wien, Oskar-Morgenstern-Platz 1,
1090 Vienna, Austria
e-mail: joachim.mahnkopf@univie.ac.at

J. W. Cogdell et al. (eds.), *Cohomology of Arithmetic Groups*,
Springer Proceedings in Mathematics & Statistics 245,
https://doi.org/10.1007/978-3-319-95549-0_5

using an analogous statement for p-adic modular forms ("*p-adic Gouvea–Mazur conjecture*").

In another direction, following a suggestion by R. Taylor, Buzzard gave an elementary proof of boundedness of dimension of the slope subspace $\mathcal{M}_k(\Gamma_0(pN))^\beta$ as the weight k varies (cf. [7]). This follows from local constancy of dimension of slope subspaces but Buzzard's proof is very different: he does not make use of methods from rigid analytic geometry, overconvergent modular forms or p-adic spectral theory but only uses elementary methods and even gave an explicit bound for the dimension of the slope subspaces.

An analogue of the Gouvea–Mazur Conjecture for slope subspaces of cohomology groups of arithmetic subgroups in reductive algebraic groups having discrete series and with coefficients in a finite dimensional irreducible representation L_λ of highest weight λ is proven in [17].

In this article we want to begin looking at the higher rank case of the p-adic Gouvea–Mazur Conjecture (we note that in case $\mathbf{GL}_2$ this conjecture has been proven in [21]). Thus, we replace $\mathcal{M}_k(\Gamma_0(pN))$ by the cohomology group of an arithmetic subgroup of a semi simple algebraic group with coefficients in certain p-adic Verma modules and we want to prove bounds for the dimension of the slope subspaces which are independent of the coefficient system. This is an analogue of the result of [7] in the p-adic (higher rank) case. The proof uses p-adic spectral theory but apart from that, similar to the proof in [7], is more of an elementary nature.

To be more precise we introduce some notation. We fix a prime $p \in \mathbb{N}$. We let $\mathbf{G}/\mathbb{Q}$ be a semi simple connected algebraic group defined over $\mathbb{Q}$ which contains a maximal torus $\mathbf{T}/\mathbb{Q}$ which splits over $\mathbb{Q}_p$. We let $\Gamma \subseteq \mathbf{G}(\mathbb{Q})$ be an arithmetic subgroup which is contained in the Iwahori subgroup $\mathcal{I}^1$ of $\mathbf{G}(\mathbb{Q}_p)$. We denote by $\mathfrak{g}$ the complexification of the Lie algebra of $\mathbf{G}$ and we select a Chevalley basis in $\mathfrak{g}$. This defines $\mathbb{Z}$-structures $\mathfrak{g}(\mathbb{Z})$ resp. $\mathfrak{h}(\mathbb{Z})$ resp. $\mathcal{U}_\mathbb{Z}$ on $\mathfrak{g}$ resp. on the Cartan subalgebra $\mathfrak{h} \leq \mathfrak{g}$ resp. on the universal enveloping algebra $\mathcal{U}$ of $\mathfrak{g}$. We let $K/\mathbb{Q}_p$ be a finite extension with ring of integers $\mathcal{O}$ and we denote by $\mathfrak{h}^*_\Gamma(\mathcal{O})$ the set of all linear forms $\lambda : \mathfrak{h}(K) = \mathfrak{h}(\mathbb{Z}) \otimes K \to K$ which are defined over $\mathcal{O}$, i.e. which satisfy $\lambda(\mathfrak{h}_\Gamma(\mathbb{Z})) \subseteq \mathcal{O}$; here, $\mathfrak{h}_\Gamma(\mathbb{Z}) \leq \mathfrak{h}(K)$ is a certain $\mathbb{Z}$-lattice depending on $\mathbf{G}/\mathbb{Q}_p$; e.g. if $\mathbf{G}/\mathbb{Q}_p$ is simply connected then $\mathfrak{h}_\Gamma(\mathbb{Z}) = \mathfrak{h}(\mathbb{Z})$ (cf. Sect. 2.2.3). For any $\lambda \in \mathfrak{h}^*_\Gamma(\mathcal{O})$ we denote by V_λ the Verma module of highest weight λ, i.e. V_λ is a $\mathcal{U}_{\mathfrak{g}(K)}$-module, hence, a $\mathfrak{g}(K)$-module and we denote by $\hat{V}_\lambda$ the completion of V_λ w.r.t. a certain norm on V_λ. Using basic p-adic Lie theory we integrate $\hat{V}_\lambda$ to a representation of the sub semigroup D of $\mathbf{G}(\mathbb{Q}_p)$ which is generated by $\mathcal{I}^1$ and the semi group $\mathbf{T}(\mathbb{Q}_p)^{--}$ which consists of all $t \in \mathbf{T}(\mathbb{Q}_p)$ satisfying $v_p(\alpha(t)) < 0$ for all $\alpha \in \Delta$. We then fix an element $h \in \mathbf{T}(\mathbb{Q})^{++}$ (hence, h is regular) and denote by $\mathrm{D}_h \leq \mathbf{G}(\mathbb{Q}_p)$ the sub semigroup generated by $\mathcal{I}^1$ and h. We set $\Delta = \{g \in \mathbf{G}(\mathbb{Q}) : g \in \mathrm{D}_h\}$. Any $\zeta \in \Delta\, (\subseteq \mathrm{D}^{-1})$ defines a Hecke operator $T_\zeta = \Gamma\zeta\Gamma \in \mathcal{H}(\Gamma\backslash\Delta/\Gamma)$ which acts on $H^i(\Gamma, \hat{V}_\lambda)$ and $H^i(\Gamma, \hat{V}_\lambda)$ has slope decompositions w.r.t. T_ζ (the slope $\leq \beta$ subspace $H^i(\Gamma, \hat{V}_\lambda)^{\leq\beta}$, $\beta \in \mathbb{Q}_{\geq 0}$, coincides with the slope $\leq \beta$-subspace of the Hausdorff quotient $\tilde{H}^i(\Gamma, \hat{V}_\lambda)$ of $H^i(\Gamma, \hat{V}_\lambda)$ which is a Banach space and on which T_ζ acts as a compact operator; cf. 6.6 Remark).

To formulate our result we denote by g_i the number of i-simplices in a triangulation of the Borel–Serre compactification $\Gamma\backslash\bar{X}$ of $\Gamma\backslash X$, $\sigma = \max_{\alpha\in\Phi^+}\mathrm{ht}(\alpha)$ is the maximal height of a positive root and $s = |\Phi^+|$ is the number of positive roots. Also, we assume that $p > \sigma + 2$. We then obtain the following bound on the dimension of the slope-β subspaces $H^i(\Gamma, \hat{V}_\lambda)^{\leq\beta}$ of $H^i(\Gamma, \hat{V}_\lambda)$.

Theorem (cf. 6.6 Corollary.) *For all $\beta \in \mathbb{Q}_{\geq 0}$, all weights $\lambda \in \mathfrak{h}^*_\Gamma(\mathcal{O})$, and all i we have*

$$\dim H^i(\Gamma, \hat{V}_\lambda)^{\leq\beta} \leq \mathrm{m}\beta^s + \mathrm{n};$$

here, $\mathrm{m} = 11\frac{g_i}{s}(\frac{\sigma(p-1)}{p-2-\sigma})^s \in \mathbb{R}_{\geq 0}$ and $\mathrm{n} \in \mathbb{N}$ is an integer which only depends on g_i and s. In particular, m and n only depend on Γ (and, hence, on $\mathbf{G}$ and p) and i, i.e. they do not depend on λ.

Coefficient systems $\mathcal{D}_k$ depending on a p-adic parameter k have been constructed and their cohomology has been investigated by Ash-Stevens in [1] ("overconvergent cohomology"); these investigations have been taken up by Urban in [20]. The construction of Ash-Stevens is different as they construct the modules $\mathcal{D}_k$ directly on the group as spaces of certain distributions. In contrast we integrate a Verma module on the Lie algebra to a representation of the group using basic p-adic Lie theory; in particular, this reduces part of the analysis of the modules $\hat{V}_\lambda$ to calculations in the Lie algebra $\mathfrak{g}(K)$.

2 p-Adic Verma Modules

2.1 The p-Adic Complex Numbers

We select a prime $p \in \mathbb{N}$. We denote by $\mathbb{C}_p$ the p-adic complex numbers and by $|\cdot|$ the norm on $\mathbb{C}_p$ normalized by $|p| = p^{-1}$. We denote by $K/\mathbb{Q}_p$ a finite extension which is contained in $\mathbb{C}_p$. The norm $|\cdot|$ induces via restriction a norm on K, v_p denotes the corresponding valuation on K (i.e. $v_p(p) = 1$) and $\mathcal{O} = \mathcal{O}_K$ is the ring of integers in K.

2.2 Lie Algebras

2.2.1 Notations

Let $\mathfrak{g}$ be a finite dimensional complex semi simple Lie algebra of rank ℓ. We denote by $\mathfrak{h}$ a Cartan subalgebra in $\mathfrak{g}$ and by $\Phi = \Phi(\mathfrak{g}, \mathfrak{h})$ the set of roots of $\mathfrak{g}$ w.r.t. to $\mathfrak{h}$. We choose a basis Δ of Φ and we denote by Φ^+ the set of positive roots. For each root $\alpha \in \Phi$ we write $\mathfrak{g}(\alpha)$ for the corresponding root subspace and we denote by $h_\alpha \in \mathfrak{h}$

the coroot corresponding to $\alpha \in \Phi$. We select elements $x_\alpha \in \mathfrak{g}(\alpha)$, $\alpha \in \Phi$, such that $\{h_\alpha, \alpha \in \Delta, x_\beta, \beta \in \Phi\}$ is a *Chevalley basis* of $\mathfrak{g}$. We denote by $\mathfrak{h}^*$ the vector space of all $\mathbb{C}$-linear mappings $\lambda : \mathfrak{h} \to \mathbb{C}$ and we denote by $\Gamma_{\text{ad}} \leq \mathfrak{h}^*$ resp. $\Gamma_{\text{sc}} \leq \mathfrak{h}^*$ the root resp. the weight lattice of $\mathfrak{g}$; hence, $\Gamma_{\text{ad}} \leq \Gamma_{\text{sc}}$.

We denote by $\mathfrak{b}$ resp. $\mathfrak{b}^-$ the Borel subgroup corresponding to Δ resp. to $-\Delta$ and by $\mathfrak{n}$ resp. $\mathfrak{n}^-$ the unipotent radical of $\mathfrak{b}$ resp. of $\mathfrak{b}^-$. By $\mathcal{U} = \mathcal{U}_\mathfrak{g}$ resp. $\mathcal{U}_\mathfrak{b}$ resp. $\mathcal{U}_{\mathfrak{b}^-}$ resp. $\mathcal{U}^+ = \mathcal{U}_\mathfrak{n}$ resp. $\mathcal{U}^- = \mathcal{U}_{\mathfrak{n}^-}$ resp. $\mathcal{U}^\circ = \mathcal{U}_\mathfrak{h}$ we understand the universal enveloping algebra of $\mathfrak{g}$ resp. $\mathfrak{b}$ resp. $\mathfrak{b}^-$ resp. $\mathfrak{n}$ resp. $\mathfrak{n}^-$ resp. $\mathfrak{h}$.

Finally, $\mu \leq \lambda$, $\lambda, \mu \in \mathfrak{h}^*$, means that $\lambda - \mu$ is a linear combination of positive roots with non negative coefficients. For any $\lambda \in \mathfrak{h}^*$ we define a relative height function

$$\text{ht}_\lambda : \{\mu : \mu \leq \lambda\} \to \mathbb{N}_0$$

by $\text{ht}_\lambda(\mu) = \text{ht}(\lambda - \mu)$; here, $\text{ht} = \text{ht}_\Delta$ is the height function corresponding to Δ linearly extended to Γ_{ad}.

2.2.2 $\mathbb{Z}$-Structures

The above choice of a Chevalley basis of $\mathfrak{g}$ determines $\mathbb{Z}$-forms $\mathfrak{g}(\mathbb{Z}) = \bigoplus_{\alpha\in\Phi} \mathbb{Z}x_\alpha \oplus \bigoplus_{\alpha\in\Delta} \mathbb{Z}h_\alpha$ resp. $\mathfrak{h}(\mathbb{Z}) = \bigoplus_{\alpha\in\Delta} \mathbb{Z}h_\alpha$ resp. $\mathfrak{n}(\mathbb{Z}) = \bigoplus_{\alpha\in\Phi^+} \mathbb{Z}x_\alpha$ resp. $\mathfrak{b}(\mathbb{Z}) = \bigoplus_{\alpha\in\Delta} \mathbb{Z}h_\alpha \oplus \bigoplus_{\alpha\in\Phi^+} \mathbb{Z}x_\alpha$ of $\mathfrak{g}$ resp. $\mathfrak{h}$ resp. $\mathfrak{n}$ resp. $\mathfrak{b}$ and we set $\mathfrak{g}(R) = \mathfrak{g}(\mathbb{Z}) \otimes R$, $\mathfrak{h}(R) = \mathfrak{h}(\mathbb{Z}) \otimes R$, $\mathfrak{n}(R) = \mathfrak{n}(\mathbb{Z}) \otimes R$ and $\mathfrak{b}(R) = \mathfrak{b}(\mathbb{Z}) \otimes R$, R any $\mathbb{Z}$-module. We fix an ordering $\Phi^+ = \{\alpha_1, \ldots, \alpha_s\}$ of the set of positive roots and we set

$$X_+^\mathbf{n} = \frac{x_{\alpha_1}^{n_1}}{n_1!} \cdots \frac{x_{\alpha_s}^{n_s}}{n_s!} \in \mathcal{U}^+, \quad \text{and} \quad X_-^\mathbf{n} = \frac{x_{-\alpha_1}^{n_1}}{n_1!} \cdots \frac{x_{-\alpha_s}^{n_s}}{n_s!} \in \mathcal{U}^-,$$

where $\mathbf{n} = (n_1, \ldots, n_s) \in \mathbb{N}_0^s$. Moreover, for $\mathbf{n} \in \mathbb{N}_0^\Delta$ we set

$$H^\mathbf{n} = \prod_{\alpha\in\Delta} \binom{h_\alpha}{n_\alpha} \in \mathcal{U}^\circ.$$

The lattice $\bigoplus_{\mathbf{n}_1,\mathbf{n}_2\in\mathbb{N}_0^s,\mathbf{n}_0\in\mathbb{N}_0^\Delta} \mathbb{Z}X_-^{\mathbf{n}_1} H^{\mathbf{n}_0} X_+^{\mathbf{n}_2}$ resp. $\bigoplus_{\mathbf{n}\in\mathbb{N}_0^s} \mathbb{Z}X_-^\mathbf{n}$ resp. $\bigoplus_{\mathbf{n}\in\mathbb{N}_0^\Delta} \mathbb{Z}H^\mathbf{n}$ then is a $\mathbb{Z}$-form $\mathcal{U}_\mathbb{Z}$ of $\mathcal{U}$ resp. a $\mathbb{Z}$-form $\mathcal{U}_\mathbb{Z}^-$ of $\mathcal{U}^-$ resp. a $\mathbb{Z}$-form $\mathcal{U}_\mathbb{Z}^\circ$ of $\mathcal{U}^\circ$ (as associative algebras). We set $\mathcal{U}_R = \mathcal{U}_\mathbb{Z} \otimes R$, $\mathcal{U}_R^- = \mathcal{U}_\mathbb{Z}^- \otimes R$ and $\mathcal{U}_R^\circ = \mathcal{U}_\mathbb{Z}^\circ \otimes R$ (R a $\mathbb{Z}$-algebra).

The inclusion $\mathfrak{g}(\mathbb{Z}) \subseteq \mathfrak{g}(K) \subseteq \mathcal{U}_{\mathfrak{g}(K)}$ induces a morphism $\mathcal{U}_{\mathfrak{g}(\mathbb{Z})} \to \mathcal{U}_{\mathfrak{g}(K)}$, hence, a morphism $\mathcal{U}_{\mathfrak{g}(\mathbb{Z})} \otimes K \to \mathcal{U}_{\mathfrak{g}(K)}$ which is an isomorphism by PBW Theorem. The inclusion $\mathcal{U}_{\mathfrak{g}(\mathbb{Z})} \subseteq \mathcal{U}_\mathbb{Z}$ $(\subseteq \mathcal{U})$ induces an injective morphism $\mathcal{U}_{\mathfrak{g}(\mathbb{Z})} \otimes K \hookrightarrow \mathcal{U}_\mathbb{Z} \otimes K = \mathcal{U}_K$ which also is surjective because $\mathcal{U}_{\mathfrak{g}(\mathbb{Z})}$ contains the elements x_β, $\beta \in \Phi$ and h_α, $\alpha \in \Delta$ which generate $\mathcal{U}_K$ as K-algebra

$$\begin{array}{ccc} \mathcal{U}_{\mathbb{Z}} \otimes K & & \\ \cup & \overset{\cong}{\searrow} & \\ \mathcal{U}_{\mathfrak{g}(\mathbb{Z})} \otimes K & \cong & \mathcal{U}_{\mathfrak{g}(K)}. \end{array}$$

2.2.3 $\mathbb{Z}$-Structures on $\mathfrak{h}$

We select a $\mathbb{Z}$-lattice $\Gamma \leq \mathfrak{h}^*$ satisfying $\Gamma_{\mathrm{ad}} \leq \Gamma \leq \Gamma_{\mathrm{sc}}$. We select a $\mathbb{Z}$-basis $(\xi_i)_{i=1,\ldots,\ell}$ of Γ and we denote by $(\mathrm{h}_i)_{i=1,\ldots,\ell}$ the basis of $\mathfrak{h}$ which is dual to $(\xi_i)_i$. We define

$$\mathfrak{h}_\Gamma(\mathbb{Z}) = \{H \in \mathfrak{h} : \lambda(H) \in \mathbb{Z} \text{ for all } \lambda \in \Gamma\};$$

hence, $\mathfrak{h}(\mathbb{Z}) = \mathfrak{h}_{\Gamma_{\mathrm{sc}}}(\mathbb{Z}) \leq \mathfrak{h}_\Gamma(\mathbb{Z}) \leq \mathfrak{h}_{\Gamma_{\mathrm{ad}}}(\mathbb{Z})$ and $\mathfrak{h}_\Gamma(\mathbb{Z}) = \bigoplus_i \mathbb{Z}\mathrm{h}_i$, i.e. $(\mathrm{h}_i)_i$ is a $\mathbb{Z}$-basis of $\mathfrak{h}_\Gamma(\mathbb{Z})$. For any $\mathbb{Z}$-algebra R we set

$$\mathfrak{h}_\Gamma(R) = R \otimes \mathfrak{h}_\Gamma(\mathbb{Z}) = \bigoplus_i R\mathrm{h}_i,$$

hence, $\mathfrak{h}_\Gamma(\mathbb{Z}_p)$ is a lattice in $\mathfrak{h}_\Gamma(\mathbb{Q}_p)$. The inclusion $\mathfrak{h}(\mathbb{Z}) = \bigoplus_{\alpha \in \Delta} \mathbb{Z}h_\alpha \subseteq \mathfrak{h}_\Gamma(\mathbb{Z}) = \bigoplus_i \mathbb{Z}h_i$ induces an inclusion

$$\mathfrak{h}(K) = \mathfrak{h}(\mathbb{Z}) \otimes K \hookrightarrow \mathfrak{h}_\Gamma(\mathbb{Z}) \otimes K,$$

which is an isomorphism of K-vector spaces, hence, $\mathfrak{h}_\Gamma(\mathbb{Z}) \leq \mathfrak{h}(K)$ is a $\mathbb{Z}$-structure on $\mathfrak{h}(K) = \mathfrak{h}(\mathbb{Z}) \otimes K$.

We look at dual spaces. We set

$$\mathfrak{h}^*_\Gamma(\mathbb{Z}) := \{\lambda \in \mathfrak{h}^* : \lambda(\mathrm{h}_i) \in \mathbb{Z} \text{ for all } i = 1, \ldots, \ell\},$$

hence, $\mathfrak{h}^*_\Gamma(\mathbb{Z}) = \Gamma = \bigoplus_i \mathbb{Z}\xi_i$. We set $\mathfrak{h}^*_\Gamma(R) = \mathfrak{h}^*_\Gamma(\mathbb{Z}) \otimes R$ (R a $\mathbb{Z}$-algebra). We denote by $\mathfrak{h}^*(K)$ the K-vector space of all K-linear mappings $\lambda : \mathfrak{h}(K) \to K$. We obtain isomorphisms of K-modules

$$K \otimes \mathfrak{h}^*_\Gamma(\mathbb{Z}) \to \{\lambda : \mathfrak{h}_\Gamma(\mathbb{Z}) \to K,\ \lambda \text{ is } \mathbb{Z}\text{-linear}\} \overset{\mathrm{res}}{\leftarrow} \mathfrak{h}^*(K),$$

where the first isomorphism is given by mapping $\sum_i \alpha_i \otimes \xi_i$ to $\sum_i \alpha_i \xi_i|_{\mathfrak{h}_\Gamma(\mathbb{Z})}$ and the second is given by restricting mappings from $\mathfrak{h}(K)$ to $\mathfrak{h}_\Gamma(\mathbb{Z})$. We identify $K \otimes \mathfrak{h}^*_\Gamma(\mathbb{Z}) = \mathfrak{h}^*(K)$, hence, $\mathfrak{h}^*_\Gamma(\mathbb{Z}) \leq \mathfrak{h}^*(K)$ is a lattice and

$$\mathfrak{h}^*_\Gamma(\mathcal{O}) = \{\lambda \in \mathfrak{h}^*(K) : \lambda(\mathrm{h}_i) \in \mathcal{O} \text{ for all } i = 1, \ldots, \ell\},$$

i.e. $\mathfrak{h}^*_\Gamma(\mathcal{O})$ consists of elements $\lambda \in \mathfrak{h}^*(K)$ satisfying $\lambda(\mathfrak{h}_\Gamma(\mathbb{Z})) \subseteq \mathcal{O}$. In particular, any root $\alpha \in \Gamma = \mathfrak{h}^*_\Gamma(\mathbb{Z})$ yields an element $\alpha \in \mathfrak{h}^*(K)$ which leaves $\mathfrak{h}_\Gamma(\mathbb{Z})$ invariant and which is the K-linear extension of $\alpha|_{\mathfrak{h}_\Gamma(\mathbb{Z})}$ (α even leaves $\mathfrak{h}_{\Gamma_{\mathrm{ad}}}(\mathbb{Z})$ invariant).

2.2.4 Notation

There is a morphism of semi groups $\mathbb{N}_0^s \to \bigoplus_{i=1}^s \mathbb{N}_0\alpha_i \subseteq \Gamma_{\rm sc}$ which sends $\mathbf{n} = (n_1, \dots, n_s)$ to $\sum_{i=1}^s n_i\alpha_i$. To facilitate notation we sometimes also understand by $\mathbf{n} \in \mathbb{N}_0^s$ the corresponding element $\sum_i n_i\alpha_i \in \Gamma_{\rm sc}$; expressions like "$\lambda - \mathbf{n}$" or "$\mathbf{n} - m\alpha$" where $\mathbf{n} \in \mathbb{N}_0^s$ and $\lambda, \alpha \in \mathfrak{h}^*$ or $\in \mathfrak{h}^*(K)$ are to be understood in this sense. In particular, we define the height of $\mathbf{n} \in \mathbb{N}_0^s$ as $\mathrm{ht}(\mathbf{n}) = \mathrm{ht}(\sum_i n_i\alpha_i) = \sum_i n_i \mathrm{ht}(\alpha_i)$.

2.3 *Verma Modules*

2.3.1 Norm on Verma Modules

We let $\lambda \in \mathfrak{h}^*(K)$. We define a structure of $\mathfrak{b}(K)$-module on the 1-dimensional K vector space $K\mathrm{v}_\lambda$ by $\mathfrak{n}(K)\mathrm{v}_\lambda = 0$ and $H\mathrm{v}_\lambda = \lambda(H)\mathrm{v}_\lambda$ for all $H \in \mathfrak{h}(K)$. This yields a $\mathcal{U}_{\mathfrak{b}(K)}$-module $K\mathrm{v}_\lambda$ and we denote by

$$V_\lambda = V_\lambda(K) := \mathcal{U}_{\mathfrak{g}(K)} \otimes_{\mathcal{U}_{\mathfrak{b}(K)}} K\mathrm{v}_\lambda$$

the Verma module of highest weight λ, i.e. V_λ is a $\mathcal{U}_{\mathfrak{g}(K)} = \mathcal{U}_K$-module with $X \in \mathcal{U}_K$ acting on the first factor of a tensor. We identify $1 \otimes \mathrm{v}_\lambda$ with v_λ and obtain $V_\lambda = \mathcal{U}_{\mathfrak{g}(K)} 1 \otimes \mathrm{v}_\lambda = \mathcal{U}_{\mathfrak{g}(K)}\mathrm{v}_\lambda$, i.e. $\mathrm{v}_\lambda \in V_\lambda$ is a maximal vector in V_λ. Since $\mathcal{U}_K = \mathcal{U}_K^- \mathcal{U}_K^\circ \mathcal{U}_K^+$ we obtain as $\mathcal{U}_K^-$-modules

$$V_\lambda = \mathcal{U}_K^- \mathrm{v}_\lambda = \bigoplus_{\mathbf{n} \in \mathbb{N}_0^s} K X_-^{\mathbf{n}} \mathrm{v}_\lambda.$$

We note that $hX_-^{\mathbf{n}}\mathrm{v}_\lambda = (\lambda - \mathbf{n})(h)\mathrm{v}_\lambda$, $h \in \mathfrak{h}(K)$; here, $\alpha \in \Gamma$ is viewed as element in $\mathfrak{h}^*(K)$ (cf. Sect. 2.2.3)

We put $\Pi_\lambda^V = \{\lambda - \mathbf{n},\ \mathbf{n} \in \mathbb{N}_0^s\}$, i.e. Π_λ^V is the set of weights appearing in V_λ.

The $\mathcal{U}_K$-module V_λ yields a representation

$$\rho_\lambda^V : \mathcal{U}_K \to \mathrm{End}(V_\lambda).$$

We set $\sigma = \max_{\alpha \in \Phi^+} \mathrm{ht}(\alpha)$ and we select a rational number $\epsilon \in \mathbb{Q}$ such that

$$\frac{1}{p-1} < \epsilon < \frac{p-2}{p-1}\frac{1}{\sigma}. \tag{1}$$

In particular, from now on we must assume that

$$p > \sigma + 2. \tag{2}$$

We define a norm $|\cdot| = |\cdot|_\epsilon = |\cdot|_{\epsilon,\lambda}$ on V_λ as follows. If $v = \sum_{\mathbf{n}} a_{\mathbf{n}} X_-^{\mathbf{n}} \mathrm{v}_\lambda \in V_\lambda$, $a_{\mathbf{n}} \in K$, then we set

$$|v| = \sup_{\mathbf{n}\in\mathbb{N}_0^s} |a_{\mathbf{n}}| p^{-\epsilon \mathrm{ht}(\mathbf{n})}.$$

Remark The first of the inequalities in (1) will ensure that $\rho_\lambda^V(t_\alpha x_\alpha)$ where $\alpha \in \Phi^-$ and $t_\alpha \in \mathcal{O}$ has norm less than or equal to $\Theta := p^{-1/(p-1)}$ and the second inequality will ensure that $\rho_\lambda^V(t_\alpha x_\alpha)$ for $\alpha \in \Phi^+$ and $t_\alpha \in \mathcal{O}$ divisible by p has norm less than or equal to Θ. Hence, the exponential will converge for both these elements (cf. Sects. 2.3.2 and 2.5).

We denote by $\hat{V}_\lambda$ the completion of V_λ w.r.t. the norm $|\cdot|$, i.e. $(\hat{V}_\lambda, |\cdot|)$ is a K-Banach space. Explicitly, $\hat{V}_\lambda$ is given as follows. We let $\widehat{\bigoplus}_{\mathbf{n}\in\mathbb{N}_0^s} K X_-^{\mathbf{n}} \mathrm{v}_\lambda \leq \prod_{\mathbf{n}\in\mathbb{N}_0^s} K X_-^{\mathbf{n}} \mathrm{v}_\lambda$ be the subspace consisting of all sequences $(a_{\mathbf{n}} X_-^{\mathbf{n}} \mathrm{v}_\lambda)_{\mathbf{n}}$ satisfying the following property: for all $\delta > 0$ there are only finitely many $\mathbf{n} \in \mathbb{N}_0^s$ such that

$$|a_{\mathbf{n}} X_-^{\mathbf{n}} \mathrm{v}_\lambda| = |a_{\mathbf{n}}| p^{-\epsilon \mathrm{ht}(\mathbf{n})} > \delta. \tag{3}$$

The space $\widehat{\bigoplus}_{\mathbf{n}\in\mathbb{N}_0^s} K X_-^{\mathbf{n}} \mathrm{v}_\lambda$ then is complete w.r.t. $|\cdot|$ and is the closure of the set $\{\delta_{\mathbf{n}},\ \mathbf{n} \in \mathbb{N}_0^s\}$ where $\delta_{\mathbf{n}}$ is the sequence with value 1 at $X_-^{\mathbf{n}} \mathrm{v}_\lambda$ and value 0 else (cf. [18], 3.H Example, p. 49; we identify $K X_-^{\mathbf{n}} \mathrm{v}_\lambda$ with K). In particular, $\hat{V}_\lambda = \widehat{\bigoplus}_{\mathbf{n}\in\mathbb{N}_0^s} K X_-^{\mathbf{n}} \mathrm{v}_\lambda$ is the completion of V_λ.

After eventually enlargening K there is an element $\kappa \in K$ such that

$$v_p(\kappa) = \epsilon. \tag{4}$$

Thus, any element in $v \in V_\lambda$ can be normalized by multiplying with an element in K to have absolute value 1 and the Banach space $\hat{V}_\lambda$ is orthonormalizable and the vectors

$$\mathrm{w}_{\mathbf{n}} = \kappa^{-\mathrm{ht}(\mathbf{n})} X_-^{\mathbf{n}} \mathrm{v}_\lambda, \quad \mathbf{n} \in \mathbb{N}_0^s,$$

form an ONB in $\hat{V}_\lambda$.

We also denote by $|\cdot| = |\cdot|_\epsilon = |\cdot|_{\epsilon,\lambda}$ the norm on $\mathrm{End}(\hat{V}_\lambda)$ defined by $|\varphi| = \sup_{v\in\hat{V}_\lambda} \frac{|\varphi(v)|}{|v|}$.

2.3.2 Estimate on Norms

We set $\Theta = p^{-\frac{1}{p-1}}$. We let λ be an element in $\mathfrak{h}_\Gamma^*(\mathcal{O}) \leq \mathfrak{h}^*(K)$.

Lemma *(1) For all $\alpha \in \Phi^+$ we have*

$$|\rho_\lambda^V(x_\alpha)| \leq p^{\epsilon \mathrm{ht}(\alpha)};$$

in particular, $|\rho_\lambda^V(p x_\alpha)| < \Theta$.

(2) For all $\alpha \in \Phi^-$ we have

$$|\rho_\lambda^V(x_\alpha)| \leq p^{-\epsilon \mathrm{ht}(-\alpha)};$$

in particular, $|\rho_\lambda^V(x_\alpha)| < \Theta$.

(3) For all $H \in \mathfrak{h}_\Gamma(\mathcal{O})$ *we have*

$$|\rho_\lambda^V(H)| \leq 1;$$

in particular, $|\rho_\lambda^V(pH)| < \Theta$.

Proof (1) Let $\alpha \in \Phi^+$. Using 3.1.3 Lemma in [16] we find

$$x_\alpha X_-^{\mathbf{n}} = \sum_{\substack{\mathbf{m}_-,\mathbf{m}_+ \in \mathbb{N}_0^s \\ \gamma \in \Delta}} \zeta_{\mathbf{m}_-,\gamma,\mathbf{m}_+} X_-^{\mathbf{m}_-} h_\gamma X_+^{\mathbf{m}_+} + \sum_{\mathbf{m}_-,\mathbf{m}_+ \in \mathbb{N}_0^s} \zeta_{\mathbf{m}_-,\mathbf{m}_+} X_-^{\mathbf{m}_-} X_+^{\mathbf{m}_+},$$

where $\zeta_{\mathbf{m}_-,\gamma,\mathbf{m}_+}, \zeta_{\mathbf{m}_-,\mathbf{m}_+} \in \mathbb{Z}$. Since $X_+^{\mathbf{m}_+}$ annihilates v_λ if $\mathbf{m}_+ \neq 0$ and $h_\gamma \mathrm{v}_\lambda = \lambda(h_\gamma)\mathrm{v}_\lambda$ we obtain

$$\begin{aligned} x_\alpha X_-^{\mathbf{n}} \mathrm{v}_\lambda &= \sum_{\substack{\mathbf{m}_- \in \mathbb{N}_0^s \\ \gamma \in \Delta}} \zeta_{\mathbf{m}_-,\gamma,0} X_-^{\mathbf{m}_-} \lambda(h_\gamma)\mathrm{v}_\lambda + \sum_{\mathbf{m}_- \in \mathbb{N}_0^s} \zeta_{\mathbf{m}_-,0} X_-^{\mathbf{m}_-} \mathrm{v}_\lambda \\ &= \sum_{\mathbf{m}_- \in \mathbb{N}_0^s} \zeta_{\mathbf{m}_-} X_-^{\mathbf{m}_-} \mathrm{v}_\lambda, \end{aligned}$$

where $\zeta_{\mathbf{m}_-} = \sum_{\gamma \in \Delta} \lambda(h_\gamma)\zeta_{\mathbf{m}_-,\gamma,0} + \zeta_{\mathbf{m}_-,0} \in \mathcal{O}$ (note that $\lambda(h_\gamma) \in \mathcal{O}$ because $\lambda \in \mathfrak{h}_\Gamma^*(\mathcal{O})$). Since $x_\alpha X_-^{\mathbf{n}} \mathrm{v}_\lambda$ has weight $\lambda - (\mathbf{n} - \alpha)$ all coefficients $\zeta_{\mathbf{m}}$ with $\mathbf{m} \neq \mathbf{n} - \alpha$, i.e. $\sum_{i=1}^s m_i \alpha_i \neq \sum_{i=1}^s n_i \alpha_i - \alpha$, vanish. This yields

$$|x_\alpha X_-^{\mathbf{n}} \mathrm{v}_\lambda| = \sup\{|\zeta_{\mathbf{m}}| p^{-\epsilon \mathrm{ht}(\mathbf{m})},\ \mathbf{m} \in \mathbb{N}_0^s\} \leq \sup\{p^{-\epsilon \mathrm{ht}(\mathbf{n})} p^{\epsilon \mathrm{ht}(\alpha)},\ \mathbf{m} \in \mathbb{N}_0^s\} = p^{\epsilon \mathrm{ht}(\alpha)} p^{-\epsilon \mathrm{ht}(\mathbf{n})}.$$

If $v = \sum_{\mathbf{n} \in \mathbb{N}_0^s} c_{\mathbf{n}} X_-^{\mathbf{n}} \mathrm{v}_\lambda$, $c_{\mathbf{n}} \in K$, is an arbitrary element in V_λ then

$$\begin{aligned} |x_\alpha v| &= |\sum_{\mathbf{n} \in \mathbb{N}_0^s} c_{\mathbf{n}} x_\alpha X_-^{\mathbf{n}} \mathrm{v}_\lambda| \leq \sup_{\mathbf{n} \in \mathbb{N}_0^s} |c_{\mathbf{n}}| |x_\alpha X_-^{\mathbf{n}} \mathrm{v}_\lambda| \\ &\leq \sup_{\mathbf{n} \in \mathbb{N}_0^s} |c_{\mathbf{n}}| p^{\epsilon \mathrm{ht}(\alpha)} p^{-\epsilon \mathrm{ht}(\mathbf{n})} \\ &= p^{\epsilon \mathrm{ht}(\alpha)} |v|, \end{aligned}$$

hence,

$$|\rho_\lambda^V(x_\alpha)| \leq p^{\epsilon \mathrm{ht}(\alpha)}.$$

Since $\mathrm{ht}(\alpha) \leq \sigma$ and $\epsilon < \frac{1}{\sigma}\frac{p-2}{p-1}$ this implies $|\rho_\lambda^V(px_\alpha)| \leq p^{\epsilon \mathrm{ht}(\alpha) - 1} < p^{-\frac{1}{p-1}}$.

(2) Let $\alpha \in \Phi^-$. Since $x_\alpha, X_-^{\mathbf{n}} \in \mathcal{U}_{\mathcal{O}}^-$ we obtain

$$x_\alpha X_-^{\mathbf{n}} \mathrm{v}_\lambda = \sum_{\mathbf{m}\in\mathbb{N}_0^s} \zeta_{\mathbf{m}} X_-^{\mathbf{m}} \mathrm{v}_\lambda$$

where $\zeta_{\mathbf{m}} \in \mathcal{O}$ and $\zeta_{\mathbf{m}} = 0$ if $\mathbf{m} \neq \mathbf{n} - \alpha$, i.e. $\sum_i m_i\alpha_i \neq \sum_i n_i\alpha_i - \alpha$. This yields

$$|x_\alpha X_-^{\mathbf{n}} \mathrm{v}_\lambda| = \sup\{|\zeta_{\mathbf{m}}| p^{-\epsilon \mathrm{ht}(\mathbf{m})},\ \mathbf{m} \in \mathbb{N}_0^s\} \le p^{-\epsilon \mathrm{ht}(-\alpha)} p^{-\epsilon \mathrm{ht}(\mathbf{n})}.$$

If $v = \sum_{\mathbf{n}\in\mathbb{N}_0^s} c_{\mathbf{n}} X_-^{\mathbf{n}} \mathrm{v}_\lambda$, $c_{\mathbf{n}} \in K$, is an arbitrary element in V_λ then

$$|x_\alpha v| \le p^{-\epsilon \mathrm{ht}(-\alpha)} |v|,$$

hence,

$$|\rho_\lambda^V(x_\alpha)| \le p^{-\epsilon \mathrm{ht}(-\alpha)}.$$

Since $\mathrm{ht}(-\alpha) \ge 1$ and $\epsilon > \frac{1}{p-1}$ this yields $|\rho_\lambda^V(x_\alpha)| < p^{-\frac{1}{p-1}}$.

(3) Let $H \in \mathfrak{h}_\Gamma(\mathcal{O})$. Since $\lambda \in \mathfrak{h}_\Gamma^*(\mathcal{O})$ we obtain $\lambda(H) \in \mathcal{O}$ and since $\Delta \subset \Gamma_{\mathrm{sc}}$ we obtain $\sum_i n_i\alpha_i(h_\gamma) \in \mathbb{Z}$ for all $\gamma \in \Delta$. Hence, $HX_-^{\mathbf{n}} \mathrm{v}_\lambda = (\lambda - \mathbf{n})(H) X_-^{\mathbf{n}} \mathrm{v}_\lambda \in \mathcal{O}X_-^{\mathbf{n}} \mathrm{v}_\lambda$ and we find $|HX_-^{\mathbf{n}} \mathrm{v}_\lambda| \le |X_-^{\mathbf{n}} \mathrm{v}_\lambda|$ for all $\mathbf{n} \in \mathbb{N}_0^s$. Hence, $|\rho_\lambda^V(H)| \le 1$. Since $p > \sigma + 2 > 2$ this implies $|p\rho_\lambda^V(H)| < p^{-\frac{1}{p-1}}$. Thus, the proof of the Lemma is complete.

Remark We set

$$G_\Theta = G_\Theta^{\mathrm{End}(V_\lambda)} = \{X \in \mathrm{End}(V_\lambda) : |X| < \Theta\}.$$

2.3.2 Lemma implies that $|\rho_\lambda^V(X)| \in G_\Theta^{\mathrm{End}(V_\lambda)}$ for all elements X of the form

$$X = t_\alpha x_\alpha \text{ where } t_\alpha \in p\mathbb{Z}_p \text{ if } \alpha > 0 \text{ and } t_\alpha \in \mathbb{Z}_p \text{ if } \alpha < 0 \text{ or } X = t_i \mathrm{h}_i,\ i = 1, \ldots, \ell, \text{ where } t_i \in p\mathbb{Z}_p. \tag{5}$$

2.4 *The Normed Lie Algebra* $\mathfrak{g}(K)$

2.4.1 Notations

We select a $\mathbb{Z}$-lattice $\Gamma \le \mathfrak{h}^*$ satisfying $\Gamma_{\mathrm{ad}} \le \Gamma \le \Gamma_{\mathrm{sc}}$. As in Sect. 2.2.3 we select a $\mathbb{Z}$-basis $(\xi_i)_{i=1,\ldots,\ell}$ of Γ and we denote by $(\mathrm{h}_i)_{i=1,\ldots,\ell}$ the basis of $\mathfrak{h}$ which is dual to $(\xi_i)_i$. Thus, $\mathfrak{h}_\Gamma(\mathbb{Z}) = \bigoplus_i \mathbb{Z}\mathrm{h}_i$ is a lattice in $\mathfrak{h}(K)$.

2.4.2 Norm on $\mathfrak{g}(K)$

The elements x_α, $\alpha \in \Phi$, $\mathrm{h}_1, \ldots \mathrm{h}_\ell$ form a K-basis of $\mathfrak{g}(K)$. We define a norm $|\cdot| = |\cdot|_\Gamma$ on $\mathfrak{g}(K)$ by

$$|\sum_{\alpha\in\Phi} a_\alpha x_\alpha + \sum_{i=1}^{\ell} b_i \mathrm{h}_i| = \sup\{|a_\alpha| p^{\epsilon\mathrm{ht}(\alpha)}, \alpha \in \Phi, |b_i|, i = 1, \ldots, \ell\}$$

$(a_\alpha, b_i \in K)$. We note that $\sup_i |b_i| = \sup_{\mu\in\Gamma} |\mu(\sum_i b_i \mathrm{h}_i)|$, hence, the definition of the norm does not depend on the choice of the basis $(\xi_i)_i$.

Lemma $(\mathfrak{g}(K), |\cdot|)$ *is a complete normed Lie Algebra (cf. [6], Sect. 2.8.2, Definition 1)*

Proof We set $u_\alpha = p^{\epsilon\mathrm{ht}(\alpha)} x_\alpha$, $\alpha \in \Phi$. The elements $\{u_\alpha, \alpha \in \Phi, \mathrm{h}_i, i = 1, \ldots, \ell\}$ form an ONB of $\mathfrak{g}(K)$. It is sufficient to show that $|[X, Y]| \leq |X||Y| = 1$ for any elements X, Y of the ONB. 25.2 Theorem, p. 147 in [14] or Sect. 1.2, p. 3 in [2] show that

$$[x_\alpha, x_\beta] = \begin{cases} 0 & \alpha+\beta \notin \Phi,\ \alpha+\beta \neq 0 \\ \zeta x_{\alpha+\beta} & \alpha+\beta \in \Phi \\ \sum_{\gamma\in\Delta} \zeta_\gamma h_\gamma = \sum_i \zeta_i' \mathrm{h}_i & \alpha+\beta = 0 \end{cases}$$

where $\zeta, \zeta_\gamma \in \mathbb{Z}$ and, hence, $\zeta_i' \in \mathbb{Z}$. Hence, we obtain $|[u_\alpha, u_\beta]| \leq 1$. Since $[\mathrm{h}_i, x_\alpha] = \alpha(\mathrm{h}_i) x_\alpha$ where $\alpha(\mathrm{h}_i) \in \mathbb{Z}$ (because $\mathrm{h}_i \in \mathfrak{h}_\Gamma(\mathbb{Z})$) we obtain $|[\mathrm{h}_i, u_\alpha]| \leq 1$. Finally, since $|[\mathrm{h}_i, \mathrm{h}_j]| = 0$ the proof is complete.

Taking $K = \mathbb{Q}_p$ we see in particular that $\mathfrak{g}(\mathbb{Q}_p)$ is a normed $\mathbb{Q}_p$-Lie algebra. We set

$$G_\Theta = G_\Theta^{\mathfrak{g}(\mathbb{Q}_p)} = \{X \in \mathfrak{g}(\mathbb{Q}_p) : |X| < \Theta\}.$$

We note that $X \in G_\Theta^{\mathfrak{g}(\mathbb{Q}_p)}$ for all elements X as in Eq. (5).

2.4.3 The Representation $\hat{\rho}_\lambda$

We look at the representation $\rho_\lambda^V : \mathfrak{g}(K) \to \mathrm{End}(V_\lambda)$. The definition of the norm on $\mathfrak{g}(K)$ and 2.3.2 Lemma imply that $|\rho_\lambda^V| \leq 1$. In particular, any $\rho_\lambda^V(X)$, $X \in \mathfrak{g}(K)$, is a bounded, hence, continuous operator on V_λ and therefore extends to a continuous operator $\hat{\rho}_\lambda^V(X) = \hat{\rho}_\lambda(X)$ on $\hat{V}_\lambda$. We obtain a representation

$$\begin{aligned} \hat{\rho}_\lambda = \hat{\rho}_\lambda^V : \mathfrak{g}(K) &\to \mathrm{End}(\hat{V}_\lambda) \\ X &\mapsto \hat{\rho}_\lambda(X). \end{aligned}$$

Since $|\hat{\rho}_\lambda| \leq |\rho_\lambda^V| \leq 1$ and since all elements in Eq. (5) have norm $< \Theta$ we obtain

$$|\hat{\rho}_\lambda| \leq 1 \text{ and } |\hat{\rho}_\lambda(X)| < \Theta \text{ for all } X \text{ as in Eq. (5).} \tag{6}$$

2.4.4 The Submodule $\hat{V}_{\lambda,\mathcal{O}}$

We let $\lambda \in \mathfrak{h}_\Gamma^*(\mathcal{O})$. We define the sub $\mathcal{O}$-module

$$\hat{V}_{\lambda,\mathcal{O}} = \{v \in \hat{V}_\lambda : |v| \leq 1\} \leq \hat{V}_\lambda.$$

We set

$$\hat{V}_{\lambda,\mathcal{O}}(\mu) = \bigoplus_{\substack{\mathbf{n}\in\mathbb{N}_0^s \\ \sum_i n_i\alpha_i = \lambda-\mu}} \mathcal{O}\mathrm{w}_{\mathbf{n}}.$$

Lemma *(1)* $\hat{V}_{\lambda,\mathcal{O}} \leq V_\lambda$ *is a closed sub* $\mathcal{O}$*-module.*
(2) $\hat{V}_{\lambda,\mathcal{O}} = \widehat{\bigoplus}_{\mathbf{n}\in\mathbb{N}_0^s}\mathcal{O}\mathrm{w}_{\mathbf{n}} = \widehat{\bigoplus}_{\mu\leq\lambda}\hat{V}_{\lambda,\mathcal{O}}(\mu)$.
(3) $\hat{V}_{\lambda,\mathcal{O}}$ *is invariant under the elements* $\hat{\rho}_\lambda(X)$ *where* X *runs through all elements as in Eq. (5).*

Proof (1) obvious. (2) Since $|\mathrm{w}_{\mathbf{n}}| = 1$ we know that $\mathrm{w}_{\mathbf{n}} \in \hat{V}_{\lambda,\mathcal{O}}$. To prove the reverse inclusion let $v \in \hat{V}_{\lambda,\mathcal{O}}$. Since $(\mathrm{w}_{\mathbf{n}})_{\mathbf{n}}$ is an ONB we can write $v = \sum_{\mathbf{n}} a_{\mathbf{n}}\mathrm{w}_{\mathbf{n}}$, $a_{\mathbf{n}} \in K$, and since $|v| = 1$ this implies $|a_{\mathbf{n}}| \leq 1$, hence, $a_{\mathbf{n}} \in \mathcal{O}$ for all $\mathbf{n} \in \mathbb{N}_0^s$. The second equality is immediate by the definition of $\hat{V}_{\lambda,\mathcal{O}}$. (3) Since $|\hat{\rho}_\lambda(X)| < \Theta \leq 1$ for all elements X as in Eq. (5) (cf. Eq. (6)) we obtain that any $\hat{\rho}_\lambda^V(X)$, X as in Eq. (5), leaves $\hat{V}_{\lambda,\mathcal{O}}$ invariant.

2.4.5 The Subalgebra $\mathfrak{i}$

We denote by $\mathfrak{i} = \mathfrak{i}(\Gamma) \leq \mathfrak{g}(\mathbb{Q}_p)$ the sub $\mathbb{Z}_p$-Lie algebra which is generated as Lie algebra by the elements $t_\alpha x_\alpha$, where $\alpha \in \Phi$ and $t_\alpha \in \mathbb{Z}_p$ if $\alpha < 0$ and $t_\alpha \in p\mathbb{Z}_p$ if $\alpha > 0$ and the elements pH where $H \in \mathfrak{h}_\Gamma(\mathbb{Z}_p)$. The results obtained so far can be reformulated in terms of the $\mathbb{Z}_p$-Lie algebra $\mathfrak{i}$. In a strict sense we will not need this but since it is very convenient to do so we record the result.

Lemma *(1) As* $\mathbb{Z}_p$*-module* $\mathfrak{i}$ *is the direct sum*

$$\mathfrak{i} = \bigoplus_{\alpha\in\Phi^-} \mathbb{Z}_p x_\alpha \oplus \bigoplus_{\alpha\in\Phi^+} \mathbb{Z}_p p x_\alpha \oplus p\mathfrak{h}_\Gamma(\mathbb{Z}_p).$$

In particular, the elements X *as in Eq. (5) are a* K*-basis of* $\mathfrak{i}$.

(2) The norm of any $X \in \mathfrak{i}$ *is bounded by* $|X| < \Theta$.
(3) For any $X \in \mathfrak{i}$ *the norm of* $\hat{\rho}_\lambda^V(X)$ *is bounded by* $|\hat{\rho}_\lambda^V(X)| < \Theta$.
(4) $\mathfrak{i} = G_\Theta^{\mathfrak{g}(\mathbb{Q}_p)}$.
(5) $\hat{V}_{\lambda,\mathcal{O}}$ *is invariant under* $\mathfrak{i}$ *(acting via* $\hat{\rho}_\lambda$*).*

Proof (1) "$\subseteq$": We first show that the right hand side of the equality in the Lemma is a sub $\mathbb{Z}_p$-Lie algebra of $\mathfrak{g}(\mathbb{Q}_p)$. Since $\{x_\alpha,\ \alpha \in \Phi,\ h_\beta, \beta \in \Delta\}$ is a Chevalley basis (cf. e.g. [2], Sect. 1.2, p. 3) we know that $[x_\alpha, x_\beta] \in \mathbb{Z}x_{\alpha+\beta}$ if $\alpha + \beta \neq 0$ is a root and $[x_\alpha, x_\beta] \in \mathfrak{h}(\mathbb{Z}) \subseteq \mathfrak{h}_\Gamma(\mathbb{Z})$ if $\alpha = -\beta$. Hence, the bracket of any two elements $t_\alpha x_\alpha$, $t_\beta x_\beta$ where t_α is contained in $\mathbb{Z}_p$ if $\alpha < 0$ and contained in $p\mathbb{Z}_p$ if

$\alpha > 0$ and analogously for t_β again is contained in $\mathfrak{i}$. We let $H \in \mathfrak{h}_\Gamma(\mathbb{Z}_p)$, hence, $\beta(H) \in \mathbb{Z}_p$ for all $\beta \in \Phi$. We obtain $[pH, x_\beta] = p\beta(H)x_\beta \in p\mathbb{Z}_p x_\beta \subseteq \mathfrak{i}$. Since the bracket of any two elements in $\mathfrak{h}_\Gamma(\mathbb{Z}_p)$ vanishes we see that $\mathfrak{i}$ is closed under the bracket. Since the right hand side contains all generators of $\mathfrak{i}$ the inclusion follows. Since the inclusion "$\supseteq$" is obvious the proof is complete.

(2) Any $X \in \mathfrak{i}$ is a $\mathbb{Z}_p$- linear combination of the element x_α with $\alpha < 0$, px_α with $\alpha > 0$ and $p\mathrm{h}_i$, $i = 1, \ldots, \ell$, all of which have norm less than Θ.

(3) Follows from (2) because $|\hat\rho_\lambda^V| \le 1$.

(4) "$\supseteq$": We let $X = \sum_{\alpha\in\Phi} c_\alpha x_\alpha + \sum_i c_i \mathrm{h}_i \in \mathfrak{g}(\mathbb{Q}_p)$ with $|X| < \Theta$ ($c_\alpha, c_i \in \mathbb{Q}_p$). The definition of the norm then implies $|c_i| < \Theta$ for all i and since $\Theta < 1$ this implies $c_i \in p\mathbb{Z}_p$. Moreover, for any $\alpha \in \Phi$ we obtain $p^{\epsilon \mathrm{ht}(\alpha)}|c_\alpha| < \Theta$. We set $v_\alpha = v_p(c_\alpha)$ and obtain $v_\alpha > \epsilon\mathrm{ht}(\alpha) + \frac{1}{p-1}$. We distinguish cases. Case 1: $\alpha > 0$. Since $\epsilon > \frac{1}{p-1}$ we find $\epsilon\mathrm{ht}(\alpha) + \frac{1}{p-1} > \frac{1}{p-1} + \frac{1}{p-1} > 0$, hence, $v_\alpha > 0$ and $c_\alpha \in p\mathbb{Z}_p$. Case 2: $\alpha < 0$. Since $\mathrm{ht}(\alpha) \ge -\sigma$ and $\epsilon < \frac{1}{\sigma}\frac{p-2}{p-1}$ we find $\epsilon\mathrm{ht}(\alpha) + \frac{1}{p-1} \ge \epsilon(-\sigma) + \frac{1}{p-1} > -\sigma\frac{1}{\sigma}\frac{p-2}{p-1} + \frac{1}{p-1} = -\frac{p-3}{p-1} > -1$. Thus, $c_\alpha \in \mathbb{Z}_p$ and (1) implies that $X \in \mathfrak{i}$. "$\subseteq$": follows from (2)

(5) Since $|\hat\rho_\lambda(X)| < \Theta \le 1$ for all $X \in \mathfrak{i}$ by (3) we obtain that any $\hat\rho_\lambda(X)$, $X \in \mathfrak{i}$, leaves $\hat V_{\lambda,\mathcal{O}}$ invariant. Thus, the proof of the Lemma is complete.

2.5 Exponential Mapping

We set

$$G_\Theta = G_\Theta^{\mathrm{End}(\hat V_\lambda)} = \{X \in \mathrm{End}(\hat V_\lambda) : |X| < \Theta\}.$$

Reference [6], Sect. 2.8.4 Proposition 4, p. 175 implies that the exponential defines an analytic isomorphism

$$\begin{array}{cccc} & \mathrm{End}(\hat V_\lambda) & & \mathrm{Aut}(\hat V_\lambda) \\ & \cup & & \cup \\ \exp: & G_\Theta & \to & 1 + G_\Theta \\ & X & \mapsto & \sum_{i\ge 0}\frac{X^i}{i!}. \end{array} \tag{7}$$

The associative algebra $\mathrm{End}(\hat V_\lambda)$ with the bracket $[X, Y] = XY - YX$ is a normed Lie-Algebra. We denote by $h : G_\Theta \times G_\Theta \to \mathrm{End}(\hat V_\lambda)$ the Hausdorff function, i.e. h is represented by the Hausdorff series $H(X, Y)$ which converges on G_Θ (cf. [6], Sect. 2.8.3 Proposition 2, p. 173) and which satisfies

$$\exp(X)\exp(Y) = \exp(h(X, Y))$$

for all $X, Y \in G_\Theta$ (cf. [6], Sect. 2.8.4 Proposition 5, p. 176). Since $\mathfrak{g}(K)$ is a normed Lie Algebra and $\hat{\rho}^V_\lambda$ is a Lie Algebra morphism with $|\hat{\rho}^V_\lambda| \leq 1$, i.e. $\hat{\rho}^V_\lambda(G_\Theta^{\mathfrak{g}(K)}) \subseteq G_\Theta^{\mathrm{End}(\hat{V}_\lambda)}$ and since H is a sum of iterated Lie brackets in X and Y we find $h(\hat{\rho}^V_\lambda(X), \hat{\rho}^V_\lambda(Y)) = \hat{\rho}^V_\lambda(h(X, Y))$ for all $X, Y \in G_\Theta^{\mathfrak{g}(K)}$. Hence,

$$\exp(\hat{\rho}^V_\lambda(X)) \exp(\hat{\rho}^V_\lambda(Y)) = \exp(h(\hat{\rho}^V_\lambda(X), \hat{\rho}^V_\lambda(Y))) = \exp(\hat{\rho}^V_\lambda(h(X, Y))) \qquad (8)$$

for all $X, Y \in G_\Theta^{\mathfrak{g}(K)}$. In particular, this holds for all $X, Y \in \mathfrak{i} \subseteq G_\Theta^{\mathfrak{g}(K)}$.

2.6 *Finite Dimensional Representations*

2.6.1 Norm on Finite Dimensional Representations

We let (ρ_0, L_{ρ_0}) be a faithful complex finite dimensional representation of $\mathfrak{g}$. We write $\rho_0 = \bigoplus_i \rho_{\lambda_i}$ where $(\rho_{\lambda_i}, L_{\lambda_i})$ is the irreducible $\mathfrak{g}$-module of highest weight $\lambda_i \in \Gamma_{\mathrm{sc}}$. We set $L_{\lambda_i}(\mathbb{Z}) = \mathcal{U}_\mathbb{Z} v_{\lambda_i} = \sum_\mathbf{n} \mathbb{Z} X_-^\mathbf{n} v_{\lambda_i}$ (almost all summands vanish) where $v_{\lambda_i} \in L_{\lambda_i}$ is a highest weight vector. Thus, $L_{\lambda_i}(\mathbb{Z})$ is an admissible, i.e. $\mathcal{U}_\mathbb{Z}$-invariant lattice in L_{λ_i} and we set $L_{\lambda_i}(R) = L_{\lambda_i}(\mathbb{Z}) \otimes R$, R any $\mathbb{Z}$-algebra. We obtain a representation $\rho_{\lambda_i} : \mathcal{U}_{\mathbb{Q}_p} \to \mathrm{End}(L_{\lambda_i}(\mathbb{Q}_p))$ and, hence, a semi simple representation

$$\rho_0 : \mathcal{U}_{\mathbb{Q}_p} \to \mathrm{End}(L_{\rho_0}(\mathbb{Q}_p))$$

where $L_{\rho_0}(\mathbb{Q}_p) = \bigoplus_i L_{\lambda_i}(\mathbb{Q}_p)$ and a corresponding Lie Algebra representation of $\mathfrak{g}(\mathbb{Q}_p)$.

Depending on ρ_0 we introduce the following notation

- $\Gamma = \Gamma_{\rho_0} = \bigcup_i \Gamma_{\rho_{\lambda_i}}$ $(\subseteq \Gamma_{\mathrm{sc}})$ is the weight lattice of ρ_0
- $\mathfrak{i} = \mathfrak{i}(\Gamma)$ $(\supseteq p\mathfrak{h}_\Gamma(\mathbb{Z}_p))$ (cf. Sect. 2.4.5).

We let $\lambda \in \Gamma = \Gamma_{\rho_0}$ be an arbitrary dominant weight and we denote by Π_λ the set of weights of L_{ρ_λ}. We select a (finite) subset $M_\lambda \subseteq \mathbb{N}_0^s$ such that $L_\lambda(\mathbb{Z}_p) = \bigoplus_{\mathbf{n} \in M_\lambda} \mathbb{Z}_p X_-^\mathbf{n} v_\lambda$, hence,

$$L_\lambda(\mathbb{Q}_p) = \bigoplus_{\mathbf{n} \in M_\lambda} \mathbb{Q}_p X_-^\mathbf{n} v_\lambda = \bigoplus_{\mu \in \Pi_\lambda} \bigoplus_{\substack{\mathbf{n} \in M_\lambda \\ \lambda - \sum_i n_i \alpha_i = \mu}} \mathbb{Q}_p X_-^\mathbf{n} v_\lambda.$$

Let $v = \sum_{\mathbf{n} \in M_\lambda} c_\mathbf{n} X_-^\mathbf{n} v_\lambda \in L_\lambda(\mathbb{Q}_p)$ $(c_\mathbf{n} \in \mathbb{Q}_p)$; we define a norm on $L_\lambda(\mathbb{Q}_p)$ by setting

$$|v| = \sup_{\mathbf{n} \in M_\lambda} |c_\mathbf{n}| p^{-\epsilon \mathrm{ht}(\mathbf{n})} = \sup_{\mu \in \Pi_\lambda} p^{-\epsilon \mathrm{ht}_\lambda(\mu)} \sup_{\substack{\mathbf{n} \in M_\lambda \\ \lambda - \sum_i n_i \alpha_i = \mu}} |c_\mathbf{n}|. \qquad (9)$$

Let $L_\lambda(\mathbb{Z}_p) = \bigoplus_{\mathbf{n} \in N_\lambda} \mathbb{Z}_p X_-^\mathbf{n} v_\lambda$ for some finite subset $N_\lambda \subseteq \mathbb{N}_0^s$; then $\{X_-^\mathbf{n} v_\lambda, \mathbf{n} \in M_\lambda, \lambda - \sum_i n_i \alpha_i = \mu\}$ and $\{X_-^\mathbf{n} v_\lambda, \mathbf{n} \in N_\lambda, \lambda - \sum_i n_i \alpha_i = \mu\}$ are $\mathbb{Z}_p$-basis of

$\bigoplus_{\substack{\mathbf{n}\in M_\lambda \\ \lambda-\sum_i n_i\alpha_i=\mu}} \mathbb{Z}_p X_-^{\mathbf{n}} v_\lambda = \bigoplus_{\substack{\mathbf{n}\in N_\lambda \\ \lambda-\sum_i n_i\alpha_i=\mu}} \mathbb{Z}_p X_-^{\mathbf{n}} v_\lambda$. Thus, the coefficients of any element in $\bigoplus_{\substack{\mathbf{n}\in M_\lambda \\ \lambda-\sum_i n_i\alpha_i=\mu}} \mathbb{Q}_p X_-^{\mathbf{n}} v_\lambda$ w.r.t. these two basis are related by a matrix in $\mathbf{GL}_n(\mathbb{Z}_p)$ where $n = \dim \bigoplus_{\substack{\mathbf{n}\in M_\lambda \\ \lambda-\sum_i n_i\alpha_i=\mu}} \mathbb{Q}_p X_-^{\mathbf{n}} v_\lambda$, hence, Eq. (9) implies $\sup_{\mathbf{n}\in M_\lambda} |c_{\mathbf{n}}| p^{-\epsilon \mathrm{ht}(\mathbf{n})} = \sup_{\mathbf{n}\in N_\lambda} |c_{\mathbf{n}}| p^{-\epsilon \mathrm{ht}(\mathbf{n})}$, i.e. $|v|$ does not depend on the choice of the subset $M_\lambda \subseteq \mathbb{N}_0^s$.

We let $\alpha \in \Phi$. Since $x_\alpha X_-^{\mathbf{n}} v_\lambda = \sum_{\mathbf{m}\in M_\lambda} d_{\mathbf{m}} X_-^{\mathbf{m}} v_\lambda$ where $d_{\mathbf{m}} \in \mathbb{Z}_p$ and $\mathbf{m} \in M_\lambda$ runs over elements satisfying $\lambda - \mathbf{m} = \lambda - \mathbf{n} + \alpha$ we obtain $|x_\alpha X_-^{\mathbf{n}} v_\lambda| \le p^{\epsilon \mathrm{ht}(\alpha)} p^{-\epsilon \mathrm{ht}(\mathbf{n})}$, hence, for any $v = \sum_{\mathbf{n}\in M_\lambda} c_{\mathbf{n}} X_-^{\mathbf{n}} v_\lambda \in L_\lambda(\mathbb{Q}_p)$, $c_{\mathbf{n}} \in \mathbb{Q}_p$, we find

$$|\rho_\lambda(x_\alpha)(v)| \le \sup_{\mathbf{n}\in M_\lambda} |c_{\mathbf{n}}| p^{\epsilon \mathrm{ht}(\alpha)} p^{-\epsilon \mathrm{ht}(\mathbf{n})} = p^{\epsilon \mathrm{ht}(\alpha)} |v|.$$

Similarly, for $H \in \mathfrak{h}_\Gamma(\mathbb{Z}_p)$ we find $|Hv| \le |v|$ (note that $\lambda - \mathbf{n} \in \Gamma$, hence, $(\lambda - \mathbf{n})(H) \in \mathbb{Z}_p$). Thus,

$$|\rho_\lambda(x_\alpha)| \le p^{\epsilon \mathrm{ht}(\alpha)} \ (\alpha \in \Phi) \quad \text{and} \quad |\rho_\lambda(H)| \le 1 \ (H \in \mathfrak{h}_\Gamma(\mathbb{Z}_p))$$

and as in Sects. 2.3.2 and 2.4.3 we deduce

$$|\rho_\lambda(X)| < \Theta \ \text{ for all } X \text{ as in Eq. (5)} \quad \text{and} \quad |\rho_\lambda| \le 1$$

(the norm on $\mathfrak{g}(\mathbb{Q}_p)$ has been defined in Sect. 2.4.2).

We return to the situation from the beginning of Sect. 2.6 and we let $v = \sum_i v_i \in L_{\rho_0}(\mathbb{Q}_p)$, $v_i \in L_{\lambda_i}(\mathbb{Q}_p)$; we define a norm on $L_{\rho_0}(\mathbb{Q}_p)$ by

$$|v| = \sup_i |v_i|.$$

Since $\lambda_i \in \Gamma$ for all i we obtain from the above

$$|\rho_0(X)| < \Theta \ \text{ for all } X \text{ as in Eq. (5)} \quad \text{and} \quad |\rho_0| \le 1.$$

Equivalently, the first of the above inequalities holds for all $X \in \mathfrak{i}$.

2.6.2 Exponential Mapping

We set $G_\Theta = G_\Theta^{\mathrm{End}(L_{\rho_0}(\mathbb{Q}_p))} = \{X \in \mathrm{End}(L_{\rho_0}(\mathbb{Q}_p)) : |X| < \Theta\}$. Reference [6], II.8.4 Proposition 4, p. 175 implies that the exponential

$$\begin{array}{cccc} & \mathrm{End}(L_{\rho_0}(\mathbb{Q}_p)) & & \mathrm{Aut}(L_{\rho_0}(\mathbb{Q}_p)) \\ & \cup & & \cup \\ \exp: & G_\Theta & \overset{\exp}{\to} & 1 + G_\Theta \end{array} \tag{10}$$

defines an analytic isomorphism.

3 Chevalley Groups and p-Adic Verma Modules

3.1 Semi Simple Algebraic Groups and Chevalley Groups

We let $\mathbf{G}/\mathbb{Q}$ be a connected semi simple algebraic group defined over $\mathbb{Q}$ which contains a maximal torus $\mathbf{T}/\mathbb{Q}$ which is split over $\mathbb{Q}_p$. Hence, $\mathbf{G}$ is $\mathbb{Q}_p$-split and there is a Chevalley group scheme $\mathbf{G}_{\rho_0}/\mathbb{Z}_p$ attached to the faithful finite dimensional complex representation $\rho_0 : \mathfrak{g} \to \mathrm{End}(L_{\rho_0})$ of the finite dimensional complex semi simple Lie algebra $\mathfrak{g}$ such that $\mathbf{G}/\mathbb{Q}_p = \mathbf{G}_{\rho_0}/\mathbb{Q}_p$.

We recall the definition and some basic properties of the Chevalley group scheme $\mathbf{G}_{\rho_0}/\mathbb{Z}_p$. We use the notations introduced in Sect. 1 for the Lie algebra $\mathfrak{g}$ and the representation ρ_0 appearing in the definition of $\mathbf{G}_{\rho_0}$. E.g. $\mathfrak{h} \leq \mathfrak{g}$ is a Cartan subalgebra, $\Phi = \Phi(\mathfrak{g}, \mathfrak{h})$ is the root system of $\mathfrak{g}$ w.r.t. $\mathfrak{h}$, Δ is a basis of Φ and $\mathfrak{g}(\mathbb{Z})$ is a $\mathbb{Z}$-form of $\mathfrak{g}$ determined by a *Chevalley basis* $\{x_\alpha,\ \alpha \in \Phi,\ h_\alpha,\ \alpha \in \Delta\}$ of $\mathfrak{g}$. Moreover,

- $L_{\rho_0}(\mathbb{Z})$ is a $\mathcal{U}_{\mathbb{Z}}$-invariant lattice in L_{ρ_0} and we set $L_{\rho_0}(R) = L_{\rho_0}(\mathbb{Z}) \otimes R$ for a $\mathbb{Z}$-algebra R; hence, $L_{\rho_0}(R)$ is a $\mathcal{U}_R$ and a $\mathfrak{g}(R)$-module
- $\Gamma = \Gamma_{\rho_0}$ $(\subseteq \Gamma_{\mathrm{sc}})$ is the weight lattice of ρ_0
- $\mathfrak{i} = \mathfrak{i}(\Gamma) \leq \mathfrak{g}(\mathbb{Q}_p)$ (cf. Sect. 2.4.5)
- $(\xi_i)_i$ is a basis of Γ (cf. Sect. 2.2.3)
- $(\mathrm{h}_i)_i$ is the basis of $\mathfrak{h}_\Gamma(\mathbb{Z})$ dual to $(\xi_i)_i$.

We identify $L_{\rho_0}(\mathbb{Z}) = \mathbb{Z}^n$ via a basis of $L_{\rho_0}(\mathbb{Z})$. Let T_α be an indeterminate. Since $\frac{\rho_0(x_\alpha)^i}{i!} \in \mathrm{End}(L_{\rho_0}(\mathbb{Z})) = M_n(\mathbb{Z})$ is nilpotent the series

$$\exp(T_\alpha \rho_0(x_\alpha)) = \sum_{i \geq 0} T_\alpha^i \frac{\rho_0(x_\alpha)^i}{i!} \in \mathrm{Aut}(L_{\rho_0}(\mathbb{Z}[T_\alpha])) = \mathbf{GL}_n(\mathbb{Z}[T_\alpha])$$

converges. Let R be a $\mathbb{Z}$-algebra and $t_\alpha \in R$. Specializing $T_\alpha \mapsto t_\alpha$ we obtain elements

$$x_\alpha(t_\alpha) = x_\alpha^{\rho_0}(t_\alpha) \in \mathrm{Aut}(L_{\rho_0}(R)) = \mathbf{GL}_n(R).$$

If R is an integral domain of characteristic 0, i.e. $\mathbb{Z} \subseteq R$, hence, $M_n(\mathbb{Z}) \leq M_n(R) \leq M_n(\mathrm{Quot}(R))$ is a subring and $i!$ is invertible in the field $\mathrm{Quot}(R)$ then we find

$$x_\alpha(t_\alpha) = \sum_{i \geq 0} t_\alpha^i \frac{\rho(x_\alpha)^i}{i!} = \sum_{i \geq 0} \frac{\rho(t_\alpha x_\alpha)^i}{i!}.$$

We denote by

$$G_{\rho_0,R} \leq \mathrm{Aut}(L_{\rho_0}(R))$$

the subgroup generated (as an abstract group) by the elements $x_\alpha(t_\alpha)$ where $\alpha \in \Phi$ and $t_\alpha \in R$.

From now on we will work over $\mathbb{Z}_p$. We let R be a $\mathbb{Z}_p$-algebra. We select a basis $\mathcal{B}$ of $L_{\rho_0}(\mathbb{Z}_p)$ and obtain an embedding

$$G_{\rho_0,R} \to \mathbf{GL}_n(R) = \mathrm{Aut}(L_{\rho_0}(R))$$

where $n = \dim L_{\rho_0}(\mathbb{Z})$. The group $G_{\rho_0,\bar{\mathbb{Q}}_p}$ then is the group of $\bar{\mathbb{Q}}_p$-points of a closed subgroup $\mathbf{G}_{\rho_0}/\mathbb{Q}_p \subseteq \mathbf{GL}_n/\mathbb{Q}_p$ which is defined over $\mathbb{Q}_p$ (and even defined over $\mathbb{Q}$; cf. [2], Sect. 3.3. (3), p. 15; cf. also [9], Sect. 3, p. 223) and which by our assumption equals $\mathbf{G}/\mathbb{Q}_p$. Thus, we will write $\mathbf{G}$ for $\mathbf{G}_{\rho_0}$. The canonical $\mathbb{Z}_p$-structure on $\mathbf{GL}_n$ induces a $\mathbb{Z}_p$-structure $\mathbf{G}/\mathbb{Z}_p = \mathbf{G}_{\rho_0}/\mathbb{Z}_p$ on $\mathbf{G}/\mathbb{Q}_p$ (the canonical $\mathbb{Z}$-structure on $\mathbf{GL}_n$ even induces a $\mathbb{Z}$-structure $\mathbf{G}/\mathbb{Z} = \mathbf{G}_{\rho_0}/\mathbb{Z}$ on $\mathbf{G}/\mathbb{Q}$) and on closed subgroups $\mathbf{H}$ of $\mathbf{G}/\mathbb{Q}_p$ (cf. [2], 3.4, p. 18; [9], Sect. 4, p. 224), i.e.

$$\mathbf{H}(\mathbb{Z}_p) = \mathbf{H}(\bar{\mathbb{Q}}_p) \cap \mathbf{GL}_n(\mathbb{Z}_p) \tag{11}$$

for any closed subgroup $\mathbf{H}/\mathbb{Q}_p$ of $\mathbf{G}/\mathbb{Q}_p$. In particular, $\mathbf{G}(\mathbb{Z}_p) \subseteq \mathrm{Aut}(L_{\rho_0}(\mathbb{Z}_p))$ and $G_{\rho_0,\mathbb{Z}_p} \subseteq \mathrm{Aut}(L_{\rho_0}(\mathbb{Z}_p)) \cap \mathbf{G}(\bar{\mathbb{Q}}_p) = \mathbf{G}(\mathbb{Z}_p)$.

For any $\alpha \in \Phi$ and $t \in R^*$ we define the elements $h_\alpha(t) = h_\alpha^{\rho_0}(t) \in \mathbf{G}(R)$ as $h_\alpha(t) = \mu_\alpha(\begin{pmatrix} t & \\ & t^{-1}\end{pmatrix})$ where $\mu_\alpha = \mu_\alpha^{\rho_0} : \mathbf{SL}_2/\mathbb{Z}_p \to \mathbf{G}/\mathbb{Z}_p$ is the morphism defined in [2], 3.2 (1), p. 13 and 3.4, p. 18. There is a unique $\mathbb{Q}_p$-split (it is even $\mathbb{Q}$-split) torus $\mathbf{T}/\mathbb{Q}_p$ in $\mathbf{G}/\mathbb{Q}_p$ such that $\mathbf{T}(\bar{\mathbb{Q}}_p)$ is generated by the elements $h_\alpha(t_\alpha)$ with $\alpha \in \Delta$ and $t_\alpha \in \bar{\mathbb{Q}}_p^*$ (cf. [2], Sect. 3.3 (3), p. 15). For any weight μ contained in the weight lattice Γ_{ρ_0} of ρ_0 we define a character

$$\mu^\circ : \mathbf{T}/\mathbb{Q}_p \to \mathbb{G}_m/\mathbb{Q}_p \tag{12}$$

by $\prod_{\alpha\in\Delta} h_\alpha(t_\alpha) \mapsto \prod_{\alpha\in\Delta} t_\alpha^{\langle\mu,h_\alpha\rangle}$, $t_\alpha \in \bar{\mathbb{Q}}_p$, which is defined over $\mathbb{Q}_p$ (and even over $\mathbb{Q}$; cf. [2], Sect. 3.3 (3), p. 15). Moreover, the coordinate algebra of $\mathbf{T}/\mathbb{Z}$ is generated by all rational characters of $\mathbf{T}/\mathbb{Q}$, i.e. $\mathbb{Z}[\mathbf{T}/\mathbb{Z}] = \mathbb{Z}[\chi,\ \chi \in X(\mathbf{T}/\mathbb{Q})]$ (cf. [9], Proposition 1, p. 226; "$\mathbf{T}/\mathbb{Z} = \mathcal{H}_M$" in [9]), hence, any μ°, $\mu \in \Gamma_{\rho_0}$, is defined over $\mathbb{Z}$, hence, over $\mathbb{Z}_p$. The assignment

$$\Gamma_{\rho_0} \to X(\mathbf{T}/\mathbb{Q}_p), \quad \mu \mapsto \mu^\circ \tag{13}$$

is an isomorphism of groups (cf. [2], 3.3 (3), p. 15). In particular, any root $\alpha \in \Phi$ yields a character $\alpha^\circ \in X(\mathbf{T}/\mathbb{Q}_p)$. In the following we will write more simply α instead of α°. We define $\mathbf{T}(\mathbb{Q}_p)^+$ resp. $\mathbf{T}(\mathbb{Q}_p)^{++}$ as the semigroup of all $t \in \mathbf{T}(\mathbb{Q}_p)$ such that $v_p(\alpha(t)) \geq 0$ resp. $v_p(\alpha(t)) \geq 1$ for all $\alpha \in \Delta$. Analogously, we define $\mathbf{T}(\mathbb{Q}_p)^-$ resp. $\mathbf{T}(\mathbb{Q}_p)^{--}$ as the semigroup of all $t \in \mathbf{T}(\mathbb{Q}_p)$ such that $v_p(\alpha(t)) \leq 0$ resp. $v_p(\alpha(t)) \leq 1$ for all $\alpha \in \Delta$. We note that the assignment $t \mapsto h_\alpha(t)$ is the coroot in $X_*(\mathbf{T}/\mathbb{Q}_p)$ corresponding to the root α°.

We denote by $\mathbf{B}/\mathbb{Z}_p$ resp. $\mathbf{B}^-/\mathbb{Z}_p$ the Borel subgroup of $\mathbf{G}/\mathbb{Z}_p$ corresponding to the set Φ^+ resp. Φ^- of positive resp. negative roots and $\mathbf{N}/\mathbb{Z}_p$ resp. $\mathbf{N}^-/\mathbb{Z}_p$ is the unipotent radical of $\mathbf{B}/\mathbb{Z}_p$ resp. $\mathbf{B}^-/\mathbb{Z}_p$.

We denote by $(t \mapsto \mathrm{h}_i(t))_i$ the $\mathbb{Z}$-basis of $X_*(\mathbf{T}/\mathbb{Q}_p)$ which is dual to the basis $(\xi_i^\circ)_i$ of $X(\mathbf{T}/\mathbb{Q}_p)$ under the perfect pairing $X(\mathbf{T}/\mathbb{Q}_p) \times X_*(\mathbf{T}/\mathbb{Q}_p) \to \mathbb{Z}$. In particular, for any $\mu = \sum_i z_i \xi_i \in \Gamma$, $z_i \in \mathbb{Z}$, we find

$$\mu^\circ(\mathrm{h}_j(t)) = \prod_i (\xi_i^\circ)^{z_i}(\mathrm{h}_j(t)) = t^{z_j} = t^{\mu(\mathrm{h}_j)}. \tag{14}$$

For $t' \in p\mathbb{Z}_p$ and $t := \exp t' \in 1 + p\mathbb{Z}_p$ and any $v_\mu \in L_{\rho_0}(\mathbb{Q}_p)$ of weight $\mu\,(\in \Gamma)$ we obtain

$$\exp(t'\rho_0(\mathrm{h}_i))v_\mu = \sum_k (t')^k \frac{\rho_0(\mathrm{h}_i)^k(v_\mu)}{k!} = \exp(t'\mu(\mathrm{h}_i))v_\mu = t^{\mu(\mathrm{h}_i)}v_\mu$$

(note that $|\rho_0(t'\mathrm{h}_i)| < \Theta$ and $\mu(\mathrm{h}_i) \in \mathbb{Z}$). On the other hand, $\mathrm{h}_i(t)v_\mu = \mu^\circ(\mathrm{h}_i(t))\, v_\mu = t^{\mu(\mathrm{h}_i)}v_\mu$ (cf. [2], Sect. 3.2 (1), p. 13) we deduce

$$\exp(t'\rho_0(\mathrm{h}_i)) = \mathrm{h}_i(\exp t') \quad (\text{ in Aut } (L_{\rho_0}(\mathbb{Q}_p))). \tag{15}$$

For any dominant weight $\lambda \in \Gamma_{\rho_0}$ we define a representation

$$\begin{aligned} \pi_{\lambda^\circ} : \mathbf{G}(\bar{\mathbb{Q}}_p) &\to \mathrm{Aut}(L_\lambda(\bar{\mathbb{Q}}_p)) \\ x_\alpha^{\rho_0}(t_\alpha) &\mapsto \quad x_\alpha^{\rho_\lambda}(t_\alpha) \end{aligned}$$

which is defined over $\mathbb{Z}_p$ (cf. [2], Sect. 3.2 (4), p. 14 and Sect. 3.5, p. 19).

Mod p-reduction. We denote by $\mathbf{G}_{(p)}$ or by $\mathbf{G}/\mathbb{F}_p$ the mod p-reduction of $\mathbf{G}/\mathbb{Z}_p$, i.e. $\mathbf{G}_{(p)}(\bar{\mathbb{F}}_p)$ is the set of zeros of $\mathbb{F}_p \otimes J \leq \mathbb{F}_p \otimes \mathbb{Z}_p[\mathbf{GL}_n]$ in $\mathbf{GL}_n(\bar{\mathbb{F}}_p)$ where $J \leq \mathbb{Z}_p[\mathbf{GL}_n]$ is the ideal of functions vanishing on $\mathbf{G}$. Hence, $\mathbf{G}_{(p)}$ is an affine variety defined over $\mathbb{F}_p$ (a closed subgroup of $\mathbf{GL}_n/\mathbb{F}_p$). Moreover, since $\mathbf{G}/\mathbb{Z}_p$ has good reduction (cf. [2], Sect. 3.4, p. 18) we know that $\mathbb{F}_p \otimes J$ equals the ideal consisting of all $f \in \mathbb{F}_p \otimes \mathbb{Z}_p[\mathbf{GL}_n]$ which vanish on $\mathbf{G}_{(p)}(\bar{\mathbb{F}}_p)$, hence, $\mathbb{F}_p[\mathbf{G}_{(p)}] = \mathbb{F}_p \otimes \mathbb{Z}_p[\mathbf{G}]$. The mod p-reduction $\mathbf{G}_{(p)}$ is a semi simple group defined over $\mathbb{F}_p$; cf. [2], 4.3, p. 21/22). Analogously, the mod p-reductions $\mathbf{N}_{(p)}, \mathbf{B}_{(p)}, \mathbf{N}^-_{(p)}, \mathbf{B}^-_{(p)}, \ldots$ are defined.

We denote by

$$\wp : \mathrm{Aut}(L_{\rho_0}(\mathbb{Z}_p)) \stackrel{\mathcal{B}}{=} \mathbf{GL}_n(\mathbb{Z}_p) \to \mathbf{GL}_n(\mathbb{F}_p) \stackrel{\bar{\mathcal{B}}}{=} \mathrm{Aut}(L_{\rho_0}(\mathbb{F}_p))$$

the mod p reduction map which sends (x_{ij}) to $(x_{ij} \pmod p)$ ($\bar{\mathcal{B}}$ is the basis of $L_{\rho_0}(\mathbb{F}_p) = \mathbb{F}_p \otimes L_{\rho_0}(\mathbb{Z}_p)$ induced by the basis $\mathcal{B}$ of $L_{\rho_0}(\mathbb{Z}_p)$). If $(x_{ij}) \in \mathbf{G}(\mathbb{Z}_p) \leq \mathbf{GL}_n(\mathbb{Z}_p)$ then $\wp(x)$ obviously is contained in $\mathbf{G}(\mathbb{F}_p)$, hence, $\wp$ induces a map $\mathbf{G}(\mathbb{Z}_p) \to \mathbf{G}(\mathbb{F}_p)$. If $\phi : \mathbf{G}/\mathbb{Z}_p \to \mathbf{G}'/\mathbb{Z}_p$ is a morphism we denote by $\bar{\phi} : \mathbf{G}(\mathbb{F}_p) \to \mathbf{G}'(\mathbb{F}_p)$ the mod p reduction of ϕ, i.e. $\bar{\phi} \circ \wp = \wp \circ \phi$.

3.2 Iwahori Subgroup

The Iwahori subgroup $\mathcal{I}$ of $\mathbf{G}(\mathbb{Z}_p)$ is defined as the set of all $k \in \mathbf{G}(\mathbb{Z}_p)$ such that $\wp(k) \in \mathbf{B}^-(\mathbb{F}_p)$, i.e. $\mathcal{I} = \wp^{-1}(\mathbf{B}^-(\mathbb{F}_p))$. We also define $\mathcal{I}^1 \leq \mathcal{I}$ as the subgroup consisting of all elements $k \in \mathbf{G}(\mathbb{Z}_p)$ satisfying $\wp(k) \in \mathbf{N}^-(\mathbb{F}_p)$. The group $\mathcal{I}$ resp. $\mathcal{I}^1$ decomposes

$$\mathcal{I} = (\mathcal{I} \cap \mathbf{N}(\mathbb{Q}_p))\mathbf{T}(\mathbb{Z}_p)\mathbf{N}^-(\mathbb{Z}_p)$$

resp.

$$\mathcal{I}^1 = (\mathcal{I}^1 \cap \mathbf{N}(\mathbb{Q}_p))\mathbf{T}(\mathbb{Z}_p)^1\mathbf{N}^-(\mathbb{Z}_p).$$

where $\mathbf{T}(\mathbb{Z}_p)^1 = \{t \in \mathbf{T}(\mathbb{Z}_p) : \wp(t) = 1\}$. We set

$$N^-_{\mathbb{Z}_p} = \langle x_\alpha(t_\alpha),\ \alpha \in \Phi^-, t_\alpha \in \mathbb{Z}_p\rangle, \quad N_{p\mathbb{Z}_p} = \langle x_\alpha(t_\alpha),\ \alpha \in \Phi^+, t_\alpha \in p\mathbb{Z}_p\rangle$$

and

$$T^1_\Gamma = \left\{\prod_{i=1}^{\ell} \mathrm{h}_i(t_i),\ t_i \in 1 + p\mathbb{Z}_p\right\}.$$

We note that Eq. (15) yields

$$T^1_\Gamma = \exp \rho_0(p\mathfrak{h}_\Gamma(\mathbb{Z}_p)). \tag{16}$$

Proposition *(1)* $N^-_{\mathbb{Z}_p} = \mathbf{N}^-(\mathbb{Z}_p)$
(2) $T^1_\Gamma = \mathbf{T}(\mathbb{Z}_p)^1$
(3) $N_{p\mathbb{Z}_p} = \mathcal{I}^1 \cap \mathbf{N}(\mathbb{Q}_p) = \mathcal{I} \cap \mathbf{N}(\mathbb{Q}_p)$.

In particular, the Iwahori decomposition yields $\mathcal{I}^1 = N_{p\mathbb{Z}_p} T^1_\Gamma N^-_{\mathbb{Z}_p}$, *hence,* $\mathcal{I}^1$ *is generated by the elements* $x_\alpha(t_\alpha)$ *where* $t_\alpha \in p\mathbb{Z}_p$ *if* $\alpha \in \Phi^+$ *and* $t_\alpha \in \mathbb{Z}_p$ *if* $\alpha \in \Phi^-$ *and* $\exp t_i\rho_0(\mathrm{h}_i)$, $i = 1, \ldots, \ell$, *where* $t_i \in p\mathbb{Z}_p$.

Proof (1) We fix an ordering of the positive and the negative roots as in [9], Sect. 3, p. 223. Any $n \in \mathbf{N}^\epsilon(\bar{\mathbb{Q}}_p)$, $\epsilon = \pm$, has a unique representation of the form $n = \prod_{\alpha\in\Phi^\epsilon} x_\alpha(Z_\alpha(n))$, where $Z_\alpha(n) \in \bar{\mathbb{Q}}_p$ (cf. [2], Sect. 3.2 (2), p. 14; note that $\nu_\pm$ is defined over $\mathbb{Q}$ by [2], Sect. 3.3 (3), p. 15). Then the coordinate algebra of $\mathbf{N}^\epsilon/\mathbb{Z}$ is generated by the functions $Z_\alpha : n \mapsto Z_\alpha(n)$, $\alpha \in \Phi^\epsilon$, i.e. $\mathbb{Z}[\mathbf{N}^\epsilon/\mathbb{Z}] = \mathbb{Z}[Z_\alpha, \alpha \in \Phi^\epsilon]$ (cf. [9], Proposition 1, p. 226). In particular, if $n \in \mathbf{N}^-(\mathbb{Z}_p)$ then $n = \prod_{\alpha\in\Phi^-} x_\alpha(t_\alpha)$ where $t_\alpha \in \mathbb{Z}_p$, hence, $n \in N^-_{\mathbb{Z}_p}$. Since $N^-_{\mathbb{Z}_p} \subseteq \mathbf{GL}_n(\mathbb{Z}_p) \cap \mathbf{N}^-(\bar{\mathbb{Q}}_p) = \mathbf{N}^-(\mathbb{Z}_p)$ the claim follows.

(2) "$\subseteq$": Let $t \in \mathbf{T}(\mathbb{Z}_p)^1$. We write $t = \prod_i \mathrm{h}_i(t_i)$ where $t_i \in \bar{\mathbb{Q}}_p$. Since $\mathbb{Z}[\mathbf{T}/\mathbb{Z}]$ is generated by the characters $\chi \in X(\mathbf{T}/\mathbb{Q})$ (cf. [9], Proposition 1, p. 226) we know that $\chi(t) \in 1 + p\mathbb{Z}_p$ for all $\chi \in X(\mathbf{T}/\mathbb{Q})$. In particular, if we choose $\chi = \xi_i^\circ$ we obtain $t_i \in 1 + p\mathbb{Z}_p$ for all i.

"$\supseteq$": Let $t = \prod_i \mathrm{h}_i(t_i)$ with $t \in 1 + p\mathbb{Z}_p$ for all i. For all $\mu \in \Gamma$ we obtain $\mu^\circ(t) = \prod_i t_i^{\mu(\mathrm{h}_i)} \in 1 + p\mathbb{Z}_p$ (note that $\mu(\mathrm{h}_i) \in \mathbb{Z}$) and since $\mathbb{Z}[\mathbf{T}/\mathbb{Z}]$ is generated by $X(\mathbf{T}/\mathbb{Q}) \cong \Gamma$ this implies that $t \in \mathbf{T}(\mathbb{Z}_p)$ and $\wp(t) = 1$.

(3) We prove the first equality. "$\subseteq$" We have $N_{p\mathbb{Z}_p} \subseteq N_{\mathbb{Z}_p} \subseteq \mathbf{N}(\mathbb{Z}_p)$. Let $x_\alpha(t_\alpha)$, $t_\alpha \in p\mathbb{Z}_p$, be a generator of $N_{p\mathbb{Z}_p}$. Since $\wp(x_\alpha(t_\alpha)) = x_\alpha(t_\alpha \pmod p) = x_\alpha(0) = 1$ we find $\wp(N_{p\mathbb{Z}_p}) = 1$, hence, $N_{p\mathbb{Z}_p} \subseteq \mathcal{I}^1$.

"$\supseteq$" Let $x \in \mathcal{I}^1 \cap \mathbf{N}(\mathbb{Q}_p)$. We write $x = \prod_{\alpha\in\Phi^+} x_\alpha(t_\alpha)$, $t_\alpha \in \bar{\mathbb{Q}}_p$. Since $x \in \mathcal{I}^1 \cap \mathbf{N}(\mathbb{Q}_p) \subseteq \mathbf{G}(\mathbb{Z}_p) \cap \mathbf{N}(\bar{\mathbb{Q}}_p) = \mathbf{N}(\mathbb{Z}_p)$ we know that $t_\alpha \in \mathbb{Z}_p$ for all $\alpha \in \Phi^+$. Since $\wp(x) \in \wp(\mathcal{I}) = \mathbf{B}^-(\mathbb{F}_p)$ and $\wp(x) \in \mathbf{N}(\mathbb{F}_p)$ we obtain $\wp(x) = 1$ and since $\wp(x) = \prod_{\alpha\in\Phi^+} x_\alpha(t_\alpha \pmod p)$ we deduce that $t_\alpha \equiv 0 \pmod{p\mathbb{Z}_p}$ for all $\alpha \in \Phi^+$, hence, $x \in N_{p\mathbb{Z}_p}$.

Finally, since any $t \in T_\Gamma^1$ can be written $t = \prod_i \mathrm{h}_i(t_i)$ where $t_i \in 1 + p\mathbb{Z}_p$ and since $\mathrm{h}_i(t_i) = \exp t_i' \rho_0(\mathrm{h}_i)$ where $t_i' = \log t_i \in p\mathbb{Z}_p$ by Eq. (15) the last claim follows. Thus, the Lemma is proven.

In the following Lemma we put $G_\Theta = G_\Theta^{\mathfrak{g}(\mathbb{Q}_p)}$ and $\exp : G_\Theta \to \mathrm{Aut}(L_{\rho_0}(\mathbb{Q}_p))$ is the exponential map (cf. Sect. 2.6.2).

Proposition *(1)* $\exp \rho_0(G_\Theta) \leq \mathbf{G}_{\rho_0}(\mathbb{Q}_p)$ *is a subgroup*
(2) $\mathcal{I}^1 \subseteq \exp \rho_0(G_\Theta)$
(3) $|x - 1| < \Theta$ *for all* $x \in \mathcal{I}^1$*, i.e.* $\mathcal{I}^1$ *is bounded.*

Proof (1) Let $X, Y \in G_\Theta^{\mathfrak{g}(\mathbb{Q}_p)}$. Hence, $|X|, |Y| < \Theta$ and $|\rho_0(X)|, |\rho_0(Y)| < \Theta$ because $|\rho_0| \leq 1$ (cf. Sect. 2.6.1). We obtain

$$\exp(\rho_0(X)) \exp(\rho_0(Y)) = \exp(h(\rho_0(X), \rho_0(Y))) = \exp(\rho_0(h(X, Y)) \tag{17}$$

where h denotes the Hausdorff functions on $G_\Theta^{\mathrm{End}(L_{\rho_0}(\mathbb{Q}_p))}$ and on $G_\Theta^{\mathfrak{g}(\mathbb{Q}_p)}$ which are represented by the Hausdorff series H (cf. [6] II.8.4, Proposition 4, p. 175; Sect. II.8.3, p. 175; II.8.4 Proposition 5, p. 176). Here, $h(X, Y) \in \mathfrak{g}(\mathbb{Q}_p)$ and since $|h(X, Y)| < \Theta$ (cf. [6], Sect. II.8.3, p. 174) we know that $h(X, Y) \in G_\Theta^{\mathfrak{g}(\mathbb{Q}_p)}$. Thus, $\exp(\rho_0(X)) \exp(\rho_0(Y)) \in \exp \rho_0(G_\Theta^{\mathfrak{g}(\mathbb{Q}_p)})$.

(2) Since $t_\alpha x_\alpha \in G_\Theta$ where $t_\alpha \in \mathbb{Z}_p$ if $\alpha < 0$ and $t_\alpha \in p\mathbb{Z}_p$ if $\alpha > 0$ (cf. Sect. 2.4.2) we see that $x_\alpha(t_\alpha) = \exp \rho_0(t_\alpha x_\alpha) \in \exp \rho_0(G_\Theta)$ where $t_\alpha \in \mathbb{Z}_p$ if $\alpha < 0$ and $t_\alpha \in p\mathbb{Z}_p$ if $\alpha > 0$. Since $\exp \rho_0(G_\Theta)$ is a group this implies that $\exp \rho_0(G_\Theta)$ contains $N_{p\mathbb{Z}_p}$ and $N_{\mathbb{Z}_p}^-$. Using Eq. (15) any $t \in T_\Gamma^1$ can be written $t = \prod_i \mathrm{h}_i(t_i) = \prod_i \exp(\rho_0(t_i'\mathrm{h}_i))$, $t_i' \in p\mathbb{Z}_p$. Since $t_i'\mathrm{h}_i \in G_\Theta$ if $t_i' \in p\mathbb{Z}_p$ (cf. Sect. 2.4.2) and since $\exp \rho_0(G_\Theta)$ is a group this implies that $T_\Gamma^1 \subseteq \exp \rho_0(G_\Theta)$. Thus, the group $\exp \rho_0(G_\Theta)$ contains a set of generators of $\mathcal{I}^1$.

(3) We write $x \in \mathcal{I}^1$ as $x = \exp \rho_0(X)$ with $X \in G_\Theta$. Using $|\rho_0| < 1$ (cf. Sect. 2.6.1) we obtain $|x - 1| = |\exp \rho_0(X) - 1| = |\rho_0(X)| \leq |\rho_0||X| < \Theta$ (cf. [6], Sect. II.8.4, Proposition 4, p. 175). Thus, the proof of the Lemma is complete.

3.3 Highest Weight Representation $\hat{\pi}_{\lambda,\mathcal{I}^1}$ of $\mathcal{I}^1$

3.3.1 The Representation $\hat{\pi}_{\lambda,\mathcal{I}^1}$

We select a weight $\lambda \in \mathfrak{h}^*_\Gamma(\mathcal{O})$ with corresponding Verma module $\hat{V}_\lambda$. Since $|\rho_0| \leq 1$ we know that $\rho_0(G_\Theta^{\mathfrak{g}(\mathbb{Q}_p)}) \subseteq G_\Theta^{\mathrm{End}(L_{\rho_0}(\mathbb{Q}_p))}$, hence, Eq. (10) yield an analytic isomorphism $\exp : \rho_0(G_\Theta^{\mathfrak{g}(\mathbb{Q}_p)}) \to \exp \rho_0(G_\Theta^{\mathfrak{g}(\mathbb{Q}_p)})$. Since also $|\hat{\rho}_\lambda| \leq 1$ (cf. Eq. (6)) Eq. (7) yields an analytic mapping $\exp : \hat{\rho}_\lambda(G_\Theta^{\mathfrak{g}(\mathbb{Q}_p)}) \to \mathrm{Aut}(\hat{V}_\lambda)$, i.e. we obtain the diagram

$$\begin{array}{ccccc} \mathcal{I}^1 \subseteq \exp \rho_0(G_\Theta^{\mathfrak{g}(\mathbb{Q}_p)}) & & \stackrel{\hat{\pi}_{\lambda,\mathcal{I}^1}}{\rightarrow} & & \subseteq \mathrm{Aut}(\hat{V}_\lambda) \\ \exp \uparrow \cong & & & & \uparrow \exp \\ \rho_0(G_\Theta^{\mathfrak{g}(\mathbb{Q}_p)}) & \stackrel{\rho_0}{\leftarrow} & G_\Theta^{\mathfrak{g}(\mathbb{Q}_p)} & \stackrel{\hat{\rho}_\lambda|_{\mathfrak{g}(\mathbb{Q}_p)}}{\rightarrow} & \hat{\rho}_\lambda(G_\Theta^{\mathfrak{g}(\mathbb{Q}_p)}) \\ \cap & & \cap & & \cap \\ \mathrm{End}(L_{\rho_0}(\mathbb{Q}_p)) & & \mathfrak{g}(\mathbb{Q}_p) & & \mathrm{End}(\hat{V}_\lambda(K)). \end{array} \tag{18}$$

Since ρ_0 is assumed to be faithful, i.e. $\rho_0 : \mathfrak{g}(\mathbb{Q}_p) \to \rho_0(\mathfrak{g}(\mathbb{Q}_p))$ is bijective we obtain a well defined map

$$\begin{array}{rcl} \hat{\pi}_{\lambda,\mathcal{I}^1} : \exp \rho_0(G_\Theta^{\mathfrak{g}(\mathbb{Q}_p)}) & \to & \mathrm{Aut}(\hat{V}_\lambda) \\ \exp \rho_0(X) & \mapsto & \exp \hat{\rho}_\lambda^V(X). \end{array}$$

Via restriction we obtain a representation $\hat{\pi}_{\lambda,\mathcal{I}^1} : \mathcal{I}^1 \to \mathrm{Aut}(\hat{V}_\lambda)$ (cf. 3.2 Proposition (2).

Remark (1) For any pair (α, t_α) where $\alpha \in \Phi$ and $t_\alpha \in \mathbb{Z}_p$ if $\alpha < 0$ and $t_\alpha \in p\mathbb{Z}_p$ if $\alpha > 0$ we set

$$x_\alpha^{\hat{\rho}_\lambda^V}(t_\alpha) = \exp(\hat{\rho}_\lambda^V(t_\alpha x_\alpha)) \in \mathrm{Aut}(\hat{V}_\lambda).$$

Since $x_\alpha^{\rho_0}(t_\alpha) = \exp(\rho_0(t_\alpha x_\alpha)) \in \exp \rho_0(G_\Theta^{\mathfrak{g}(\mathbb{Q}_p)})$ the definition of $\hat{\pi}_{\lambda,\mathcal{I}^1}$ yields

$$\hat{\pi}_{\lambda,\mathcal{I}^1}(x_\alpha^{\rho_0}(t_\alpha)) = x_\alpha^{\hat{\rho}_\lambda^V}(t_\alpha)$$

for all $\alpha \in \Phi$ and $t_\alpha \in \mathbb{Z}_p$ if $\alpha < 0$ and $t_\alpha \in p\mathbb{Z}_p$ if $\alpha > 0$.

(2) Since $X_n \to X$, $X_n, X \in \mathrm{End}(\hat{V}_\lambda)$ implies that $X_n(v) \to X(v)$ we obtain for all $v \in V_\lambda$ and all $t_\alpha \in p\mathbb{Z}_p$ if $\alpha > 0$ and $t_\alpha \in \mathbb{Z}_p$ if $\alpha < 0$

$$\hat{\pi}_{\lambda,\mathcal{I}^1}(x_\alpha^{\rho_0}(t_\alpha))(v) = \exp(\hat{\rho}_\lambda^V(t_\alpha x_\alpha))(v) = \sum_{i \geq 0} \left(\frac{t_\alpha^i (\hat{\rho}_\lambda^V(x_\alpha))^i(v)}{i!} \right); \tag{19}$$

analogously, for $\mathrm{h}_i(t_i) = \exp t_i' \rho_0(\mathrm{h}_i)$, $t_i \in 1 + p\mathbb{Z}_p$, $t_i' = \log t_i$.

Lemma *The map* $\hat{\pi}_{\lambda,\mathcal{I}^1} : \mathcal{I}^1 \to \mathrm{Aut}(\hat{V}_\lambda)$ *is a representation of the group* $\mathcal{I}^1$.

Proof We have to show that $\hat{\pi}_{\lambda,\mathcal{I}^1}$ is multiplicative. Let $x, y \in \mathcal{I}^1$. Since $\mathcal{I}^1 \subseteq \exp \rho_0(G_\Theta^{\mathfrak{g}(\mathbb{Q}_p)})$ there are $X, Y \in G_\Theta^{\mathfrak{g}(\mathbb{Q}_p)}$ such that $x = \exp \rho_0(X)$ and $y = \exp \rho_0(Y)$. Equation (17) yields

$$xy = \exp \rho_0(X) \exp \rho_0(Y) = \exp \rho_0(h(X, Y))$$

and Eq. (8) in Sect. 2.5 yields

$$\exp \hat{\rho}_\lambda^V(X) \exp \hat{\rho}_\lambda^V(Y) = \exp \hat{\rho}_\lambda^V(h(X, Y)).$$

This implies that $\hat{\pi}_{\lambda,\mathcal{I}^1}(xy) = \hat{\pi}_{\lambda,\mathcal{I}^1}(x)\hat{\pi}_{\lambda,\mathcal{I}^1}(y)$ and the proof is complete.

3.3.2 Weights of $\hat{\pi}_{\lambda,\mathcal{I}^1}$

Let $\lambda \in \mathfrak{h}_\Gamma^*(\mathcal{O})$, i.e. $\lambda(\mathrm{h}_i) \in \mathcal{O}$ for all $i = 1, \ldots, \ell$. We define a character

$$\lambda^\circ : \mathbf{T}(\mathbb{Z}_p)^1 \to 1 + p\mathcal{O}$$

as follows: using 3.2 Lemma (2) and Eq. (16) any $t \in \mathbf{T}(\mathbb{Z}_p)^1$ can be written $t = \exp \rho_0(H)$ with $H \in p\mathfrak{h}_\Gamma(\mathbb{Z}_p)$; we then set

$$\lambda^\circ(t) = \exp \lambda(H),$$

i.e. following diagram commutes (compare Eq. (18))

$$\begin{array}{ccccc} T_\Gamma^1 = \exp(\rho_0(p\mathfrak{h}(\mathbb{Z}_p))) & & \stackrel{\lambda^\circ}{\to} & & 1 + p\mathcal{O} \\ \exp \uparrow & & & & \uparrow \exp \\ \rho_0(p\mathfrak{h}(\mathbb{Z}_p)) & \stackrel{\rho_0}{\leftarrow} & p\mathfrak{h}_\Gamma(\mathbb{Z}_p) = \bigoplus_i p\mathbb{Z}_p \mathrm{h}_i & \stackrel{\lambda}{\to} & p\mathcal{O}. \end{array} \tag{20}$$

Remark Using Eq. (15) we obtain

$$\lambda^\circ(\mathrm{h}_i(t_i)) = \exp(\lambda(t_i'\mathrm{h}_i)) \tag{21}$$

for all $i = 1, \ldots, \ell$, $t_i = \exp(t_i') \in 1 + p\mathbb{Z}_p$ ($t_i' \in p\mathbb{Z}_p$). Let $\alpha \in \Phi$ and write $h_\alpha = \sum_i z_i \mathrm{h}_i$, $z_i \in \mathbb{Z}$. Since $\prod_i \mathrm{h}_i(t)^{z_i} v_\mu = \prod_i \mu(\mathrm{h}_i(t))^{z_i} v_\mu = \prod_i t^{z_i \mu(\mathrm{h}_i)} = t^{\mu(h_\alpha)}$ (cf. Eq. (14)) and since $h_\alpha(t) v_\mu = t^{\mu(\alpha)} v_\mu$, $t \in \mathbb{Q}_p$, $\mu \in \Gamma$, (cf. [2], Sect. 3.2 (1), p. 13) we obtain

$$h_\alpha(t) = \prod_i \mathrm{h}_i^{z_i}(t), \quad t \in \bar{\mathbb{Q}}_p. \tag{22}$$

If $\lambda \in \Gamma_{\rho_0}$ then a (algebraic) character λ° on $\mathbf{T}(\bar{\mathbb{Q}}_p)$ has been defined in Sect. 3.1. On $\mathbf{T}(\mathbb{Z}_p)^1$ the two definitions of "λ°" coincide because for any $t_\alpha = \exp(t'_\alpha) \in 1 + p\mathbb{Z}_p$ ($t'_\alpha \in p\mathbb{Z}_p$) we obtain

$$\lambda^\circ(h_\alpha(t_\alpha)) = \prod_i \lambda^\circ(\mathrm{h}_i^{z_i}(t_\alpha)) = \prod_i \exp t'_\alpha \lambda(z_i \mathrm{h}_i) = t_\alpha^{\lambda(h_\alpha)}$$

(note that $\lambda(h_\alpha) \in \mathbb{Z}$).

In the following, we will consider the restriction of $\hat{\pi}_{\lambda,\mathcal{I}^1}$ to $\mathbf{T}(\mathbb{Z}_p)^1 \subseteq \mathcal{I}^1$.

Lemma *For any $t \in \mathbf{T}(\mathbb{Z}_p)^1$ and any $\mathbf{n} \in \mathbb{N}_0^s$ we obtain*

$$\hat{\pi}_{\lambda,\mathcal{I}^1}(t)\mathrm{w}_\mathbf{n} = \lambda^\circ(t)(\mathbf{n}^\circ)^{-1}(t)\mathrm{w}_\mathbf{n}.$$

Proof We write $t = \prod_i \mathrm{h}_i(t_i)$ where $t_i = \exp(t'_i) \in 1 + p\mathbb{Z}_p$, $t'_i \in p\mathbb{Z}_p$ (cf. 3.2 Lemma (2)) and obtain:

$$\begin{aligned}
\hat{\pi}_{\lambda,\mathcal{I}^1}(t)\mathrm{w}_\mathbf{n} = \hat{\pi}_{\lambda,\mathcal{I}^1}\left(\prod_i \mathrm{h}_i(t_i)\right)\mathrm{w}_\mathbf{n} &\overset{(15)}{=} \hat{\pi}_{\lambda,\mathcal{I}^1}\left(\exp\left(\rho_0\left(\sum_i t'_i \mathrm{h}_i\right)\right)\right)\mathrm{w}_\mathbf{n} \\
&= \exp\left(\hat{\rho}_\lambda^V\left(\sum_i t'_i \mathrm{h}_i\right)\right)\mathrm{w}_\mathbf{n} \\
&= \sum_{k\geq 0} \frac{\hat{\rho}_\lambda^V\left(\sum_i t'_i \mathrm{h}_i\right)^k}{k!}\mathrm{w}_\mathbf{n} \\
&= \sum_{k\geq 0} \frac{\left((\lambda - \mathbf{n})\left(\sum_i t'_i \mathrm{h}_i\right)\right)^k}{k!}\mathrm{w}_\mathbf{n} \\
&= \exp\left((\lambda - \mathbf{n})\left(\sum_i t'_i \mathrm{h}_i\right)\right)\mathrm{w}_\mathbf{n} \\
&= \exp\left(\lambda\left(\sum_i t'_i \mathrm{h}_i\right)\right)\exp\left((-\mathbf{n})\left(\sum_i t'_i \mathrm{h}_i\right)\right)\mathrm{w}_\mathbf{n} \\
&\overset{(21)}{=} \lambda^\circ(t)(\mathbf{n}^\circ)^{-1}(t)\mathrm{w}_\mathbf{n}
\end{aligned}$$

(note that $\mathbf{n}^\circ(t) = \prod_{i=1,\ldots,s}(\alpha_i^\circ)^{n_i}(t)$, $t \in \mathbf{T}(\mathbb{Q}_p)$ and that the definition of α° in this section coincides with the one in Sect. 3.1; cf. the above Remark).

3.3.3 Norm of $\hat{\pi}_{\lambda,\mathcal{I}^1}$

Lemma *For any $k \in \mathcal{I}^1$ the operator $\hat{\pi}_{\lambda,\mathcal{I}^1}(k)$ on $\hat{V}_\lambda$ has norm less than or equal to 1. In particular, any $\hat{\pi}_{\lambda,\mathcal{I}^1}(k)$, $k \in \mathcal{I}^1$, is continuous and maps $\hat{V}_{\lambda,\mathcal{O}}$ to itself.*

Proof Let $k \in \mathcal{I}^1$. By 3.2 Proposition (2) there is $X \in G_\Theta^{\mathfrak{g}(\mathbb{Q}_p)}$ such that $k = \exp \rho_0(X)$. Since $|\exp(\varphi) - 1| \leq |\varphi|$, $\varphi \in G_\Theta^{\mathrm{End}(\hat{V}_\lambda)}$ (cf. [6], Sect. II.8.4, Proposition 4, p. 175) we obtain

$$|\hat{\pi}_{\lambda,\mathcal{I}^1}(k) - 1| = |\exp(\hat{\rho}_\lambda^V(X)) - 1| \leq |\hat{\rho}_\lambda^V(X)| < \Theta \leq 1.$$

The second to last inequality holds because $|\hat{\rho}_\lambda^V| \leq 1$ (cf. Eq. (6)) and $|X| < \Theta$. This yields $|\hat{\pi}_{\lambda,\mathcal{I}^1}(k)| \leq 1$ and the Lemma is proven.

3.4 The Representation of $\mathbf{T}(\mathbb{Q}_p)^{--}$ on Verma Modules

For the moment we let $\mathbf{T}/\mathbb{Z}_p$ be any torus which is split over $\mathbb{Z}_p$, i.e. there is an isomorphism $\Phi : \mathbf{T}/\mathbb{Z}_p \to \mathbb{G}_m^a/\mathbb{Z}_p$ for some $a \in \mathbb{N}$. This induces a bijection $\Phi : \mathbf{T}(\mathbb{Z}_p)^1 \to \mathbb{G}_m^a(\mathbb{Z}_p)^1 = (1 + p\mathbb{Z}_p)^a$. We let $t \in \mathbf{T}(\mathbb{Q}_p)$ and we denote by $\zeta \in \mathbb{Z}_p^*$ a primitive $(p-1)$-st root of unity. There are unique $m_i \in \mathbb{Z}$ and $n_i \in \mathbb{Z}/(p-1)\mathbb{Z}$ such that $\Phi(t)(p^{m_1}\zeta^{n_1}, \ldots, p^{m_a}\zeta^{n_a}) \in (1 + p\mathbb{Z}_p)^a$. We set $t^\dagger = \Phi^{-1}((p^{m_1}\zeta^{n_1}, \ldots, p^{m_a}\zeta^{n_a})) \in \mathbf{T}(\mathbb{Q}_p)$ and obtain $\Phi(tt^\dagger) \in (1 + p\mathbb{Z}_p)^a$, hence,

$$tt^\dagger \in \mathbf{T}(\mathbb{Z}_p)^1.$$

We now let $\mathbf{T}/\mathbb{Q}_p$ again denote as before the maximal $\mathbb{Q}_p$-split torus in $\mathbf{G}/\mathbb{Q}_p$ as defined in Sect. 3.1. The group $\mathbf{G}/\mathbb{Q}_p$ has a $\mathbb{Z}_p$-structure which induces a $\mathbb{Z}_p$-structure on $\mathbf{T}$ and any character $\mu^\circ \in X(\mathbf{T}/\mathbb{Q}_p)$, $\mu \in \Gamma$, is defined over $\mathbb{Z}_p$ (cf. Sect. 3.1); hence, $\mathbf{T} \cong \mathbb{G}_m^\ell$ over $\mathbb{Z}_p$.

We recall that $\mathbf{T}(\mathbb{Q}_p)^-$ resp. $\mathbf{T}(\mathbb{Q}_p)^{--}$ is the set of all $t \in \mathbf{T}(\mathbb{Q}_p)$ satisfying $v_p(\alpha(t)) \leq 0$ resp. $v_p(\alpha(t)) < 0$ for all $\alpha \in \Delta$. Let $\lambda \in \mathfrak{h}_\Gamma^*(\mathcal{O})$. We define a representation of semi groups

$$\hat{\pi}_{\lambda,\mathbf{T}} : \mathbf{T}(\mathbb{Q}_p)^- \to \mathrm{Aut}(\hat{V}_\lambda)$$

by defining for any $t \in \mathbf{T}(\mathbb{Q}_p)^-$ and $v = \sum_{\mathbf{n}\in\mathbb{N}_0^s} a_\mathbf{n} X_-^\mathbf{n} \mathrm{v}_\lambda \in \hat{V}_\lambda$ ($a_\mathbf{n} \in K$, $a_\mathbf{n} \to 0$)

$$tv = \sum_\mathbf{n} \lambda^\circ(tt^\dagger)(\mathbf{n}^\circ)^{-1}(t) a_\mathbf{n} X_-^\mathbf{n} \mathrm{v}_\lambda. \tag{23}$$

We note that $\lambda^\circ(tt^\dagger)$ has been defined in Sect. 3.3.2. Since $\lambda^\circ(tt^\dagger) \in 1 + p\mathcal{O}$ and since

$$v_p(\mathbf{n}^\circ(t^{-1})) = v_p\left(\prod_{i=1,\dots,s} \alpha_i(t^{-1})^{n_i}) = \sum_{i=1,\dots,s} n_i v_p(\alpha_i(t^{-1})\right) \geq 0 \tag{24}$$

(note that $t \in \mathbf{T}(\mathbb{Q}_p)^-$) the factor $\lambda^\circ(tt^\dagger)\mathbf{n}^\circ(t^{-1})$ has norm less than or equal to 1; thus, the sum defining tv converges. Moreover, since $(t_1t_2)^\dagger = t_1^\dagger t_2^\dagger$ the map $\hat{\pi}_{\lambda,\mathbf{T}}$ is multiplicative and, hence, defines a representation of $\mathbf{T}(\mathbb{Q}_p)^-$. If $t \in \mathbf{T}(\mathbb{Z}_p)^1$ then $t^\dagger = 1$ and we obtain

$$tv = \sum_{\mathbf{n}} \lambda^\circ(t)(\mathbf{n}^\circ)^{-1}(t) a_{\mathbf{n}} X_-^{\mathbf{n}} \mathrm{v}_\lambda. \tag{25}$$

Lemma *(1) For any $t \in \mathbf{T}(\mathbb{Q}_p)^-$ the operator $\hat{\pi}_{\lambda,\mathbf{T}}(t)$ on $\hat{V}_\lambda$ has norm less than or equal to 1, hence, it is continuous.*
(2) The operator $\hat{\pi}_{\lambda,\mathbf{T}}(t)$, $t \in \mathbf{T}(\mathbb{Q}_p)^{--}$, even is compact.

Proof (1) Let $t \in \mathbf{T}(\mathbb{Q}_p)^-$. Since the factor $\lambda^\circ(tt^\dagger)\mathbf{n}^\circ(t^{-1})$ has norm less than or equal to 1 (cf. Eq. (24)) we obtain for any $v = \sum_{\mathbf{n}} a_{\mathbf{n}} X_-^{\mathbf{n}} \mathrm{v}_\lambda \in \hat{V}_\lambda$

$$|\hat{\pi}_{\lambda,\mathbf{T}}(t)(v)| = |\sum_{\mathbf{n}} a_{\mathbf{n}} \lambda^\circ(tt^\dagger)\mathbf{n}^\circ(t^{-1}) X_-^{\mathbf{n}} \mathrm{v}_\lambda| \leq \sup_{\mathbf{n}} |a_{\mathbf{n}}| p^{-\epsilon \mathrm{ht}(\mathbf{n})} = |v|.$$

This implies $|\hat{\pi}_{\lambda,\mathbf{T}}(t)| \leq 1$.
(2) To prove that $\hat{\pi}_{\lambda,\mathbf{T}}(t)$, $t \in \mathbf{T}(\mathbb{Q}_p)^{--}$, is compact we look at its representing matrix $\mathcal{D}$ w.r.t. the ONB $(\mathrm{w}_{\mathbf{n}})_{\mathbf{n}}$. The definition of $\hat{\pi}_{\lambda,\mathbf{T}}(t)$ shows that $\mathcal{D}$ is a diagonal matrix with **n**th diagonal entry equal to $\lambda^\circ(tt^\dagger)\mathbf{n}^\circ(t^{-1})$. Since $t \in \mathbf{T}(\mathbb{Q}_p)^{--}$ Eq. (24) implies that

$$v_p(\mathbf{n}^\circ(t^{-1})) = \sum_{i=1,\dots,s} n_i v_p(\alpha_i(t^{-1})) \geq \sum_i n_i \mathrm{ht}(\alpha_i) = \mathrm{ht}(\mathbf{n}) \tag{26}$$

hence, $|\lambda^\circ(tt^\dagger)\mathbf{n}^\circ(t^{-1})| \leq p^{-\mathrm{ht}(\mathbf{n})}$. Thus, $(\lambda^\circ(tt^\dagger)\mathbf{n}^\circ(t^{-1}))_{\mathbf{n}} \to 0$ and the operator $\hat{\pi}_{\lambda,\mathbf{T}}(t)$ is compact by Sect. 3, Proposition 4, p. 72 or Corollary p. 73 in [19].

3.5 The Sub Semigroup D Generated by $\mathcal{I}^1$ and $\mathbf{T}(\mathbb{Q}_p)^-$

We note the following consequences of the Iwahori decomposition.

- For all $t_1, t_2 \in \mathbf{T}(\mathbb{Q}_p)^-$ it holds that

$$\mathcal{I}^1 t_1 \mathcal{I}^1 t_2 \mathcal{I}^1 = \mathcal{I}^1 t_1 t_2 \mathcal{I}^1. \tag{27}$$

  This follows by writing the element in the factor in the "middle $\mathcal{I}^1$" using Iwahori decomposition as $k = k^- k^\circ k^+$ with $k^- \in \mathbf{N}^-(\mathbb{Z}_p)$, $k^\circ \in \mathbf{T}(\mathbb{Z}_p)^1$, $k^+ \in \mathbf{N}(\mathbb{Q}_p) \cap \mathcal{I}^1$ and using that $\mathrm{Inn}(t_2^{-1})(\mathbf{N}(\mathbb{Q}_p) \cap \mathcal{I}^1) \subseteq \mathbf{N}(\mathbb{Q}_p) \cap \mathcal{I}^1$ and $\mathrm{Inn}(t_1)(\mathbf{N}^-(\mathbb{Z}_p)) \subseteq \mathbf{N}^-(\mathbb{Z}_p)$.

- Let $k = k^- k^\circ k^+ \in \mathcal{I}^1$ and $m = m^- m^\circ m^+ \in \mathcal{I}^1$ with $k^-, m^- \in \mathbf{N}^-(\mathbb{Z}_p)$, $k^\circ, m^\circ \in \mathbf{T}(\mathbb{Z}_p)^1$, $k^+, m^+ \in \mathbf{N}(\mathbb{Q}_p) \cap \mathcal{I}^1$. If $kt = t'm$ with $t, t' \in \mathbf{T}(\mathbb{Q}_p)^-$ then

$$t^{-1}k^+t = m^+, \qquad k^- = t'm^-(t')^{-1} \quad \text{and} \quad k^\circ t = t'm^\circ, \tag{28}$$

  i.e. $t \equiv t' \pmod{\mathbf{T}(\mathbb{Z}_p)^1}$. This follows from $k^- t k^\circ \mathrm{Inn}(t^{-1})(k^+) = \mathrm{Inn}(t')(m^-)\, t'm^\circ m^+$: the first three factors on each side are contained in $\mathbf{B}^-(\mathbb{Q}_p)$ while the last factor is contained in $\mathbf{N}(\mathbb{Q}_p)$, hence, $\mathrm{Inn}(t^{-1})(k^+) = m^+$. Analogously we obtain $k^- = \mathrm{Inn}(t')(m^-)$ and therefore $k^\circ t = t'm^\circ$.

We denote by $\mathrm{D} \leq \mathbf{G}(\mathbb{Q}_p)$ the sub semi group generated by $\mathcal{I}^1$ and $\mathbf{T}(\mathbb{Q}_p)^-$. Using Eq. (27) any $x \in \mathrm{D}$ can be written

$$x = k_1 t k_2 \tag{29}$$

where $k_1, k_2 \in \mathcal{I}^1$ and $t \in \mathbf{T}(\mathbb{Q}_p)^-$ and in this representation t is uniquely determined up to a factor in $\mathbf{T}(\mathbb{Z}_p)^1$ (cf. Eq. (28)).

Lemma *(1) For all $t \in \mathbf{T}(\mathbb{Q}_p)^-$ and $k \in \mathcal{I}^+ := \mathcal{I}^1 \cap \mathbf{N}(\mathbb{Q}_p) = N_{p\mathbb{Z}_p}$ we have*

$$\hat{\pi}_{\lambda,\mathcal{I}^1}(k)\hat{\pi}_{\lambda,\mathbf{T}}(t) = \hat{\pi}_{\lambda,\mathbf{T}}(t)\hat{\pi}_{\lambda,\mathcal{I}^1}(\mathrm{Inn}(t^{-1})(k))$$

(note that $\mathrm{Inn}(t^{-1})(k) \in \mathcal{I}^+$*).*

For all $t \in \mathbf{T}(\mathbb{Q}_p)^-$ and $k \in \mathcal{I}^- := \mathbf{N}^-(\mathbb{Z}_p) = N^-_{\mathbb{Z}_p}$ we have

$$\hat{\pi}_{\lambda,\mathbf{T}}(t)\hat{\pi}_{\lambda,\mathcal{I}^1}(k) = \hat{\pi}_{\lambda,\mathcal{I}^1}(\mathrm{Inn}(t)(k))\hat{\pi}_{\lambda,\mathbf{T}}(t)$$

(note that $\mathrm{Inn}(t)(k) \in \mathcal{I}^-$*).*

(2) For all $t \in \mathbf{T}(\mathbb{Z}_p)^1 = T^1_\Gamma$ we have

$$\hat{\pi}_{\lambda,\mathbf{T}}(t) = \hat{\pi}_{\lambda,\mathcal{I}^1}(t).$$

(3) Let $k_1 t k_2 = m_1 t' m_2 \in \mathrm{D}$ where $k_1, k_2, m_1, m_2 \in \mathcal{I}^1$ and $t, t' \in \mathbf{T}(\mathbb{Q}_p)^-$. Then

$$\hat{\pi}_{\lambda,\mathcal{I}^1}(k_1)\hat{\pi}_{\lambda,\mathbf{T}}(t)\hat{\pi}_{\lambda,\mathcal{I}^1}(k_2) = \hat{\pi}_{\lambda,\mathcal{I}^1}(m_1)\hat{\pi}_{\lambda,\mathbf{T}}(t')\hat{\pi}_{\lambda,\mathcal{I}^1}(m_2).$$

Proof (1) Let $t \in \mathbf{T}(\mathbb{Q}_p)^-$ and $k \in \mathcal{I}^+$. Since $\mathcal{I}^+ = N_{p\mathbb{Z}_p}$ any element $k \in \mathcal{I}^+$ can be written as a product of elements $x_\alpha(t_\alpha) = x_\alpha^{\rho_0}(t_\alpha)$ with $\alpha \in \Phi^+$ and $t_\alpha \in p\mathbb{Z}_p$, hence, it is sufficient to show that $\hat{\pi}_{\lambda,\mathcal{I}^1}(x_\alpha(t_\alpha))\hat{\pi}_{\lambda,\mathbf{T}}(t)\mathrm{w}_\mathbf{n} = \hat{\pi}_{\lambda,\mathbf{T}}(t)\hat{\pi}_{\lambda,\mathcal{I}^1}(\mathrm{Inn}(t^{-1})(x_\alpha(t_\alpha)))\mathrm{w}_\mathbf{n}$ for all $\alpha \in \Phi^+$, $t_\alpha \in p\mathbb{Z}_p$ and all $\mathbf{n} \in \mathbb{N}_0^s$. We calculate using continuity of $\hat{\pi}_{\lambda,\mathbf{T}}(t)$:

$$\begin{aligned}
\hat\pi_{\lambda,\mathbf{T}}(t)\hat\pi_{\lambda,\mathcal{I}^1}(\mathrm{Inn}(t^{-1})(x_\alpha(t_\alpha)))\mathbf{w_n} &= \hat\pi_{\lambda,\mathbf{T}}(t)\hat\pi_{\lambda,\mathcal{I}^1}(x_\alpha(\alpha(t^{-1})t_\alpha))\mathbf{w_n}\\
&= \hat\pi_{\lambda,\mathbf{T}}(t)\exp(\hat\rho_\lambda^V(\alpha(t^{-1})t_\alpha x_\alpha))\mathbf{w_n}\\
&= \hat\pi_{\lambda,\mathbf{T}}(t)\sum_{m\in\mathbb{N}_0}\alpha(t^{-1})^m t_\alpha^m\frac{\hat\rho_\lambda^V(x_\alpha)^m}{m!}\mathbf{w_n}\\
&= \sum_{m\in\mathbb{N}_0}\frac{\alpha(t^{-1})^m t_\alpha^m}{m!}\hat\pi_{\lambda,\mathbf{T}}(t)\hat\rho_\lambda^V(x_\alpha)^m\mathbf{w_n}\\
&= \sum_{m\in\mathbb{N}_0}\frac{\alpha(t^{-1})^m t_\alpha^m}{m!}\lambda^\circ(tt^\dagger)(\mathbf{n}-m\alpha)^\circ(t^{-1})\hat\rho_\lambda^V(x_\alpha)^m\mathbf{w_n}\\
&= \sum_{m\in\mathbb{N}_0}\frac{t_\alpha^m}{m!}\hat\rho_\lambda^V(x_\alpha)^m\hat\pi_{\lambda,\mathbf{T}}(t)\mathbf{w_n}\\
&= \exp(\hat\rho_\lambda^V(t_\alpha x_\alpha))\hat\pi_{\lambda,\mathbf{T}}(t)\mathbf{w_n}\\
&= \hat\pi_{\lambda,\mathcal{I}^1}(x_\alpha(t_\alpha))\hat\pi_{\lambda,\mathbf{T}}(t)\mathbf{w_n}.
\end{aligned}$$

Hence, the claim is proven. The case $t\in\mathbf{T}(\mathbb{Q}_p)^-$ and $k\in\mathcal{I}^-$ is analogous.

(2) This follows from 3.3.2 Lemma (note that $t^\dagger=1$).

(3) It is sufficient to show that

$$kt=t'm \tag{30}$$

with $k,m\in\mathcal{I}^1$ and $t,t'\in\mathbf{T}(\mathbb{Q}_p)^-$ implies that $\hat\pi_{\lambda,\mathcal{I}^1}(k)\hat\pi_{\lambda,\mathbf{T}}(t)=\hat\pi_{\lambda,\mathbf{T}}(t')\hat\pi_{\lambda,\mathcal{I}^1}(m)$. We write $k=k^-k^\circ k^+$ and $m=m^-m^\circ m^+$ where $k^+,m^+\in\mathcal{I}^+$, $k^\circ,m^\circ\in\mathbf{T}(\mathbb{Z}_p)^1$ and $k^-,m^-\in\mathcal{I}^-$. Thus, we have to show that

$$\hat\pi_{\lambda,\mathcal{I}^1}(k^-)\hat\pi_{\lambda,\mathcal{I}^1}(k^\circ)\hat\pi_{\lambda,\mathcal{I}^1}(k^+)\hat\pi_{\lambda,\mathbf{T}}(t)=\hat\pi_{\lambda,\mathbf{T}}(t')\hat\pi_{\lambda,\mathcal{I}^1}(m^-)\hat\pi_{\lambda,\mathcal{I}^1}(m^\circ)\hat\pi_{\lambda,\mathcal{I}^1}(m^+).$$

Using (1) and (2) this is equivalent to

$$\hat\pi_{\lambda,\mathcal{I}^1}(k^-)\hat\pi_{\lambda,\mathbf{T}}(k^\circ t)\hat\pi_{\lambda,\mathcal{I}^1}(t^{-1}k^+t)=\hat\pi_{\lambda,\mathcal{I}^1}(t'm^-t'^{-1})\hat\pi_{\lambda,\mathbf{T}}(t'm^\circ)\hat\pi_{\lambda,\mathcal{I}^1}(m^+).$$

This last equation holds by Eq. (28). Hence, the Lemma is proven.

Using the Lemma and Eq. (29) we obtain a well defined representation

$$\begin{aligned}
\hat\pi_\lambda:\ \mathrm{D} &\to \mathrm{Aut}(\hat V_\lambda)\\
k_1tk_2 &\mapsto \hat\pi_{\lambda,\mathcal{I}^1}(k_1)\hat\pi_{\lambda,\mathbf{T}}(t)\hat\pi_{\lambda,\mathcal{I}^1}(k_2)
\end{aligned}$$

To prove multiplicativity let $k_1tk_2, m_1t'm_2\in\mathrm{D}$ ($t,t'\in\mathbf{T}(\mathbb{Q}_p)^-$, $k_1,k_2,m_1,m_2\in\mathcal{I}^1$) and set $g=k_2m_1\in\mathcal{I}^1$. Writing $g=g^-g^\circ g^+$, where $g^-\in\mathcal{I}^-$, $g^\circ\in\mathbf{T}(\mathbb{Z}_p)^1$, $g^+\in\mathcal{I}^+$ and using 3.5 Lemma (1), (2) as in the proof of the 3.5 Lemma (3) multiplicativity follows.

Proposition *(1) The operator $\hat\pi_\lambda(x)$, $x\in\mathrm{D}$, has norm less than or equal to* 1. *In particular, $\hat\pi_\lambda(x)$, $x\in\mathrm{D}$, maps $\hat V_{\lambda,\mathcal{O}}$ to itself:*

$$\hat{\pi}_\lambda : \mathrm{D} \to \mathrm{Aut}(\hat{V}_{\lambda,\mathcal{O}}).$$

(2) Let $x = k_1 t k_2 \in \mathrm{D}$ with $k_1, k_2 \in \mathcal{I}^1$ and $t \in \mathbf{T}(\mathbb{Q}_p)^{--}$; then $\hat{\pi}_\lambda(x)$ is compact.

Proof (1) We write $x \in \mathrm{D}$ as $x = k_1 t k_2$ where $k_1, k_2 \in \mathcal{I}^1$ and $t \in \mathbf{T}(\mathbb{Q}_p)^-$, hence, $\hat{\pi}_\lambda(x) = \hat{\pi}_{\lambda,\mathcal{I}^1}(k_1)\hat{\pi}_{\lambda,\mathbf{T}}(t)\hat{\pi}_{\lambda,\mathcal{I}^1}(k_2)$. The claim then follows from 3.3.3 Lemma and 3.4 Lemma.

(2) As in (1) this follows from 3.3.3 Lemma and 3.4 Lemma.

4 Truncations

We keep the notations from Sect. 3; in particular, $\mathbf{G}$ is a connected semi simple algebraic groups defined over $\mathbb{Q}$ and split over $\mathbb{Q}_p$, hence, $\mathbf{G}$ is isomorphic over $\mathbb{Q}_p$ to the Chevalley group $\mathbf{G}_{\rho_0}$ attached to the faithful finite dimensional complex representation $\rho_0 : \mathfrak{g} \to \mathrm{End}(L_{\rho_0})$ and $\Gamma = \Gamma_{\rho_0}$ is the weight lattice of ρ_0

4.1 The Submodule $\hat{V}_{\lambda,\mathcal{O}}(r, \sigma)$

Let $\lambda \in \mathfrak{h}^*_\Gamma(\mathcal{O})$. In Sect. 2.4.4 we defined the sub $\mathcal{O}$-module $\hat{V}_{\lambda,\mathcal{O}} = \{v \in \hat{V}_\lambda : |v| \leq 1\}$ of $\hat{V}_\lambda$. We select an element $\omega \in K$ such that

$$v_p(\omega) = \delta := \epsilon - \frac{1}{\sigma}\frac{p-2}{p-1} \in \mathbb{Q}_{<0}, \tag{31}$$

(eventually we enlarge K). We remark that Eq. (1) implies that $\delta = \epsilon - \frac{1}{\sigma}\frac{p-2}{p-1} < 0$. For any $r \in \mathbb{N}$ we define the following sub $\mathcal{O}$-module of $V_{\lambda,\mathcal{O}}$

$$\begin{aligned} V_{\lambda,\mathcal{O}}(r,\sigma) &= \bigoplus_{\substack{\mathbf{n}\in\mathbb{N}_0^s \\ r+\delta\mathrm{ht}(\mathbf{n})\geq 0}} p^r\omega^{\mathrm{ht}(\mathbf{n})}\mathcal{O}\mathrm{w}_\mathbf{n} \oplus \bigoplus_{\substack{\mathbf{n}\in\mathbb{N}_0^s \\ r+\delta\mathrm{ht}(\mathbf{n})<0}} \mathcal{O}\mathrm{w}_\mathbf{n} \\ &= \bigoplus_{\substack{\mu\in\Pi_\lambda^V \\ r+\delta\mathrm{ht}_\lambda(\mu)\geq 0}} p^r\omega^{\mathrm{ht}_\lambda(\mu)} V_{\lambda,\mathcal{O}}(\mu) \oplus \bigoplus_{\substack{\mu\in\Pi_\lambda^V \\ r+\delta\mathrm{ht}_\lambda(\mu)<0}} V_{\lambda,\mathcal{O}}(\mu) \end{aligned}$$

and

$$\hat{V}_{\lambda,\mathcal{O}}(r,\sigma) = \text{closure of } V_{\lambda,\mathcal{O}}(r,\sigma) \text{ in } \hat{V}_\lambda.$$

Thus, $\hat{V}_{\lambda,\mathcal{O}}(r,\sigma)$ is a sub $\mathcal{O}$-module of $\hat{V}_{\lambda,\mathcal{O}}$ and consists of all elements $\sum_\mathbf{n} a_\mathbf{n}\mathrm{w}_\mathbf{n} \in \hat{V}_\lambda$ such that $a_\mathbf{n} \in \mathcal{O}$ and $v_p(a_\mathbf{n}) \geq r + \delta\mathrm{ht}(\mathbf{n})$ (i.e. $a_\mathbf{n} \in p^r\omega^{\mathrm{ht}(\mathbf{n})}\mathcal{O}$ if $r + \delta\mathrm{ht}(\mathbf{n}) \geq 0$).

Lemma *Let* $\lambda \in \mathfrak{h}^*_\Gamma(\mathcal{O})$. *Then,* $\hat{V}_{\lambda,\mathcal{O}}$ *and* $\hat{V}_{\lambda,\mathcal{O}}(r,\sigma)$ *are* $\mathcal{I}^1$*-invariant sub* $\mathcal{O}$*-modules of* $\hat{V}_\lambda$.

Proof Lemma 3.3.3 implies that $\mathcal{I}^1$ leaves $\hat{V}_{\lambda,\mathcal{O}}$ invariant. To prove invariance of $\hat{V}_{\lambda,\mathcal{O}}(r,\sigma)$ under $\mathcal{I}^1$ it suffices to show that

$$x_\alpha(t_\alpha)\mathsf{c}\mathsf{w}_\mathbf{n} = \sum_{i\geq 0} \frac{t_\alpha^i \hat{\rho}_\lambda^V(x_\alpha)^i}{i!}\mathsf{c}\mathsf{w}_\mathbf{n} \in \hat{V}_{\lambda,\mathcal{O}}(r,\sigma) \tag{32}$$

and

$$t\mathsf{c}\mathsf{w}_\mathbf{n} \in \hat{V}_{\lambda,\mathcal{O}}(r,\sigma); \tag{33}$$

here, $\mathbf{n}$ runs over all elements in $\mathbb{N}_0^s$, $\mathsf{c}=\mathsf{c}_\mathbf{n} \in p^r\omega^{\mathrm{ht}(\mathbf{n})}\mathcal{O}$ if $r+\delta\mathrm{ht}(\mathbf{n})\geq 0$ and $\mathsf{c}\in\mathcal{O}$ else, $\alpha\in\Phi$ with $t_\alpha\in p\mathbb{Z}_p$ if $\alpha>0$ and $t_\alpha\in\mathbb{Z}_p$ if $\alpha<0$ and $t\in\mathbf{T}(\mathbb{Z}_p)^1$ (these elements generate $\mathcal{I}^1$; cf. 3.2 Lemma).

We first prove Eq. (32). We look at a single summand $\frac{t_\alpha^i\hat{\rho}_\lambda^V(x_\alpha)^i}{i!}\mathsf{c}\mathsf{w}_\mathbf{n}$ appearing on the right hand side:

- Since $x_\alpha^i\mathbf{w}_\mathbf{n}$ has weight $\lambda-\mathbf{n}+i\alpha$ w.r.t. $\mathfrak{h}(K)$ we obtain

$$\frac{t_\alpha^i\hat{\rho}_\lambda^V(x_\alpha^i)}{i!}\mathsf{w}_\mathbf{n} = \sum_{\substack{\mathbf{m}\\ \sum_{i=1}^s m_i\alpha_i=\mathbf{n}-i\alpha}} a_\mathbf{m}\mathsf{w}_\mathbf{m}$$

  with $a_\mathbf{m}\in K$.
- We assume $\alpha>0$, hence, $t_\alpha\in p\mathbb{Z}_p$. Since $v_p(i!)\leq i/(p-1)$ we obtain using 2.4.3 Lemma

$$\left|\frac{t_\alpha^i\hat{\rho}_\lambda^V(x_\alpha)^i}{i!}\right| \leq \frac{p^{-i}p^{\epsilon i\mathrm{ht}(\alpha)}}{p^{-i/(p-1)}} = p^{i\left(\epsilon\mathrm{ht}(\alpha)-\frac{p-2}{p-1}\right)}.$$

  Thus, for any $\mathbf{m}$ satisfying $\sum_{i=1}^s m_i\alpha_i=\mathbf{n}-i\alpha$ we find

$$|a_\mathbf{m}|\leq|\sum_{\substack{\mathbf{k}\\ \sum_{i=1}^s k_i\alpha_i=\mathbf{n}-i\alpha}} a_\mathbf{k}\mathsf{w}_\mathbf{k}| = \left|\frac{t_\alpha^i x_\alpha^i}{i!}\mathsf{w}_\mathbf{n}\right| \leq p^{i\left(\epsilon\mathrm{ht}(\alpha)-\frac{p-2}{p-1}\right)} \leq p^{\delta\mathrm{ht}(i\alpha)} = |\omega^{\mathrm{ht}(\mathbf{m}-\mathbf{n})}|,$$

  which implies that $|\mathsf{c}a_\mathbf{m}|\leq 1$ and $|\mathsf{c}a_\mathbf{m}|\leq|p^r\omega^{\mathrm{ht}(\mathbf{m})}|$ (distinguish the cases $r+\delta\mathrm{ht}(\mathbf{n})\geq 0$ and $r+\delta\mathrm{ht}(\mathbf{n})<0$). Hence, $\frac{t_\alpha^i x_\alpha^i}{i!}\mathsf{c}\mathsf{w}_\mathbf{n}=\sum_\mathbf{m}\mathsf{c}a_\mathbf{m}\mathsf{w}_\mathbf{m}\in\hat{V}_{\lambda,\mathcal{O}}(r,\sigma)$ which shows that Eq. (32) holds.
- We assume $\alpha<0$, hence, $t_\alpha\in\mathbb{Z}_p$. Using 2.3.2 Lemma and the first inequality in Eq. (1) we obtain

$$\left|\frac{t_\alpha^i\hat{\rho}_\lambda^V(x_\alpha)^i}{i!}\mathsf{w}_\mathbf{n}\right| \leq p^{i\epsilon\mathrm{ht}(\alpha)+i/(p-1)}\leq 1.$$

  Thus, for all $\mathbf{m}$ satisfying $\mathbf{m}=\mathbf{n}-i\alpha$ we obtain

$$|a_{\mathbf{m}}| \le 1$$

which implies that $|\mathsf{c}a_{\mathbf{m}}| \le 1$ and $|\mathsf{c}a_{\mathbf{m}}| \le |\mathsf{c}| \le |p^r\omega^{\mathrm{ht}(\mathbf{n})}| \le |p^r\omega^{\mathrm{ht}(\mathbf{m})}|$ (note that $\mathrm{ht}(\alpha) < 0$). Again, this implies that $\frac{t_\alpha^i x_\alpha^i}{i!}\mathsf{c}\mathrm{w}_{\mathbf{n}} = \sum_{\mathbf{m}} \mathsf{c}a_{\mathbf{m}}\mathrm{w}_{\mathbf{m}} \in \hat{V}_{\lambda,\mathcal{O}}(r,\sigma)$, hence, Eq. (32) holds in this case too.

Finally, for any $t \in \mathbf{T}(\mathbb{Z}_p)^1$ 3.3.2 Lemma implies that $t\mathsf{c}\mathrm{w}_{\mathbf{n}} = \lambda^\circ(t)\mathbf{n}^\circ(t^{-1})\mathsf{c}\mathrm{w}_{\mathbf{n}} \in \mathcal{O}\mathsf{c}\mathrm{w}_{\mathbf{n}}$ because $\lambda^\circ(t)\mathbf{n}^\circ(t^{-1}) \in \mathcal{O}^*$), hence, Eq. (33) holds. Thus, the proof of the Lemma is complete.

4.2 Truncation of $\hat{V}_\lambda$

Definition Let $\lambda \in \mathfrak{h}_\Gamma^*(\mathcal{O})$. The truncation $\hat{\mathbf{V}}_{\lambda,\mathcal{O}}[r,\sigma]$ of $\hat{V}_\lambda$ of length r (and slope $\frac{1}{\sigma}$) is defined as

$$\hat{\mathbf{V}}_{\lambda,\mathcal{O}}[r,\sigma] := \frac{\hat{V}_{\lambda,\mathcal{O}}}{\hat{V}_{\lambda,\mathcal{O}}(r,\sigma)}.$$

Remark (1) By 4.1 Lemma $\hat{\mathbf{V}}_\lambda[r,\sigma]$ is a $\mathcal{I}^1$-module.

(2) Since σ only depends on **G** we will write more simply $\hat{V}_{\lambda,\mathcal{O}}(r)$ resp. $\hat{\mathbf{V}}_{\lambda,\mathcal{O}}[r]$ for $\hat{V}_{\lambda,\mathcal{O}}(r,\sigma)$ resp. $\hat{\mathbf{V}}_{\lambda,\mathcal{O}}[r,\sigma]$.

Lemma *For any weight $\lambda \in \mathfrak{h}_\Gamma^*(\mathcal{O})$ and any $r \in \mathbb{N}$ there is an embedding (of abelian groups)*

$$\hat{\mathbf{V}}_{\lambda,\mathcal{O}}[r] \le \bigoplus_{h=0}^{\lfloor -r/\delta \rfloor} \left(\frac{\mathcal{O}}{p^r\omega^h\mathcal{O}}\right)^{(h+1)^{s-1}}.$$

Proof We denote by N_h the number of s-tuples $\mathbf{n} = (n_1 \dots, n_s) \in \mathbb{N}_0^s$ such that

$$\mathrm{ht}(\mathbf{n}) = \sum_{i=1}^{s} n_i \mathrm{ht}(\alpha_i) = h.$$

It is easy to see that

$$N_h \le (h+1)^{s-1}. \tag{34}$$

The definition of the truncation yields

$$\hat{\mathbf{V}}_{\lambda,\mathcal{O}}[r] = \bigoplus_{\substack{h\in\mathbb{N}_0 \\ r+\delta h \ge 0}} \bigoplus_{\substack{\mathbf{n}\in\mathbb{N}_0^s \\ \mathrm{ht}(\mathbf{n})=h}} \frac{\mathcal{O}}{p^r\omega^h\mathcal{O}}\mathrm{w}_{\mathbf{n}} \cong \bigoplus_{h=0}^{\lfloor -r/\delta \rfloor} \left(\frac{\mathcal{O}}{p^r\omega^h\mathcal{O}}\right)^{N_h},$$

and together with Eq. (34) this yields the claim.

4.3 The Representation of $\mathbf{T}(\mathbb{Q}_p)^{--}$ on $\hat{V}_{\lambda,\mathcal{O}}$

Lemma *Let $h \in \mathbf{T}(\mathbb{Q}_p)^{--}$. Then*

$$\hat{\pi}_\lambda(h)\hat{V}_{\lambda,\mathcal{O}}(r) \subseteq p^r \hat{V}_{\lambda,\mathcal{O}}.$$

Proof We let $\mathbf{n} \in \mathbb{N}_0^s$ and we look at

$$\hat{\pi}_\lambda(h)\mathsf{c}_\mathbf{n}\mathsf{w}_\mathbf{n} = \lambda^\circ(hh^\dagger)\mathbf{n}^\circ(h^{-1})\mathsf{c}_\mathbf{n}\mathsf{w}_\mathbf{n},$$

where $\mathsf{c}_\mathbf{n}$ is as in the proof of 4.1 Lemma. We first assume $r + \delta\mathrm{ht}(\mathbf{n}) \geq 0$, hence, $\mathsf{c}_\mathbf{n} \in p^r\omega^{\mathrm{ht}(\mathbf{n})}\mathcal{O}$. Since $v_p(\alpha(h^{-1})) \geq 1$ for all $\alpha \in \Delta$ we know that $v_p(\mathbf{n}^\circ(h^{-1})) = v_p(\prod_{i=1,\ldots,s}\alpha_i^{n_i}(h^{-1})) \geq \mathrm{ht}(\mathbf{n})$. Hence,

$$v_p(\lambda^\circ(hh^\dagger)\mathbf{n}^\circ(h^{-1})\mathsf{c}_\mathbf{n}) \geq r + (1+\delta)\mathrm{ht}(\mathbf{n}) \geq r$$

(note that $1+\delta \geq 0$) and we obtain $\hat{\pi}_\lambda(h)\mathsf{c}_\mathbf{n}\mathsf{w}_\mathbf{n} \in p^r\mathcal{O}\mathsf{w}_\mathbf{n} \subseteq p^r\hat{V}_{\lambda,\mathcal{O}}$. Next we assume $r + \delta\mathrm{ht}(\mathbf{n}) < 0$, hence, $\mathsf{c}_\mathbf{n} \in \mathcal{O}$. We then find

$$v_p(\lambda^\circ(hh^\dagger)\mathbf{n}^\circ(h^{-1})\mathsf{c}_\mathbf{n}) \geq \mathrm{ht}(\mathbf{n}) \geq (-\delta)\mathrm{ht}(\mathbf{n}) \geq r$$

(note that $-\delta \leq 1$). Thus, $\hat{\pi}_\lambda(h)\mathsf{c}_\mathbf{n}\mathsf{w}_\mathbf{n} \in p^r\mathcal{O}\mathsf{w}_\mathbf{n} \subseteq p^r\hat{V}_{\lambda,\mathcal{O}}$ and the Lemma is proven.

5 Hecke Algebra and Cohomology

5.1 Notations

We keep the notations from the previous sections, i.e. $\mathbf{G}/\mathbb{Q}$ is a connected semi simple algebraic group defined over $\mathbb{Q}$ which contains a maximal torus $\mathbf{T}$ defined over $\mathbb{Q}$ and split over $\mathbb{Q}_p$. In particular, $\mathbf{G}/\mathbb{Q}_p$ is split, hence, isomorphic to a Chevalley group $\mathbf{G}/\mathbb{Q}_p = \mathbf{G}_{\rho_0}/\mathbb{Q}_p$ (cf. Sect. 3.1).

We select an arithmetic subgroup $\Gamma \leq \mathbf{G}(\mathbb{Q})$ satisfying $\Gamma \leq \mathcal{I}^1$.

5.2 Hecke Algebra

Also, from now on, we fix an element $h \in \mathbf{T}(\mathbb{Q})^{++}$, i.e. $v_p(\alpha(h)) \geq 1$ for all $\alpha \in \Delta$. We denote by

$$\mathrm{D}_h = \langle \mathcal{I}^1, h\rangle \leq \mathbf{G}(\mathbb{Q}_p)$$

the sub semi group generated by $\mathcal{I}^1$ and h and we set

$$\Delta_h = \{g \in \mathbf{G}(\mathbb{Q}) : g \in \mathrm{D}_h\}.$$

Hence,

$$\mathrm{D}_h = \bigcup_{e \in \mathbb{N}_0} \mathcal{I}^1 h^e \mathcal{I}^1 \tag{35}$$

(cf. Eq. (27)) and we obtain the following diagram of inclusions

$$\begin{array}{ccccc} & & \mathbf{G}(\mathbb{Q}) & & \mathbf{G}(\mathbb{Q}_p) \\ & & \cup & & \cup \\ & & & & \mathrm{D}^{-1} \\ & & & & \cup \\ h & \in & \Delta_h & \subseteq & \mathrm{D}_h \\ & & \cup & & \cup \\ & & \Gamma & \subseteq & \mathcal{I}^1. \end{array}$$

We write $\mathcal{H} = \mathcal{H}(\Gamma\backslash\Delta_h/\Gamma)$ for the Hecke algebra attached to the pair $\Gamma \subseteq \Delta_h$. Thus, $\mathcal{H}$ is a $\mathbb{Z}$-algebra with $\mathbb{Z}$-linear basis $\{T_\zeta = \Gamma\zeta\Gamma,\ \zeta \in \Delta_h\}$ and for any $\mathbb{Z}$-algebra R we set $\mathcal{H}_R = \mathcal{H} \otimes R$.

5.3 *Cohomology of* Γ

We let $\lambda \in \mathfrak{h}^*_\Gamma(\mathcal{O})$ (here, $\Gamma = \Gamma_{\rho_0}$ is the weight lattice of ρ_0). We recall that $\hat{V}_\lambda$ is a D-module (cf. Sect. 3.5), hence, it is a Γ-module, i.e. the cohomology groups

$$H^i(\Gamma, \hat{V}_\lambda)$$

are defined and the Hecke algebra $\mathcal{H}$ acts on them. For convenience we recall the definition of this action. The groups $H^i(\Gamma, \hat{V}_\lambda)$ may be defined as the cohomology groups of the standard complex $(C^i(\Gamma, \hat{V}_\lambda))_{i\geq 0}$ where $C^i(\Gamma, \hat{V}_\lambda)$ is the set of all mappings $c : \Gamma^{i+1} \to \hat{V}_\lambda$ which satisfy $c(\gamma\gamma_0, \ldots, \gamma\gamma_i) = \gamma c(\gamma_0, \ldots, \gamma_i)$, $\gamma, \gamma_0, \ldots, \gamma_i \in \Gamma$. Let $\zeta \in \Delta_h$. We fix a system of representatives $\gamma_1, \ldots, \gamma_d$ for $\zeta^{-1}\Gamma\zeta \cap \Gamma\backslash\Gamma$, hence, we obtain

$$T_\zeta = \bigcup_{i=1,\ldots,d} \Gamma\zeta\gamma_i.$$

For any $\eta \in \Gamma$ and any index i satisfying $1 \leq i \leq d$ there is an index $\eta(i)$ such that

$$\Gamma\zeta\gamma_i\eta = \Gamma\zeta\gamma_{\eta(i)},$$

In particular, there are $\rho_i(\eta) \in \Gamma, i = 1, \dots, d$, such that $\zeta\gamma_i\eta = \rho_i(\eta)\zeta\gamma_{\eta(i)}$. Let now $c \in C^k(\Gamma, \hat{V}_\lambda)$ be any cochain; we then define $T_\zeta(c)$ as the cochain $c' \in C^k(\Gamma, \hat{V}_\lambda)$, which is given by

$$c'(\eta_0, \dots, \eta_k) = \sum_{1 \le i \le d} (\zeta\gamma_i)^{-1} c(\rho_i(\eta_0), \dots, \rho_i(\eta_k)) \tag{36}$$

(cf. [15], p. 227). Thus, T_ζ defines an operator on $C^\bullet(\Gamma, \hat{V}_\lambda)$. Since the action of T_ζ commutes with the coboundary operator, T_ζ acts on cohomology with coefficients in $\hat{V}_\lambda$, i.e. T_ζ defines an element in $\mathrm{End}(H^\bullet(\Gamma, \hat{V}_\lambda))$ which is independent of the choice of the representatives $\gamma_1, \dots, \gamma_d$ (cf. [15], p. 227).

Note. The Hecke algebra acts from the right on cohomology but since later on we will mostly consider only a single Hecke operator T_ζ this will not be relevant and we therefore write the image of a cohomology class c under T as $T(c)$.
Remark Since also $\hat{V}_{\lambda,\mathcal{O}}$ is a D-module (cf. 3.5 Proposition) the coset T_ζ defines via Eq. (36) an operator on $C^i(\Gamma, \hat{V}_{\lambda,\mathcal{O}})$ and, hence, a Hecke operator

$$T_\zeta : H^i(\Gamma, \hat{V}_{\lambda,\mathcal{O}}) \to H^i(\Gamma, \hat{V}_{\lambda,\mathcal{O}}).$$

5.4 Hecke Operators on Cohomology of $\hat{V}_{\lambda,\mathcal{O}}$

Lemma 4.1 implies that $\hat{V}_{\lambda,\mathcal{O}}(r)$ is a $\mathcal{I}^1$, hence, a Γ-module and the inclusion i : $\hat{V}_{\lambda,\mathcal{O}}(r) \subseteq \hat{V}_{\lambda,\mathcal{O}}$ induces a mapping

$$i^* : H^i(\Gamma, \hat{V}_{\lambda,\mathcal{O}}(r)) \to H^i(\Gamma, \hat{V}_{\lambda,\mathcal{O}}).$$

Lemma *(1) The mapping i^* induces an injection*

$$i^* : H^i(\Gamma, \hat{V}_{\lambda,\mathcal{O}}(r))^{\mathrm{TF}} \hookrightarrow H^i(\Gamma, \hat{V}_{\lambda,\mathcal{O}})^{\mathrm{TF}},$$

where the upper index "TF" denotes the maximal torsion free quotient. In particular, we may identify $H^i(\Gamma, \hat{V}_{\lambda,\mathcal{O}}(r))^{\mathrm{TF}}$ with its image in $H^i(\Gamma, \hat{V}_{\lambda,\mathcal{O}})^{\mathrm{TF}}$ under i^*.

(2) Let $\zeta \in \Delta_h$, hence, $\zeta \in \mathcal{I}^1 h^e \mathcal{I}^1$ for some $e \in \mathbb{N}_0$ and we assume that $e \in \mathbb{N}$. The Hecke operator T_ζ induces an operator on $H^i(\Gamma, \hat{V}_{\lambda,\mathcal{O}})^{\mathrm{TF}}$ and we obtain

$$T_\zeta\big(H^i(\Gamma, \hat{V}_{\lambda,\mathcal{O}}(r))^{\mathrm{TF}}\big) \subseteq p^r H^i(\Gamma, \hat{V}_{\lambda,\mathcal{O}})^{\mathrm{TF}}.$$

Proof (1) The exact sequence of $\mathcal{O}$-modules

$$0 \to \hat{V}_{\lambda,\mathcal{O}}(r) \xrightarrow{i} \hat{V}_{\lambda,\mathcal{O}} \xrightarrow{\pi} \hat{\mathbf{V}}_{\lambda,\mathcal{O}}[r] \to 0$$

yields an exact sequence

$$H^{i-1}(\Gamma, \hat{\mathbf{V}}_{\lambda,\mathcal{O}}[r]) \to H^i(\Gamma, \hat{V}_{\lambda,\mathcal{O}}(r)) \stackrel{i^*}{\to} H^i(\Gamma, \hat{V}_{\lambda,\mathcal{O}}) \stackrel{\pi^*}{\to} H^i(\Gamma, \hat{\mathbf{V}}_{\lambda,\mathcal{O}}[r]).$$

Since $H^i(\Gamma, \hat{\mathbf{V}}_{\lambda,\mathcal{O}}[r])$ is a finite abelian group (cf. Sect. 5.5.1 below) we further obtain an exact sequence

$$0 \to H^i(\Gamma, \hat{V}_{\lambda,\mathcal{O}}(r))^{\mathrm{TF}} \stackrel{i^*}{\to} H^i(\Gamma, \hat{V}_{\lambda,\mathcal{O}})^{\mathrm{TF}} \stackrel{\pi^*}{\to} Q \to 0, \tag{37}$$

where Q is a certain subquotient of $H^i(\Gamma, \hat{\mathbf{V}}_{\lambda,\mathcal{O}}[r])$. In particular, i^* is injective.

(2) Since T_ζ is an operator on $H^i(\Gamma, \hat{V}_{\lambda,\mathcal{O}})$ it induces an operator on $H^i(\Gamma, \hat{V}_{\lambda,\mathcal{O}})^{\mathrm{TF}}$. To prove the second claim we note that $\gamma_i \in \Gamma \subseteq \mathcal{I}^1$, where γ_i are the elements appearing in Eq. (36), hence, we can write $\zeta\gamma_i = k_1 h^e k_2$ where $k_1, k_2 \in \mathcal{I}^1$. Since $\hat{\pi}_\lambda(k_i)$ leaves $\hat{V}_{\lambda,\mathcal{O}}$ and $\hat{V}_{\lambda,\mathcal{O}}(r)$ invariant (cf. 4.1 Lemma) and since

$$\hat{\pi}_\lambda(h^{-e})\hat{V}_{\lambda,\mathcal{O}}(r) \subseteq p^r \hat{V}_{\lambda,\mathcal{O}}$$

(cf. 4.3 Lemma) we obtain

$$(\zeta\gamma_i)^{-1}\hat{V}_{\lambda,\mathcal{O}}(r) \subseteq p^r \hat{V}_{\lambda,\mathcal{O}}.$$

Using the definition of the action of T_ζ (on $C^i(\Gamma, \hat{V}_{\lambda,\mathcal{O}})$) this yields

$$T_\zeta(C^i(\Gamma, \hat{V}_{\lambda,\mathcal{O}}(r))) \subseteq p^r C^i(\Gamma, \hat{V}_{\lambda,\mathcal{O}}) \; (\subseteq C^i(\Gamma, \hat{V}_{\lambda,\mathcal{O}}))$$

(here, we view $C^i(\Gamma, \hat{V}_{\lambda,\mathcal{O}}(r))$ as subset of $C^i(\Gamma, \hat{V}_{\lambda,\mathcal{O}})$, i.e. it as embedded via i^*). The last equation implies the claim, hence, the Lemma is proven.

5.5 *Cohomology and Banach Spaces*

5.5.1 Finite Free Resolutions

According to [3] the arithmetic subgroup $\Gamma \leq \mathbf{G}(\mathbb{Q})$ is a group of type FL (cf. [5], p. 218), i.e. there is a finite length resolution of the trivial $\mathbb{Z}[\Gamma]$-module $\mathbb{Z}$ (i.e. Γ acts trivially on $\mathbb{Z}$)

$$0 \to M_d \to \cdots \stackrel{\delta^2}{\to} M_1 \stackrel{\delta^1}{\to} M_0 \to \mathbb{Z} \to 0 \tag{38}$$

where each M_i is a free $\mathbb{Z}[\Gamma]$-module of finite rank (cf. [5], p. 199). The cohomology groups $H^i(\Gamma, \hat{V}_\lambda)$ may then be computed as the cohomology groups of the complex

$$D^\bullet: \quad 0 \to D^0 \stackrel{d^1}{\to} D^1 \stackrel{d^2}{\to} \cdots \to D^d \to 0$$

where

$$D^k = \mathrm{Hom}_{\mathbb{Z}[\Gamma]}(M_k, \hat{V}_\lambda), \tag{39}$$

i.e.

$$H^k(\Gamma, \hat{V}_\lambda) = H^k((\mathrm{Hom}_{\mathbb{Z}[\Gamma]}(M_i, \hat{V}_\lambda))_i).$$

Replacing $\hat{V}_\lambda$ by the Γ-module $\hat{\mathbf{V}}_{\lambda,\mathcal{O}}[r]$ we obtain in particular, that the cohomology groups $H^i(\Gamma, \hat{\mathbf{V}}_{\lambda,\mathcal{O}}[r])$ are finite abelian groups (note that $\hat{\mathbf{V}}_{\lambda,\mathcal{O}}[r]$ is a finite abelian group; cf. 4.2 Lemma).

We note that the functor "$\otimes_{\mathbb{Z}[\Gamma]} K[\Gamma] = \otimes_{\mathbb{Z}} K$" is exact, hence, the complex

$$0 \to M_d \otimes_{\mathbb{Z}[\Gamma]} K[\Gamma] \to \cdots \xrightarrow{\delta^2} M_1 \otimes_{\mathbb{Z}[\Gamma]} K[\Gamma] \xrightarrow{\delta^1} M_0 \otimes_{\mathbb{Z}[\Gamma]} K[\Gamma] \to \mathbb{Z} \otimes_{\mathbb{Z}[\Gamma]} K[\Gamma] \to 0$$

is exact, i.e. it is a finite free resolution of the $K[\Gamma]$-module $\mathbb{Z} \otimes_{\mathbb{Z}[\Gamma]} K[\Gamma] = K$ on which Γ acts trivially. Since

$$\mathrm{Hom}_{\mathbb{Z}[\Gamma]}(M_k, \hat{V}_\lambda) = \mathrm{Hom}_{K[\Gamma]}(M_k \otimes_{\mathbb{Z}[\Gamma]} K[\Gamma], \hat{V}_\lambda)).$$

the groups $H^i(\Gamma, \hat{V}_\lambda)$ can also be computed from this complex.

5.5.2 Coboundary Operator

We select a (finite) basis $(m_j)_j = (m_j^k)_j$ of M_k, which yields an isomorphism of K-vector spaces $\mathrm{Hom}_{\mathbb{Z}[\Gamma]}(M_k, \hat{V}_\lambda) \cong \hat{V}_\lambda^{g_k}$, where $g_k = \dim_{\mathbb{Z}[\Gamma]} M_k$. We define a norm $|\cdot|$ on $\mathrm{Hom}_{\mathbb{Z}[\Gamma]}(M_k, \hat{V}_\lambda)$ by

$$|c| = \sup_j |c(m_j)|, \quad c \in \mathrm{Hom}_{\mathbb{Z}[\Gamma]}(M_k, \hat{V}_\lambda).$$

Thus, $(D^k, |\cdot|) = (\mathrm{Hom}_{\mathbb{Z}[\Gamma]}(M_k, \hat{V}_\lambda), |\cdot|)$ is a K-Banach space which is isomorphic to the K-Banach space $\hat{V}_\lambda^{g_k}$ which carries the sup norm.

Lemma *The coboundary operator* $d^k : \mathrm{Hom}_{\mathbb{Z}[\Gamma]}(M_{k-1}, \hat{V}_\lambda) \to \mathrm{Hom}_{\mathbb{Z}[\Gamma]}(M_k, \hat{V}_\lambda)$ *has norm less than or equal to* 1.

Proof Let $m \in M_k$; then $\delta^k(m) \in M_{k-1}$ can be written $\delta^k(m) = \sum_j \alpha_j m_j$ where $m_j = m_j^{k-1}$ and $\alpha_j = \sum_\gamma \alpha_{j,\gamma}\gamma \in \mathbb{Z}[\Gamma]$ $(\alpha_{j,\gamma} \in \mathbb{Z})$. Let $c \in \mathrm{Hom}_{\mathbb{Z}[\Gamma]}(M_{k-1}, \hat{V}_\lambda)$. We obtain

$$|(d^k c)(m)| = |c(\delta^k(m))| = |\sum_j \alpha_j c(m_j)| \leq \max_j |\alpha_j c(m_j)|.$$

Since $|\hat{\pi}_\lambda(\gamma)| \leq 1$ (cf. 3.5 Proposition) and $\alpha_{j,\gamma} \in \mathbb{Z}$ we obtain $|\alpha_j c(m_j)| = |\sum_\gamma \alpha_{j,\gamma}\hat{\pi}_\lambda(\gamma)c(m_j)| \leq |c(m_j)|$. Hence, $|d^k(c)(m)| \leq \max_j |c(m_j)| = |c|$ and this holds for all $m \in M_k$. In particular, this holds for all $m = m_j^k$ which shows that $|d^k(c)| \le |c|$, hence, $|d^k| \le 1$ and the Lemma is proven.

5.5.3 Norm on Cohomology

We denote by Z^k resp. B^k the subspace of cocycles resp. coboundaries in $\mathrm{Hom}_{\mathbb{Z}[\Gamma]}(M_k, \hat{V}_\lambda)$. Since d^k is continuous $Z^k \le \mathrm{Hom}_{\mathbb{Z}[\Gamma]}(M_k, \hat{V}_\lambda)$ is closed, hence, Z^k also is a Banach space with norm $|\cdot|$ given by restricting the norm on $\mathrm{Hom}_{\mathbb{Z}[\Gamma]}(M_k, \hat{V}_\lambda)$ to Z^k. We denote by $\bar{B}^k \le Z^k$ the closure of the space of coboundaries B^k in Z^k; the norm on Z^k induces a norm $|\cdot|$ on the quotient

$$\tilde{H}^k(\Gamma, \hat{V}_\lambda) = \tilde{H}^k((\mathrm{Hom}_{\mathbb{Z}[\Gamma]}(M_i, \hat{V}_\lambda))_i) := Z^k/\bar{B}^k$$

which is a Banach space. On the other hand, the norm on $Z^k \le \mathrm{Hom}_{\mathbb{Z}[\Gamma]}(M_k, \hat{V}_\lambda)$ induces a semi norm $|\cdot|$ on Z^k/B^k and the quotient topology on Z^k/B^k coincides with the topology defined by this semi norm. Thus, $H^k(\Gamma, \hat{V}_\lambda) = H^k((\mathrm{Hom}_{\mathbb{Z}[\Gamma]}(M_i, \hat{V}_\lambda))_i) = Z^k/B^k$ becomes a topological vector space with topology induced by a semi norm. The Hausdorff quotient $H^k(\Gamma, \hat{V}_\lambda)/\overline{\{0\}}$ of $H^k(\Gamma, \hat{V}_\lambda)$ is isomorphic to $Z^k/\bar{B}^k = \tilde{H}^k((\mathrm{Hom}_{\mathbb{Z}[\Gamma]}(M_i, \hat{V}_\lambda))_i) = \tilde{H}^k(\Gamma, \hat{V}_\lambda)$. The quotient topology on $H^k(\Gamma, \hat{V}_\lambda)/\overline{\{0\}}$ coincides with the topology defined by the norm on $H^k(\Gamma, \hat{V}_\lambda)/\overline{\{0\}}$ which is induced by the semi norm on $H^k(\Gamma, \hat{V}_\lambda)$.

Let

$$0 \to N_t \to \cdots \xrightarrow{\delta^2} N_1 \xrightarrow{\delta^1} N_0 \to \mathbb{Z} \to 0$$

be a second finite length resolution of $\mathbb{Z}$ where each N_i is a free $\mathbb{Z}[\Gamma]$-module of finite rank. The complex

$$E^\bullet: \quad 0 \to E^0 \xrightarrow{d^1} E^1 \xrightarrow{d^2} \cdots \to E^t \to 0$$

where $E^k = \mathrm{Hom}_{\mathbb{Z}[\Gamma]}(N_k, \hat{V}_\lambda)$ also computes the cohomology $H^k(\Gamma, \hat{V}_\lambda)$. There are mappings of complexes

$$\Phi_\bullet : M_\bullet \to N_\bullet \quad \text{and} \quad \Psi : N_\bullet \to M_\bullet$$

which lift the identity on $\mathbb{Z}$, hence, they are inverse to each other modulo $\mathbb{Z}[\Gamma]$-homotopy equivalence. We obtain morphisms of complexes $\Phi^\bullet$ and $\Psi^\bullet$ where

$$\begin{aligned} \Psi^k : \mathrm{Hom}_{\mathbb{Z}[\Gamma]}(M_k, \hat{V}_\lambda) &\to \mathrm{Hom}_{\mathbb{Z}[\Gamma]}(N_k, \hat{V}_\lambda) \\ c &\mapsto c \circ \Psi_k \end{aligned}$$

and analogously for Φ^k. We select (finite) $\mathbb{Z}[\Gamma]$-basis $(m_i)_i$ of M_k and $(n_i)_i$ of N_k and we write $\Psi_k(n_i) = \sum_j z_{ij} m_j$, $z_{ij} \in \mathbb{Z}[\Gamma]$. For any $c \in \mathrm{Hom}_{\mathbb{Z}[\Gamma]}(M_k, \hat{V}_\lambda)$ and all i we find

$$|\Psi^k(c)(n_i)| = |c(\Psi_k(n_i))| = |c\left(\sum_j z_{ij}m_j\right)| = |\sum_j z_{ij}c(m_j)| \leq \max_j |c(m_j)| = |c|$$

(note that $|\hat{\pi}_\lambda(\gamma)| \leq 1$ for all $\gamma \in \Gamma \subseteq \mathcal{I}^1$ by 3.5 Proposition, hence, $|z_{ij}| \leq 1$). Thus, $|\Psi^k(c)| \leq |c|$ which implies $|\Psi^k| \leq 1$, i.e. Ψ^k is continuous. Analogously, we see that Φ^k is continuous, hence, Ψ^k and Φ^k induce homeomorphisms on cohomology which are inverse to each other. In particular, the topology on $H^i(\Gamma, \hat{V}_\lambda)$ does not depend on the choice of the finite free resolution (38).

5.5.4 Compact Hecke Operators

Since $\hat{V}_\lambda$ and $\hat{V}_{\lambda,\mathcal{O}}$ are left modules for Δ_h^{-1} ($\subseteq$ D) (cf. 3.5 Proposition), following [20], Sect. 4.2.5–4.2.10, p. 1738 ff. an action of Hecke operators $T_\zeta = \Gamma\zeta\Gamma, \zeta \in \Delta_h$, on cochains $D^k = \mathrm{Hom}_{\mathbb{Z}[\Gamma]}(M_k, \hat{V}_\lambda)$ and $\mathrm{Hom}_{\mathbb{Z}[\Gamma]}(M_k, \hat{V}_{\lambda,\mathcal{O}})$ is defined.

Lemma *For any $\zeta \in \Delta_h$ the Hecke operator T_ζ acting on $\mathrm{Hom}_{\mathbb{Z}[\Gamma]}(M_k, \hat{V}_\lambda)$ has norm less than or equal to* 1. *If $\zeta \in \mathcal{I}^1 h^e \mathcal{I}^1$ with $e \geq 1$ (cf. Eq. (35)) then T_ζ even is a compact operator.*

Proof We recall from [20], Sect. 4.2.6, p. 1739 that the Hecke operator $T_\zeta = \Gamma\zeta\Gamma$ acting on D^k is given as

$$\mathrm{Cor}^{\Gamma}_{\Gamma\cap\zeta^{-1}\Gamma\zeta} \circ [\zeta] \circ \mathrm{Res}^{\Gamma}_{\Gamma\cap\zeta\Gamma\zeta^{-1}};$$

All three mappings appearing as factors in the above definition of T_ζ on cochains are compositions of mappings of the following two forms (cf. [20], p. 1738/1739):

(1) Let $\Gamma, \Gamma' \leq \mathbf{G}(\mathbb{Q})$ be arithmetic subgroups with $\Gamma, \Gamma' \leq \mathcal{I}^1$. Let $M_\bullet$ resp. $M'_\bullet$ be a finite $\mathbb{Z}[\Gamma]$ resp. $\mathbb{Z}[\Gamma']$-resolution of $\mathbb{Z}$ by modules which are free of finite rank. Attached to a compatible pair (ϕ, f) of mappings $\phi : \Gamma \to \Gamma'$ and $f : \hat{V}_\lambda \to \hat{V}_\lambda$ a map

$$A : \mathrm{Hom}_{\mathbb{Z}[\Gamma']}(M'_\bullet, \hat{V}_\lambda) \stackrel{\phi^*_\bullet \otimes f}{\to} \mathrm{Hom}_{\mathbb{Z}[\Gamma]}(M_\bullet, \hat{V}_\lambda)$$

is defined in [20], Sect. 4.2.5, p. 1738; here, $\phi_\bullet : M_\bullet \to M'_\bullet$ is a morphism of $\mathbb{Z}$-modules (it is compatible with ϕ; cf. [20], p. 1738). The restriction $\mathrm{Res}^{\Gamma'}_{\Gamma}$, $\Gamma \subseteq \Gamma'$, is attached to the pair $(\phi, f) = (\mathrm{incl.}, \mathrm{id}_{\hat{V}_\lambda})$, the mapping $[\zeta]$ is attached to the pair $(\phi, f) = (\mathrm{Inn}(\zeta), \hat{\pi}_\lambda(\zeta^{-1}))$ (cf. [20], p. 1739; note that in [20], $\hat{V}_\lambda$ is viewed as a right $\mathcal{I}^1$-module via $v\gamma = \gamma^{-1}v, \gamma \in \mathcal{I}^1$; cf. [20], p. 1738, 1741). Thus, in these cases we have $|f| \leq 1$ (cf. 3.5 Proposition). As for the corestriction $\mathrm{Cor}^{\Gamma}_{\Gamma'}$, $\Gamma' \subseteq \Gamma$, it is the composition of two mappings, the first one being of the form

$$A' : \mathrm{Hom}_{\mathbb{Z}[\Gamma']}(M'_\bullet, \hat{V}_\lambda) \stackrel{(\tau, \mathrm{id}_{\hat{V}_\lambda})}{\to} \mathrm{Hom}_{\mathbb{Z}[\Gamma']}(M_\bullet, \hat{V}_\lambda)$$

where $\tau : M_\bullet \to M'_\bullet$ is Γ'-equivariant morphism. We want to show that $|A| \leq 1$. We let $(m_i)_i$ resp. $(m'_i)_i$ be a $\mathbb{Z}[\Gamma]$ resp. $\mathbb{Z}[\Gamma']$-basis of M_k resp. M'_k. We write $\phi_k(m_i) = \sum_j z_{ij} m'_j$, $z_{ij} \in \mathbb{Z}[\Gamma']$, and obtain

$$|A(c)(m_i)| = |f(c(\phi_k(m_i)))| \leq |c\left(\sum_j z_{ij} m'_j\right)| = |\sum_j z_{ij} c(m'_j)| \leq \max_j |c(m'_j)| = |c|;$$

here, we have used that $|f| \leq 1$ and $|z_{ij}| \leq 1$ (cf. 3.5 Proposition). Hence, $|A(c)| \leq |c|$ which shows that $|A| \leq 1$. Similarly, we obtain $|A'| \leq 1$.

(2) Let $\Gamma' \leq \Gamma$ be a subgroup of finite index and write $\Gamma = \bigcup_i \gamma_i \Gamma'$, i.e. $(\gamma_i)_i$ is a system of representatives for Γ/Γ'. The second factor of $\mathrm{Cor}^{\Gamma}_{\Gamma'}$ is of the form

$$\begin{aligned} B : \mathrm{Hom}_{\mathbb{Z}[\Gamma']}(M'_\bullet, \hat{V}_\lambda) &\to \quad \mathrm{Hom}_{\mathbb{Z}[\Gamma]}(M_\bullet, \hat{V}_\lambda) \\ c \quad &\mapsto \{x \mapsto \textstyle\sum_i \gamma_i (c(\gamma_i^{-1} x))\}. \end{aligned}$$

We want to show that also $|B| \leq 1$. We note that M_k also is a free $\mathbb{Z}[\Gamma']$-module with a finite basis $(k_l)_l$ (which defines a norm on $\mathrm{Hom}_{\mathbb{Z}[\Gamma']}(M_k, \hat{V}_\lambda)$). We write $\gamma_i^{-1} m_j = \sum_l z_{ijl} k_l$ where $z_{ijl} \in \mathbb{Z}[\Gamma']$. We obtain

$$|B(c)(m_j)| = |\sum_i \gamma_i c(\gamma_i^{-1} m_j)| \leq \max_i |c(\gamma_i^{-1} m_j)| = \max_i |c\left(\sum_l z_{ijl} k_l\right)| \leq \max_{i,l} |c(k_l)| = |c|,$$

hence, again we obtain $|B| \leq 1$. Thus, T_ζ has norm less than or equal to 1

In particular, $\mathrm{Cor}^{\Gamma}_{\Gamma\cap\zeta^{-1}\Gamma\zeta}$ and $\mathrm{Res}^{\Gamma}_{\Gamma\cap\zeta\Gamma\zeta^{-1}}$ are continuous. Since $f : \hat{V}_\lambda \to \hat{V}_\lambda$, $v \mapsto \zeta^{-1} v$, is compact if $\zeta \in \mathcal{I}^1 h^e \mathcal{I}^1$ with $e \geq 1$ (cf. 3.5 Proposition) Sect. 4.2.9, p. 1740 in [20] implies that $[\zeta]$ is compact, hence, T_ζ is compact. Thus the Lemma is proven.

6 The Characteristic Series

6.1 Characteristic Series

As before $K/\mathbb{Q}_p$ is a finite extension with ring of integers $\mathcal{O}$ and local prime $\varpi \in \mathcal{O}$ and we denote by v_ϖ the valuation on K normalized by $v_\varpi(\varpi) = 1$. We let V be a K-Banach space with ONB $(w_i)_{i\in I}$ and T a compact operator on V. We denote by $\mathrm{Fr}_T(X) = \det(1 - XT) = c_0 + c_1 X + c_2 X^2 + \cdots \in K[[X]]$ the Fredholm determinant of T. We assume that $|T| \leq 1$, hence, the representing matrix of T w.r.t. $(w_i)_{i\in I}$ has entries in $\mathcal{O}$ and $\mathrm{Fr}_T(X) \in \mathcal{O}[[X]]$. We let $\mathcal{N}_T = \mathcal{N}(\mathrm{Fr}_T)$ be the Newton polygon of the power series Fr_T, i.e. $\mathcal{N}_T$ is the lower convex envelope of the points $(i, v_p(c_i))$, $i = 0, 1, 2, 3, \ldots$ (see e.g. [10], p. 207/8 for a precise definition of the Newton polygon). Thus, $\mathcal{N}_T$ is a polygon of infinite length (with finitely or infinitely many segments); in particular, if Fr_T is a polynomial then $\mathcal{N}_T$ has finitely many segments and ends with a vertical line. We will also view the polygon $\mathcal{N}_T$ as a piecewise linear function

$$\mathcal{N}_T : [0, m) \to \mathbb{R}_{\geq 0}$$

where $m = m_T$ is the degree of Fr_T ($m < \infty$ precisely if Fr_T is a polynomial).

6.2 An Estimate on Coefficients of Characteristic Series

The following Proposition extends the estimate on the Newton polygon in [7] to the case of compact operators on Banach spaces.

Proposition *Let T be a compact operator on the orthonormalizable K-Banach space V with $|T| \leq 1$. We set $V_{\mathcal{O}} = \{v \in V : |v| \leq 1\}$ and we assume that there is a sub $\mathcal{O}$-module $V'_{\mathcal{O}} \leq V_{\mathcal{O}}$ such that*

1. *$V_{\mathcal{O}}/V'_{\mathcal{O}}$ is a finite abelian group, hence, we can write (as $\mathcal{O}$-modules)*

$$V_{\mathcal{O}}/V'_{\mathcal{O}} = \bigoplus_{i=1}^{k} \mathcal{O}/\varpi^{a_i}\mathcal{O},$$

   where $a_1 \geq a_2 \geq \cdots \geq a_k > 0$. We set $a_i = 0$ if $i > k$.
2. *$T(V'_{\mathcal{O}}) \subseteq \varpi^r V_{\mathcal{O}}$ for some $r \in \mathbb{N}$ with $r \geq a_1$.*

Let $\mathrm{Fr}_T = \det(1 - XT) = \sum_{i\geq 0} c_i X^i$ be the Fredholm determinant of T. Then for all $i \geq 0$ the coefficient c_i satisfies

$$v_{\varpi}(c_i) \geq \sum_{h=1}^{i} r - a_h.$$

Proof Since T is compact we know that $T = \lim_n T_n$, where T_n, $n \in \mathbb{N}$, is a continuous operator on V of finite rank. We let $U_n \leq V$ be a finite dimensional sub K-vector space which contains the image of T_n, hence, $T_n(U_n) \subseteq U_n$. Since $U_n \leq V$ is finite dimensional it is closed, hence, U_n is a (sub) Banach space and since the valuation on K is discrete there is a (finite) ONB $(e_i)_{i\in I}$ of U_n. Since $T_n \to T$ and $|T| \leq 1$ we know that $|T_n| \leq 1$ for almost all $n \in \mathbb{N}$ and removing finitely many of the T_n we may assume that $|T_n| \leq 1$ for all $n \in \mathbb{N}$. We set $U_{n,\mathcal{O}} = U_n \cap V_{\mathcal{O}} \leq U_n$. The $\mathcal{O}$-sub module $U_{n,\mathcal{O}}$ satisfies the following properties.

- $U_{n,\mathcal{O}}$ has $\mathcal{O}$-basis $(e_i)_{i\in I}$; in particular, $U_n = U_{n,\mathcal{O}} \otimes K$.
- $U_{n,\mathcal{O}}$ is complemented in $V_{\mathcal{O}}$.

Proof. Since V has an ONB, [18], Theorem 5.16, ι, p. 179 or [18], p. 180 implies that the finite orthonormal set $(e_i)_{i\in I}$ can be extended to an ONB $(e_i)_{i\in J}$ of V and the closure of the $\mathcal{O}$-span of the basis vectors e_i, $i \in J - I$ is a complement.

- $T_n(U_{n,\mathcal{O}}) \subseteq U_n \cap V_{\mathcal{O}} = U_{n,\mathcal{O}}$.
- We set

$$U'_{n,\mathcal{O}} = U_{n,\mathcal{O}} \cap V'_{\mathcal{O}} = U_n \cap V'_{\mathcal{O}} \leq U_{n,\mathcal{O}}.$$

We first claim that if n is sufficiently large, i.e. $|T - T_n|$ is sufficiently small then

$$T_n(V'_{\mathcal{O}}) \subseteq \varpi^r V_{\mathcal{O}}. \tag{40}$$

To prove this let $v \in V'_{\mathcal{O}}$. Hence, $T_n(v) = T(v) + (T_n - T)(v)$. The first summand is contained in $\varpi^r V_{\mathcal{O}}$ by assumption. We write $(T_n - T)(v) = \sum_{i\in J} \alpha_i e_i$, $\alpha_i \in K$. If n is sufficiently large so that $|T - T_n| \leq |\varpi^r|$ we see that $|\alpha_i| \leq |\varpi^r|$ for all i, hence, $\alpha_i \in \varpi^r \mathcal{O}$. Hence, we obtain $(T_n - T)(v) \in \varpi^r V_{\mathcal{O}}$ which proves Eq. (40).

Removing finitely many T_n we may assume that Eq. (40) holds for all n. In particular, we then obtain for all n

$$T_n(U'_{n,\mathcal{O}}) \subseteq T_n(U_{n,\mathcal{O}}) \cap T_n(V'_{\mathcal{O}}) \subseteq U_{n,\mathcal{O}} \cap \varpi^r V_{\mathcal{O}} \subseteq \varpi^r U_{n,\mathcal{O}};$$

for the last inclusion note that $U_{n,\mathcal{O}} \leq V_{\mathcal{O}}$ is complemented.

We denote by

$$\det(1 - XT_n|_{U_n}) = c_0^{(n)} + c_1^{(n)} X + c_2^{(n)} X^2 + \cdots + c_m^{(n)} X^m$$

the characteristic polynomial of $T_n|_{U_n}$. Proposition 7, (c), (d), p. 75 in [19] shows that

$$\lim_n c_i^{(n)} = c_i.$$

Since $U_{n,\mathcal{O}}$ is a finitely generated free $\mathcal{O}$-module there is a $\mathcal{O}$-basis $(f_j)_{j=1,\ldots,m}$ of $U_{n,\mathcal{O}}$ such that

$$U'_{n,\mathcal{O}} = \bigoplus_{i=1}^{m'} \mathcal{O}\varpi^{b_j} f_j,$$

where $m' \leq m$ and $b_1 \geq b_2 \geq \cdots \geq b_{m'} \geq 0$. Since

$$U_{n,\mathcal{O}}/U'_{n,\mathcal{O}} \hookrightarrow V_{\mathcal{O}}/V'_{\mathcal{O}}$$

where $V_{\mathcal{O}}/V'_{\mathcal{O}}$, hence, $U_{n,\mathcal{O}}/U'_{n,\mathcal{O}}$ is finite by assumption we obtain $m' = m$ and

$$b_1 \leq a_1, b_2 \leq a_2, \ldots, b_m \leq a_m,$$

where we have set $a_i = 0$ if $i > k$.

We denote by $\mathcal{D}=(n_{ij})_{i,j=1,\dots,m}$ the matrix representing T_n w.r.t. the basis $(f_j)_j$ of U_n, i.e. $T_n(f_j)=\sum_i n_{ij}f_i$. Since $T_n(U'_{n,\mathcal{O}})\subseteq\varpi^r U_{n,\mathcal{O}}$ we know that $T_n(\varpi^{b_j}f_j)\in\varpi^r U_{n,\mathcal{O}}$, hence, $T_n(f_j)\in\varpi^{r-b_j}U_{n,\mathcal{O}}$ (note that $b_j\le r$), i.e. $\varpi^{r-b_j}|n_{ij}$ for all i and all j. Since $b_j\le a_j$ we obtain $\varpi^{r-a_j}|n_{ij}$ for all $i,j\le m$ (note that $a_j\le r$), i.e. the jth column of $\mathcal{D}$ is divisible by ϖ^{r-a_j}. This implies

$$c_i^{(n)}\in\varpi^{\sum_{h=1}^i r-a_h}\mathcal{O}$$

for all $i\le m$. Since $c_i=\lim_n c_i^{(n)}$ this yields

$$v_\varpi(c_i)\ge\sum_{h=1}^i r-a_h.$$

Hence, the Proposition is proven.

6.3 Slope Decomposition

We keep the notations from the previous sections, i.e. $\mathbf{G}/\mathbb{Q}$ is a connected semi simple algebraic group containing a maximal torus $\mathbf{T}/\mathbb{Q}$ which is split over $\mathbb{Q}_p$; hence, $\mathbf{G}/\mathbb{Q}_p\cong\mathbf{G}_{\rho_0}/\mathbb{Q}_p$ for some faithful finite dimensional complex representation ρ_0 of the complex Lie Algebra $\mathfrak{g}$. We selected an arithmetic subgroup $\Gamma\le\mathbf{G}(\mathbb{Q})$. Also we fixed $h\in\mathbf{T}(\mathbb{Q})^{++}$ and we defined $\Delta_h=\{g\in\mathbf{G}(\mathbb{Q}):\ g\in\mathrm{D}_h\}$ (cf. Sect. 5.2). We let $\zeta\in\Delta_h$ and $\lambda\in\mathfrak{h}^*_\Gamma(\mathcal{O})$ (here, $\Gamma=\Gamma_{\rho_0}=$ weight lattice of ρ_0). Thus, $\zeta\in\mathcal{I}^1h^e\mathcal{I}^1$ for some $e\in\mathbb{N}_0$ and we assume that $e\ge 1$; 5.5.4 Lemma then implies that $T_\zeta=\Gamma\zeta\Gamma$ is a compact operator on the Hausdorff quotient $\tilde{H}^i(\Gamma,\hat{V}_\lambda)$ of $H^i(\Gamma,\hat{V}_\lambda)$ with $|T_\zeta|\le 1$.

We denote by $\tilde{\mathrm{Fr}}^i_\lambda(X)\in\mathcal{O}[[X]]$ resp. $\tilde{\mathcal{N}}^i_\lambda=\mathcal{N}(\tilde{\mathrm{Fr}}^i_\lambda(X))$ the Fredholm determinant of $T_\zeta:\ \tilde{H}^i(\Gamma,\hat{V}_\lambda)\to\tilde{H}^i(\Gamma,\hat{V}_\lambda)$ resp. the Newton polygon of $\tilde{\mathrm{Fr}}^i_\lambda(X)$. Since $\tilde{\mathrm{Fr}}^i_\lambda(X)$ has constant coefficient c_0 equal to 1 the Newton polygon $\tilde{\mathcal{N}}^i_\lambda$ starts in $(0,0)$ and is monotonically increasing (not necessarily strictly monotonically increasing). In particular, all slopes of $\tilde{\mathcal{N}}^i_\lambda$ are non negative.

Let $\beta\in\mathbb{Q}_{\ge 0}$ and let $\beta_1,\dots,\beta_k$ be the slopes of $\tilde{\mathcal{N}}^i_\lambda$ which are smaller than or equal to β. We denote by $\tilde{\mathcal{N}}^{i,\le\beta}_\lambda$ the portion of $\tilde{\mathcal{N}}^i_\lambda$ which consists of the segments of $\tilde{\mathcal{N}}^i_\lambda$ which have slopes $\beta_1,\dots,\beta_k$. The Fredholm determinant then factorizes

$$\tilde{\mathrm{Fr}}^i_\lambda=\tilde{\mathrm{Fr}}^{i,\le\beta}_\lambda\tilde{\mathrm{Fr}}^{i,>\beta}_\lambda$$

where $\tilde{\mathrm{Fr}}^{i,\le\beta}_\lambda$ is a polynomial whose degree is the sum of the lengths of the projections to the x-axis of the segments of slopes $\beta_1,\dots,\beta_k$ and which satisfies $\mathcal{N}(\tilde{\mathrm{Fr}}^{i,\le\beta}_\lambda)=\tilde{\mathcal{N}}^{i,\le\beta}_\lambda$ (cf. [10], Prop. 6.5.10, p. 216). Theorem 2.3.8, p. 1708 in [20] then yields a slope-$\leq \beta$ decomposition

$$\tilde{H}^i(\Gamma, \hat{V}_\lambda) = \tilde{H}^i(\Gamma, \hat{V}_\lambda)^{\leq\beta} \oplus \tilde{H}^i(\Gamma, \hat{V}_\lambda)^{>\beta},$$

i.e. both summands are T_ζ-invariant, $\tilde{\mathrm{Fr}}_\lambda^{i,\leq\beta*}(T_\zeta)$ is invertible on $\tilde{H}^i(\Gamma, \hat{V}_\lambda)^{>\beta}$ (here, $P^*(X) := X^{\deg P} P(1/X)$, P a polynomial) and the Fredholm determinant of T_ζ on $\tilde{H}^i(\Gamma, \hat{V}_\lambda)^{\leq\beta}$ is $\tilde{\mathrm{Fr}}_\lambda^{i,\leq\beta}$. We call $\tilde{H}^i(\Gamma, \hat{V}_\lambda)^{\leq\beta}$ the slope $\leq \beta$ subspace of $\tilde{H}^i(\Gamma, \hat{V}_\lambda)$ w.r.t. to T_ζ (cf. [20], Corollary 2.3.3, p. 1707). Since $\mathcal{N}(\tilde{\mathrm{Fr}}_\lambda^{i,\leq\beta}) = \tilde{\mathcal{N}}_\lambda^{i,\leq\beta}$ we obtain that the dimension of the slope $\leq \beta$ subspace $\tilde{H}^i(\Gamma, \hat{V}_\lambda)^{\leq\beta}$ equals the sum of the lengths of the projections to the x-axis of the segments of $\tilde{\mathcal{N}}_\lambda^i$ which are of slope less than or equal to β.

6.4 Unit Ball in Cohomology

We set

$$\tilde{H}^i(\Gamma, \hat{V}_\lambda)_{\mathcal{O}} := \{\omega \in \tilde{H}^i(\Gamma, \hat{V}_\lambda) : |\omega| \leq 1\} \leq \tilde{H}^i(\Gamma, \hat{V}_\lambda). \tag{41}$$

Lemma *The mapping of $\mathcal{O}$-modules*

$$\begin{aligned} \Phi : H^i(\Gamma, \hat{V}_{\lambda,\mathcal{O}}) &\to \tilde{H}^i(\Gamma, \hat{V}_\lambda)_{\mathcal{O}} \\ c + B^i(\hat{V}_{\lambda,\mathcal{O}}) &\mapsto c + \bar{B}^i(\hat{V}_\lambda) \end{aligned}$$

is well defined, surjective and commutes with Hecke operators. Here, $Z^i(X)$ resp. $B^i(X)$ is the space of cocycles resp. coboundaries in $\mathrm{Hom}_{\mathbb{Z}[\Gamma]}(M_i, X)$, $X = \hat{V}_{\lambda,\mathcal{O}}, \hat{V}_\lambda$ *and $\bar{B}^i(\hat{V}_\lambda)$ is the closure of $B^i(\hat{V}_\lambda)$ in $Z^i(\hat{V}_\lambda)$.*

Proof The inclusion of $\mathbb{Z}[\Gamma]$-modules $\hat{V}_{\lambda,\mathcal{O}} \subseteq \hat{V}_\lambda$ induces a mapping in cohomology

$$\begin{aligned} f_1 : H^i(\Gamma, \hat{V}_{\lambda,\mathcal{O}}) &\to H^i(\Gamma, \hat{V}_\lambda) \\ c + B^i(\hat{V}_{\lambda,\mathcal{O}}) &\mapsto c + B^i(\hat{V}_\lambda). \end{aligned}$$

Composing with the projection

$$\begin{aligned} f_2 : H^i(\Gamma, \hat{V}_\lambda) &\to \tilde{H}^i(\Gamma, \hat{V}_\lambda) \\ c + B^i(\hat{V}_\lambda) &\mapsto c + \bar{B}^i(\hat{V}_\lambda) \end{aligned}$$

we obtain the mapping

$$\begin{aligned} \Phi = f_2 \circ f_1 : H^i(\Gamma, \hat{V}_{\lambda,\mathcal{O}}) &\to \tilde{H}^i(\Gamma, \hat{V}_\lambda) \\ c + B^i(\hat{V}_{\lambda,\mathcal{O}}) &\mapsto c + \bar{B}^i(\hat{V}_\lambda). \end{aligned}$$

(1) We selected a (finite) $\mathbb{Z}[\Gamma]$-basis $(m_j)_j$ of M_i which yields a norm on $\mathrm{Hom}_{\mathbb{Z}[\Gamma]}(M_i, \hat{V}_\lambda)$ and, hence, on $\tilde{H}^i(\Gamma, \hat{V}_\lambda)$ (cf. 5.5.2). Let $c + B^i(\hat{V}_{\lambda,\mathcal{O}}) \in H^i(\Gamma, \hat{V}_{\lambda,\mathcal{O}})$, $c \in Z^i(\hat{V}_{\lambda,\mathcal{O}})$. Then,

$$|\Phi(c + B^i(\hat{V}_{\lambda,\mathcal{O}}))| = \inf_{b \in \bar{B}^i(\hat{V}_\lambda)} |c + b| \leq |c| = \sup_j |c(m_j)| \leq 1,$$

where the last inequality holds because c has its values in $\hat{V}_{\lambda,\mathcal{O}}$. Thus, $|\Phi(c + B^i(\hat{V}_{\lambda,\mathcal{O}}))| \leq 1$ and Φ induces a well defined mapping

$$\Phi : H^i(\Gamma, \hat{V}_{\lambda,\mathcal{O}}) \to \tilde{H}^i(\Gamma, \hat{V}_\lambda)_\mathcal{O}. \tag{42}$$

(2) We show that Φ in Eq. (42) is surjective. Let $c + \bar{B}^i(\hat{V}_\lambda) \in \tilde{H}^i(\Gamma, \hat{V}_\lambda)_\mathcal{O}$, $c \in Z^i(\hat{V}_\lambda)$. Then

$$|c + \bar{B}^i(\hat{V}_\lambda)| = \inf_{b \in \bar{B}^i(\hat{V}_\lambda)} |c + b| \leq 1,$$

hence, for any $\epsilon > 0$ there is $b \in \bar{B}^i(\hat{V}_\lambda)$ such that $|c + b| < 1 + \epsilon$. Since the valuation on $\mathrm{Hom}_{\mathbb{Z}[\Gamma]}(M_i, \hat{V}_\lambda)$ is discrete this implies $|c + b| \leq 1$ if ϵ is chosen sufficiently small. We obtain $|(c + b)(m_j)| \leq 1$ for all j and this implies that $|(c + b)(m)| \leq 1$ for all $m = \sum_j z_j m_j \in M_i$ (because $z_j \in \mathbb{Z}[\Gamma]$ has norm less than or equal to 1; cf. 3.5 Proposition). The cocycle $c + b$ therefore has its values in $\hat{V}_{\lambda,\mathcal{O}}$, i.e. $c + b \in \mathrm{Hom}_{\mathbb{Z}[\Gamma]}(M_i, \hat{V}_{\lambda,\mathcal{O}})$. Since $b \in \bar{B}^i(\hat{V}_\lambda) \subseteq Z^i(\hat{V}_\lambda)$ we see that $c + b \in Z^i(\hat{V}_\lambda)$, hence, $c + b \in Z^i(\hat{V}_{\lambda,\mathcal{O}})$. Since $\Phi(c + b + B^i(\hat{V}_{\lambda,\mathcal{O}})) = c + b + \bar{B}^i(\hat{V}_\lambda) = c + \bar{B}^i(\hat{V}_\lambda)$ we finally obtain that Φ is surjective.

(3) Since $T(c + \bar{B}^i(\hat{V}_\lambda)) := T(c) + \bar{B}^i(\hat{V}_\lambda)$, $c \in Z^i(\hat{V}_\lambda)$, the map f_2 obviously commutes with the action of Hecke operators. Since f_1 is Hecke equivariant we obtain that Φ is Hecke equivariant and the Lemma is proven.

6.5 The Newton Polygon of T_ζ on Cohomology

We recall that $g_i = \mathrm{rk}_{\mathbb{Z}[\Gamma]} M_i$ (cf. Sect. 5.5.2). Also, we recall that we have set $\delta = \epsilon - \frac{1}{\sigma}\frac{p-2}{p-1} \in \mathbb{Q}_{<0}$ and $\omega \in K$ is an element with $v_p(\omega) = \delta$ (cf. Eq. (31) in Sect. 4.1). Moreover, we select $\zeta \in \Delta_h$ and we assume that $\zeta \in \mathcal{I}^1 h^e \mathcal{I}^1$ for some $e \geq 1$ (cf. Eq. (35)); this yields a compact operator $T_\zeta = \Gamma\zeta\Gamma$ on $\tilde{H}^i(\Gamma, \hat{V}_\lambda)$ of norm less than or equal to 1 (cf. 5.5.4 Lemma). By $B_s(X)$ we denote the sth Bernoulli polynomial.

Theorem *For all weights $\lambda \in \mathfrak{h}_\Gamma^*(\mathcal{O})$ and all $i \in \mathbb{N}_0$ the Newton polygon $\tilde{\mathcal{N}}_\lambda^i$ of T_ζ acting on $\tilde{H}^i(\Gamma, \hat{V}_\lambda)$ lies above the piecewise linear function $f_\infty : \mathbb{R}_{\geq 0} \to \mathbb{R}_{\geq 0}$ which starts in $(0, 0)$ and has slope 0 in the interval*

$$0 \leq x \leq g_i \frac{B_s(2) - B_s(1)}{s}$$

and slope $j(-\delta)$ in the interval

$$g_i \frac{B_s(j+1) - B_s(1)}{s} \leq x \leq g_i \frac{B_s(j+2) - B_s(1)}{s}, \quad j = 1, 2, \ldots$$

Equivalently, f_∞ may be defined as the piecewise linear function joining the points

$$(0,0),\ P_j = \left(g_i \frac{B_s(j+2) - B_s(1)}{s}, g_i(-\delta) \sum_{h=0}^{j} h(h+1)^{s-1} \right), \quad j = 0, 1, 2, \ldots.$$

Proof We set $g = g_i$ and proceed in steps.

Step 1.

We select a number $r \in \mathbb{N}$. The short exact sequence

$$0 \to \hat{V}_{\lambda,\mathcal{O}}(r) \to \hat{V}_{\lambda,\mathcal{O}} \to \hat{\mathbf{V}}_{\lambda,\mathcal{O}}[r] \to 0$$

induces an exact sequence

$$\cdots \to H^{i-1}(\Gamma, \hat{\mathbf{V}}_{\lambda,\mathcal{O}}[r]) \to H^i(\Gamma, \hat{V}_{\lambda,\mathcal{O}}(r)) \to H^i(\Gamma, \hat{V}_{\lambda,\mathcal{O}}) \to H^i(\Gamma, \hat{\mathbf{V}}_{\lambda,\mathcal{O}}[r]) \to \cdots$$

and, hence, an exact sequence

$$0 \to H^i(\Gamma, \hat{V}_{\lambda,\mathcal{O}}(r))^{\mathrm{TF}} \to H^i(\Gamma, \hat{V}_{\lambda,\mathcal{O}})^{\mathrm{TF}} \to Q \to 0$$

where Q is a subquotient of the finite $\mathcal{O}$-module $H^i(\Gamma, \hat{\mathbf{V}}_{\lambda,\mathcal{O}}[r])$ (cf. Eq. (37)). Hence,

$$\frac{H^i(\Gamma, \hat{V}_{\lambda,\mathcal{O}})^{\mathrm{TF}}}{H^i(\Gamma, \hat{V}_{\lambda,\mathcal{O}}(r))^{\mathrm{TF}}} \quad \text{is a subquotient of } H^i(\Gamma, \hat{\mathbf{V}}_{\lambda,\mathcal{O}}[r]).$$

Moreover,

$$H^i(\Gamma, \hat{\mathbf{V}}_{\lambda,\mathcal{O}}[r]) \quad \text{is a subquotient of} \quad \mathrm{Hom}_{\mathbb{Z}[\Gamma]}(M_i, \hat{\mathbf{V}}_{\lambda,\mathcal{O}}[r]) \cong \hat{\mathbf{V}}_{\lambda,\mathcal{O}}[r]^g.$$

By 4.2 Lemma we know that

$$\hat{\mathbf{V}}_{\lambda,\mathcal{O}}[r] \leq \bigoplus_{h=0}^{\lfloor -r/\delta \rfloor} \left(\frac{\mathcal{O}}{p^r \omega^h \mathcal{O}} \right)^{(h+1)^{s-1}},$$

hence,

$$H^i(\Gamma, \hat{\mathbf{V}}_{\lambda,\mathcal{O}}[r]) \quad \text{is a subquotient of} \quad \bigoplus_{h=0}^{\lfloor -r/\delta \rfloor} \left(\frac{\mathcal{O}}{p^r \omega^h \mathcal{O}} \right)^{g(h+1)^{s-1}}.$$

Moreover, the mapping Φ (cf. 6.4 Lemma) together with 5.4 Lemma (1) induces a surjective morphism

$$\Phi: H^i(\Gamma, \hat{V}_{\lambda,\mathcal{O}})^{\mathrm{TF}}/H^i(\Gamma, \hat{V}_{\lambda,\mathcal{O}}(r))^{\mathrm{TF}} \to \tilde{H}^i(\Gamma, \hat{V}_\lambda)_{\mathcal{O}}/\Phi(H^i(\Gamma, \hat{V}_{\lambda,\mathcal{O}}(r))^{\mathrm{TF}}).$$

Altogether we obtain that

$$\tilde{H}^i(\Gamma, \hat{V}_\lambda)_{\mathcal{O}}/\Phi(H^i(\Gamma, \hat{V}_{\lambda,\mathcal{O}}(r))^{\mathrm{TF}}) \quad \text{is connected to} \quad \bigoplus_{h=0}^{\lfloor -r/\delta \rfloor} \left(\frac{\mathcal{O}}{p^r \omega^h \mathcal{O}}\right)^{g(h+1)^{s-1}} \tag{43}$$

via a chain of successive subquotients. Thus, $\tilde{H}^i(\Gamma, \hat{V}_\lambda)_{\mathcal{O}}/\Phi(H^i(\Gamma, \hat{V}_{\lambda,\mathcal{O}}(r))^{\mathrm{TF}})$ is a finite $\mathcal{O}$-module and we denote by

$$\varpi^{b_1}, \varpi^{b_2}, \ldots, \varpi^{b_k}, \qquad b_1 \geq b_2 \geq \cdots \geq b_k > 0$$

its elementary divisors. The right hand side of equation (43) has elementary divisors

$$(p^r, \ldots, p^r, p^r\omega, \ldots, p^r\omega, p^r\omega^2, \ldots, p^r\omega^2, \ldots, p^r\omega^{\lfloor -r/\delta \rfloor}, \ldots, p^r\omega^{\lfloor -r/\delta \rfloor})$$

which have p-adic valuations (w.r.t. v_p)

$$(a_i)_{i=1}^l = (r, \ldots, r, r+\delta, \ldots, r+\delta, r+2\delta, \ldots, r+2\delta, \ldots, r+\lfloor \frac{-r}{\delta} \rfloor \delta, \ldots, r+\lfloor \frac{-r}{\delta} \rfloor \delta),$$

with $p^r\omega^j$ and, hence, $r + j\delta$ appearing $g(j+1)^{s-1}$-times. Equation (43) implies

$$(\mathrm{e}r \geq)\ \mathrm{e}a_1 \geq b_1, \mathrm{e}a_2 \geq b_2, \ldots, \mathrm{e}a_k \geq b_k, \tag{44}$$

where e is the ramification degree of $K/\mathbb{Q}_p$. We set $b_i = 0$ if $i > k$ and $a_i = 0$ if $i > l$, hence, $\mathrm{e}a_i \geq b_i$ for all $i = 1, 2, 3, \ldots$.

Step 2.

Since Φ is Hecke equivariant 5.4 Lemma implies that

$$\begin{aligned} T_\varsigma(\Phi(H^i(\Gamma, \hat{V}_{\lambda,\mathcal{O}}(r))^{\mathrm{TF}})) &= \Phi(T_\varsigma(H^i(\Gamma, \hat{V}_{\lambda,\mathcal{O}}(r))^{\mathrm{TF}})) \\ &\subseteq \Phi(p^r H^i(\Gamma, \hat{V}_{\lambda,\mathcal{O}})^{\mathrm{TF}}) \\ &\subseteq \varpi^{\mathrm{e}r} \tilde{H}^i(\Gamma, \hat{V}_\lambda)_{\mathcal{O}}. \end{aligned}$$

We write the Fredholm determinant of T_ς on $\tilde{H}^i(\Gamma, \hat{V}_\lambda)$ as $\tilde{\mathrm{Fr}}^i_\lambda(X) = \sum_{i=0}^\infty c_i X^i \in \mathcal{O}[[X]]$. Thus, 6.2 Proposition applied to T_ς acting on $V = \tilde{H}^i(\Gamma, \hat{V}_\lambda)$ (with "$V'_{\mathcal{O}} = \Phi(H^i(\Gamma, \hat{V}_{\lambda,\mathcal{O}}(r))^{\mathrm{TF}})$") together with Eq. (44) yields

$$v_{\varpi}(c_j) \geq \sum_{h=1}^{j} \mathrm{er} - b_h \geq \sum_{h=1}^{j} \mathrm{er} - ea_h$$

for all $j = 0, 1, 2, \ldots$. Hence, we obtain

$$v_p(c_j) \geq \sum_{h=1}^{j} r - a_h \tag{45}$$

for all $j = 0, 1, 2, \ldots$

Step 3.

We denote by $f_r : \mathbb{R}_{\geq 0} \to \mathbb{R}$ the piecewise linear function which starts in $(0, 0)$ and has slope $0 = 0\delta$ in the interval $0 \leq x \leq g = g1^{s-1}$, slope $j(-\delta)$ in the interval $g\sum_{h=1}^{j} h^{s-1} < x \leq g\sum_{h=1}^{j+1} h^{s-1}$, $j = 1, 2, 3, \ldots, \lfloor -r/\delta \rfloor$ and slope r in the range $x > g\sum_{h=1}^{\lfloor -r/\delta \rfloor + 1} h^{s-1}$. Equation (45) then implies that

$$v_p(c_j) \geq f_r(j) \tag{46}$$

for all $j = 0, 1, 2, 3, \ldots$. We note that Eq. (46) holds for all $r \in \mathbb{N}$ because r was arbitrary in $\mathbb{N}$. Since

$$\sum_{h=1}^{j} h^{s-1} = \frac{B_s(j+1) - B_s(1)}{s} \tag{47}$$

for all $j = 0, 1, 2, 3, \ldots$ we see that the piecewise linear functions f_r and f_∞ coincide in $0 \leq x \leq g\sum_{h=1}^{\lfloor -r/\delta \rfloor + 1} h^{s-1}$, i.e. $f_\infty(j) = f_r(j)$ if r is sufficiently large compared to j. Thus, Eq. (46) yields

$$v_p(c_j) \geq f_\infty(j)$$

for all $j = 0, 1, 2, \ldots$. Since f_∞ is convex this implies that $\tilde{\mathcal{N}}_\lambda^i$ lies above f_∞.

Step 4.

Since the interval where f_∞ has slope $j(-\delta)$ is $g\sum_{h=1}^{j} h^{s-1} \leq x \leq g\sum_{h=1}^{j+1} h^{s-1}$, $j = 0, 1, 2, \ldots$ (cf. Eq. (47)) and this has length $g(j+1)^{s-1}$ we obtain that f_∞ has value $g(-\delta)\sum_{h=0}^{j} h(h+1)^{s-1}$ at the point $x = g\sum_{h=0}^{j}(h+1)^{s-1} = g\sum_{h=1}^{j+1} h^{s-1}$ $= g\frac{B_s(j+2) - B_s(1)}{s}$. Hence, f_∞ satisfies the alternative description as given in the theorem. This concludes the proof of the Theorem.

Corollary *The Newton polygon $\tilde{\mathcal{N}}_\lambda^i$ lies above the piecewise linear function $f_\infty^* : \mathbb{R}_{\geq 0} \to \mathbb{R}_{\geq 0}$ which joins the points*

$$(0,0),\ Q_j = \left(g_i \frac{B_s(j+2) - B_s(1)}{s}, g_i(-\delta)\frac{B_{s+1}(j+1) - B_{s+1}(1)}{s+1}\right),\quad j = 0, 1, 2, \ldots$$

Proof Using Eq. (47) we obtain that the second coordinate of P_j is bigger than or equal to $g_i(-\delta)\sum_{h=0}^{j} h^s = g_i(-\delta)\sum_{h=1}^{j} h^s = g_i(-\delta)\frac{B_{s+1}(j+1)-B_{s+1}(1)}{s+1}$, which is the second coordinate of Q_j. Hence, the claim follows.

6.6 Bound for the Dimension of Slope Subspaces

Corollary *For all $\beta \in \mathbb{Q}_{\geq 0}$, all weights $\lambda \in \mathfrak{h}^*_\Gamma(\mathcal{O})$, and all i we have*

$$\dim \tilde{H}^i(\Gamma, \hat{V}_\lambda)^{\leq \beta} \leq \mathrm{m}\beta^s + \mathrm{n};$$

here, $\mathrm{m} = \mathrm{m}_\Gamma = 11\frac{g_i}{s}(\frac{\sigma(p-1)}{p-2-\sigma})^s \in \mathbb{R}_{\geq 0}$ and $\mathrm{n} = \mathrm{n}_\Gamma \in \mathbb{N}$ is an integer which only depends on g_i and s (cf. Eq. (51) below); in particular, m and n only depend on Γ (and, hence, on **G** *and p) and i, i.e. they do not depend on λ and ζ.*

Proof (*We very closely follow the proof of 3.10 Corollary in* [17].) Let $h : \mathbb{R}_{\geq 0} \to \mathbb{R}$ be any function such that $f^*_\infty(x) \geq h(x)$ for all $x \geq 0$ and let $(d(\varepsilon), y)$ with $d(\varepsilon) > 0$ be an intersection point of h and the function $w_\varepsilon : x \mapsto (\beta + \varepsilon)x$ where $\varepsilon > 0$. 6.5 Corollary yields

$$(\beta + \varepsilon)x > \tilde{\mathcal{N}}^i_\lambda(x) \geq h(x)$$

for all $x \in (0, \dim \tilde{H}^i(\Gamma, \hat{V}_\lambda)^{\leq \beta}]$ and we deduce that $d(\varepsilon) \geq \dim \tilde{H}^i(\Gamma, \hat{V}_\lambda)^{\leq \beta}$. Thus, $d(\varepsilon)$ is an upper bound for the dimension of the slope $\leq \beta$-subspace.

We want to explicitly determine such a bound $d(\varepsilon)$ and for this we define a lower bound h for f^*_∞ as follows. Since B_s is a polynomial of degree s and leading coefficient 1 there is a natural number $M = M(s) \in \mathbb{N}$ such that

$$x_s(j) := g_i \frac{B_s(j+2) - B_s(1)}{s} \leq 2^{\frac{1}{s+1}} \frac{g_i}{s} j^s \tag{48}$$

and

$$y_s(j) := g_i(-\delta)\frac{B_{s+1}(j+1) - B_{s+1}(1)}{s+1} \geq \left(\frac{1}{2}\right)^{\frac{1}{s}} \frac{(-\delta)g_i}{s+1} j^{s+1} \tag{49}$$

for all $j \geq M$. We define the function

$$h : [x_s(M), \infty) \to \mathbb{R}_{\geq 0}, \quad x \mapsto cx^{\frac{s+1}{s}},$$

where $c = 4^{-\frac{1}{s}} g_i^{-\frac{1}{s}} s^{\frac{s+1}{s}} \frac{1}{(s+1)}(-\delta) \in \mathbb{R}_{>0}$. We note that $x_s(M) > 0$ by Eq. (47). We then obtain for all $j \geq M$ (note that $x_s(j) \geq x_s(M)$ by Eq. (47))

$$h(x_s(j)) = h\left(g_i \frac{B_s(j+2) - B_s(1)}{s}\right) \overset{(48)}{\leq} c\left(2^{\frac{1}{s+1}} \frac{g_i}{s} j^s\right)^{\frac{s+1}{s}}$$
$$= c 2^{\frac{1}{s}} \left(\frac{g_i}{s}\right)^{\frac{s+1}{s}} j^{s+1}$$
$$= \left(\frac{1}{2}\right)^{\frac{1}{s}} \frac{g_i(-\delta)}{s+1} j^{s+1} \overset{(49)}{\leq} y_s(j).$$

Since f^*_∞ is the piecewise linear function connecting the points $Q_j = (x_s(j), y_s(j))$, $j \in \mathbb{N}_0$ and $(0,0)$, and since h passes below the points Q_j, $j \geq M$, and is convex, i.e. the set $\{(x, y),\ y \geq h(x)\}$ is convex this implies that $h(x) = cx^{\frac{s+1}{s}} \leq f^*_\infty(x)$ for all $x \geq x_s(M)$. We extend h to a function $h : \mathbb{R}_{\geq 0} \to \mathbb{R}$ by setting $h(x) = f^*_\infty(x)$ for $x \in [0, x_s(M)]$ and $h(x) = cx^{\frac{s+1}{s}}$ if $x > x_s(M)$, hence, $f^*_\infty(x) \geq h(x)$ for all $x \in [0, \infty)$ (note that this redefines the value at $x_s(M)$ of h). Since h starts with a segment of slope 0 which has positive length g_i (cf. definition of f^*_∞ in 6.5 Corollary) and since h jumps downwards at $x_s(M)$ we see that for all $\varepsilon > 0$ the functions h and $x \mapsto (\beta + \varepsilon)x$ always intersect in a point $(d(\varepsilon), y)$ with $d(\varepsilon) > 0$ and this point satisfies $d(\varepsilon) \leq \max((\frac{\beta+\varepsilon}{c})^s, x_s(M))$. Since $(\beta/c)^s, x_s(M) \geq 0$ are positive we obtain

$$\dim \tilde{H}^i(\Gamma, \hat{V}_\lambda)^{\leq \beta} \leq \max((\beta/c)^s, x_s(M)) \leq \left(\frac{\beta}{c}\right)^s + x_s(M).$$

Since further

$$\left(\frac{1}{c}\right)^s = 4 g_i s^{-(s+1)} (s+1)^s (-\delta)^{-s} = 4 g_i s^{-1} \left(1 + \frac{1}{s}\right)^s (-\delta)^{-s} \leq 4 \frac{g_i}{s} (-\delta)^{-s} \mathbf{e} \tag{50}$$

where $\mathbf{e} = 2.718\ldots$ (hence, $4\mathbf{e} \leq 11$) the claim of the Corollary holds with $\mathrm{m}_\delta = 11 \frac{g_i}{s}(-\delta)^{-s}$ in place of m and $\mathrm{n} = x_s(M)$. The claim still holds if we replace n with any larger number and since Eq. (48) implies that $x_s(M) \leq 2^{\frac{1}{s+1}} \frac{g_i}{s} M^s$ the claim in particular holds with

$$\mathrm{n} = \lceil 2^{\frac{1}{s+1}} \frac{g_i}{s} M^s \rceil + 1 \in \mathbb{N}. \tag{51}$$

Moreover, Eq. (1) and $p > \sigma + 2$ implies that $(-\delta)$ may take any value which is strictly smaller than $\frac{1}{\sigma}\frac{p-2}{p-1} - \frac{1}{p-1} = \frac{p-2-\sigma}{\sigma(p-1)}$, hence, m_δ may take any value which is strictly bigger than $11 \frac{g_i}{s}(\frac{\sigma(p-1)}{p-2-\sigma})^s$. Thus, the claim holds with $\mathrm{m} = 11 \frac{g_i}{s}(\frac{\sigma(p-1)}{p-2-\sigma})^s$ and the Corollary is proven.

Remark We note that $\mathrm{m} \geq 11 \frac{g_i}{s} \sigma^s$ and $\lim_{p \to \infty} \mathrm{m} = 11 \frac{g_i}{s} \sigma^s$.

Remark (cf. [20], Sect. 2.3.10, p. 1709/1710). We denote by $Z^i(M, \hat{V}_\lambda)$ resp. $B^i(M, \hat{V}_\lambda)$ the space of cocycles resp. coboundaries in $C^i(M, \hat{V}_\lambda) = \mathrm{Hom}_{\mathbb{Z}[\Gamma]}(M_i, \hat{V}_\lambda)$. The compact operator T_ζ on cochains (cf. 5.5.4 Lemma) induces a compact operator on the closed subspace $Z^{i-1}(M, \hat{V}_\lambda)$ which thus has slope decompositions;

hence, the quotient (Banach) space $C^{i-1}(M,\hat V_\lambda)/Z^{i-1}(M,\hat V_\lambda)$ has slope decompositions (cf. [20], 2.3.5 Corollary, p. 1707). Since $C^{i-1}(M,\hat V_\lambda)/Z^{i-1}(M,\hat V_\lambda)\cong B^i(M,\hat V_\lambda)$ via d^{i-1} the quotient $H^i(\Gamma,\hat V_\lambda)=Z^i(M,\hat V_\lambda)/B^i(M,\hat V_\lambda)$ has slope-β decompositions. The proof of Proposition 2.3.9, p. 1709 in [20] implies that $\overline{B^i(M,\hat V_\lambda)}^{\leq\beta}=\overline{B^i(M,\hat V_\lambda)^{\leq\beta}}=B^i(M,\hat V_\lambda)^{\leq\beta}$, hence, 2.3.5 Corollary in [20] implies that $H^i(\Gamma,\hat V_\lambda)^{\leq\beta}=\tilde H^i(\Gamma,\hat V_\lambda)^{\leq\beta}$ for all $\beta\in\mathbb{Q}_{\geq 0}$. The bound in 6.6 Corollary therefore also holds for the dimension of the slope subspaces of $H^i(\Gamma,\hat V_\lambda)$.

Index of Symbols

2.1. $p, K, \mathcal{O}, |\cdot|, v_p$
2.2.1. $\mathfrak{g}, \mathfrak{h}, \Phi=\Phi(\mathfrak{g},\mathfrak{h}), \Delta, h_\alpha, x_\alpha, \Gamma_{\rm ad}, \Gamma_{\rm sc}; \mathfrak{n}, \mathfrak{b}, \mathfrak{n}^-, \mathfrak{b}^-, \mathcal{U}, \mathcal{U}^+, \mathcal{U}^-, \mathcal{U}^\circ$
2.2.2. $s, X^{\mathbf{n}}_-, X^{\mathbf{n}}_+, H^{\mathbf{n}}$
2.2.3. $\xi_i, \mathrm{h}_i, \mathfrak{h}_\Gamma(\mathbb{Z}), \mathfrak{h}^*_\Gamma(\mathcal{O})$
2.2.4. $\mathrm{ht}(\mathbf{n})$
2.3.1. $V_\lambda, \mathrm{v}_\lambda, \Pi^V_\lambda, \rho^V_\lambda, \sigma, \epsilon, |\cdot|=|\cdot|_{\epsilon,\lambda}, \kappa, \mathrm{w}_{\mathbf{n}}$,
2.3.2. $\Theta, G_\Theta=G_\Theta^{\mathrm{End}(V_\lambda)}$
2.4.2. $|\cdot|=|\cdot|_\Gamma$
2.4.3. $\hat\rho_\lambda=\hat\rho^V_\lambda$
2.4.4. $\hat V_{\lambda,\mathcal{O}}$
2.4.5. $\mathrm{i}=\mathrm{i}(\Gamma)$
2.5. $G_\Theta=G_\Theta^{\mathrm{End}(\hat V_\lambda)}$
2.6.1. $(\rho_0, L_{\rho_0}), \Gamma=\Gamma_{\rho_0}$ ($\Gamma_{\rm ad}\leq\Gamma\leq\Gamma_{\rm sc}$ is a lattice), $|\cdot|$
2.6.2. $G_\Theta=G_\Theta^{\mathrm{End}(L_{\rho_0}(\mathbb{Q}_p))}$
3.1. $\mathbf{G}, \mathbf{T}, \mathbf{G}_{\rho_0}, x_\alpha(t_\alpha)=x^{\rho_0}_\alpha(t_\alpha), G_{\rho_0,R}, h_\alpha(t_\alpha), \mu^\circ, \mathbf{T}^{+(+)}, \mathbf{T}^{-(-)}, \mathbf{B}, \mathbf{N}, \mathbf{B}^-, \mathbf{N}^-$, $\mathrm{h}_i(t), \pi_{\lambda^\circ}$
3.2. $\mathcal{I}, \mathcal{I}^1, \mathbf{T}(\mathbb{Z}_p)^1, N^-_{\mathbb{Z}_p}, N_{p\mathbb{Z}_p}, T^1_\Gamma$
3.3.1. $\hat\pi_{\lambda,\mathcal{I}^1}$
3.3.2. λ°
3.4. $t^\dagger, \hat\pi_{\lambda,\mathbf{T}}$
3.5. $\mathrm{D}, \hat\pi_\lambda$
4.1. $\omega, \delta, \hat V_{\lambda,\mathcal{O}}(r,\sigma)=\hat V_{\lambda,\mathcal{O}}(r)$,
4.2. $\hat{\mathbf{V}}_{\lambda,\mathcal{O}}[r,\sigma]=\hat{\mathbf{V}}_{\lambda,\mathcal{O}}[r]$
5.1. Γ (arithmetic subgroup of $\mathbf{G}(\mathbb{Q})$; same symbol as in 2.6.1 but different meaning)
5.2. h, D_h, Δ_h (sub semi group of $\mathbf{G}(\mathbb{Q})$; $\Delta=$ basis of root system; cf. 2.2.1), $\mathcal{H}=\mathcal{H}(\Gamma\backslash\Delta_h\Gamma), T_\zeta=\Gamma\zeta\Gamma$
5.5.1. $M_k, D^\bullet=(D^k)_k$,
5.5.2. $g_i, |\cdot|$
5.5.3. $\tilde H^i(\Gamma,\hat V_\lambda)$
6.1. $\varpi, v_\varpi, \mathrm{Fr}_T, \mathcal{N}_T$
6.3. $\tilde{\mathrm{Fr}}^i_\lambda, \tilde{\mathcal{N}}^i_\lambda, \tilde{\mathcal{N}}^{i,\leq\beta}_\lambda, \tilde{\mathrm{Fr}}^{i,\leq\beta}_\lambda, \tilde H^i(\Gamma,\hat V_\lambda)^{\leq\beta}$
6.4. $\tilde H^i(\Gamma,\hat V_\lambda)_{\mathcal{O}}$
6.5. $B_s, f_\infty, f^*_\infty$
6.6. $H^i(\Gamma,\hat V_\lambda)^{\leq\beta}$

Acknowledgements I am grateful to the referee for pointing out errors and for helpful remarks. I am thankful for the invitation to participate in a conference in honor of Prof. Schwermer's 66th birthday.

References

1. Ash, A., Stevens, G.: p-adic deformations of arithmetic cohomology, preprint (2008)
2. Borel, A.: Properties and linear representations of Chevalley groups. Seminar on Algebraic Groups and Related Finite Groups. LNM, vol. 131. Springer, Berlin (1970)
3. Borel, A., Serre, J.-P.: Corners and arithmetic groups. Comment. Math. Helv. **48**, 436–491 (1974)
4. Bewersdorff, J.: Eine Lefschetzsche Fixpunktformel für Hecke Operatoren, Bonner Mathematische Schriften, vol. 164 (1985)
5. Brown, K.S.: Cohomology of Groups. GTM, vol. 87. Springer, New York (1982)
6. Bourbaki, N.: Lie Groups and Lie Algebras, Chap. 1–3. Springer, Berlin (1989)
7. Buzzard, K.: Families of modular forms. J. de Théorie de Nombres de Bordeaux **13**, 43–52 (2001)
8. Buzzard, K., Calegari, F.: A counter-example to the Gouvea–Mazur conjecture. Comptes rendus - Mathématiques **338**(10), 751–753 (2004)
9. Chevalley, C.: Certains schémas de groupes semi simples. Semin. Bourbaki **219**, 219–234 (1961)
10. Gouvêa, F.: p-Adic Numbers. Springer, Berlin (1993)
11. Gouvêa, F., Mazur, B.: Families of modular eigenforms. Math. Comput. **58**(198), 793–805 (1992)
12. Hida, H.: On p-adic Hecke algebras for GL2 over totally real fields. Ann. Math. **128**, 295–384 (1988)
13. Hida, H.: Elementary Theory of Eisenstein Series and L-Functions. London Mathematical Society. Cambridge University Press, Cambridge (1993)
14. Humphreys, J.: Introduction to Lie Algebras and Representation Theory. Springer, New York (1972)
15. Kuga, M., Parry, W., Sah, C.: Group cohomology and Hecke operators. In: Hano, J., et al. (eds.) Manifolds and Lie groups, pp. 223–266. Birkhäuser, Boston (1981)
16. Mahnkopf, J.: On truncation of irreducible representations of Chevalley groups. J. Number Theory **133**, 3149–3182 (2013)
17. Mahnkopf, J.: Local constancy of dimension of slope subspaces of automorphic forms. Pac. J. Math. **289**(2), 317–380 (2017)
18. vanRooij, A.C.M.: Non-archimedean Functional Analysis. Marcel Dekker, New York (1978)
19. Serre, J.-P.: Endomorphismes complétement continus des espaces de Banach p-adiques. Publ. Math. I.H.E.S **12**, 69–85 (1962)
20. Urban, E.: Eigenvarieties for reductive groups. Ann. Math. **174**, 1685–1784 (2011)
21. Wan, T.: Dimension variation of classical and p-adic modular forms. Invent. Math. **133**, 449–463 (1998)

Cohomological Representations and Functorial Transfer from Classical Groups

A. Raghuram and Makarand Sarnobat

Abstract We study whether the property of being cohomological is preserved under Langlands functoriality for the transfer of tempered representations from real classical Lie groups to an appropriate general linear group.

Keywords Cohomological representations · Langlands functoriality

1 Introduction

The main aim of this article is to study the cohomological properties of representations of $\mathrm{GL}(n, \mathbb{R})$ and $\mathrm{GL}(n, \mathbb{C})$ which are obtained via Langlands transfers of tempered representations of classical groups. This purely local quest has a global motivation: the project started with the following observation that was made in [9, Theorem 3.2]. Let $G = \mathrm{GL}(2)/F$ for a number field F. Let μ be a strongly pure dominant integral weight for G and let M_μ be the finite-dimensional representation of G with highest weight μ. Suppose that π is a cuspidal automorphic representation of G that is cohomological with respect to $M_\mu^{\vee}$, i.e., $\pi_\infty \otimes M_\mu^{\vee}$ has nontrivial relative Lie algebra cohomology (equivalently, that the finite part π_f contributes to the cohomology of a locally symmetric space of G with coefficients in the sheaf associated to the contragredient representation $M_\mu^{\vee}$ of M_μ). Let's denote this as $\pi \in \mathrm{Coh}(G, \mu^{\vee})$. For an integer $r \geq 1$, suppose that $\mathrm{Sym}^r(\pi)$, which is the Langlands transfer corresponding to the L-group homomorphism $\mathrm{Sym}^r : \mathrm{GL}_2(\mathbb{C}) \to \mathrm{GL}_{r+1}(\mathbb{C})$, is cuspidal on $\mathrm{GL}(r+1)/F$. We can transfer the weight μ to a weight $\mathrm{Sym}^r(\mu)$ for $\mathrm{GL}(r+1)/F$ and [9, Theorem 3.2] says that

A. Raghuram (✉) · M. Sarnobat
Indian Institute of Science Education and Research, Dr. Homi Bhabha Road, Pashan, Pune 411008, India
e-mail: raghuram@iiserpune.ac.in

M. Sarnobat
e-mail: makarand.sarnobat@students.iiserpune.ac.in

J. W. Cogdell et al. (eds.), *Cohomology of Arithmetic Groups*,
Springer Proceedings in Mathematics & Statistics 245,
https://doi.org/10.1007/978-3-319-95549-0_6

$$\pi \in \mathrm{Coh}(\mathrm{GL}(2), \mu^{\vee}) \implies \mathrm{Sym}^r(\pi) \in \mathrm{Coh}(\mathrm{GL}(r+1), \mathrm{Sym}^r(\mu)^{\vee}). \quad (1.1)$$

We might say that the property of being cohomological is preserved under Langlands transfer and in fact the coefficient system also transfers in a nice way; indeed, if $\mu = (\mu_\tau)_{\tau:F\to\mathbb{C}}$ with $\mu_\tau = (a_\tau, b_\tau)$ then $\mathrm{Sym}^r(\mu) = (\mathrm{Sym}^r(\mu)_\tau)_{\tau:F\to\mathbb{C}}$ with $\mathrm{Sym}^r(\mu)_\tau = (ra_\tau, (r-1)a_\tau + b_\tau, \dots, a_\tau + (r-1)b_\tau, rb_\tau)$. Such a result was then used to study the arithmetic of symmetric power L-functions attached to π. The reader is also referred to [10] where there is a general discussion involving Langlands functoriality, cohomological representations, and applications to the special values of L-functions.

The question we address in this article is to what extent we can generalize the implication (1.1) if we start with a classical group G and transfer to an appropriate $\mathrm{GL}(N)$. The local archimedean results obtained in this article apply to questions of proving nonvanishing of cuspidal cohomology as studied in [1]. The main result of this article is Theorem 5.6. As a sampling, for the introduction, consider a dominant integral weight $\lambda = (\lambda_1, \dots, \lambda_n)$ for the symplectic group $\mathrm{Sp}(2n, \mathbb{R})$. Denote the inclusion of its dual group $\mathrm{SO}(2n+1, \mathbb{C})$ into $\mathrm{GL}(2n+1, \mathbb{C})$ as ι. Define a transfered weight $\iota(\lambda) = (\lambda_1, \dots, \lambda_n, 0, -\lambda_n, \dots, -\lambda_1)$. First, one observes that the transfer of finite-dimensional representations is well-behaved: $\iota(M_\lambda) = M_{\iota(\lambda)}$. Now, we let G be any group in the inner class of $\mathrm{Sp}(2n, \mathbb{R})$ and let $\pi_\lambda \in \mathrm{Coh}(G, \lambda^{\vee})$ (henceforth a purely local notation) be a discrete series representation. Then one proves that $\iota(\pi_\lambda) \in \mathrm{Coh}(\mathrm{GL}(2n+1, \mathbb{R}), \iota(\lambda)^{\vee})$ generalizing (1.1) for the case of symplectic groups. Moreover, if π is a tempered representation of G, which transfers to a cohomological and necessarily tempered representation of $\mathrm{GL}(2n+1, \mathbb{R})$, then from an easy classification of such representations of general linear groups, one concludes that π is a discrete series representation. We need some care in the even orthogonal case where we need certain half-integral twists; for example, in this case we have $\iota(M_\lambda) = M_{\iota(\lambda)} \otimes |\,|^{-1/2}$, and further $\iota(\pi(\lambda)) \otimes |\,|^{1/2} \in \mathrm{Coh}(\mathrm{GL}(2n, \mathbb{R}), \iota(\lambda)^{\vee})$. The reader should consult Theorem 5.6 for the details.

One of the earliest instances of studying functorial behaviour of cohomological representations is to be found in the paper of Labesse and Schwermer [7] where (1.1) is proved for $r = 2$ and is used to construct nontrivial cuspidal cohomology classes for SL(3) over certain number fields. It's indeed a great pleasure for the authors to dedicate this article to Joachim Schwermer on the occasion of his 60th birthday!

2 Background and Notations

2.1 The Groups

All the groups considered here are real Lie groups. Let's recall the definitions of the classical Lie groups that will be considered in this article. Let I_n be the $n \times n$ identity matrix. Define $J_n :=$ anti-diag$(1, \dots, 1)$, i.e., $J_n(i, j) = \delta_{i,n-j+1}$, and let $J'_{2n} :=$ anti-diag$(J_n, -J_n)$. Let $p + q = n$; we will often assume that $p \geq q \geq 1$.

Let $I_{p,q} = \mathrm{diag}(1, \ldots, 1, -1, \ldots, -1)$ be the $n \times n$ diagonal matrix with p many 1's and q many -1's. Put $J_{p,q} = \text{anti-diag}(J_q, I_{p-q}, J_q)$. Let ${}^tA$ and $A^* = \overline{{}^tA}$ denote the transpose and the conjugate-transpose of A, respectively; where, conjugation is either in $\mathbb{C}$ or in $\mathbb{H}$ as the case might be. Define the classical groups:

$$\begin{aligned}
\mathrm{Sp}(2n, \mathbb{R}) &:= \{A \in \mathrm{GL}(2n, \mathbb{R}) \ : \ {}^tA \, J'_{2n} \, A \ = \ J'_{2n}\}, \\
\mathrm{Sp}(p, q) &:= \{A \in \mathrm{GL}(n, \mathbb{H}) \ : \ A^* \, J_{p,q} \, A \ = \ J_{p,q}\}, \\
\mathrm{SO}(p, q) &:= \{A \in \mathrm{SL}(n, \mathbb{R}) \ : \ {}^tA \, J_{p,q} \, A \ = \ J_{p,q}\}, \\
\mathrm{U}(p, q) &:= \{A \in \mathrm{GL}(p+q, \mathbb{C}) \ : \ A^* \, J_{p,q} \, A \ = \ J_{p,q}\}, \\
\mathrm{SO}^*(2n) &:= \left\{A \in \mathrm{SU}(n, n) \ : \ {}^tA \begin{pmatrix} 0 & I_n \\ I_n & 0 \end{pmatrix} A = \begin{pmatrix} 0 & I_n \\ I_n & 0 \end{pmatrix}\right\},
\end{aligned}$$

where in the last definition, $\mathrm{SU}(n, n) = \mathrm{U}(n, n) \cap \mathrm{SL}(2n, \mathbb{C})$.

Let G be a real reductive group (and often it will stand for a simple classical Lie group as above). We denote the Lie algebra of G by $\mathfrak{g}$ and its complexification by $\mathfrak{g}_{\mathbb{C}}$. Let K be a maximal compact subgroup of G, and T a maximal torus in G. Let $\mathfrak{k}_{\mathbb{C}}$ (resp., $\mathfrak{t}_{\mathbb{C}}$) be the complexification of the Lie algebra of K (resp., T). Let $\Phi = \Phi(\mathfrak{g}_{\mathbb{C}}, \mathfrak{t}_{\mathbb{C}})$ be the set of roots of $\mathfrak{t}_{\mathbb{C}}$ in $\mathfrak{g}_{\mathbb{C}}$, $\Phi^+ = \Phi^+(\mathfrak{g}_{\mathbb{C}}, \mathfrak{t}_{\mathbb{C}})$ a positive system, and Δ the corresponding base of Φ consisting of indecomposables in Φ^+. Let ρ stand for half the sum of positive roots in $\Phi(\mathfrak{g}_{\mathbb{C}}, \mathfrak{t}_{\mathbb{C}})$. If G is the real points of an algebraic group $\mathcal{G}/\mathbb{R}$ (as will be the case in all the groups that we consider) then we let B stand for the Borel subgroup of $\mathcal{G}(\mathbb{C})$ corresponding to Φ^+.

2.2 *Representations of G*

(As a general reference, see [3].) Let (π, V) be an irreducible representation of G. We say that (π, V) is *admissible* if V, as a representation of K, breaks up as a sum of irreducible representations such that each isotypic component is finite-dimensional, i.e., if $V|_K = \oplus_{\sigma \in \hat{K}} m_\sigma \sigma$, then all the m_σ's are finite. Given a unitary representation (π, V) of G, the subspace of K-finite smooth vectors is a $(\mathfrak{g}, K)$-module which completely determines the representation π. Let $X(G)$ be the group of all continuous homomorphisms of G to $\mathbb{R}^\times$. An irreducible admissible representation (π, V) of G is called *tempered* if for each $v, w \in V_0$, there exists a constant C such that

$$|\langle \pi(g)v, w \rangle| \leq C \, \Xi(g)$$

for all $g \in G$, where V_0 is the corresponding $(\mathfrak{g}, K)$-module, Ξ the Harish-Chandra's function and ${}^\circ G = \bigcap_{\chi \in X(G)} \ker |\chi|$. A weight $\lambda \in \mathfrak{t}_{\mathbb{C}}^*$ is called dominant integral if $\langle \lambda, \alpha \rangle$ is a non-negative integer for all $\alpha \in \Phi^+(\mathfrak{g}_{\mathbb{C}}, \mathfrak{t}_{\mathbb{C}})$. Isomorphism classes of irreducible finite-dimensional representations of $\mathcal{G}(\mathbb{C})$ are in one-to-one correspondence with the dominant integral weights. We will denote the irreducible representation of $\mathcal{G}(\mathbb{C})$ with highest weight λ by M_λ; its restriction to G will also be denoted M_λ.

2.3 Representations with Cohomology

(See [3].) Let V be a $\mathfrak{g}$ module. Let $\mathfrak{k}$ be a subalgebra of $\mathfrak{g}$. Then define the following chain complex: $C^q(\mathfrak{g}, \mathfrak{k}; V) = \mathrm{Hom}_{\mathfrak{k}}(\wedge^q(\mathfrak{g}/\mathfrak{k}), V)$, where the action of $\mathfrak{k}$ on $\wedge^q(\mathfrak{g}/\mathfrak{k})$ is induced by the adjoint representation. The map $d : C^q \to C^{q+1}$ is defined by

$$df(x_0, \ldots, x_q) = \sum_i (-1)^i x_i \cdot f(x_0, \ldots, \hat{x_i}, \ldots, x_q) + \sum_{i<j} (-1)^{i+j} f([x_i, x_j], x_0, \ldots, \hat{x_i}, \ldots, \hat{x_j}, \ldots, x_q). \tag{2.1}$$

The cohomology groups of this chain complex are denoted by $H^i(\mathfrak{g}, \mathfrak{k}; V)$.

We shall now introduce the category in which we will be working for the rest of the paper. Let $\mathfrak{k}$ be reductive in $\mathfrak{g}$. Let V be a $\mathfrak{k}$-module. For the sake of completeness we include some definitions though they are very standard. A vector $v \in V$ is called $\mathfrak{k}$-*finite* if $\mathcal{U}(\mathfrak{k}) \cdot v$ is a finite-dimensional subspace of V. A $\mathfrak{k}$-module V is called *locally $\mathfrak{k}$-finite* if every element of V is $\mathfrak{k}$-finite. Equivalently, V is locally $\mathfrak{k}$-finite if every finite-dimensional subspace of V is contained in a $\mathfrak{k}$-invariant finite-dimensional subspace. A $\mathfrak{g}$-module V is called a $(\mathfrak{g}, \mathfrak{k})$-*module* if V is locally $\mathfrak{k}$-finite and V is semisimple as a $\mathfrak{k}$-module. A $(\mathfrak{g}, \mathfrak{k})$-module is called *admissible* if the isotypic components for $\mathfrak{k}$ are all finite-dimensional. We will work with the category of $(\mathfrak{g}, \mathfrak{k})$-modules. For a connected Lie group G, and K a connected compact subgroup of G, with $\mathfrak{g}$ and $\mathfrak{k}$ the respective Lie algebras, every $(\mathfrak{g}, K)$-module is a $(\mathfrak{g}, \mathfrak{k})$-module. A $(\mathfrak{g}, \mathfrak{k})$-module V is called *cohomological* if there exists a finite-dimensional $(\mathfrak{g}, \mathfrak{k})$-module M_λ such that $H^\bullet(\mathfrak{g}, \mathfrak{k}; V \otimes M_\lambda)$ is non-zero. When (π, V) is a representation of a real Lie group G, π is called cohomological if π is cohomological as a $(\mathfrak{g}, \mathfrak{k})$-module. In this case we write $\pi \in \mathrm{Coh}(G, \lambda)$.

2.4 The Langlands Dual Group

Let G be a connected reductive algebraic group over $\mathbb{R}$ with a maximal torus T defined over $\mathbb{R}$. Let $X^*(T) = \mathrm{Hom}(T, \mathbb{G}_m)$ and $X_*(T) = \mathrm{Hom}(\mathbb{G}_m, T)$ be the group of characters and co-characters, respectively. The root datum of G is given by $(X^*(T), \Phi, X_*(T), \Phi^\wedge)$, where Φ is the set of roots of G with respect to T, and $\Phi^\wedge$ the set of co-roots. The connected component of the Langlands dual of G is the connected complex reductive group ${}^LG^\circ$ whose root datum is $(X_*(T), \Phi^\wedge, X^*(T), \Phi)$. See [2, Chap. I] for the definition of the dual group ${}^LG$ and for more details.

2.5 The Weil Group and the Langlands Transfer

Let $W_{\mathbb{R}}$ be the Weil group of $\mathbb{R}$. Then $W_{\mathbb{R}} = \mathbb{C}^{\times} \cup j\mathbb{C}^{\times}$, where $j^2 = -1$ and $jzj^{-1} = \bar{z}$, the complex conjugate of z. (See [6].) For a real reductive group G, let $\Pi(G)$ denote the set of all equivalence classes of irreducible and admissible representations of G, and let $\Phi(G)$ denote the set of all equivalence classes of admissible homomorphisms $W_{\mathbb{R}}$ to the Langlands dual ${}^LG$ of G. (See [2, Sect. 8.2].) Then, to every $\phi \in \Phi(G)$, Langlands attached a finite non-empty subset (called the L-packet of ϕ) of $\Pi(G)$ with some desired properties. (See [8] or [2].) Conversely, to every element of $\Pi(G)$ is associated an element of $\Phi(G)$ called its Langlands parameter. Now, suppose we have an L-group homomorphism $\alpha : {}^LG \to {}^LH$, then composing an element of $\Phi(G)$ with α we obtain an element of $\Phi(H)$, which in turn corresponds to a finite subset of $\Pi(H)$. The representations thus obtained will be called the transfers of the representation of G which we started with. We will be considering the situation where $\alpha = \iota$ is an inclusion from the L-group of a classical group G to the L-group of an appropriate general linear group. It's well-known that for GL_n the L-packets are singletons; in this case we will denote the transfer of a representation π by $\iota(\pi)$. The specific cases that will be considered in this article are the following:

G	${}^LG^\circ$	Transferred to
$\mathrm{Sp}(2n, \mathbb{R})$	$\mathrm{SO}(2n+1, \mathbb{C})$	$\mathrm{GL}(2n+1, \mathbb{R})$
$\mathrm{Sp}(p, q),\ p+q=n$	$\mathrm{SO}(2n+1, \mathbb{C})$	$\mathrm{GL}(2n+1, \mathbb{R})$
$\mathrm{SO}(p, q),\ p+q=2n+1$	$\mathrm{Sp}(2n, \mathbb{C})$	$\mathrm{GL}(2n, \mathbb{R})$
$\mathrm{SO}(p, q),\ p+q=2n$	$\mathrm{SO}(2n, \mathbb{C})$	$\mathrm{GL}(2n, \mathbb{R})$
$\mathrm{SO}^*(2n)$	$\mathrm{SO}(2n, \mathbb{C})$	$\mathrm{GL}(2n, \mathbb{R})$
$\mathrm{U}(p, q),\ p+q=n$	$\mathrm{GL}(n, \mathbb{C})$	$\mathrm{GL}(n, \mathbb{C})$

3 Transfer of Tempered Representations

In this section we will compute the transfer of tempered representations of the groups G (mentioned in the table above), to representations of $\mathrm{GL}(N, \mathbb{R})$ or $\mathrm{GL}(N, \mathbb{C})$ as the case might be. A representation $\pi \in \Pi(G)$ is tempered if and only if the Langlands parameter corresponding to π is bounded in ${}^LG^\circ$ (see [8]). Now, when we transfer such a representation to an appropriate $\mathrm{GL}(N)$ using the inclusion of ${}^LG^\circ$ to $\mathrm{GL}(N, \mathbb{C})$, then it's clear that if the image of $W_{\mathbb{R}}$ is bounded in ${}^LG^\circ$, then it is also bounded in $\mathrm{GL}(N, \mathbb{C})$. Thus, *the transfer of a tempered representation of G is a tempered representation of* $\mathrm{GL}(N, \mathbb{R})$ *or* $\mathrm{GL}(N, \mathbb{C})$.

3.1 Tempered Cohomological Representation of $\mathrm{GL}(n, \mathbb{R})$

Let $\lambda = (\lambda_1, \ldots, \lambda_n)$ be a dominant integral weight for $\mathrm{GL}(n, \mathbb{R})$; then $\lambda_j \in \mathbb{Z}$; $\lambda_1 \geq \cdots \geq \lambda_n$. We also suppose that λ is a *pure* weight, i.e., there exists an integer w, called the purity weight of λ such that $\lambda_i + \lambda_{n-i+1} = \mathsf{w}$. Define an n-tuple of integers $\ell = \ell(\lambda) = (\ell_1, \ldots, \ell_n)$ by $\ell = 2\lambda + 2\rho_n - \mathsf{w}$, where ρ_n is half the sum of positive roots of $\mathrm{GL}(n, \mathbb{R})$. Then $\ell_i = 2\lambda_i + n - 2i + 1 - \mathsf{w}$ for $1 \leq i \leq n$. Let's note the parity condition:

$$\ell_i \equiv n + 1 - \mathsf{w} \pmod 2. \tag{3.1}$$

When n is even, let P be the standard parabolic subgroup of type $(2, 2, \ldots, 2)$. Then define

$$J_\lambda = \mathrm{Ind}_{P(\mathbb{R})}^{\mathrm{GL}(n,\mathbb{R})}(D(\ell_1)|det|^{\frac{\mathsf{w}}{2}} \otimes \cdots \otimes D(\ell_{n/2})|det|^{\frac{\mathsf{w}}{2}}).$$

We know that J_λ is irreducible, essentially tempered and cohomological with respect to the finite-dimensional representation $M_\lambda^\vee$. When n is odd, we let P be the standard parabolic subgroup of type $(2, 2, \ldots, 2, 1)$, and for any sign character $\epsilon : \mathbb{R}^\times \to \{\pm 1\}$, define

$$J_\lambda^\epsilon = \mathrm{Ind}_{P(\mathbb{R})}^{\mathrm{GL}(n,\mathbb{R})}(D(\ell_1)|det|^{\frac{\mathsf{w}}{2}} \otimes \cdots \otimes D(\ell_{(n-1)/2})|det|^{\frac{\mathsf{w}}{2}} \otimes \epsilon|det|^{\frac{\mathsf{w}}{2}}).$$

We know that J_λ^ϵ is irreducible, essentially tempered and cohomological with respect to the finite-dimensional representation $M_\lambda^\vee$. For a pure weight λ, if $\pi \in \mathrm{Coh}(\mathrm{GL}(n, \mathbb{R}), \lambda^\vee)$ is essentially tempered, then $\pi = J_\lambda$ if n is even, and $\pi = J_\lambda^\epsilon$ for some ϵ if n is odd.

Remark 3.1 Since temperedness is captured by boundedness of the parameter, it's easy to see that the representation J_λ (or J_λ^ϵ) is tempered if and only if the purity weight $\mathsf{w} = 0$. When n is odd we do not have any conditions on ϵ.

3.2 Tempered Cohomological Representation of $\mathrm{GL}(n, \mathbb{C})$

A dominant-integral weight for $\mathrm{GL}(n, \mathbb{C})$, as a real Lie group, is of the form $\lambda = ((\lambda_1, \ldots, \lambda_n), (\lambda_1^*, \ldots, \lambda_n^*))$ with $\lambda_j, \lambda_j^* \in \mathbb{Z}$, $\lambda_1 \geq \cdots \geq \lambda_n$, $\lambda_1^* \geq \cdots \geq \lambda_n^*$. We say λ is pure if there exists an integer w such that $\lambda_i + \lambda_{n-i+1}^* = \mathsf{w}$. For such a pure weight, define a string of half-integers $a = (a_1, \ldots, a_n) = \lambda + \rho_n$ and $b = (b_1, \ldots, b_n) = \mathsf{w} - \lambda - \rho_n$. Now, define the representation J_λ induced from the Borel subgroup $B(\mathbb{C})$ of $\mathrm{GL}(n, \mathbb{C})$ as

$$J_\lambda = \mathrm{Ind}_{B(\mathbb{C})}^{\mathrm{GL}(n,\mathbb{C})}(z^{a_1}\bar{z}^{b_1} \otimes \cdots \otimes z^{a_n}\bar{z}^{b_n}).$$

We know that J_λ is irreducible, essentially tempered and cohomological with respect to the finite-dimensional representation $M_\lambda^\vee$. For a pure weight λ, if $\pi \in \mathrm{Coh}(\mathrm{GL}(n, \mathbb{C}), \lambda^\vee)$ is essentially tempered, then $\pi = J_\lambda$.

Remark 3.2 J_λ is tempered if and only if the purity weight $\mathsf{w} = 0$.

4 Transfer of Finite-Dimensional Representations

We will compute the transfers of finite-dimensional representations of a split classical group G (from the table in Sect. 2.5) to the appropriate $\mathrm{GL}(N, \mathbb{R})$ or $\mathrm{GL}(N, \mathbb{C})$. Towards this we need some preparation. Given the natural inclusion $\iota : {}^LG^\circ \to \mathrm{GL}_N(\mathbb{C})$, we may take the dual torus ${}^LT^\circ$ to be contained in ${}^LD^\circ$ the diagonal matrices in $\mathrm{GL}_N(\mathbb{C})$, where D is the diagonal torus in $\mathrm{GL}_N(\mathbb{R})$ or $\mathrm{GL}_N(\mathbb{C})$.

4.1 Some Notation Concerning Classical Groups

We will fix a torus for the split groups of the types B_n, C_n, D_n and also for the maximally split unitary groups. In each of these cases, we then describe a root system corresponding to the chosen torus, along with a base and set of positive roots, and compute the half sum of positive roots ρ.

(1) For type B_n: The split group in this case is $\mathrm{SO}(n+1, n)$. The diagonal torus T consists of all matrices of the form

$$\mathrm{diag}(t_1, \ldots, t_n, 1, t_n^{-1}, \ldots, t_1^{-1}),$$

with $t_j \in \mathbb{R}^\times$. The complexified Lie algebra of $\mathrm{SO}(n+1, n)$ is $\mathfrak{so}_{2n+1}$, and that of T, denoted $\mathfrak{t}_\mathbb{C}$, has matrices of the form $\mathrm{diag}(a_1, \ldots, a_n, 0, -a_n, \ldots, -a_1)$, where $a_j \in \mathbb{C}$. Denote by e_i, the character on the torus which picks out a_i. Then we fix the following data:

- $\Phi(\mathfrak{g}_\mathbb{C}, \mathfrak{t}_\mathbb{C}) = \{\pm e_i \pm e_j : i < j\} \cap \{\pm e_i\}$.
- The base $\Delta = \{e_1 - e_2, e_2 - e_3, \ldots, e_{n-1} - e_n, e_n\}$.
- The set of positive roots $\Phi^+ = \{e_i \pm e_j\} \cap \{e_i\}$.
- The half sum of positive roots $\rho = (n - \frac{1}{2})e_1 + (n - \frac{3}{2})e_2 + \cdots + \frac{1}{2}e_n$. We will also write ρ as the n-tuple $(n - \frac{1}{2}, n - \frac{3}{2}, \ldots, \frac{1}{2})$.

(2) For type C_n: The split group in this case is $\mathrm{Sp}(2n, \mathbb{R})$. The diagonal torus T consists of all matrices of the form:

$$\mathrm{diag}(t_1, \ldots, t_n, t_n^{-1}, \ldots, t_1^{-1}),$$

with $t_j \in \mathbb{R}^\times$. The complexified Lie algebra of $\mathrm{Sp}(2n, \mathbb{R})$ is denoted $\mathfrak{sp}_{2n}$, and that of T, denoted $\mathfrak{t}_\mathbb{C}$, has matrices of the form $\mathrm{diag}(a_1, \ldots, a_n, -a_n, \ldots, -a_1)$ with $a_j \in \mathbb{C}$. Denote by e_i, the character on the torus which picks out a_i. Then we fix the following data:

- $\Phi(\mathfrak{g}_\mathbb{C}, \mathfrak{t}_\mathbb{C}) = \{\pm e_i \pm e_j : i < j\} \cap \{\pm 2e_i\}$.
- The base $\Delta = \{e_1 - e_2, e_2 - e_3, \ldots, e_{n-1} - e_n, 2e_n\}$.
- The set of positive roots $\Phi^+ = \{e_i \pm e_j\} \cap \{2e_i\}$.
- The half sum of positive roots $\rho = ne_1 + (n-1)e_2 + \cdots + 1e_n$. We will also write ρ as the n-tuple $(n, n-1, \ldots, 1)$.

(3) For type D_n: The split group in this case is $\mathrm{SO}(n, n)$. The diagonal torus T consists of all matrices of the form:

$$\mathrm{diag}(t_1, \ldots, t_n, t_n^{-1}, \ldots, t_1^{-1}),$$

with $t_j \in \mathbb{R}^\times$. The complexified Lie algebra of $\mathrm{SO}(2n, \mathbb{R})$ is denoted $\mathfrak{so}_{2n}$, and that of T, denoted $\mathfrak{t}_\mathbb{C}$, has matrices of the form $\mathrm{diag}(a_1, \ldots, a_n, -a_n, \ldots, -a_1)$ with $a_j \in \mathbb{C}$. Denote by e_i, the character on the torus which picks out a_i. Then we fix the following data:

- $\Phi(\mathfrak{g}_\mathbb{C}, \mathfrak{t}_\mathbb{C}) = \{\pm e_i \pm e_j : i < j\}$.
- The base $\Delta = \{e_1 - e_2, e_2 - e_3, \ldots, e_{n-1} - e_n, e_{n-1} + e_n\}$.
- The set of positive roots $\Phi^+ = \{e_i \pm e_j\}$.
- The half sum of positive roots $\rho = (n-1)e_1 + (n-2)e_2 + \cdots + 1e_{n-1}$. We will also write ρ as the n-tuple $(n-1, n-2, \ldots, 1, 0)$.

(4) The unitary groups: When n is even take the unitary group as $\mathrm{U}(\frac{n}{2}, \frac{n}{2})$, and let $p = n/2$. For a maximal torus T of $\mathrm{U}(\frac{n}{2}, \frac{n}{2})$ take all matrices of the form

$$\{\mathrm{diag}(t_1, \ldots, t_p, \bar{t}_p^{-1}, \ldots, \bar{t}_1^{-1}) : t_i \in \mathbb{C}^\times\}.$$

Its Lie algebra $\mathfrak{t}$ has matrices of the form $\mathrm{diag}(a_1, \ldots, a_p, -\bar{a}_p, \ldots, -\bar{a}_1)$ with $a_j \in \mathbb{C}$. When n is odd we consider $\mathrm{U}(\frac{n+1}{2}, \frac{n-1}{2})$, and put $p = (n+1)/2$ and $q = (n-1)/2$. For T in $\mathrm{U}(\frac{n+1}{2}, \frac{n-1}{2})$ take all matrices of the form

$$\{\mathrm{diag}(t_1, \ldots, t_{p-1}, t_p, \bar{t}_{p-1}^{-1}, \ldots, \bar{t}_1^{-1}) : t_i \in \mathbb{C}^\times \text{ for } 1 \le i \le p-1, \text{ and } t_p \in S^1\},$$

hence $\mathfrak{t}$ has matrices of the form $\mathrm{diag}(a_1, \ldots, a_{p-1}, a_p, -\bar{a}_{p-1}, \ldots, -\bar{a}_1)$ with $a_j \in \mathbb{C}$. In either case, denote by e_i, the character on the torus which picks out the ith entry of the diagonal matrix. We have:

- $\Phi(\mathfrak{g}_\mathbb{C}, \mathfrak{t}_\mathbb{C}) = \{e_i - e_j : i \neq j,\ 1 \le i, j \le n\}$.
- The base $\Delta = \{e_1 - e_2, e_2 - e_3, \ldots, e_{n-1} - e_n\}$.
- The set of positive roots $\Phi^+ = \{e_i - e_j : i < j\}$.

- The half sum of positive roots $\rho = (\frac{n-1}{2})e_1 + (\frac{n-3}{2})e_2 + \cdots + (\frac{1-n}{2})e_n$. We will also write ρ as the n-tuple $(\frac{n-1}{2}, \frac{n-3}{2}, \ldots, \frac{1-n}{2})$.

In all the above cases (1)–(4), a weight $\lambda \in \mathfrak{t}_{\mathbb{C}}^*$, will be written as an n-tuple $\lambda = (\lambda_1, \ldots, \lambda_n)$ and this will stand for $\lambda_1 e_1 + \cdots + \lambda_n e_n$.

4.2 *Parametrizing the Highest Weights*

The algebraic irreducible finite-dimensional representations of $\mathrm{Sp}(2n, \mathbb{C})$, $\mathrm{SO}(n, \mathbb{C})$, $\mathrm{GL}(n, \mathbb{C})$ are classified by the highest weights of the corresponding Lie algebra $\mathfrak{g}_{\mathbb{C}}$. For the groups under consideration, the highest weights are as follows:

(1) $\mathrm{Sp}(2n, \mathbb{C})$: $\lambda = (\lambda_1, \ldots, \lambda_n)$; $\lambda_j \in \mathbb{Z}$ and $\lambda_1 \geq \lambda_2 \geq \cdots \geq \lambda_n \geq 0$.
(2) $\mathrm{SO}(2n+1, \mathbb{C})$: $\lambda = (\lambda_1, \ldots, \lambda_n)$; $\lambda_j \in \mathbb{Z}$ and $\lambda_1 \geq \lambda_2 \geq \cdots \geq \lambda_n \geq 0$.
(3) $\mathrm{SO}(2n, \mathbb{C})$: $\lambda = (\lambda_1, \ldots, \lambda_n)$; $\lambda_j \in \mathbb{Z}$ and $\lambda_1 \geq \lambda_2 \geq \cdots \geq |\lambda_n|$.
(4) $\mathrm{GL}(n, \mathbb{C})$: $\lambda = (\lambda_1, \ldots, \lambda_n)$; $\lambda_j \in \mathbb{Z}$ and $\lambda_1 \geq \lambda_2 \geq \cdots \geq \lambda_n$.
(4') $\mathrm{GL}(n, \mathbb{C})$ as a real group: $\lambda = ((\lambda_1, \ldots, \lambda_n), (\lambda_1^*, \ldots, \lambda_n^*))$; $\lambda_i, \lambda_i^* \in \mathbb{Z}$ and $\lambda_1 \geq \lambda_2 \geq \cdots \geq \lambda_n$ as well as $\lambda_1^* \geq \lambda_2^* \geq \cdots \geq \lambda_n^*$.

4.3 *Transfers of the Finite-Dimensional Representations*

Now we proceed to compute the transfers of the finite-dimensional representations of the split real forms of the groups in Sect. 4.2. We denote by M_λ the finite-dimensional representation of (a classical group) G which is obtained by restriction of the finite-dimensional representation of the corresponding complex group with highest weight λ. Let $\phi(M_\lambda) : W_{\mathbb{R}} \to {}^L G$ be the corresponding Langlands parameter. Note that such an M_λ is obtained as the Langlands quotient of the induced representation $\mathrm{Ind}_B^{G(\mathbb{C})}(\otimes_{i=1}^n \chi_{\lambda_i} |\cdot|^{\rho_i})$, where ρ_i is the ith coefficient of ρ when written as an n-tuple, and $\chi_{\lambda_i}(x) = x^{\lambda_i}$. Here, $\mathrm{Ind}_B^{G(\mathbb{C})}$ denotes the normalized parabolic induction from the Borel subgroup B to $G(\mathbb{C})$, and $\otimes_{i=1}^n \chi_{\lambda_i} |\cdot|^{\rho_i}$ is as usual a character of the torus $T(\mathbb{C})$ inflated up to a character of B by making it trivial on the unipotent radical of B.

4.3.1 From $\mathbf{Sp}(2n, \mathbb{R})$ to $\mathbf{GL}(2n+1, \mathbb{R})$

For $G = \mathrm{Sp}(2n, \mathbb{R})$, ${}^L G^\circ = \mathrm{SO}(2n+1, \mathbb{C})$. The parameter $\phi(M_\lambda)$ is given by: $\phi(M_\lambda)(z) = (z\bar{z})^{\lambda+\rho}$, i.e.,

$$\phi(M_\lambda)(z) = \begin{pmatrix} (z\bar{z})^{\lambda_1+n} & & & & & & \\ & \ddots & & & & & \\ & & (z\bar{z})^{\lambda_n+1} & & & & \\ & & & 1 & & & \\ & & & & (z\bar{z})^{-\lambda_n-1} & & \\ & & & & & \ddots & \\ & & & & & & (z\bar{z})^{-\lambda_1-n} \end{pmatrix},$$

and

$$\phi(M_\lambda)(j) = \mathrm{diag}((-1)^{\lambda_1}, \ldots, (-1)^{\lambda_n}, 1, (-1)^{-\lambda_n}, \ldots, (-1)^{-\lambda_1}).$$

The transfer of M_λ to $\mathrm{GL}(2n+1, \mathbb{R})$ is the unique irreducible quotient of

$$\mathrm{Ind}_{B_{2n+1}(\mathbb{R})}^{\mathrm{GL}(2n+1,\mathbb{R})}(\otimes_{i=1}^{n} \chi_{\lambda_i} |\cdot|^{n-i+1} \otimes 1 \otimes_{i=n+2}^{2n+1} \chi_{-\lambda_{2n+2-i}} |\cdot|^{n-i+1}).$$

(Here and elsewhere, B_m will denote the Borel subgroup of GL_m of all upper-triangular matrices.) This representation is the finite-dimensional representation of $\mathrm{GL}(2n+1, \mathbb{R})$ with highest weight $\iota(\lambda) = (\lambda_1, \ldots, \lambda_n, 0, -\lambda_n, \ldots, -\lambda_1)$. Summarising, we have:

Proposition 4.1 *For $G = \mathrm{Sp}(2n, \mathbb{R})$, the transfer of M_λ to $\mathrm{GL}(2n+1, \mathbb{R})$ is the finite-dimensional representation $M_{\iota(\lambda)}$ with highest weight $\iota(\lambda) = (\lambda_1, \ldots, \lambda_n, 0, -\lambda_n, \ldots, -\lambda_1)$.*

4.3.2 From $\mathrm{SO}(n+1, n)$ to $\mathrm{GL}(2n, \mathbb{R})$

We have ${}^L G^\circ = \mathrm{Sp}(2n, \mathbb{C})$, for $G = \mathrm{SO}(n+1, n)$. The Langlands parameter $\phi(M_\lambda)$ is given by:

$$\begin{aligned} \phi(M_\lambda)(z) &= \mathrm{diag}(a_1, \ldots, a_n, a_n^{-1}, \ldots, a_1^{-1}), \text{ where } a_i = (z\bar{z})^{\lambda_i+n-\frac{2i-1}{2}}, \text{ and} \\ \phi(M_\lambda)(j) &= \mathrm{diag}((-1)^{\lambda_1}, \ldots, (-1)^{\lambda_n}, (-1)^{\lambda_n}, \ldots, (-1)^{\lambda_1}). \end{aligned}$$

The resulting representation of $\mathrm{GL}(2n, \mathbb{R})$ is the unique irreducible quotient of

$$\mathrm{Ind}_{B_{2n}(\mathbb{R})}^{\mathrm{GL}(2n,\mathbb{R})}(\chi_{\lambda_1}|\cdot|^{\frac{2n-1}{2}} \otimes \cdots \otimes \chi_{\lambda_1}|\cdot|^{\frac{1}{2}} \otimes \chi_{-\lambda_n}|\cdot|^{-\frac{1}{2}} \otimes \cdots \otimes \chi_{-\lambda_1}|\cdot|^{-\frac{2n-1}{2}}),$$

which one knows to be the finite-dimensional representation of $\mathrm{GL}(2n, \mathbb{R})$ with highest weight $\iota(\lambda) = (\lambda_1, \ldots, \lambda_n, -\lambda_n, \ldots, -\lambda_1)$. We have:

Proposition 4.2 *For $G = \mathrm{SO}(n+1, n)$ the transfer of M_λ to $\mathrm{GL}(2n, \mathbb{R})$ is the finite-dimensional representation $M_{\iota(\lambda)}$ with highest weight $\iota(\lambda) = (\lambda_1, \ldots, \lambda_n, -\lambda_n, \ldots, -\lambda_1)$.*

4.3.3 From SO(n, n) to GL($2n, \mathbb{R}$)

We have ${}^LG^\circ = \mathrm{SO}(2n, \mathbb{C})$. (See [4].) The parameter for M_λ is:

$$\phi(M_\lambda)(z) = \mathrm{diag}(a_1, \ldots, a_n, a_n^{-1}, \ldots, a_1^{-1}), \text{ where } a_i = (z\bar{z})^{\lambda_i+n-i},\ 1 \le i \le n, \text{ and}$$
$$\phi(M_\lambda)(j) = \mathrm{diag}((-1)^{\lambda_1}, \ldots, (-1)^{\lambda_n}, (-1)^{\lambda_n}, \ldots, (-1)^{\lambda_1}).$$

Then the transfer of M_λ is the unique irreducible quotient of

$$\mathrm{Ind}_{B_{2n}(\mathbb{R})}^{\mathrm{GL}(2n,\mathbb{R})}(\chi_{\lambda_1}|\cdot|^{n-1} \otimes \chi_{\lambda_2}|\cdot|^{n-2} \otimes \cdots \otimes \chi_{\lambda_n} \otimes \chi_{-\lambda_n} \otimes \cdots \otimes \chi_{-\lambda_1}|\cdot|^{1-n}).$$

One sees (after, if necessary, switching χ_{λ_n} and $\chi_{-\lambda_n}$), that this representation is the same as

$$\iota(M_\lambda) = M_{\iota(\lambda)} \otimes |\cdot|^{-\frac{1}{2}},$$

where the $M_{\iota(\lambda)}$ is the finite-dimensional representation of $\mathrm{GL}(2n, \mathbb{R})$ with highest weight $\iota(\lambda) = (\lambda_1, \ldots, |\lambda_n|, -|\lambda_n|, \ldots, -\lambda_1)$. For $\lambda = (\lambda_1, \ldots, \lambda_{n-1}, \lambda_n)$, define $\lambda^\# = (\lambda_1, \ldots, \lambda_{n-1}, -\lambda_n)$. Then the representations M_λ and $M_{\lambda^\#}$ transfer to the same representation of $\mathrm{GL}(2n, \mathbb{R})$, since we can conjugate by an element of $\mathrm{SO}(2n, \mathbb{C})$ to switch the n-th and $(n+1)$-th entries of a diagonal $2n \times 2n$-matrix. We have the following

Proposition 4.3 *For $G = \mathrm{SO}(n, n)$ the transfer of M_λ to a representation of $\mathrm{GL}(2n, \mathbb{R})$ is $M_{\iota(\lambda)} \otimes |\cdot|^{-\frac{1}{2}}$, where $M_{\iota(\lambda)}$ is the finite-dimensional representation with highest weight $\iota(\lambda) = (\lambda_1, \ldots, \lambda_{n-1}, |\lambda_n|, -|\lambda_n|, -\lambda_{n-1}, \ldots, -\lambda_1)$.*

4.4 *Unitary Groups to* GL($n, \mathbb{C}$)

Similar to the case of even orthogonal groups, we need to subdivide into two cases depending on the parity of n. Recall that a dominant integral weight of $\lambda = ((\lambda_1, \ldots, \lambda_n), (\lambda_1^*, \ldots, \lambda_n^*))$ with $\lambda_i, \lambda_i^* \in \mathbb{Z}$ and $\lambda_i \ge \lambda_{i+1}, \lambda_i^* \ge \lambda_{i+1}^*$.

4.4.1 n Even

In this case we transfer from the unitary group $\mathrm{U}(\frac{n}{2}, \frac{n}{2})$. The Langlands dual group is $\mathrm{GL}(n, \mathbb{C}) \rtimes \mathbb{Z}/2\mathbb{Z}$ and ${}^LG^\circ = \mathrm{GL}(n, \mathbb{C})$. The absolute rank of $U(\frac{n}{2}, \frac{n}{2})$ is n. Let M_λ be the finite-dimensional representation of $\mathrm{U}(\frac{n}{2}, \frac{n}{2})$ with highest weight $\lambda = (\lambda_1, \ldots, \lambda_n)$. Let $p = \frac{n}{2}$. The restriction of the Langlands parameter $\phi(M_\lambda)$ to $\mathbb{C}^\times$ is given by

$$\phi(M_\lambda)(z) = \mathrm{diag}(a_1, \ldots, a_p, \bar{a}_p^{-1}, \ldots, \bar{a}_1^{-1}),$$

where, for $1 \le i \le p$, we have

$$a_i = z^{\lambda_i+\rho_i}\bar{z}^{-\lambda_{n-i+1}-\rho_{n-i+1}}.$$

We will transfer representations of $U(\frac{n}{2},\frac{n}{2})$ to representations of $GL(n,\mathbb{C})$ using stable base change (see [5]). The dual for $U(\frac{n}{2},\frac{n}{2})$ as noted above is $GL(n,\mathbb{C})\rtimes\mathbb{Z}/2\mathbb{Z}$, where the action of θ, the non-trivial element of $\mathbb{Z}/2\mathbb{Z}$, is given by $g\mapsto\Phi_n{}^tg^{-1}\Phi_n^{-1}$ where $(\Phi_n)_{ij}=(-1)^{i-1}\delta_{i,n-j+1}$. Now, ${}^LR_{\mathbb{C}\backslash\mathbb{R}}GL_n=GL(n,\mathbb{C})\times GL(n,\mathbb{C})\rtimes\mathbb{Z}/2\mathbb{Z}$, where the semidirect product acts by interchanging the two copies of $GL(n,\mathbb{C})$. Define the map BC as:

$$BC: {}^LU(\frac{n}{2},\frac{n}{2})\hookrightarrow {}^LR_{\mathbb{C}\backslash\mathbb{R}}GL_n(\mathbb{C}),\quad (g,1)\mapsto(g,\theta(g),1),\quad (g,\theta)\mapsto(g,\theta(g),\theta).$$

Note that $R_{\mathbb{C}\backslash\mathbb{R}}GL_n(\mathbb{C})=GL(n,\mathbb{C})\times GL(n,\mathbb{C})$. Hence the parameter obtained by stable base change gives us a representation of $GL(n,\mathbb{C})\times GL(n,\mathbb{C})$. Such a transfered representation of $GL(n,\mathbb{C})\times GL(n,\mathbb{C})$ determines a representation of $GL(n,\mathbb{C})$ with the Langlands parameter obtained by projecting on the first component. Similarly, a representation π of $GL(n,\mathbb{C})$, with Langlands parameter $\Phi(\pi)$, determines a representation of $GL(n,\mathbb{C})\times GL(n,\mathbb{C})$ with parameter $(\Phi(\pi),\theta(\Phi(\pi)))$. This gives us a way to go back and forth between representations of $GL(n,\mathbb{C})$ and $GL(n,\mathbb{C})\times GL(n,\mathbb{C})$, hence transferring a representation of $U(\frac{n}{2},\frac{n}{2})$ to $GL(n,\mathbb{C})$.

Thus the representation of $GL(n,\mathbb{C})$ obtained by stable base-change of the finite-dimensional representation M_λ has inducing data

$$z^{\lambda_1+\frac{n-1}{2}}\bar{z}^{-\lambda_n+\frac{n-1}{2}}\otimes\cdots\otimes z^{\lambda_p+\frac{1}{2}}\bar{z}^{-\lambda_{p+1}+\frac{1}{2}}\otimes z^{\lambda_{p+1}-\frac{1}{2}}\bar{z}^{-\lambda_p-\frac{1}{2}}\otimes\cdots\otimes z^{\lambda_n-\frac{n-1}{2}}\bar{z}^{-\lambda_1-\frac{n-1}{2}}.$$

The transfered representation of $GL(n,\mathbb{C})$ is the finite-dimensional representation with highest weight $\iota(\lambda)=((\lambda_1,\ldots,\lambda_n),(-\lambda_n,\ldots,-\lambda_1))$ which has purity weight 0. A low-dimensional example might help the reader to see the finer details:

Example 4.4 Consider $G=U(2,2)$. Take a dominant integral weight $\lambda=(\lambda_1,\ldots,\lambda_4)$ and recall that $\lambda_1\geq\lambda_2\geq\lambda_3\geq\lambda_4$. The torus of $U(2,2)$ has the form $T=\{\mathrm{diag}(a_1,a_2,\bar{a}_2^{-1},\bar{a}_1^{-1}) : a_i\in\mathbb{C}^\times\}$. The Langlands parameter for M_λ on $\mathbb{C}^\times$ is

$$\begin{pmatrix} z^{\lambda_1+\frac{3}{2}}\bar{z}^{-\lambda_4+\frac{3}{2}} & & & \\ & z^{\lambda_2+\frac{1}{2}}\bar{z}^{-\lambda_3+\frac{1}{2}} & & \\ & & z^{\lambda_3-\frac{1}{2}}\bar{z}^{-(\lambda_2+\frac{1}{2})} & \\ & & & z^{\lambda_4-\frac{3}{2}}\bar{z}^{-(\lambda_1+\frac{3}{2})} \end{pmatrix}.$$

This is the parameter for the finite-dimensional representation of $GL(4,\mathbb{C})$ with highest weight $\iota(\lambda)=((\lambda_1,\ldots,\lambda_4),(-\lambda_4,\ldots,-\lambda_1))$.

4.4.2 n Odd

In this case we transfer from the unitary group $\mathrm{U}(\frac{n+1}{2}, \frac{n-1}{2})$. The Langlands dual group is $\mathrm{GL}(n, \mathbb{C}) \rtimes \mathbb{Z}/2\mathbb{Z}$ and ${}^LG^\circ = \mathrm{GL}(n, \mathbb{C})$. Let M_λ be the finite-dimensional representation of $\mathrm{U}(\frac{n+1}{2}, \frac{n-1}{2})$ with highest weight $\lambda = (\lambda_1, \ldots, \lambda_n)$. We will use stable base change as in the previous case to obtain a representation of $\mathrm{GL}(n, \mathbb{C})$. For convenience let $p = \frac{n+1}{2}$ and $q = \frac{n-1}{2}$. Observe that $p = q + 1$. Then the Langlands parameter $\phi(M_\lambda)$ is given by:

$$\phi(M_\lambda)(z) = \mathrm{diag}(a_1, \ldots, a_q, a_p, \bar{a}_q^{-1}, \ldots, \bar{a}_1^{-1}),$$

where

$$a_i = z^{\lambda_i+\rho_i} \bar{z}^{-\lambda_{n-i+1}-\rho_{n-i+1}}.$$

Thus we observe that the transfered representation has inducing data:

$$z^{\lambda_1+\frac{n-1}{2}} \bar{z}^{-\lambda_n+\frac{n-1}{2}} \otimes \cdots \otimes z^{\lambda_q+1} \bar{z}^{-\lambda_{n-q+1}+1} \otimes z^{\lambda_p} \bar{z}^{-\lambda_p} \otimes z^{\lambda_{n-q+1}-1} \bar{z}^{-\lambda_q-1} \otimes \cdots \otimes z^{\lambda_n-\frac{n-1}{2}} \bar{z}^{-\lambda_1-\frac{n-1}{2}},$$

and, as before, it is finite-dimensional with highest weight $\iota(\lambda) = ((\lambda_1, \ldots, \lambda_n), (\lambda_1^*, \ldots, \lambda_n^*))$ which has purity 0. The reader should work out a low-dimensional example like $\mathrm{U}(2, 1)$ as in Example 4.4.

4.5 *Result*

To conclude, we summarise the results of this section in the following table:

G	Transfered to	Highest weight $\iota(\lambda)$ of $\iota(M_\lambda)^*$
$\mathrm{Sp}(2n, \mathbb{R})$	$\mathrm{GL}(2n+1, \mathbb{R})$	$(\lambda_1, \ldots, \lambda_n, 0, -\lambda_n, \ldots, -\lambda_1)$
$\mathrm{SO}(n+1, n)$	$\mathrm{GL}(2n, \mathbb{R})$	$(\lambda_1, \ldots, \lambda_n, -\lambda_n, \ldots, -\lambda_1)$
$\mathrm{SO}(n, n)$	$\mathrm{GL}(2n, \mathbb{R})$	$(\lambda_1, \ldots, \lvert\lambda_n\rvert, -\lvert\lambda_n\rvert, \ldots, -\lambda_1)$
$\mathrm{U}(\frac{n}{2}, \frac{n}{2})$	$\mathrm{GL}(n, \mathbb{C})$	$((\lambda_1, \ldots, \lambda_n)(-\lambda_n, \ldots, -\lambda_1))$
$\mathrm{U}(\frac{n+1}{2}, \frac{n-1}{2})$	$\mathrm{GL}(n, \mathbb{C})$	$((\lambda_1, \ldots, \lambda_n)(-\lambda_n, \ldots, -\lambda_1))$

In the table, $\iota(M_\lambda)^*$ is an *ad hoc* notation to denote $\iota(M_\lambda)$ in all cases except the case of $\mathrm{SO}(n, n)$ where $\iota(M_\lambda)^* = \iota(M_\lambda) \otimes |\cdot|^{\frac{1}{2}}$.

5 Transfer of Discrete Series Representations

Let G be a semisimple Lie group. Let K be a maximal compact subgroup of G. It is well-known that the group G has discrete series representations if and only if the ranks of G and K are equal. Henceforth we assume G has this property. We will very briefly review some important notations and properties of discrete series representations of G and refer the reader to [3, Chap. II, Sect. 5.1] for more details. Let W and W_K be the Weyl groups of G and K, respectively. For the root system $\Phi = \Phi(\mathfrak{g}_{\mathbb{C}}, \mathfrak{t}_{\mathbb{C}})$ (see Sect. 2.1), let $P(\Phi)$ stand for the set of weights; then $P(\Phi)$ is a lattice in $i\mathfrak{t}^*$. An element Λ of $P(\Phi)$ is called regular if for all $\alpha \in \Delta$ we have $\langle \Lambda, \alpha \rangle > 0$. The equivalence classes of discrete series representations of G correspond canonically and bijectively to the orbits of W_K in the set of regular elements in $P(\Phi)$. Let ω_Λ be the class of representations associated to a regular element $\Lambda \in P(\Phi)$. The elements of ω_Λ have infinitesimal character χ_Λ. Put $\lambda = \Lambda - \rho$. The following theorem is well-known (see [3, Theorem II.5.3]):

Theorem 5.1 *Let $(\pi, V) \in \omega_\Lambda$ and $\lambda = \Lambda - \rho$. Let V_K be the underlying $(\mathfrak{g}, K)$-module of K-finite vectors in V. Let μ be a dominant integral weight for G. If $\mu \neq \Lambda - \rho$, then $H^\bullet(\mathfrak{g}, \mathfrak{k}; V_K \otimes M_\mu^\vee) = 0$. Furthermore, $\dim(H^i(\mathfrak{g}, \mathfrak{k}; V_K \otimes M_\lambda^\vee)) = \delta_{i,q}$, where $q = \dim(G/K)/2$ is the middle-dimension.*

For $\pi_\lambda \in \omega_\Lambda$ we write $\pi_\lambda \in \mathrm{Coh}(G, \lambda^\vee)$. (The choice of π_λ in ω_Λ will not matter in what follows.) In this section we will calculate the transfers of the discrete series representations of the classical groups to an appropriate general linear group, checking whether the transfered representations are cohomological or not. Further, if they are cohomological, then we identify the finite-dimensional coefficient system with respect to which they are cohomological. To begin, we review the discrete series representations of $\mathrm{GL}(2, \mathbb{R})$.

5.1 Parameters of Discrete Series for GL(2, ℝ)

A discrete series representation of $\mathrm{GL}(2, \mathbb{R})$ is parametrized by a positive integer l. We denote the corresponding representation by D_l. Consider the representation $D_{l,t} := D_l \otimes |\cdot|^t$ of $\mathrm{GL}(2, \mathbb{R})$. The Langlands parameter of $D_{l,t}$ is given by a homomorphism $\phi(D_{l,t}) : W_{\mathbb{R}} \to \mathrm{GL}(2, \mathbb{C})$, which is

$$\phi(D_{l,t})(z) = \begin{pmatrix} (z\bar{z})^t (\frac{z}{\bar{z}})^{\frac{l}{2}} & 0 \\ 0 & (z\bar{z})^t (\frac{z}{\bar{z}})^{-\frac{l}{2}} \end{pmatrix}, \quad \phi(D_{l,t})(j) = \begin{pmatrix} 0 & (-1)^l \\ 1 & 0 \end{pmatrix}.$$

5.2 $\mathrm{Sp}(p, q)$ to $\mathrm{GL}(2n+1, \mathbb{R})$

Let $G = \mathrm{Sp}(p, q)$ with $p + q = n$ and $p \geq q$. Then $\rho = ne_1 + (n-1)e_2 + \cdots + e_n$. For a dominant integral $\lambda = (\lambda_1, \lambda_2, \ldots, \lambda_n) \in X^*(T)$, take a discrete series representation π_λ of $\mathrm{Sp}(p, q)$ corresponding to $\Lambda = \lambda + \rho$. Note that all the various $\mathrm{Sp}(p, q)$ are in the same inner class along with $\mathrm{Sp}(2n, \mathbb{R})$. Hence they have the same Langlands dual group; recall that ${}^LG^\circ$ is $\mathrm{SO}(2n+1, \mathbb{C})$. The Langlands parameter $\phi(\pi_\lambda)$ (which is a map from $W_\mathbb{R}$ to $\mathrm{SO}(2n+1, \mathbb{C})$) is given by:

$$\begin{aligned}
\phi(\pi_\lambda)(z) &= \mathrm{diag}(a_1, \ldots, a_n, 1, a_n^{-1}, \ldots, a_1^{-1}), \quad a_i = (z/\bar{z})^{\lambda_i + n - i + 1}, \\
\phi(\pi_\lambda)(j) &= \text{anti-diag}((-1)^{2(\lambda_1+n)}, \ldots, (-1)^{2(\lambda_n+1)}, 1, \ldots, 1) = \text{anti-diag}(1, \ldots, 1).
\end{aligned}$$

As in Sect. 4, π_λ transfers to a representation of $\mathrm{GL}(2n+1, \mathbb{R})$, the inducing data of which is

$$D_{2(\lambda_1+n)} \times D_{2(\lambda_2+n-1)} \times \cdots \times D_{2(\lambda_n+1)} \times 1.$$

Note that the ith and the $(n-i+1)$th entry of the Langlands parameter combine to give us the two-dimensional parameter for the discrete series in the inducing data. From Sect. 3 we know that this representation of $\mathrm{GL}(2n+1, \mathbb{R})$ is tempered and cohomological with respect to M_μ, where μ is

$$\begin{aligned}
\mu &= \frac{(2(\lambda_1+n), \ldots, 2(\lambda_n+1), 0, -2(\lambda_n+1), \ldots, -2(\lambda_1+n))}{2} - \rho_{2n+1} \\
&= (\lambda_1, \ldots, \lambda_n, 0, -\lambda_n, \ldots, \lambda_1) \;=\; \iota(\lambda).
\end{aligned}$$

Hence, we have the following proposition:

Proposition 5.2 *Suppose π_λ is a discrete series representation of* $\mathrm{Sp}(p, q)$, *corresponding to the weight $\lambda = (\lambda_1, \lambda_2, \ldots, \lambda_n)$. Then*

$$\pi_\lambda \in \mathrm{Coh}(\mathrm{Sp}(p, q), \lambda^\vee) \;\; \textit{and} \;\; \iota(\pi_\lambda) \in \mathrm{Coh}(\mathrm{GL}(2n+1, \mathbb{R}), \iota(\lambda)^\vee),$$

where, $\iota(\lambda) = (\lambda_1, \ldots, \lambda_n, 0, -\lambda_n, \ldots, -\lambda_1) = \iota(\lambda)^\vee$.

5.3 *From* $\mathrm{SO}(p, q)$ *with* $p + q = 2n + 1$ *to* $\mathrm{GL}(2n, \mathbb{R})$

Since $p + q$ is odd, all the groups $\mathrm{SO}(p, q)$ are in the same inner class. Thus, they have the same Langlands dual group and ${}^LG^\circ = \mathrm{Sp}(2n, \mathbb{C})$. Also, $\rho = (n - \frac{1}{2})e_1 + (n - \frac{3}{2})e_2 + \cdots + \frac{1}{2}e_n$. Let π_λ be a discrete series representation of G, where G is any of the groups in the inner class of $\mathrm{SO}(2n+1)$. Then, the Langlands parameter $\phi(\pi_\lambda)$ is given by:

$$\begin{aligned}\phi(\pi_\lambda)(z) &= \mathrm{diag}(a_1,\ldots,a_n,a_n^{-1},\ldots,a_1^{-1}),\quad a_i=(z/\bar{z})^{\lambda_i+\frac{2n+1-2i}{2}},\\ \phi(\pi_\lambda)(j) &= \text{anti-diag}((-1)^{2(\lambda_1+n-\frac{1}{2})},\ldots,(-1)^{2(\lambda_n+\frac{1}{2})},1,\ldots,1).\end{aligned}$$

This transfers to a representation of $\mathrm{GL}(2n,\mathbb{R})$ with inducing data

$$D_{2(\lambda_1+(n-\frac{1}{2}))}\times D_{2(\lambda_2+(n-\frac{3}{2}))}\times\cdots\times D_{2(\lambda_n+\frac{1}{2})}.$$

Once again, from Sect. 3 we know that this representation of $\mathrm{GL}(2n,\mathbb{R})$ is tempered, and cohomological with respect to the finite-dimensional representation having highest weight $\iota(\lambda)=(\lambda_1,\ldots,\lambda_n,-\lambda_n,\ldots,-\lambda_1)$. Thus we have the following proposition:

Proposition 5.3 *Suppose* π_λ *is a discrete series representation of* $\mathrm{SO}(p,q)$. *Then*

$$\pi_\lambda\in\mathrm{Coh}(\mathrm{SO}(p,q),\lambda^\vee)\ \textit{and}\ \iota(\pi_\lambda)\in\mathrm{Coh}(\mathrm{GL}(2n,\mathbb{R}),\iota(\lambda)^\vee),$$

where, $\iota(\lambda)=(\lambda_1,\ldots,\lambda_n,0,-\lambda_n,\ldots,-\lambda_1)=\iota(\lambda)^\vee$.

5.4 *Even* SO(p,q), $p+q=2n$ *to* GL$(2n,\mathbb{R})$

In this case, there are two inner classes:

- $\mathrm{SO}(p,q)$ such that p and q are both even, and $\mathrm{SO}^*(2n)$,
- $\mathrm{SO}(p,q)$ such that p and q are both odd.

Note that we are considering discrete series representations and we need that the rank of G and rank of K have to be equal. In the case of even orthogonal groups this translates to the condition that pq is even. Thus forcing p and q both to be even. So we are going to assume that we are dealing with the first inner class only. We also have $\rho=(n-1)e_1+(n-2)e_2+\cdots+1e_{n-1}$. Now there are two cases to be considered here:

- n is even.
- n is odd.

When n is even the inner class having discrete series representation contains $\mathrm{SO}(n,n)$, whereas when n is odd, the class contains $\mathrm{SO}(n+1,n-1)$. The Langlands dual groups are $\mathrm{SO}(2n,\mathbb{C})$ and $\mathrm{O}(2n,\mathbb{C})$ respectively. For simplicity, we will denote any group from the inner class of $\mathrm{SO}(n,n)$ as SO_{2n} and any group from the inner class of $\mathrm{SO}(n+1,n-1)$ as SO'_{2n}. Let π_λ be a discrete series representation of SO_{2n} or SO'_{2n}. Then, the Langlands parameter $\phi(\pi_\lambda)$ is given by:

$$\begin{aligned}\phi(\pi_\lambda)(z) &= \mathrm{diag}(a_1,\ldots,a_n,a_n^{-1},\ldots,a_1^{-1}),\quad a_i=(z/\bar{z})^{\lambda_i+n-i},\\ \phi(\pi_\lambda)(j) &= \text{anti-diag}((-1)^{2(\lambda_1+n-1)},\ldots,(-1)^{2\lambda_n},1,\ldots,1)\ =\ \text{anti-diag}(1,\ldots,1).\end{aligned}$$

The transfered representation $\iota(\pi_\lambda)$ of $\mathrm{GL}(2n, \mathbb{R})$ has inducing data

$$D_{2(\lambda_1+n-1)} \otimes \cdots \otimes D_{2|\lambda_n|}.$$

Note that the transfered representation is not cohomological because of the parity condition in (3.1); however, $\iota(\pi_\lambda) \otimes |\cdot|^{\frac{1}{2}}$ is cohomological with respect to the finite-dimensional representation $M^\vee_{\iota(\lambda)}$, where $\iota(\lambda) = (\lambda_1, \ldots, |\lambda_n|, -|\lambda_n|, \ldots, -\lambda_1)$. Thus we have:

Proposition 5.4 *Suppose π_λ is a discrete series representation of $G = \mathrm{SO}_{2n}$ or $G = \mathrm{SO}'_{2n}$. Then*

$$\pi_\lambda \in \mathrm{Coh}(G, \lambda^\vee) \text{ and } \iota(\pi_\lambda) \otimes |\cdot|^{\frac{1}{2}} \in \mathrm{Coh}(\mathrm{GL}(2n, \mathbb{R}), \iota(\lambda)^\vee),$$

where, $\iota(\lambda) = (\lambda_1, \ldots, |\lambda_n|, -|\lambda_n|, \ldots, -|\lambda_1|) = \iota(\lambda)^\vee$.

This may also be rewritten as $\iota(\pi_\lambda) \in \mathrm{Coh}(\mathrm{GL}(2n, \mathbb{R}), \iota(M_\lambda)^\vee)$.

The last assertion follows from Sect. 4.3.3.

5.5 $\mathrm{U}(p, q)$, $p + q = n$ to $\mathrm{GL}(n, \mathbb{C})$

Let $p \geq q$. All the groups $\mathrm{U}(p, q)$ are inner forms of each other. Thus they have the same Langlands dual group which is $\mathrm{GL}(n, \mathbb{C}) \rtimes \mathbb{Z}/2\mathbb{Z}$, where the action of θ, the non-trivial element of $\mathbb{Z}/2\mathbb{Z}$, is given by $g \mapsto \Phi_n\, {}^t g^{-1} \Phi_n^{-1}$. Here, $\rho = (\frac{n-1}{2})e_1 + (\frac{n-3}{2})e_2 + \cdots + (-\frac{n-1}{2})e_n$. Let π_λ be a discrete series representation of $\mathrm{U}(p, q)$. Then, the Langlands parameter $\phi(\pi_\lambda)$ is given by:

$$\phi(\pi_\lambda)(z) = \mathrm{diag}(a_1, \ldots, a_n), \quad a_i = (z/\bar{z})^{\lambda_i + \frac{n-2i+1}{2}}.$$

The representation of $\mathrm{GL}(n, \mathbb{C})$ obtained by transfering π_λ via stable base change is:

$$\mathrm{Ind}_{B_n(\mathbb{C})}^{\mathrm{GL}(n,\mathbb{C})} \left((\tfrac{z}{\bar{z}})^{a_1} \otimes \cdots \otimes (\tfrac{z}{\bar{z}})^{a_n} \right),$$

where $a_i = \lambda_i + (n - 2i + 1)/2$. From Sect. 3.2, this representation of $\mathrm{GL}(n, \mathbb{C})$ is tempered, and cohomological with respect to the dual of the finite-dimensional representation with highest weight $(\lambda, \lambda^\vee) := ((\lambda_1, \ldots, \lambda_n), (-\lambda_n, \ldots, -\lambda_1))$ which is exactly $\iota(\lambda)$, the weight of the finite-dimensional representation transferred from $\mathrm{U}(\frac{n}{2}, \frac{n}{2})$ or $\mathrm{U}(\frac{n+1}{2}, \frac{n-1}{2})$, as might be the case.

Let's note that the weight $\iota(\lambda)$ does not give a self-dual coefficient system as in the cases of transfers from the other classical groups we have dealt with, however, as one might expect for stable base change from unitary groups, it gives a conjugate-self-dual representation. For $\lambda = (\lambda_1, \ldots, \lambda_n)$ with $\lambda_1 \geq \cdots \geq \lambda_n$, suppose M_λ denotes the irreducible representation of the complex (algebraic) group $\mathrm{GL}(n, \mathbb{C})$ then we have

$$M_{\iota(\lambda)} \;=\; M_\lambda \otimes \overline{M_{\lambda^\vee}} \;=\; M_\lambda \otimes \overline{M_\lambda^\vee},$$

which is now an algebraic representation of the real group $\mathrm{GL}(n,\mathbb{C})$. Clearly, $M_{\iota(\lambda)}^\vee = M_\lambda^\vee \otimes \overline{M_\lambda} = M_{\lambda^\vee} \otimes \overline{M_\lambda}$. Hence, $M_{\iota(\lambda)}^\vee$ is the representation corresponding to the weight $(\lambda^\vee, \lambda)$. Therefore, for the identity $M_{\iota(\lambda)}^\vee = M_{\iota(\lambda)^\vee}$ to hold, if we define:

$$\iota(\lambda)^\vee = (\lambda, \lambda^\vee)^\vee := (\lambda^\vee, \lambda),$$

then we have:

Proposition 5.5 *Suppose π_λ is a discrete series representation of $\mathrm{U}(p,q)$. Then*

$$\pi_\lambda \in \mathrm{Coh}(\mathrm{U}(p,q), \lambda^\vee) \;\; \textit{and} \;\; \iota(\pi_\lambda) \in \mathrm{Coh}(\mathrm{GL}(n,\mathbb{C}), \iota(\lambda)^\vee),$$

where, $\iota(\lambda) = ((\lambda_1, \ldots, \lambda_n), (-\lambda_n, \ldots, -\lambda_1)) = (\lambda, \lambda^\vee)$.

5.6 The Main Results

We summarize the results of all the above into the following theorem:

Theorem 5.6 *(1) Let $\lambda = (\lambda_1, \ldots, \lambda_n)$ be a dominant integral weight for $\mathrm{Sp}(2n,\mathbb{R})$. Then, as representations of $\mathrm{GL}(2n+1,\mathbb{R})$, we have:*

$$\iota(M_\lambda) = M_{\iota(\lambda)}, \quad \iota(\lambda) = (\lambda_1, \ldots, \lambda_n, 0, -\lambda_n, \ldots, -\lambda_1) = \iota(\lambda)^\vee.$$

Let G be in the inner class of $\mathrm{Sp}(2n,\mathbb{R})$, and if π_λ is a discrete series representation of G in $\mathrm{Coh}(G, \lambda^\vee)$, then

$$\iota(\pi_\lambda) \in \mathrm{Coh}(\mathrm{GL}(2n+1,\mathbb{R}), \iota(\lambda)^\vee).$$

(2) Let $\lambda = (\lambda_1, \ldots, \lambda_n)$ be a dominant integral weight of $\mathrm{SO}(n+1,n)$. Then, as representations of $\mathrm{GL}(2n,\mathbb{R})$, we have:

$$\iota(M_\lambda) = M_{\iota(\lambda)}, \quad \iota(\lambda) = (\lambda_1, \ldots, \lambda_n, -\lambda_n, \ldots, -\lambda_1) = \iota(\lambda)^\vee.$$

Let G be in the inner class of $\mathrm{SO}(n+1,n)$, and if π_λ is a discrete series representation of G in $\mathrm{Coh}(G, \lambda^\vee)$, then

$$\iota(\pi_\lambda) \in \mathrm{Coh}(\mathrm{GL}(2n,\mathbb{R}), \iota(\lambda)^\vee).$$

(3) Let $\lambda = (\lambda_1, \ldots, \lambda_n)$ be a dominant integral weight of $G = \mathrm{SO}(n,n)$. Then, as representations of $\mathrm{GL}(2n,\mathbb{R})$ we have:

$$\iota(M_\lambda) = M_{\iota(\lambda)} \otimes |\cdot|^{-\frac{1}{2}}, \quad \iota(\lambda) = (\lambda_1, \ldots, \lambda_{n-1}, |\lambda_n|, -|\lambda_n|, -\lambda_{n-1}, \ldots, -\lambda_1).$$

If $G = \mathrm{SO}_{2n}$ *or* SO'_{2n}, *and* π_λ *is a discrete series representation of* G *in* $\mathrm{Coh}(G, \lambda)$, *then*

$$\iota(\pi_\lambda) \otimes |\cdot|^{\frac{1}{2}} \in \mathrm{Coh}(\mathrm{GL}(2n, \mathbb{R}), \iota(\lambda)^\vee),$$

which may also be written as

$$\iota(\pi_\lambda) \in \mathrm{Coh}(\mathrm{GL}(2n, \mathbb{R}), \iota(M_\lambda)^\vee).$$

(4) *Let* $\lambda = (\lambda_1, \ldots, \lambda_n)$ *be a dominant integral weight for* $\mathrm{U}(\frac{n}{2}, \frac{n}{2})$ *or* $\mathrm{U}(\frac{n+1}{2}, \frac{n-1}{2})$. *Then, as representations of* $\mathrm{GL}(n, \mathbb{C})$, *we have:*

$$\iota(M_\lambda) = M_{\iota(\lambda)}, \quad \iota(\lambda) = ((\lambda_1, \ldots, \lambda_n), (-\lambda_n, \ldots, -\lambda_1)).$$

Let G *be in the inner class of* $\mathrm{U}(\frac{n}{2}, \frac{n}{2})$ *or* $\mathrm{U}(\frac{n+1}{2}, \frac{n-1}{2})$ *depending on the parity of* n, *and if* π_λ *is a discrete series representation of* G *in* $\mathrm{Coh}(G, \lambda^\vee)$, *then*

$$\iota(\pi_\lambda) \in \mathrm{Coh}(\mathrm{GL}(n, \mathbb{C}), \iota(\lambda)^\vee).$$

(5) *For any of the above groups in* (1) − (4), *the transfer of a tempered representation* π *is cohomological if and only if* π *is a discrete series representation of* G.

Acknowledgements We thank Dipendra Prasad for his interest in the results of this project, and for helpful tutorials on Langlands parameters. We also thank the referee for a very careful reading and helpful comments.

References

1. Bhagwat, C., Raghuram, A.: Endoscopy and Cohomology. Bull. Iranian Math. Soc. **43**(4), 317–335 (2017)
2. Borel, A.: Automorphic L-functions, Automorphic forms, Representations and L-functions (Proceedings of Symposia in Pure Mathematics, Oregon State University, Corvallis, Oregon, 1977), Part 2, XXXIII, pp. 27–61. American Mathematical Society, Providence (1979)
3. Borel, A., Wallach, N.: Continuous Cohomology, Discrete Subgroups and Representations of Reductive Groups. Mathematical Surveys and Monographs, vol. 67, 2nd edn, xviii+260 pp. American Mathematical Society, Providence (2000). ISBN: 0-8218-0851-6
4. Gross, B.H., Reeder, M.: From Laplace to Langlands via representations of orthogonal groups. Bull. Am. Math. Soc. (N.S.) **43**(2), 163–205 (2006)
5. Kim, H.H., Krishnamurthy, M.: Stable base change lift from unitary groups to GL(n). IMRP Int. Math. Res. Pap. **1**, 1–52 (2005)
6. Knapp, A.: Local Langlands Correspondence - The Archemedean case, Motives. Proceedings in Symposia of Pure Mathematics, vol. 55, Part II, pp. 393–410. Seattle, WA 1991 (1994)
7. Labesse, J.-P., Schwermer, J.: On liftings and cusp cohomology of arithmetic groups. Invent. Math. **83**, 383–401 (1986)
8. Langlands, R.P.: On the classification of irreducible representations of real algebraic groups. Representation Theory and Harmonic Analysis on Semi-Simple Lie Groups. Mathematical

Surveys and Monographs, vol. 31, pp. 101–170. American Mathematical Society, Providence (1989)
9. Raghuram, A.: Critical values of Rankin-Selberg L-functions for GL(n) $\times$ $GL(n-1)$ and the symmetric cube L-functions for GL(2). Forum Math. **28**(3), 457–489 (2016)
10. Raghuram, A., Shahidi, F.: Funtoriality and special values of L-functions. In: Gan, W.T., Kudla, S., Tschinkel, Y. (eds.) Eisenstein Series and Applications. Progress in Mathematics, vol. 258, pp. 271–294. Birkhäuser Boston, Boston (2008)

Congruence Link Complements—A 3-Dimensional Rademacher Conjecture

M. D. Baker and A. W. Reid

Dedicated to Joachim Schwermer on the occasion of his 66th birthday.

Abstract In this article we discuss a 3-dimensional version of a conjecture of Rademacher concerning genus 0 congruence subgroups of $\mathrm{PSL}(2, \mathbb{Z})$. We survey known results, as well as including some new results that make partial progress on the conjecture.

Keywords Link complement · Bianchi group · Congruence subgroup

1 Introduction

Let k be a number field with ring of integers R_k. A subgroup $\Gamma < \mathrm{PSL}(2, R_k)$ is called a *congruence subgroup* if there exists an ideal $I \subset R_k$ so that Γ contains the *principal congruence group*:

$$\Gamma(I) = \ker\{\mathrm{PSL}(2, R_k) \to \mathrm{PSL}(2, R_k/I)\},$$

where $\mathrm{PSL}(2, R_k/I) = \mathrm{SL}(2, R_k/I)/\{\pm Id\}$. The largest ideal I for which $\Gamma(I) < \Gamma$ is called the *level* of Γ. For convenience, if $n \in \mathbb{Z} \subset R_k$, we will denote the principal R_k-ideal $< n >$ simply by n.

Reid was supported in part by an NSF grant, and The Wolfensohn Fund administered through I.A.S.

M. D. Baker
IRMAR, Université de Rennes 1, 35042 Rennes Cedex, France
e-mail: mark.baker@univ-rennes1.fr

A. W. Reid (✉)
Department of Mathematics, Rice University, Houston, TX 77005, USA
e-mail: alan.reid@rice.edu

J. W. Cogdell et al. (eds.), *Cohomology of Arithmetic Groups*,
Springer Proceedings in Mathematics & Statistics 245,
https://doi.org/10.1007/978-3-319-95549-0_7

For a variety of reasons (geometric, number theoretic and topological), congruence subgroups are perhaps the most studied class of arithmetic subgroups. For example, in the case when $R_k = \mathbb{Z}$, and $\Gamma < \mathrm{PSL}(2, \mathbb{Z})$, the genus, number of cone points and number of cusps of $\mathbb{H}^2/\Gamma$ has been well-studied. This was perhaps best articulated in a conjecture of Rademacher which posited that there are only finitely many congruence subgroups $\Gamma < \mathrm{PSL}(2, \mathbb{Z})$ of genus 0 (i.e. when $\mathbb{H}^2/\Gamma$ has genus 0). The proof of this was completed by Denin in a sequence of papers [16–18]. Different proofs of this (actually of a slightly stronger version of this result) were also given by Thompson [40] and Zograf [44]. Indeed, in these two papers it is proved that there are only finitely congruence subgroups of $\mathrm{PSL}(2, \mathbb{Z})$ of any fixed genus.

The complete enumeration of congruence subgroups of genus 0 was completed in [15], where 132 groups are listed. The list of torsion-free congruence subgroups of genus 0 was completed in [35] (there are 33 and the levels are all of the form $2^a 3^b 5^c 7^d$ with $a \leq 5, b \leq 3, c \leq 2$ and $d \leq 1$, with 2^5 being the largest level). As is easy to see (and we describe this in Sect. 4 below) of those 33, only 4 are principal congruence subgroups (of levels 2, 3, 4 and 5).

Turning to dimension 3, let d be a square-free positive integer, let O_d denote the ring of integers in $\mathbb{Q}(\sqrt{-d})$, and let Q_d denote the Bianchi orbifold $\mathbb{H}^3/\mathrm{PSL}(2, O_d)$. A non-compact finite volume hyperbolic 3-manifold X is called *arithmetic* if X and Q_d are commensurable, that is to say they share a common finite sheeted cover (see [30, Chapter 8] for more on this). If N is a closed orientable 3-manifold and $L \subset N$ a link, then L is called *arithmetic* if $N \backslash L$ is an arithmetic hyperbolic 3-manifold.

In his list of problems in his Bulletin of the AMS article [42], Thurston states as Question 19:

Find topological and geometric properties of quotient spaces of arithmetic subgroups of $\mathrm{PSL}(2, \mathbb{C})$. *These manifolds often seem to have special beauty.*

For example, many of the key examples in the development of the theory of geometric structures on 3-manifolds (e.g. the figure-eight knot complement, the Whitehead link complement, the complement of the Borromean rings and the Magic manifold) are arithmetic.

The "beauty" referred to by Thurston is captured particularly well by congruence manifolds (which includes all of the above examples); i.e. manifolds $M = \mathbb{H}^3/\Gamma$ where Γ is congruence. Similarly, a manifold $\mathbb{H}^3/\Gamma$ is called principal congruence if $\Gamma = \Gamma(I)$ for some ideal I. As above we will also refer to a link $L \subset N$ as congruence (resp. principal congruence) if the manifold $N \backslash L$ is so.

In this paper we will survey what is known about the following question, which as we discuss below, can be viewed as the most natural generalization of the Rademacher conjecture to dimension 3.

Question 1.1 Are there only finitely many congruence link complements in S^3?

As noted above the work of Zograf shows that there are only finitely many congruence surfaces of any fixed genus. Thus a far reaching generalization of this to dimension 3 is the following question:

Question 1.2 Let N be a fixed closed orientable 3-manifold. Are there only finitely many congruence link complements in N?

We finish the introduction with a brief discussion of the remainder of the paper. In Sect. 3 we describe early progress on Question 1.1 in the context of the "Cuspidal Cohomology Problem". This was given particular impetus in work of Schwermer (with Fritz Grunewald) in [24]. In Sect. 4–6 we survey recent work of the authors ([7–9]), which together with the work of Goerner [21, 22] (which completed the enumeration in the cases of $d = 1, 3$) answers a question of Thurston about the complete list of principal congruence link complements asked in an email to the authors in 2009:

"Although there are infinitely many arithmetic link complements, there are only finitely many that come from principal congruence subgroups. Some of the examples known seem to be among the most general (given their volume) for producing lots of exceptional manifolds by Dehn filling, so I'm curious about the complete list."

To that end we discuss the proof of the following theorem (the final details of which will appear in work of the authors with M. Goerner in [9]).

Theorem 1.3 *There are* 48 *principal congruence link complements in* S^3. *The values of* d *and the levels* I *are listed in Table 1 in Sect. 5.*

Note that since links with at least 2 components are not generally determined by their complements (see [23]), one cannot just say "finitely many principal congruence links".

Also in Sect. 4 we describe what is known about Question 1.2 for principal congruence link complements as well as Question 1.1 for principal congruence link complements in S^3 arising from other maximal orders. In Sect. 7 we discuss some levels where Question 1.1 can be answered positively, and in Sect. 8 we finish with some comments and speculations.

2 Preliminaries

Let h_d denote the class number of $\mathbb{Q}(\sqrt{-d})$. Then as is well-known the quotient orbifold $Q_d = \mathbb{H}^3/\mathrm{PSL}(2, O_d)$ has h_d cusps. Apart from the cases of $d = 1, 3$, the cusp ends of the Q_d all have the form $T^2 \times [0, \infty)$ where T^2 is the 2-torus. When $d = 1$, the cusp end is $[S, \infty)$ where S is the Euclidean 2-orbifold which is a 2-sphere with 4 cone points of cone angle π, and when $d = 3$, the cusp end is $[B, \infty)$ where B is the Euclidean 2-orbifold which is a 2-sphere with 3 cone points of cone angle $2\pi/3$.

Let $\Gamma \leq \mathrm{PSL}(2, O_d)$ be a finite index subgroup. Then

- A *cusp*, $[c]$, of Γ is a Γ-orbit of points in $\mathbb{P}^1(\mathbb{Q}(\sqrt{-d}))$.
- A *peripheral subgroup* of Γ for $[c]$ is a maximal parabolic subgroup, $P_x < \Gamma$, fixing $x \in [c]$. Note that if $y \in [c]$, then P_x and P_y are conjugate; hence a peripheral subgroup for $[c]$ is determined up to conjugacy.

- A *set of peripheral subgroups for* Γ is the choice of one peripheral subgroup for each cusp of Γ.

We will use the term *cusp* to mean $[c]$, a choice of point x in $[c]$, as well as the end of $\mathbb{H}^3/\Gamma$ corresponding to $[c]$. Which one is meant should be clear from the context.

Recall that if $p \in \mathbb{Z}$ then p is called *inert* if the O_d-ideal $< p >$ remains prime, and p is said to *split* if the O_d-ideal $< p > = \mathcal{P}_1\mathcal{P}_2$ (and also say that $\mathcal{P}_i$ is a split prime for $i = 1, 2$). If $I \subset O_d$ is an ideal then $N(I) = |O_d/I|$ denotes the norm of the ideal I.

3 The Cuspidal Cohomology Problem

Let Γ be a non-cocompact Kleinian (resp. Fuchsian) group acting on $\mathbb{H}^3$ (resp. $\mathbb{H}^2$) with finite co-volume, and set $X = \mathbb{H}^n/\Gamma$ with $n = 2, 3$. Let $\mathcal{U}(\Gamma)$ denote the subgroup of Γ generated by parabolic elements of Γ. Note that $\mathcal{U}(\Gamma)$ is visibly a normal subgroup of Γ, and we may define:

$$V(X) = V(\Gamma) = (\Gamma/\mathcal{U}(\Gamma))^{\text{ab}} \otimes_{\mathbb{Z}} \mathbb{Q}.$$

Then the subspace of $H_1(X, \mathbb{Q})$ which defines the (degree 1) cuspidal cohomology of X (or Γ) can be identified with $V(\Gamma)$.

Setting $r(\Gamma) = \dim_{\mathbb{Q}}(V(\Gamma))$, we see that in the case when Γ is a torsion-free Fuchsian group $r(\Gamma) = 0$ if and only if the underlying space of X is a punctured S^2, which in turn is equivalent to Γ being generated by parabolic elements.

When Γ is Kleinian and $X \cong S^3 \setminus L$, then $r(\Gamma) = 0$ and Γ is generated by parabolic elements. Of course, in dimension 3, other closed manifolds provide examples of link complements $\mathbb{H}^3/\Gamma \cong \Sigma \setminus L$ satisfying $r(\Gamma) = 0$, e.g. when Σ is an integral homology 3-sphere. It is also not the case that being generated by parabolic elements forces the link complement to be contained in S^3 (see [22]).

In the setting of the Bianchi groups, $\mathrm{PSL}(2, O_d)$, *The Cuspidal Cohomology Problem* posed in the 1980's asked which Bianchi groups have $r(\mathrm{PSL}(2, O_d)) = 0$. Building on work of many people (see for example [24, 25, 34]), the solution of the Cuspidal Cohomology Problem was provided by Vogtmann [43] who determined the list of all values of d (see Theorem 3.1 below) with $r(Q_d) = 0$. In particular, this provided the list of those d for which Q_d can have a cover homeomorphic to an arithmetic link complement in S^3. Moreover in [5] it was shown that for every such d there does exist an arithmetic link complement. We summarize this discussion in the following result:

Theorem 3.1 *Q_d is covered by an arithmetic link complement in S^3 if and only if*

$$d \in \{1, 2, 3, 5, 6, 7, 11, 15, 19, 23, 31, 39, 47, 71\}.$$

4 Finiteness of Principal Congruence Link Complements

In this section we discuss the proof of finiteness of principal congruence link complements; the case of dimension 2 is well-known, and a proof in dimension 3 is given in [7] (see also [8]).

4.1 The Case of PSL(2, $\mathbb{Z}$)

Proposition 4.1 *The only principal congruence subgroups of* PSL$(2, \mathbb{Z})$ *of genus* 0 *have level* $n = 2, 3, 4, 5$.

Proof An easy argument shows that $\Gamma(n)$ is torsion-free for all $n \geq 2$. The proof is completed using the following straightforward observations:
(1) $\Gamma(n)$ has genus zero if and only if $\Gamma(n)$ is generated by parabolic elements.
(2) Let $T(n)$ denote the cyclic subgroup of $\Gamma(n)$ fixing the cusp at ∞; i.e. $T(n)$ is the cyclic group generated by $\begin{pmatrix} 1 & n \\ 0 & 1 \end{pmatrix}$. Hence the normal closure $N(n)$ of $T(n)$ in PSL$(2, \mathbb{Z})$ is a subgroup of $\Gamma(n)$. It is easy to see that $\Gamma(n)$ is generated by parabolic elements if and only $N(n) = \Gamma(n)$.
(3) Now note that PSL$(2, \mathbb{Z})/N(n)$ is isomorphic to the $(2, 3, n)$ triangle group, and this is finite if and only if n is as in the statement of Proposition 4.1. □

4.2 The Bianchi Groups

Although we will also use the idea of proof of Proposition 4.1, the case of dimension 3 requires additional technology, and handled somewhat differently; since, amongst other reasons, there are cases where the class number is greater than 1. Following [7, 8], we will appeal to systole bounds. Recall that if M is a finite volume orientable hyperbolic 3-manifold, the *systole of* M is the length of the shortest closed geodesic in M, and will be denoted by sys(M). The following is proved in [8, Lemma 4.1] (using [1] and the improvement in [33]). The solution to the Cuspidal Cohomology Problem yields only finitely many d to consider, and so the finiteness of the number of principal congruence link complements easily follows from Lemma 4.2. We include the proof since we will refer to it later.

Lemma 4.2 *Suppose that* $I \subset O_d$ *is an ideal such that* $\mathbb{H}^3/\Gamma(I)$ *is homeomorphic to a link complement in* S^3. *Then* $N(I) < 39$.

Proof If $\gamma \in \Gamma(I)$ is a hyperbolic element, its complex length is $\ell(\gamma) = \ell_0(\gamma) + i\theta(\gamma)$, where $\ell_0(\gamma)$ is the translation length of γ and $\theta(\gamma)$ is the angle incurred in translating along the axis of γ by distance $\ell_0(\gamma)$. Now, as is well-known

$\cosh(\ell(\gamma)/2) = \pm\mathrm{tr}\,(\gamma)/2$, and so we get the following inequality for $\ell_0(\gamma)$: $|\mathrm{tr}\,(\gamma)|/2 \leq \cosh(\ell_0(\gamma)/2)$. With the systole bound given in [33] of 7.171646..., the argument of [1] used in [7] can be reworked and gives:

$$|\mathrm{tr}\,(\gamma)/2| \leq \cosh(7.1717/2) \leq 18.1 \text{ and so } |\mathrm{tr}\,(\gamma)| < 37.$$

From [7, Lemma 2.5] (see also [8, Lemma 4.1]) we have $\mathrm{tr}\,(\gamma) \pm 2 \in I^2$, and so the bound on $|\mathrm{tr}\,(\gamma)|$ given above implies that $N(I) < 39$. □

Remark 4.3 The proof of Lemma 4.2 actually shows more: if $\Gamma(I)$ is a link group, then there exists $x \in I$ such that $|x|^2 < 39$. This is relevant in the case when $h_d > 1$, since there are ideals I of norm less than 39 for which no such element exists and hence $\Gamma(I)$ is not a link group.

Finally in this section we discuss the case of principal congruence link complements in arbitrary closed orientable 3-manifolds. The first remark is that in the case when N is a closed orientable 3-manifold that does not support a metric of negative curvature, then the systole bound from [1] (or the improvement in [33]) and the argument of Lemma 4.2 also proves finiteness for principal congruence link complements $N\backslash L$.

To discuss the hyperbolic case, let $X = N\backslash L$; then the dimension of $V(X)$ can be shown to be bounded above by $\dim(H_2(N;\mathbb{Q}))$, whilst on the other hand, it is known that (see [24]) as $d \to \infty$ the dimension of $V(Q_d)$ goes to infinity.

Thus we deduce the following corollary of this discussion.

Corollary 4.4 *Suppose that N is a fixed closed orientable 3-manifold. If $N\backslash L = \mathbb{H}^3/\Gamma(I)$, then there are at most finitely many d such that $I \subset O_d$.*

Thus as in the case of S^3, it remains to control the levels in these finitely many d. This was done by Lakeland and Leininger [28] using a more refined analysis of systole bounds.

Theorem 4.5 *Let N be a closed orientable 3-manifold. Then there are only finitely many principal congruence link complements in N.*

4.3 Other Maximal Orders

The Bianchi groups can be thought of as arising from the maximal order $M(2, O_d) \subset M(2,\mathbb{Q}(\sqrt{-d}))$. In the case when h_d is even, there are maximal orders $\mathcal{O} \subset M(2,\mathbb{Q}(\sqrt{-d}))$ that are not $\mathrm{GL}(2,\mathbb{Q}(\sqrt{-d}))$-conjugate to $M(2, O_d)$ and whose group of elements of norm one thereby define arithmetic Kleinian groups $\Gamma^1_{\mathcal{O}}$ commensurable with $\mathrm{PSL}(2, O_d)$, with the same co-volume as $\mathrm{PSL}(2, O_d)$, but not conjugate to $\mathrm{PSL}(2, O_d)$ (see below and [30] for further details). A version of the cuspidal cohomology problem was solved for these groups, and the following additional values of d (beyond those listed in Theorem 3.1) provide groups $\Gamma^1_{\mathcal{O}}$ with $r(\Gamma^1_{\mathcal{O}}) = 0$: namely $d \in \{10, 14, 35, 55, 95, 119\}$. In addition, for $d \in \{5, 6, 15, 39\}$ (which do arise in Theorem 3.1) groups $\Gamma^1_{\mathcal{O}}$ with $\mathcal{O} \neq M(2, O_d)$ were also shown to have $r(\Gamma^1_{\mathcal{O}}) = 0$ (see [13]).

Turning to link groups, it is known that link groups do appear as subgroups of finite index in some of these additional groups $\Gamma^1_{\mathcal{O}}$; namely in the cases of $d \in \{5, 6, 10, 15, 35, 39, 55\}$ ([6, 37, 38]). One can also make sense of the notion of congruence subgroups of the groups $\Gamma^1_{\mathcal{O}}$, however nothing is known about the existence of (principal) congruence link complements or more generally Question 1.1 in this setting. However, we can establish finiteness of principal congruence link complements in this broader sense. Before stating the result carefully we recall the construction of these other types of maximal orders. To that end, let $J \subset O_d$ be an ideal and define $\mathcal{O}_d(J)$ by:

$$\mathcal{O}_d(J) = \left\{ \begin{pmatrix} a & b \\ c & d \end{pmatrix} : a, d \in O_d, c \in J, b \in J^{-1} \right\},$$

where $J^{-1} = \{x \in \mathbb{Q}(\sqrt{-d}) : xy \in O_d, \forall y \text{ in } J\}$ is the inverse ideal.

It follows from [30, Chapter 6.7] that the *type number of maximal orders* of $M(2, \mathbb{Q}(\sqrt{-d}))$ (which is the number of distinct $\mathrm{GL}(2, \mathbb{Q}(\sqrt{-d}))$-conjugacy classes of maximal orders) is one when h_d is odd and when h_d is even it is equal to $t_d = |H_d/H_d^{(2)}|$ where H_d is the class group and $H_d^{(2)}$ the group generated by squares of elements in H_d. In particular, in the case when h_d is even, we may find a finite collection of ideals $J_1, \ldots, J_{t_d}$ so that every maximal order of $M(2, \mathbb{Q}(\sqrt{-d}))$ is $\mathrm{GL}(2, \mathbb{Q}(\sqrt{-d}))$-conjugate to one of $\mathcal{O}_d(J_i)$. By convention we take $J_1 = O_d$ and identify the maximal order $\mathcal{O}_d(J_1) = M_2(O_d)$.

Following the notation above, the maximal orders $\mathcal{O}_d(J_i)$ give rise to arithmetic Kleinian groups $\Gamma^1_{\mathcal{O}_d(J_i)}$ commensurable with $\mathrm{PSL}(2, O_d)$. Note that from the previous discussion and [30, Theorem 9.2.2] every arithmetic link group is conjugate into some group $\Gamma^1_{\mathcal{O}_d(J_i)}$

For an ideal $I \subset O_d$ with $I \neq J_i$ for $i = 1, \ldots, t_d$, we can construct the principal congruence subgroup $\Gamma_i(I)$ by reducing modulo I (we refer the reader to [30, Chapter 6.6] for more details about this). Call $\Gamma < \Gamma^1_{\mathcal{O}_d(J_i)}$ congruence if $\Gamma > \Gamma_i(I)$ for some $i = 1, \ldots, t_d$ and $I \subset O_d$.

Theorem 4.6 *There are only finitely many principal congruence link complements arising as* $\mathbb{H}^3/\Gamma_i(I)$.

Proof The proof follows the idea in the proof of Lemma 4.2. From above, there is a finite list of possible values of d, and moreover, we need only consider those d for which h_d is even—which reduces to $d \in \{5, 6, 10, 14, 15, 35, 39, 55, 95, 119\}$. Since the type number is finite, it follows, as before, that it remains to bound the norm of the ideal I. Following the proof of [7, Lemma 2.5] if $\gamma \in \Gamma_i(I)$ is a hyperbolic element then it has the form

$$\gamma = \begin{pmatrix} \pm 1 + a & b \\ c & \pm 1 + d \end{pmatrix}$$

where $a, d \in I$, $c \in I.J$ and $b \in I.J^{-1}$. Following the argument in the proof of [7, Lemma 2.5] we deduce that $\pm(a + d) = -ad + bc$, but this time we have $bc \in IJ \cdot IJ^{-1} \subset I^2$. Regardless, we can still deduce that $\operatorname{tr} \gamma \equiv \pm 2 \bmod I^2$. Given this we can complete the proof as in Lemma 4.2 to deduce that $N(I) < 39$. □

5 Techniques to Determine the List

Table 1 below gives the complete list of 48 pairs (d, I) describing all principal congruence subgroups $\Gamma(I) < \mathrm{PSL}(2, O_d)$ such that $\mathbb{H}^3/\Gamma(I)$ is a link complement in S^3. Note that if $\Gamma(I)$ determines a principal congruence link group, so does $\Gamma(\bar{I})$.

Comparing with Theorem 3.1, the reader will note that the cases of $d = 6$ and $d = 39$ do not occur in Table 1. Indeed, as shown in [8], there are no principal congruence link complements for these values of d, although we do show in [8] that there are congruence link complements.

To determine the list of those d and levels I, there are two main issues. First, proving that those on the list are link complements in S^3, and secondly, eliminating those that are not on this list. We refer the reader to [7–9] for further details. Sample computations are given in Sect. 6.

We also note that these principal congruence link complements in S^3 also give rise to many more examples of principal congruence link complements in other closed

Table 1 All principal congruence link groups

(1)	$d = 1$: $I \in \{2, < 2 \pm i >, < (1 \pm i)^3 >, 3, < 3 \pm i >, < 3 \pm 2i >, < 4 \pm i >\}$
(2)	$d = 2$: $I \in \{2, < 1 \pm \sqrt{-2} >, < 1 \pm 2\sqrt{-2} >, < 2 \pm \sqrt{-2} >, < 3 \pm \sqrt{-2} >\}$
(3)	$d = 3$: $I \in \{2, 3, < (5 \pm \sqrt{-3})/2 >, < 3 \pm \sqrt{-3} >, < (7 \pm \sqrt{-3})/2 >,$ $< 4 \pm \sqrt{-3} >, < (9 \pm \sqrt{-3})/2 >\}$
(4)	$d = 5$: $I = < 3, (1 \pm \sqrt{-5}) >$
(5)	$d = 7$: $I \in \{< (1 \pm \sqrt{-7})/2 >, 2, < (3 \pm \sqrt{-7})/2 >, < \sqrt{-7} >, < 1 \pm \sqrt{-7} >,$ $< (-5 \pm \sqrt{-7})/2 >, < 2 \pm \sqrt{-7} >, < (7 \pm \sqrt{-7})/2 >, < (1 \pm 3\sqrt{-7})/2 >\}$
(6)	$d = 11$: $I \in \{< (1 \pm \sqrt{-11})/2 >, < (3 \pm \sqrt{-11})/2 >, < (5 \pm \sqrt{-11})/2 >\}$
(7)	$d = 15$: $I \in \{< 2, (1 \pm \sqrt{-15})/2 >, < 3, (3 \pm \sqrt{-15})/2 >, < 4, (1 \pm \sqrt{-15})/2 >,$ $< 5, (5 \pm \sqrt{-15})/2 >, < 6, (-3 \pm \sqrt{-15})/2 >\}$
(8)	$d = 19$: $I = < (1 \pm \sqrt{-19})/2 >$
(9)	$d = 23$: $I \in \{< 2, (1 \pm \sqrt{-23})/2 >, < 3, (1 \pm \sqrt{-23})/2 >, < 4, (-3 \pm \sqrt{-23})/2 >\}$
(10)	$d = 31$: $I \in \{< 2, (1 \pm \sqrt{-31})/2 >, < 4, (1 \pm \sqrt{-31})/2 >, < 5, (3 \pm \sqrt{-31})/2 >\}$
(11)	$d = 47$: $I \in \{< 2, (1 \pm \sqrt{-47})/2 >, < 3, (1 \pm \sqrt{-47})/2 >, < 4, (1 \pm \sqrt{-47})/2 >\}$
(12)	$d = 71$: $I = < 2, (1 \pm \sqrt{-71})/2 >$.

non-hyperbolic 3-manifolds. For example, the case of $(7, (1 \pm \sqrt{-7})/2)$ gives a principal congruence link complement known as the Magic manifold. This manifold has many "exceptional" (i.e. non-hyperbolic) Dehn surgeries, which thereby gives closed 3-manifolds N containing a link L with $N \backslash L \cong \mathbb{H}^3/\Gamma(< (1 \pm \sqrt{-7})/2 >)$; for example infinitely many small Seifert fibered spaces, and infinitely many torus bundles over the circle admitting SOL geometry (see [31]).

5.1 Establishing $\mathbb{H}^3/\Gamma(I)$ is a Link Complement

To establish the principal congruence link groups in Table 1, we typically invoke the following strategy, which closely mimics that used in the case of $\mathrm{PSL}(2, \mathbb{Z})$. Note that only in a small number of cases do we have explicit links, indeed some of these explicit links in S^3 were in the literature (e.g. in [2, 4, 26, 41]) and were shown to give principal congruence link complements directly (see [7, 8] for further details).

(1) If $\mathbb{H}^3/\Gamma(I) \cong S^3 \backslash L$, then $\Gamma(I)$ is generated by parabolic elements.
(2) $\mathbb{H}^3/\Gamma(I) \cong S^3 \backslash L$ if and only if $\Gamma(I)$ can be trivialized by setting one parabolic from each cusp of $\Gamma(I)$ equal to 1.

Some further commentary: If $\mathbb{H}^3/\Gamma(I) \cong S^3 \backslash L$, then for each component L_i of L, there is a meridian curve x_i so that Dehn filling $S^3 \backslash L$ along the totality of these curves gives S^3. Thus, trivializing the corresponding parabolic elements $[x_i]$ in $\Gamma(I)$ gives the trivial group. Conversely, given Perelman's resolution of the Geometrization Conjecture, if $\Gamma(I)$ can be trivialized by setting one parabolic from each cusp of $\Gamma(I)$ equal to 1, then $\mathbb{H}^3/\Gamma(I)$ is homeomorphic to a link complement in S^3.

To check whether $\Gamma(I)$ is generated by parabolic elements we can proceed as follows.

Let $\Gamma(I) < \mathrm{PSL}(2, O_d)$, and let P_i be the peripheral subgroup of $\mathrm{PSL}(2, O_d)$ fixing the cusp $c_i \in \mathbb{Q}(\sqrt{-d}) \cup \{\infty\}$ for $i = 1, \ldots, h_d$. Set $P_i(I) = \mathrm{P}_i \cap \Gamma(I)$ to be the peripheral subgroup of $\Gamma(I)$ fixing c_i. Let $N_d(I)$ denote the normal closure in $\mathrm{PSL}(2, O_d)$ of $\{P_1(I), \ldots, P_{h_d}(I)\}$. Note that $N_d(I) < \Gamma(I)$ since $\Gamma(I)$ is a normal subgroup of $\mathrm{PSL}(2, O_d)$. It is clear that $\Gamma(I)$ is generated by parabolic elements if and only if $N_d(I) = \Gamma(I)$. Now we can try to use Magma [14] (using presentation from [32] and [39]) to test whether $\Gamma(I) = N_d(I)$. However, this does not always succeed in allowing us to decide on way or the other, and additional methods are required (we refer the reader to [7, 8] for further details).

To execute the second part of the strategy described above, we need to find parabolic elements in $\Gamma(I)$, one for each cusp, so that trivializing these elements trivializes the group. From above, we obtain a partial set $S = \{P_1(I), \ldots, P_{h_d}(I)\}$ of peripheral subgroups for $\Gamma(I)$. To obtain a full set of peripheral subgroups for $\Gamma(I)$ we need to add certain conjugates of the $P_i(I)$ to the partial set S.

Next, given this full set of peripheral subgroups for $\Gamma(I)$, we choose one parabolic from each of these peripheral subgroups and use Magma to check that trivializing these elements trivializes $\Gamma(I)$. This choice of parabolic elements involves trial and error. It is worth emphasizing that finding these peripheral subgroups and expressing them in terms of generators for a presentation of $\mathrm{PSL}(2, O_d)$ was a highly non-trivial exercise.

5.2 *Eliminating* $\mathbb{H}^3/\Gamma(I)$ *as a Link Complement*

We begin by discussing the case when $I =< n >$ and $n \in \mathbb{Z}$. To that end we recall the following result from [3] that places severe restrictions on the list of possible d's.

Theorem 5.1 *If* $h_d > 1$, *and* $\Gamma(n) < \mathrm{PSL}(2, O_d)$, *then* $\mathbb{H}^3/\Gamma(n)$ *is not homeomorphic to a link complement in* S^3.

This result, together with the discussion in [7, Section 4.1] allows one to deduce:

Corollary 5.2 *Suppose that* $\Gamma(n) < \mathrm{PSL}(2, O_d)$ *and* $\mathbb{H}^3/\Gamma(n)$ *is homeomorphic to a link complement in* S^3. *Then* $d \in \{1, 2, 3, 7, 11, 19\}$ *and* $n \in \{2, 3, 4, 5\}$.

Since we are reduced to the case when $h_d = 1$ some pairs (d, n) can be eliminated quite quickly using Magma. If P denotes the peripheral subgroup of $\Gamma(n)$ fixing ∞, we show the normal closure $< P > \; \neq \Gamma(n)$. Table 2 below shows the cases for which this works. In this table, N is a normal subgroup of $\mathrm{PSL}(2, O_d)$ that contains the group $< P >$. Note the orders in the final two columns are different, so we can conclude that $< P > \neq \Gamma(n)$ in each case.

To handle the remaining integral levels a combination of methods are used, we summarize these in the following combination of results from [7].

Proposition 5.3 *Assume that* $d \in \{2, 7, 11, 19\}$, *then* $\Gamma(p) < \mathrm{PSL}(2, O_d)$ *is not a link group in the following two cases:*

- p *is an inert prime in* O_d.
- $p \geq 5$ *splits in* O_d.

Using this we can eliminate the following cases.

Table 2 Dealing with some rational integer levels

d	n	Order of $\mathrm{PSL}(2, O_d)/N$	Order of $\mathrm{PSL}(2, O_d/I)$
2	3	2304	288
3	4	3840	1920
7	3	1080	360
11	2	120	60
11	4	7680	1920

Corollary 5.4 *For* $(d, n) \in \{(2,5), (7,3), (7,5), (11,2), (11,5), (19,2), (19,3), (19,5)\}$, *the groups* $\Gamma(n)$ *are not link groups.*

Finally, using cuspidal cohomology calculations (see [7, Section 4]), one can show.

Proposition 5.5 *For those pairs* (d, n) *listed below, the principal congruence subgroups* $\Gamma(n)$ *satisfy* $r(\Gamma(n)) \neq 0$.

$$\{(1,4), (1,5), (2,4), (3,5), (7,4), (11,3), (19,4)\}.$$

Other levels: In [8], a strengthening of Theorem 5.1 proved useful, namely:

Theorem 5.6 *If* $h_d > 1$, *then* $\Gamma(n) < \mathrm{PSL}(2, O_d)$ *satisfies* $r(\Gamma(n)) \neq 0$ *in the following cases (using the notation introduced earlier to indicate d and the level):*

$$(23,3), (23,5), (31,2), (47,2), (47,3), (71,2), (71,3).$$

Putting this together, in [8] it is shown that:

Corollary 5.7 *Let* $d \in \{23, 31, 47, 71\}$, $I \subset O_d$ *an ideal and* $p = 2, 3, 5$. *Suppose that* (d, p) *is as in Theorem 5.6 and* I *is divisible by* $< p >$. *Then* $\Gamma(I)$ *has non-trivial cuspidal cohomology. In particular* $\mathbb{H}^3/\Gamma(I)$ *is not homeomorphic to a link complement in* S^3.

Most of the eliminations relied on Magma computations once again and we refer the reader to Sect. 6 for some sample calculations. However, one particular case $(2, < 1 + 3\sqrt{-2} >)$ proved particularly stubborn, and we required additional help in the form of programs that can compute automatic structures on groups to eliminate this case (see [9] for details).

The strategy more generally is this (which is an extension of some of the ideas used for dealing with some of the integral levels). Using presentations for the thc Bianchi groups, as well as matrix representatives for the peripheral subgroups P_i of $\mathrm{PSL}(2, O_d)$ (as above), we identify the peripheral subgroups $P_i(I)$ for $i = 1, \ldots, h_d$, and consider the quotient group $B_d(I) = \mathrm{PSL}(2, O_d)/N_d(I)$.

If $\Gamma(I)$ is a link group then $B_d(I)$ is a finite group with order equal to $|\mathrm{PSL}(2, O_d/I)|$. Hence if $B_d(I)$ is infinite or has order greater than $|\mathrm{PSL}(2, O_d/I)|$, then $\Gamma(I)$ cannot be a link group. We can input $B_d(I)$ in the Magma routines as:

$$B_d(I) =< \mathrm{PSL}(2, O_d) | P_1(I) = \cdots = P_{h_d}(I) = 1 >$$

that is by adding the peripheral subgroups $P_i(I)$ to the relations of $\mathrm{PSL}(2, O_d)$. We distinguish two cases:
Case 1: $B_d(I)$ *is a finite group but has order larger than* $|\mathrm{PSL}(O_d/I)|$.
Case 2: $B_d(I)$ *has a finite index subgroup with "large" abelianization, i.e. of very large order compared to the size of* $\mathrm{PSL}(2, O_d/I)$ *or infinite.*

In particular, in either case we can deduce that $\Gamma(I)$ cannot be a link group. Also note that if $B_d(I)$ is infinite or of order larger than $|\mathrm{PSL}(2, O_d/J)|$ for an ideal $J \subset I$, then so is $B_d(J)$, and hence $\Gamma(J)$ is also not a link group.

6 Sample Calculations

6.1 Establishing Principal Congruence Link Complements

We include two examples which are representative of the methods used, one from [7] and one from [8].

Case of $d = 1, I = 3$: We will sketch some of the ideas from [7] to show that the principal congruence subgroup $\Gamma(3) < \mathrm{PSL}(2, O_1)$ is a twenty component link group. To that end, from [39], $\mathrm{PSL}(2, O_1)$ has the following presentation:

$$\mathrm{PSL}(2, O_1) =< a, \ell, t, u \mid \ell^2 = (t\ell)^2 = (u\ell)^2 = (a\ell)^2 = a^2 = (ta)^3 = (ua\ell)^3 = 1, [t, u] = 1 >,$$

where $a = \begin{pmatrix} 0 & -1 \\ 1 & 0 \end{pmatrix}$, $t = \begin{pmatrix} 1 & 1 \\ 0 & 1 \end{pmatrix}$, $u = \begin{pmatrix} 1 & i \\ 0 & 1 \end{pmatrix}$ and $\ell = \begin{pmatrix} i & 0 \\ 0 & -i \end{pmatrix}$ (with the obvious abuse of notation between SL and PSL). The peripheral subgroup $P < \Gamma(3)$ fixing ∞ in this case is $< t^3, u^3 >$. A Magma routine from [7] is included below and shows $< P >= \Gamma(3)$.

We include some additional preamble before the Magma computation. It can be checked that $\Gamma(3)$ is a normal subgroup of $\mathrm{PSL}(2, O_1)$ of index 360 and the peripheral subgroup $P_\infty < \mathrm{PSL}(2, O_1)$ fixing ∞ maps to a group of order 18. Hence $\mathbb{H}^3/\Gamma(3)$ has 20 cusps.

In this case (as with others in [7, 8]) it is helpful to work with an intermediate subgroup $\Gamma(3) < \Gamma < \mathrm{PSL}(2, O_1)$, where Γ is defined to be the group $< \Gamma(3), \delta >= \Gamma(3).\delta$ where $\delta = atu^{-1}$. As shown by Magma, $[\Gamma : \Gamma(3)] = 5$, and so we may deduce that the cover $\mathbb{H}^3/\Gamma(3) \to \mathbb{H}^3/\Gamma$ is a regular 5-fold cyclic cover with $\mathbb{H}^3/\Gamma$ having four cusps, and each cusp of $\mathbb{H}^3/\Gamma(3)$ projecting one-to-one to a cusp of $\mathbb{H}^3/\Gamma$.

As was alluded to above, to determine appropriate parabolic elements is a somewhat tedious but straightforward computation. Briefly, in the case at hand, the four parabolic fixed points ∞, ± 1 and $1 - i$ (the set of which we denote by S) can be shown all to be mutually inequivalent under the action of Γ. In addition, the following parabolic elements in Γ fix the points in S:

$$S' = \{t^3u^3, tat^3u^{-3}at^{-1}, t^{-1}au^3at, u^{-1}tau^3at^{-1}u\}.$$

These can be easily shown to be primitive parabolic elements in Γ.

Magma now shows that the normal closure of S' in Γ is $\Gamma(3)$. Since the parabolic elements listed above represent inequivalent cusps of $\mathbb{H}^3/\Gamma$, if we now perform Dehn filling on $\mathbb{H}^3/\Gamma$ along the curves corresponding to these parabolic elements, the normal closure computation shows that we obtain a group of order 5. Since these are primitive parabolic elements, this group is the fundamental group of a closed 3-manifold, namely some lens space L (by Geometrization). Hence we deduce that $\mathbb{H}^3/\Gamma$ is a 4 component link in L with fundamental group of order 5. From above we can compatibly fill the cusps of $\mathbb{H}^3/\Gamma(3) \to \mathbb{H}^3/\Gamma$ resulting in a 5-fold cover $N \to L$, and so $N \cong S^3$ as required.

Magma routine for $\Gamma(3)$

```
G<a,l,t,u>:=Group<a,l,t,u|l^2,a^2,(t*l)^2,(u*l)^2,(a*l)^2,(t*a)^3,(u*a*l)^3,
(t,u)>;
h:=sub<G|t^3,u^3>;
n:=NormalClosure(G,h);
print Index(G,n);
\\360
print AbelianQuotientInvariants(n);
\\[ 0, 0, 0, 0, 0, 0, 0, 0, 0, 0, 0, 0, 0, 0, 0, 0, 0, 0, 0, 0 ]
d:=sub<G|n,a*t*u^-1>;
print Index(G,d);
\\72
print AbelianQuotientInvariants(d);
\\[ 5, 0, 0, 0, 0 ]
d1:=sub<d|t^3*u^3,t*a*t^3*u^-3*a*t^-1,t^-1*a*u^3*a*t,u^-1*t*a*u^3*a*t^-1*u>;
d2:=NormalClosure(d,d1);
print Index(d,d2);
\\5
d2 eq n;
\\true
```

Case of $d = 15$, $I =< 2, \omega_{15} >$: Set $\omega_{15} = \frac{1+\sqrt{-15}}{2}$, and $I =< 2, \omega_{15} >$, an ideal of norm 2. From [39], a presentation for $\mathrm{PSL}(2, O_{15})$ is given by:

$$\mathrm{PSL}(2, O_{15}) =< a, t, u, c \,|\, a^2 = (ta)^3 = \mathrm{ucuatu}^{-1}c^{-1}u^{-1}a^{-1}t^{-1} = 1, [t, u] = [a, c] = 1 >,$$

where $a = \begin{pmatrix} 0 & -1 \\ 1 & 0 \end{pmatrix}$, $t = \begin{pmatrix} 1 & 1 \\ 0 & 1 \end{pmatrix}$, $u = \begin{pmatrix} 1 & \omega_{15} \\ 0 & 1 \end{pmatrix}$ and $c = \begin{pmatrix} 4 & 1-2\omega_{15} \\ 2\omega_{15}-1 & 4 \end{pmatrix}$. Since $h_{15} = 2$, the quotient orbifold Q_{15} has two cusps, and equivalence classes can be taken to be ∞ and $\frac{1-\sqrt{-5}}{2}$ with cusp stabilizers $P_1 =< t, u >$, and $P_2 =< tb, tu^{-1}ct^{-1} >$. Since $N(I) = 2$, $[\mathrm{PSL}(2, O_{15}) : \Gamma(I)] = 6$ and it is easy to see that $\mathbb{H}^3/\Gamma(I)$ has 6 cusps. Now it can be shown that (in the notation above) $P_1(I) =< t^2, u >$ and $P_2(I) =< uca, (c^{-1}au^{-1}c^{-1}u^{-1}ta)^2 >$. Since $\begin{pmatrix} 1 & -1 \\ 1 & 0 \end{pmatrix} = ta$, conjugating $P_1(I)$ and $P_2(I)$ by the elements $\{Id, ta, (ta)^2\}$ gives a set of 6 peripheral subgroups for $\Gamma(I)$. Now, we choose one element from each of these 6 peripheral subgroups:

$$\{t^2, (ta)u(ta)^{-1}, (ta)^2u(ta)^{-2}, uca, (ta)uca(ta)^{-1}(ta)^2(c^{-1}au^{-1}c^{-1}u^{-1}ta)^2(ta)^{-2}\}$$

In the Magma routine, Q denotes the quotient of $\Gamma(I)$ by the normal closure of these 6 parabolic elements, and Magma calculates that $Q =< 1 >$ which shows that $\Gamma(I)$ is trivialized by setting these 6 elements equal to 1. Thus $\Gamma(< 2, \omega_{15} >)$ is indeed a 6 component link group.

```
G<a,c,t,u>:=Group<a,c,t,u|a^2,(t*a)^3,u*c*u*a*t*u^-1*c^-1*u^-1*a*t^-1, (t,u),(a,c)>;
H:=sub<G|t^2,u,(c^-1*a*u^-1*c^-1*u^-1*t*a)^2,u*c*a>;
N:=NormalClosure(G,H);
print Index(G,N);
6
\\
Q:=quo<N|t^2,(t*a)*u*(t*a)^-1,(t*a)^2*u*(t*a)^-2,
u*c*a,(t*a)*u*c*a*(t*a)^-1,(t*a)^2*(c^-1*a*u^-1*c^-1*u^-1*t*a)^2*(t*a)^-2>;
print Order(Q);
1
\\
```

6.2 Eliminating Principal Congruence Link Complements

We now give some sample calculations which illustrate the discussion in Sect. 5.1. From [7], Magma rules out the following values of d and levels. Here P denotes the subgroup fixing ∞ in $\Gamma(I)$ for the ideal I in question.
(i) $|\mathrm{PSL}(2, O_1/ < 5+i >)| = 6552$, and $< P >$ is contained in a normal subgroup of index 46800.
(ii) $|\mathrm{PSL}(2, O_2/ < 4+\sqrt{-2} >)| = 1944$, and $< P >$ is contained in a normal subgroup of index 2654208.
(iii) $|\mathrm{PSL}(2, O_2/ < 2+3\sqrt{-2} >)| = 3960$, and $< P >$ is contained in a normal subgroup of index 36432.
(iv) $|\mathrm{PSL}(2, O_7/ < 3+\sqrt{-7} >)| = 1152$, and $< P >$ is contained in a normal subgroup of index 4608.

We next consider the example from [8] with $d = 6$ (so the class number is 2) and $I =< 11, 4+\sqrt{-6} >$. The norm of I is 11 and so $\mathrm{PSL}(2, O_6/I)$ has order 660. The Magma routine included below shows that the quotient group of $\mathrm{PSL}(2, O_6)$ obtained by quotienting by the normal closure of the parabolic subgroups $P_1(I)$ and $P_2(I)$ (as described above) provides a group denoted $B < a, t, u, b, c >$ in the Magma routine that has order which is too large since it has a finite index subgroup with an abelian quotient group of very large order. The relevant information needed is a presentation for $\mathrm{PSL}(2, O_6)$ (from [39]) given below, and the generators for the groups $P_1(I) =< t^{11}, t^4u >$ and $P_2(I) =< (tb)^{11}, (tb)^4(cu)^{-1} >$.

$$\mathrm{PSL}(2, O_6) =< a, t, u, b, c \mid a^2 = b^2 = (ta)^3 = (atb)^3 = (\mathrm{atubu}^{-1})^3 =$$

$$t^{-1}ctubu^{-1}c^{-1}b^{-1} = 1, [t, u] = [a, c] = 1 >,$$

where $a = \begin{pmatrix} 0 & -1 \\ 1 & 0 \end{pmatrix}$, $t = \begin{pmatrix} 1 & 1 \\ 0 & 1 \end{pmatrix}$, $u = \begin{pmatrix} 1 & \sqrt{-6} \\ 0 & 1 \end{pmatrix}$,
$b = \begin{pmatrix} -1-\sqrt{-6} & 2-\sqrt{-6} \\ 2 & 1+\sqrt{-6} \end{pmatrix}$, and $c = \begin{pmatrix} 5 & -2\sqrt{-6} \\ 2\sqrt{-6} & 5 \end{pmatrix}$.

Magma Routine for $I =< 11, 4 + \omega_6 >$:

```
B<a,t,u,b,c>:=Group<a,t,u,b,c|a^2,b^2,(t*a)^3,(a*t*b)^3, (a*t*u*b*u^-1)^3,
t^-1*c*t*u*b*u^-1*c^-1*b^-1, (t,u),(a,c),
t^11,t^4*u,(t*b)^11,(t*b)^4*(c*u)^-1>;
L:=LowIndexNormalSubgroups(B,660);
print #L;
2
\\
print Index(B,L[2]'Group);
660
\\
print AbelianQuotientInvariants(L[2]'Group);
[ 4, 120, 120, 120, 120, 120, 120, 120, 120, 120, 120 ]
\\
```

7 Congruence Link Complements

We now discuss what is known in the direction of Question 1.1. In the subsections below we prove the following theorem for which it will be convenient to recall the following.

Definition 7.1 Let $I \subset O_d$ be an ideal and let $\Gamma_0(I) = \left\{ \begin{pmatrix} a & b \\ c & d \end{pmatrix} \in \mathrm{PSL}(2, O_d) | \right.$ $\left. c \equiv 0 \bmod \mathrm{I} \right\}$ and let $\Gamma_1(I) = \left\{ \begin{pmatrix} a & b \\ c & d \end{pmatrix} \in \mathrm{PSL}(2, O_d) | c \equiv 0 \bmod I, a, d \equiv 1 \, mod \, \mathrm{I} \right\}$.

Theorem 7.2 *There are only finitely many link groups $\Gamma < \mathrm{PSL}(2, O_d)$ such that $\Gamma(I) < \Gamma$ in the following cases:*

(i) $\Gamma = \Gamma_1(I)$.
(ii) $I = \mathcal{P}$ *is a prime ideal.*
(iii) $I = \mathcal{P}_1^{n_1} \ldots \mathcal{P}_k^{n_k} \subset O_d$, *where* $h_d = 1$ *and the ideals* $\mathcal{P}_i$ *are split primes of norm* p_i, *such that* $p_i \neq p_j$ *for* $i \neq j$.

Before commencing with the proof, we note that whenever $N(I) \geq 5$ the group $\Gamma_1(I)$ is torsion-free. Moreover, in [8] examples of link groups arising as $\Gamma_1(I)$ are given when $d = 6, 39$.

7.1 *Proof of Part (i)*

We begin with some definitions (recall the proof of Lemma 4.2).

Definition 7.3 A group Γ has a small (resp. large) systole if its systole is at most (resp. greater than) 7.171646...

The proof of Lemma 4.2 shows that if Γ is a link group it has a small systole. Part (1) of Theorem 7.2 will follow immediately from our next lemma (since there are only finitely many d we need only bound the norm of the ideal I).

Lemma 7.4 *Assume that Γ is generated by parabolic elements and has a small systole, then $\Gamma = \Gamma_1(I)$ for only finitely many ideals $I \subset O_d$.*

Proof Suppose that $A \in \Gamma$ is a hyperbolic element whose translation length achieves the systole bound. Since $\Gamma = \Gamma_1(I)$ then tr $(A) \pm 2 \in I$. But then this, coupled with the argument in the proof of Lemma 4.2, now shows that $N(I) \leq 39^2$. □

7.2 *Proof of Part (ii)*

Since there are only finitely many d, there are only finitely many ramified primes, and so we can assume that either $\mathcal{P}$ is a split prime or $\mathcal{P} =< p >$ and p is inert. We can also assume that $\mathcal{P}$ does not divide 2. In the case of a split prime, $\mathrm{PSL}(2, O_d/\mathcal{P}) \cong \mathrm{PSL}(2, \mathbb{F}_p)$ and in the inert prime case $\mathrm{PSL}(2, O_d/\mathcal{P}) \cong \mathrm{PSL}(2, \mathbb{F}_{p^2})$, where $\mathbb{F}_p$ (resp. $\mathbb{F}_{p^2}$) is the field of p elements (resp. p^2 elements). Since Γ is generated by parabolic elements, its image in $\mathrm{PSL}(2, O_d/\mathcal{P})$ is generated by parabolic elements, and it follows from the classification of subgroups of $\mathrm{PSL}(2, \mathbb{F}_p)$ and $\mathrm{PSL}(2, \mathbb{F}_{p^2})$ that this image group is conjugate into the image of $\Gamma_1(\mathcal{P})$ or it generates an isomorphic copy of $\mathrm{PSL}(2, \mathbb{F}_p) < \mathrm{PSL}(2, \mathbb{F}_{p^2})$ (see [19]), which only occurs in the case when p is inert. Now in the former case, we deduce that $\Gamma_1(\mathcal{P})$ has a small systole, and so we are done by Part (1). In the latter case we can argue as follows.

The orders of $\mathrm{PSL}(2, \mathbb{F}_p)$ and $\mathrm{PSL}(2, \mathbb{F}_{p^2})$ are $p(p^2 - 1)/2$ and $p^2(p^4 - 1)/2$ respectively. Hence the index $[\mathrm{PSL}(2, O_d) : \Gamma] = p(p^2 + 1)$. Now Γ is a link group and so in particular is torsion-free. Note that since -1 is not a square modulo 3, it follows that 3 does not divide $p^2 + 1$ for any prime p. Since $\mathrm{PSL}(2, O_d)$ has elements of order 3, then unless $p = 3$ we deduce that 3 does not divide the index $[\mathrm{PSL}(2, O_d) : \Gamma]$ which is impossible if Γ is torsion-free. But there are only finitely many possible primes $\mathcal{P}$ that can divide 3 and this proves finiteness. □

7.3 *Proof of Part (iii)*

Definition 7.5 An ideal I will be referred to as large if $N(I) \geq 39^2$. Otherwise it is called small.

A variation of the proof Lemma 7.4 proves Lemma 7.6 (below) on noting that if $\mathcal{P}$ is a split prime ideal and $A \in \Gamma_0(\mathcal{P}^n)$ is a product of parabolic elements in $\Gamma_0(\mathcal{P}^n)$, then $tr(A) = \pm 2$ modulo $\mathcal{P}^n$.

Lemma 7.6 *Suppose that $\mathcal{P} \subset O_d$ is a split prime, and assume that Γ is generated by parabolic elements and has a small systole. Then Γ cannot be a subgroup of $\Gamma_0(\mathcal{P}^n)$ if $\mathcal{P}^n$ is large.*

The proof of Theorem 7.2(3) will be completed using the following propositions whose proofs are included below. Henceforth, we assume that Γ has a small systole and is generated by parabolic elements; as noted earlier, both of these properties hold for link groups.

Proposition 7.7 *If $\Gamma(\mathcal{P}_1^{n_1} \ldots \mathcal{P}_k^{n_k}) < \Gamma$, then all the primes $\mathcal{P}_i$ can be assumed to be small.*

Given this we can assume that the ideals $\mathcal{P}_i$ are small.

Proposition 7.8 *Suppose that $\Gamma(\mathcal{P}_1^{n_1} \ldots \mathcal{P}_k^{n_k}) < \Gamma$, and that $m_1 \geq 2$ is the smallest integer for which $\mathcal{P}_1^{m_1}$ is large. Then n_1 can be chosen to be less than $8m_1$.*

In fact the role of $\mathcal{P}_1$ is not important here, and the same argument proves.

Corollary 7.9 *Suppose that $\Gamma(\mathcal{P}_1^{n_1} \ldots \mathcal{P}_k^{n_k}) < \Gamma$, and that $m_i \geq 2$ is the smallest integer for which $\mathcal{P}_i^{m_i}$ is large. Then each n_i can be chosen to be less than $8m_i$.*

The proof of Theorem 7.2 is now complete, since there are only finitely many values of d, only finitely many split prime ideals of norm less than 39^2 and, for each of these ideals, the exponent is bounded using Corollary 7.9. □

Proof of Proposition 7.7 The proof of Proposition 7.7 will follow immediately from the claim below. For given this claim, if there is a prime that is not small, then after possibly relabelling this prime by $\mathcal{P}_1$, we can lower the level of the principal congruence subgroup contained in Γ. We can now repeat this process, but since Γ is assumed congruence, this must stop at some $\Gamma(I)$ where I is an ideal that can only be a product of powers of small primes. □

Claim: *Suppose that $\Gamma(\mathcal{P}_1^{n_1} \ldots \mathcal{P}_k^{n_k}) < \Gamma$ and that $\mathcal{P}_1$ is large. Then $\Gamma(\mathcal{P}_2^{n_2} \ldots \mathcal{P}_k^{n_k}) < \Gamma$.*

The proof of the claim will require some additional lemmas.

Lemma 7.10 *(i) Any matrix $M \in \mathrm{PSL}(2, O_d)$ is equivalent to a matrix $M' \in \mathrm{PSL}(2, \mathbb{Z})$ modulo $\Gamma(\mathcal{P}_1^{m_1} \ldots \mathcal{P}_k^{m_k})$. Furthermore, if $M \in \Gamma(\mathcal{P}_1^{m_1} \ldots \mathcal{P}_k^{m_k})$, where $0 \leq m_i \leq n_i$ then $M' \in \Gamma(p_1^{m_1} \ldots p_k^{m_k}) < \mathrm{PSL}(2, \mathbb{Z})$.*
(ii) If M is a parabolic matrix, then M' can be chosen to be parabolic as well.

Proof Since $O_d/\mathcal{P}_1^{m_1} \ldots \mathcal{P}_k^{m_k} \cong \mathbb{Z}/(p_1^{m_1} \ldots p_k^{m_k})\mathbb{Z}$, we have that $\mathrm{PSL}(2, O_d/\mathcal{P}_1^{m_1} \ldots \mathcal{P}_k^{m_k}) \cong \mathrm{PSL}(2, \mathbb{Z}/(p_1^{m_1} \ldots p_k^{m_k})\mathbb{Z})$ and (i) follows easily.

Since $h_d = 1$, all parabolic elements of $\mathrm{PSL}(2, O_d)$ are conjugate into the peripheral subgroup fixing ∞. Hence $M = A\begin{pmatrix}1 & b\\ 0 & 1\end{pmatrix}A^{-1}$ for some $b \in O_d$. This in turn is equivalent modulo $\Gamma(\mathcal{P}_1^{n_1}\dots\mathcal{P}_k^{n_k})$ to $A'\begin{pmatrix}1 & n\\ 0 & 1\end{pmatrix}(A')^{-1}$ where $A' \in \mathrm{PSL}(2, \mathbb{Z})$ and $n \in \mathbb{Z}$. This proves (ii). □

Lemma 7.11 *Let* $\Gamma = <\Gamma(I), M_1, M_2>< \mathrm{PSL}(2, O_d)$ *where* M_1, M_2 *are parabolic matrices such that* $M_1 \equiv \begin{pmatrix}1 & 1\\ 0 & 1\end{pmatrix} \bmod \mathcal{P}_1^{n_1}$ *and* $M_1 \equiv \begin{pmatrix}1 & 0\\ 0 & 1\end{pmatrix} \bmod \mathcal{P}_2^{n_2}\dots\mathcal{P}_k^{n_k}$, *while* $M_2 \equiv \begin{pmatrix}1 & 0\\ 1 & 1\end{pmatrix} \bmod \mathcal{P}_1^{n_1}$ *and* $M_2 \equiv \begin{pmatrix}1 & 0\\ 0 & 1\end{pmatrix} \bmod \mathcal{P}_2^{n_2}\dots\mathcal{P}_k^{n_k}$. *Then* $\Gamma(\mathcal{P}_2^{n_2}\dots\mathcal{P}_k^{n_k}) < \Gamma$.

Proof Consider the map

$$\mathrm{PSL}(2, O_d) \to \mathrm{P}(SL(2, O_d/I)) \cong \mathrm{P}\big(\mathrm{SL}(2, O_d/\mathcal{P}_1^{n_1}) \times \mathrm{SL}(2, O_d/\mathcal{P}_2^{n_2}\dots\mathcal{P}_k^{n_k})\big)$$

Now note that $\Gamma(\mathcal{P}_2^{n_2}\dots\mathcal{P}_k^{n_k})$ maps to $P\big(\mathrm{SL}(2, O_d/\mathcal{P}_1^{n_1}) \times Id\big)$ and that the matrices $\begin{pmatrix}1 & 1\\ 0 & 1\end{pmatrix}, \begin{pmatrix}1 & 0\\ 1 & 1\end{pmatrix}$ generate $\mathrm{SL}(2, O_d/\mathcal{P}_1^{n_1})$. □

Before commencing with the proof of the claim it will be convenient to introduce the following definition.

Definition 7.12 A parabolic M has $\mathcal{P}$-level n if $n \geq 0$ is the largest integer for which $M \in \Gamma(\mathcal{P}^n)$.

Proof of Claim Since Γ is generated by parabolics, it can be expressed as

$$\Gamma = <\Gamma(I), M_1, \dots, M_l>$$

where the M_i are parabolic matrices. We can assume that the M_i are in $\mathrm{PSL}(2, \mathbb{Z})$ by Lemma 7.10. Renumbering if necessary, let M_1 be of smallest $\mathcal{P}_1$-level among the M_i. Replacing Γ by a $\mathrm{PSL}(2, \mathbb{Z})$-conjugate, we can further assume that M_1 fixes ∞, and hence $M_1 = \begin{pmatrix}1 & b\\ 0 & 1\end{pmatrix}$ for some $b \in O_d$. Now $(b, p_1) = 1$ else all the M_i would have $\mathcal{P}_1$-level ≥ 1 and hence $\Gamma < \Gamma_0(\mathcal{P}_1)$, which contradicts Lemma 7.6.

Note that Γ must contain a second parabolic matrix, $M_2 = \begin{pmatrix}1+a & b\\ c & 1-a\end{pmatrix}$, not in $\Gamma_0(\mathcal{P}_1)$; hence $(c, p_1) = 1$. Replacing M_1, M_2 by suitable powers, we can further assume that $M_1 \equiv M_2 \equiv \begin{pmatrix}1 & 0\\ 0 & 1\end{pmatrix} \bmod p_2^{n_2}\dots p_k^{n_k}$ and that $M_1 \equiv \begin{pmatrix}1 & 1\\ 0 & 1\end{pmatrix}$, $M_2 \equiv \begin{pmatrix}1+a & -a^2\\ 1 & 1-a\end{pmatrix} \bmod p_1^{n_1}$.

Since the product $M_1{}^{-a}M_2M_1{}^{a} \equiv \begin{pmatrix}1 & 0\\ 1 & 1\end{pmatrix} \bmod p_1^{n_1}$, we have that a subgroup of Γ satisfies the hypotheses of Lemma 7.11; thus $\Gamma(\mathcal{P}_2^{n_2}\dots\mathcal{P}_k^{n_k}) < \Gamma$. □

Remark 7.13 The above proof shows that if $\Gamma(\mathcal{P}_1^{n_1}) < \Gamma$, then $\Gamma = \mathrm{PSL}(2, O_d)$, which is not possible for Γ a (torsion-free) link group.

Proof of Proposition 7.8: The following lemmas will be used in the proof of Proposition 7.8.

Lemma 7.14 *Let* $M_1 = \begin{pmatrix} 1 & p^r \\ 0 & 1 \end{pmatrix}$, $M_2 = \begin{pmatrix} 1 & 0 \\ p^r & 1 \end{pmatrix}$, *and* $M_3 = \begin{pmatrix} 1+p^r & -p^r \\ p^r & 1-p^r \end{pmatrix}$.
Then
(i) $\Gamma(\mathcal{P}^r) = < \Gamma(\mathcal{P}^{2r}), M_1, M_2, M_3 >$.
(ii) $\Gamma(\mathcal{P}^r) = < \Gamma(\mathcal{P}^s), M_1, M_2, M_3 >$ *for* $s > r > 1$.

Proof By Lemma 7.10 any matrix of $\Gamma(\mathcal{P}^r)$ is equivalent modulo $\Gamma(\mathcal{P}^{2r})$ to a matrix in $\mathrm{PSL}(2,\mathbb{Z})$ of the form $\begin{pmatrix} 1+ap^r & bp^r \\ cp^r & 1+dp^r \end{pmatrix}$ which is congruent modulo p^{2r} to the product $M_1^{b+a} M_2^{c-a} M_3^a$. This proves (i).

Part (ii) is an easy consequence of (i) as we now explain. Indeed, replacing s by $s+1$ if necessary, we can assume that s is even. Then, after taking suitable powers of M_1, M_2, and M_3, part (i) gives $\Gamma(\mathcal{P}^{s/2}) \subset \; < \Gamma(\mathcal{P}^s), M_1, M_2, M_3 >$. Repeating this process, we obtain $\Gamma(\mathcal{P}^k) \subset \; < \Gamma(\mathcal{P}^s), M_1, M_2, M_3 >$ for $k \le 2r$. □

Lemma 7.15 *Let* $A = \begin{pmatrix} 1 & p^\alpha \\ 0 & 1 \end{pmatrix}$, $B = \begin{pmatrix} 1+p^\beta & -p^{2\beta-\gamma} \\ p^\gamma & 1-p^\beta \end{pmatrix}$, *such that* $0 \le \alpha < r$, $4r \le \gamma \le 5r$, *and* $4r \le \beta$. *Then* $\Gamma = < \Gamma(\mathcal{P}^{8r}), A, B >$ *is of* $\mathcal{P}$*-level* $\le 7r$.

Proof In what follows we will work modulo $\mathcal{P}^{8r}$. Since A, B are in $\mathrm{PSL}(2,\mathbb{Z})$, this is the same as modulo p^{8r}.

By Lemma 7.14, it suffices to show that Γ contains the matrices $C = \begin{pmatrix} 1 & p^{7r} \\ 0 & 1 \end{pmatrix}$, $D = \begin{pmatrix} 1 & 0 \\ p^{7r} & 1 \end{pmatrix}$, and $E = \begin{pmatrix} 1+p^{7r} & -p^{7r} \\ p^{7r} & 1-p^{7r} \end{pmatrix}$. Note that $F = CD^{-1}E \equiv \begin{pmatrix} 1+p^{7r} & 0 \\ 0 & 1-p^{7r} \end{pmatrix}$.

Now $C \in \Gamma$ since C is a power of A. Note also that $ABA^{-1}B^{-1} \equiv \begin{pmatrix} 1+p^{\alpha+\gamma} & -2p^{\alpha+\beta} - p^{2\alpha+\gamma} \\ 0 & 1-p^{\alpha+\gamma} \end{pmatrix}$ and multiplying this matrix on the left by a power of A yields $G = \begin{pmatrix} 1+p^{\alpha+\gamma} & 0 \\ 0 & 1-p^{\alpha+\gamma} \end{pmatrix}$, with $4r \le \alpha+\gamma < 6r$. Hence $G \in \Gamma$ which implies that F is also in Γ since it is a power of G modulo $\mathcal{P}^{8r}$. Thus it remains to show that $D \in \Gamma$.

Consider the matrix B. If $6r \le \beta \le 8r$ then $B^{p^{8r-\beta}} \equiv \begin{pmatrix} 1 & 0 \\ p^{\gamma+8r-\beta} & 1 \end{pmatrix}$, where $\gamma + 8r - \beta \le 7r$; hence $D \in \Gamma$. Now suppose that $4r \le \beta < 6r$. Since $4r \le \alpha + \gamma < 6r$, replacing B and G by powers we can assume that the diagonals of B and G are the same, and that $\beta < 6r$ while $\gamma < 7r$. Thus $G^{-1}B \equiv \begin{pmatrix} 1 & -p^{2\beta-\gamma} + p^{3\beta-\gamma} \\ p^\gamma & 1 \end{pmatrix}$, so that $\begin{pmatrix} 1 & 0 \\ p^\gamma & 1 \end{pmatrix} \equiv \begin{pmatrix} 1 & p^\gamma \\ 0 & 1 \end{pmatrix} G^{-1}B \in \Gamma$, which implies that $D \in \Gamma$. □

Lemma 7.16 *Let $m \geq 2$ be the smallest integer for which $\mathcal{P}$ is large. If $\Gamma(\mathcal{P}^{8m}) < \Gamma$, then $\Gamma(\mathcal{P}^{7m}) < \Gamma$ as well.*

Proof Since Γ is generated by parabolic elements, it can be written as $\Gamma = < \Gamma(\mathcal{P}^{8m}), M_1, \ldots, M_l >$, with the M_i parabolic. By Lemma 7.15, it suffices to show that Γ contains the matrices $A = \begin{pmatrix} 1 & p^\alpha \\ 0 & 1 \end{pmatrix}$, $B = \begin{pmatrix} 1+p^\beta & -p^{2\beta-\gamma} \\ p^\gamma & 1-p^\beta \end{pmatrix}$, such that $0 \leq \alpha < m$, $4m \leq \gamma \leq 5m$, and $4m \leq \beta$.

As in the proof of the Claim, we can assume that the M_i are in $\mathrm{PSL}(2, \mathbb{Z})$ and that $M_1 = \begin{pmatrix} 1 & bp^\alpha \\ 0 & 1 \end{pmatrix}$ is of smallest $\mathcal{P}$-level among the M_i. Thus $0 \leq \alpha < m$. Since $(b, p) = 1$, we can assume (taking a power of M_1 and calculating mod p^{8m})) that Γ contains $A = \begin{pmatrix} 1 & p^\alpha \\ 0 & 1 \end{pmatrix}$.

Now Γ must contain a second parabolic element, $M_2 = \begin{pmatrix} 1+ap^\beta & bp^{2\beta-\gamma} \\ cp^\gamma & 1-ap^\beta \end{pmatrix}$ with $(a, p) = (b, p) = (c, p) = 1$ that is not in $\Gamma_0(\mathcal{P}^m)$. Thus $0 \leq \alpha \leq \gamma < m$. Suppose first that M_2 fixes the cusp at 0, so that $M_2 = \begin{pmatrix} 1 & 0 \\ cp^\gamma & 1 \end{pmatrix}$. Thus Γ contains the matrix $D = \begin{pmatrix} 1 & 0 \\ p^\delta & 1 \end{pmatrix}$ with $4m \leq \delta \leq 5m$. Since $D(ADA^{-1}D^{-1}) \equiv B = \begin{pmatrix} 1+p^{\alpha+\delta} & -p^{2\alpha+\delta} \\ p^\delta & 1-p^{\alpha+\delta} \end{pmatrix}$, it follows that $\Gamma(\mathcal{P}^{7m}) < \Gamma$ by Lemma 7.15.

Finally, consider the case $M_2 = \begin{pmatrix} 1+ap^\beta & bp^{2\beta-\gamma} \\ cp^\gamma & 1-ap^\beta \end{pmatrix}$ fixing a cusp other than 0 or ∞. Replacing M_2 by a power, we can assume that $M_2 = \begin{pmatrix} 1+p^\beta & p^{2\beta-\gamma} \\ p^\gamma & 1-p^\beta \end{pmatrix}$ such that $4m \leq \gamma \leq 5m$ and $4m \leq \beta$. Thus, again by Lemma 7.15, $\Gamma(\mathcal{P}^{7m}) < \Gamma$ as required. □

Lemma 7.17 *Let $m \geq 2$ be the smallest integer for which $\mathcal{P}$ is large. If $\Gamma(\mathcal{P}^n) < \Gamma$ and $n \geq 8m$, then $\Gamma(\mathcal{P}^{n-m}) < \Gamma$ as well.*

Proof Let $s = (n-8m)/2$ (resp. $(n-8m+1)/2$) if $n - 8m$ is even (resp. odd). As in the proof of Lemma 7.16, one shows that Γ contains the matrices $A = \begin{pmatrix} 1 & p^\alpha \\ 0 & 1 \end{pmatrix}$, $B = \begin{pmatrix} 1+p^\beta & -p^{2\beta-\gamma} \\ p^\gamma & 1-p^\beta \end{pmatrix}$, such that $0 \leq \alpha < m$, $4m+s \leq \gamma \leq 5m+s$, and $4m+s \leq \beta$. A slight modification of the proof of Lemma 7.15 gives $\Gamma(\mathcal{P}^{n-m}) < \Gamma$. □

We can now complete the proof of Proposition 7.8. With the notation of Proposition 7.8, if $\Gamma(\mathcal{P}^n) < \Gamma$ and $n \geq 8m$ then Lemma 7.17 implies $\Gamma(\mathcal{P}^{n-m}) < \Gamma$. Repeating this argument until the exponent of $\mathcal{P}$ is less than $8m$ proves the proposition in this case.

Now suppose that $\Gamma(\mathcal{P}_1^{n_1} \ldots \mathcal{P}_k^{n_k}) < \Gamma$, for $k > 1$. Following the proof of Lemma 7.16, Γ contains the matrices $M_1 = \begin{pmatrix} 1 & bp_1^\alpha \\ 0 & 1 \end{pmatrix}$ and $M_2 =$

$\begin{pmatrix} 1+ap_1^{\beta} & dp_1^{2\beta-\gamma} \\ cp_1^{\gamma} & 1-ap_1^{\beta} \end{pmatrix}$. Furthermore, by taking powers, we can assume that $M_1 \equiv M_2 \equiv \begin{pmatrix} 1 & 0 \\ 0 & 1 \end{pmatrix} \bmod p_2^{n_2} \dots p_k^{n_k}$. The proof of Lemma 7.16 then shows that Γ contains matrices A, B such that $A \equiv \begin{pmatrix} 1 & p^{\alpha} \\ 0 & 1 \end{pmatrix} \bmod p_1^{n_1}$ and $A \equiv \begin{pmatrix} 1 & 0 \\ 0 & 1 \end{pmatrix} \bmod p_2^{n_2} \dots p_k^{n_k}$ while $B \equiv \begin{pmatrix} 1+p^{\beta} & -p^{2\beta-\gamma} \\ p^{\gamma} & 1-p^{\beta} \end{pmatrix} \bmod p_1^{n_1}$ and $B \equiv \begin{pmatrix} 1 & 0 \\ 0 & 1 \end{pmatrix} \bmod p_2^{n_2} \dots p_k^{n_k}$. Also, $0 \le \alpha < m$, $4m+s \le \gamma \le 5m+s$, and $4m+s \le \beta$. Thus it follows that $\Gamma(\mathcal{P}_1^{n_1-m_1} \mathcal{P}_2^{n_2} \dots \mathcal{P}_k^{n_k}) < \Gamma$. Repeating the argument until the exponent of $\mathcal{P}_1$ is less than $8m_1$ completes the proof of Proposition 7.8. □

8 Final Comments and Speculations

We finish with some discussion about possible approaches to Question 1.1.

8.1 Spectral Gap

An important property of congruence manifolds is that they admit a spectral gap; i.e. there exists a number $C > 0$ (conjectured to be 1 in dimension 3) so that if $M = \mathbb{H}^3/\Gamma$ (or $\mathbb{H}^2/\Gamma$) is any congruence manifold, then the first non-zero eigenvalue of the Laplacian on M, denoted $\lambda_1(M)$, satisfies $\lambda_1(M) > C$.

The argument of [44] to prove the finiteness result in dimension 2 for congruence surfaces of genus 0 discussed in Sect. 1 is to play off the spectral gap for congruence manifolds in dimension 2, together with a result proved in [44] that says that for a sequence of genus 0 manifolds with increasing numbers of punctures we must have $\lambda_1 \to 0$.

Thus a natural question is whether there exists a "Zograf type result" in dimension 3. The answer to this in general is no since Lackenby and Souto (unpublished) have shown that there exists a sequence of hyperbolic link complements in S^3 (say M_n) with $\mathrm{Vol}(M_n) \to \infty$ and a constant $C_1 > 0$ such that $\lambda_1(M_n) > C_1$.

On the other hand there are classes of links known for which sequences as above do not arise (see [20, 27]). In particular, the result below follows from [27].

Theorem 8.1 (Lackenby) *There are only finitely many alternating links in S^3 whose complements are congruence link complements.*

8.2 Torsion

If $L \subset S^3$ is a link of n components, then $H_1(S^3 \backslash L) \cong \mathbb{Z}^n$, so that an "easy" way to exclude a congruence subgroup $\Gamma < \mathrm{PSL}(2, O_d)$ from being a link group is to prove the existence of torsion in $H_1(\Gamma, \mathbb{Z})$. Indeed, this was used in [7] to rule out certain levels as allowable for a principal congruence link complement (using computer computations of M. H. Sengun). A recent emerging theme, both in low-dimensional topology and in automorphic forms, is that "*sequences of congruence subgroups should develop torsion in* H_1". More precisely, in the light of the results, numerics and conjectures in [11, 12, 29, 36], a reasonable conjecture might be the following (see also [10, Conjecture 6.1]):

Conjecture 8.2 Let $\{\Gamma_n\}$ be a sequence of congruence subgroups of $\mathrm{PSL}(2, O_d)$ with $\mathrm{Vol}(\mathbb{H}^3/\Gamma_n) \to \infty$. Then:

$$\frac{\log |\mathrm{Tor}(H_1(\Gamma_n, \mathbb{Z}))|}{[\mathrm{PSL}(2, O_d) : \Gamma_n]} \to \frac{1}{6\pi}\mathrm{Vol}(Q_d) \text{ as } n \to \infty.$$

A positive answer to Conjecture 8.2 would of course establish the finiteness stated in Questions 1.1 and 1.2.

Acknowledgements The work contained in this paper was developed over multiple visits to the University of Texas by the first author, and the Université de Rennes 1 by the second author. We also wish to thank the Université Paul Sabatier, the Max Planck Institüt, Bonn, I.C.T.P. Trieste and The Institute for Advanced Study for their support and hospitality as this work unfolded over several years. We also wish to thank several people without whose help we would not have been able to complete this work: M. D. E. Conder, D. Holt, M. Goerner, E. O'Brien, A. Page, M. H. Sengun and A. Williams. We also thank I. Agol, N. Hoffman and P. Sarnak for useful conversations on topics related to this work. We are also very grateful to the referee for their comments and helpful suggestions.

References

1. Adams, C.C., Reid, A.W.: Systoles of hyperbolic 3-manifolds. Math. Proc. Camb. Philos. Soc. **128**, 103–110 (2000)
2. Baker, M.D.: Link complements and quadratic imaginary number fields, Ph.D. thesis M.I.T. (1981)
3. Baker, M.D.: Link complements and the homology of arithmetic subgroups of PSL $(2, C)$, I.H.E.S. preprint (1982)
4. Baker, M.D.: Link complements and integer rings of class number greater than one. In: Topology '90, pp. 55–59. Ohio State University, Mathematical Research Institute Publications 1, de Gruyter (1992)
5. Baker, M.D.: Link complements and the Bianchi modular groups. Trans. Am. Math. Soc. **353**, 3229–3246 (2001)
6. Baker, M.D.: Commensurability classes of arithmetic link complements. J. Knot Theory Ramif. **10**, 943–957 (2001)

7. Baker, M.D., Reid, A.W.: Principal congruence link complements. Ann. Fac. Sci. Toulouse **23**, 1063–1092 (2014)
8. Baker, M.D., Reid, A.W.: Principal congruence links: Class number greater than 1. Exp. Math. (2017). https://doi.org/10.1080/10586458.2017.1289867
9. Baker, M.D., Goerner, M., Reid, A.W.: All principal congruence link complements, in preparation
10. Bergeron, N.: Torsion homology growth in arithmetic groups, to appear in the proceedings of the 7th European Congress of Mathematics
11. Bergeron, N., Venkatesh, A.: The asymptotic growth of torsion homology for arithmetic groups. J. Inst. Math. Jussieu **12**, 391–447 (2013)
12. Bergeron, N., Sengun, M.H., Venkatesh, A.: Torsion homology growth and cycle complexity of arithmetic manifolds. Duke Math. J. **165**, 1629–1693 (2016)
13. Blume-Nienhaus, J.: Lefschetzzahlen fur Galois-Operationen auf der Kohomologie arithmetischer Gruppen, Ph.D. thesis, Universität Bonn, Bonn Math. Publications, 230 (1991)
14. Bosma, W., Cannon, J., Playoust, C.: The Magma algebra system. I. The user language. J. Symb. Comput. **24**, 235–265 (1997)
15. Chua, K.S., Lang, M.L., Yang, Y.: On Rademachers conjecture: congruence subgroups of genus zero of the modular group. J. Algebr. **277**, 408–428 (2004)
16. Dennin, J.B.: Fields of modular functions of genus 0. Ill. J. Math. **15**, 442–455 (1971)
17. Dennin, J.B.: Subfields of $K(2^n)$ of genus 0. Ill. J. Math. **16**, 502–518 (1972)
18. Dennin, J.B.: The genus of subfields of $K(p^n)$. Ill. J. Math. **18**, 246–264 (1974)
19. Dixon, L.E.: Linear Groups, with an Exposition of the Galois Field Theory. Dover Publications, New York (1958)
20. Futer, D., Kalfagianni, E., Purcell, J.: On diagrammatic bounds of knot volumes and spectral invariants. Geom. Dedicata **147**, 115–130 (2010)
21. Goerner, M.: Visualizing regular tessellations: principal congruence links and equivariant morphisms from surfaces to 3-manifold, Ph.D. thesis, U. C. Berkeley (2011)
22. Goerner, M.: Regular tessellation link complements. Exp. Math. **24**, 225–246 (2015)
23. Gordon, C.McA: Links and their complements. Topology and Geometry: Commemorating SISTAG. Contemporary Mathematics, vol. 314, pp. 71–82. American Mathematical Society, Providence (2002)
24. Grunewald, F., Schwermer, J.: Arithmetic quotients of hyperbolic 3- space, cusp forms and link complements. Duke Math. J. **48**, 351–358 (1981)
25. Grunewald, F., Schwermer, J.: A non-vanishing theorem for the cuspidal cohomology of SL_2 over imaginary quadratic integers. Math. Ann. **258**, 183–200 (1981)
26. Hatcher, A.: Hyperbolic structures of arithmetic type on some link complements. J. Lond. Math. Soc. **27**, 345–355 (1983)
27. Lackenby, M.: Spectral geometry, link complements and surgery diagrams. Geom. Dedicata **147**, 191–206 (2010)
28. Lakeland, G., Leininger, C.: Systoles and Dehn surgery for hyperbolic 3- manifolds. Algebr. Geomet. Topol. **14**, 1441–1460 (2014)
29. Le, T.: Growth of homology torsion in finite coverings and hyperbolic volume, to appear Annales de l'Institut Fourier
30. Maclachlan, C., Reid, A.W.: The Arithmetic of Hyperbolic 3-Manifolds. Graduate Texts in Mathematics, vol. 219. Springer, Berlin (2003)
31. Martelli, B., Petronio, C.: Dehn filling of the "Magic" 3-manifold. Commun. Anal. Geom. **14**, 969–1026 (2006)
32. Page, A.: Computing arithmetic Kleinian groups. Math. Comput. **84**, 2361–2390 (2015)
33. Palaparthi, S.: Closed geodesic lengths in hyperbolic link complements in S^3. Int. J. Pure Appl. Math. **83**, 45–53 (2013)
34. Rohlfs, J.: On the cuspidal cohomology of the Bianchi modular groups. Math. Z. **188**, 253–269 (1985)
35. Sebbar, A.: Classification of torsion-free genus zero congruence subgroups. Proc. Am. Math. Soc. **129**, 2517–2527 (2001)

36. Sengun, M.H.: On the integral cohomology of Bianchi groups. Exp. Math. **20**, 487–505 (2011)
37. Stephan, J.: Complémentaires d'entrelacs dans S^3 et ordres maximaux des algèbres de quaternions. C. R. Acad. Sci. Paris Sér. I Math. **322**, 685–688 (1996)
38. Stephan, J.: On arithmetic hyperbolic links. J. Knot Theory Ramif. **8**, 373–389 (1999)
39. Swan, R.G.: Generators and relations for certain special linear groups. Adv. Math. **6**, 1–77 (1971)
40. Thompson, J.G.: A finiteness theorem for subgroups of PSL(2, $\mathbb{R}$). Proc. Symp. Pure Math. **37**, 533–555 (1980). A.M.S. Publications
41. Thurston, W.P.: The Geometry and Topology of 3-Manifolds, Princeton University mimeographed notes (1979)
42. Thurston, W.P.: Three-dimensional manifolds, Kleinian groups and hyperbolic geometry. Bull. Am. Math. Soc. **6**, 357–381 (1982)
43. Vogtmann, K.: Rational homology of Bianchi groups. Math. Ann. **272**, 399–419 (1985)
44. Zograf, P.: A spectral proof of Rademacher's conjecture for congruence subgroups of the modular group. J. Reine Angew. Math. **414**, 113–116 (1991)

Topological Realisations of Absolute Galois Groups

Robert A. Kucharczyk and Peter Scholze

Abstract Let F be a field of characteristic 0 containing all roots of unity. We construct a functorial compact Hausdorff space X_F whose profinite fundamental group agrees with the absolute Galois group of F, i.e. the category of finite covering spaces of X_F is equivalent to the category of finite extensions of F. The construction is based on the ring of rational Witt vectors of F. In the case of the cyclotomic extension of $\mathbb{Q}$, the classical fundamental group of X_F is a (proper) dense subgroup of the absolute Galois group of F. We also discuss a variant of this construction when the field is not required to contain all roots of unity, in which case there are natural Frobenius-type automorphisms which encode the descent along the cyclotomic extension.

Keywords Galois groups · Fundamental groups · Witt vectors

2010 Mathematics Subject Classification 12F10 · 11R32 · 14F35

1 Introduction

This paper grew out of an attempt to understand whether certain constructions in p-adic Hodge theory could potentially have analogues over number fields. One important technique in p-adic Hodge theory is the possibility to relate Galois groups of p-adic fields with Galois groups or fundamental groups of more geometric objects. Some sample results of this type are the following.

R. A. Kucharczyk
Département Mathematik, ETH Zürich, Rämistrasse 101, 8092 Zürich, Switzerland
e-mail: robert.kucharczyk@math.ethz.ch

P. Scholze (✉)
Mathematisches Institut, Rheinische Friedrich-Wilhelms-Universität Bonn,
Endenicher Allee 60, 53115 Bonn, Germany
e-mail: scholze@math.uni-bonn.de

J. W. Cogdell et al. (eds.), *Cohomology of Arithmetic Groups*,
Springer Proceedings in Mathematics & Statistics 245,
https://doi.org/10.1007/978-3-319-95549-0_8

Theorem 1.1 (Fontaine–Wintenberger, [1]) *Let K be the cyclotomic extension $\mathbb{Q}_p(\zeta_{p^\infty})$ of $\mathbb{Q}_p$. Then the absolute Galois group of K is isomorphic to the absolute Galois group of $\mathbb{F}_p((t))$.*

Theorem 1.2 (Fargues–Fontaine, Weinstein, [2]) *There is a natural 'space' Z defined over $\mathbb{C}_p$ whose geometric fundamental group is the absolute Galois group of $\mathbb{Q}_p$. Formally, Z is the quotient of a 1-dimensional punctured perfectoid open unit disc by a natural action of $\mathbb{Q}_p^\times$.*

One can regard both of these theorems as instances of the general 'tilting' philosophy, [3], which relates objects of mixed characteristic with objects of equal characteristic, the latter of which have a more geometric flavour. An important feature of the tilting procedure is that it only works for 'perfectoid' objects; in the case of fields, this is related to the need to pass to the cyclotomic extension, or a similar 'big' field. Another common feature is the critical use of (p-typical) Witt vectors.

In looking for a global version of these results, one is thus led to consider a 'global' version of the Witt vectors, and the standard objects to consider are the big Witt vectors. Recall that for any commutative ring A, the ring of big Witt vectors $\mathrm{W}(A)$ can be identified with the set $1 + tA[[t]]$ of power series with constant coefficient 1, where addition of Witt vectors corresponds to multiplication of power series. The multiplication is subtler to write down, and is essentially determined by the rule that the product of $1 - at$ and $1 - bt$ is given by $1 - abt$. In particular, there is a multiplicative map $A \to \mathrm{W}(A)$, $a \mapsto [a] = 1 - at$, called the Teichmüller map.

In general, there is a map of algebras $\mathrm{W}(A) \to \prod_n A$ called the ghost map, where the product runs over all integers $n \geq 1$. If A is a $\mathbb{Q}$-algebra, the ghost map is an isomorphism, so that in particular for a field F of characteristic 0, $\mathrm{W}(F) = \prod_n F$ is just an infinite product of copies of F.

One thus cannot expect $\mathrm{W}(F)$ to have a rich structure. However, work on the K-theory of endomorphisms, [4], suggested to look at the following subring of $\mathrm{W}(F)$, called the ring of rational Witt vectors.[1]

Definition 1.3 Let A be a commutative ring. The *rational Witt vectors* over A are the elements of

$$\mathrm{W}_{\mathrm{rat}}(A) = \left\{ \frac{1 + a_1 t + \ldots + a_n t^n}{1 + b_1 t + \ldots + b_m t^m} \,\middle|\, a_i, b_j \in A \right\} \subset \mathrm{W}(A).$$

It is not hard to see that $\mathrm{W}_{\mathrm{rat}}(A)$ actually forms a subring of $\mathrm{W}(A)$. The Teichmüller map $A \to \mathrm{W}(A)$ factors over $\mathrm{W}_{\mathrm{rat}}(A)$.

Now let F be a field of characteristic 0 containing all roots of unity, and fix once and for all an embedding $\iota\colon \mathbb{Q}/\mathbb{Z} \hookrightarrow F^\times$; this 'bigness' hypothesis will be important for the construction, and all constructions will depend on ι. We also fix the standard embedding

[1] It is this connection, as well as the observation that the Dennis trace map from K-theory to topological Hochschild homology factors canonically over the K-theory of endomorphisms, that led the second author to consider the rational Witt vectors.

$$\exp\colon \mathbb{Q}/\mathbb{Z} \hookrightarrow \mathbb{C}^\times, \quad x \mapsto e^{2\pi i x}.$$

Definition 1.4 Let X_F be the set of ring maps $\mathrm{W}_{\mathrm{rat}}(F) \to \mathbb{C}$ whose restriction along $\mathbb{Q}/\mathbb{Z} \overset{\iota}{\hookrightarrow} F^\times \xrightarrow{[\cdot]} \mathrm{W}_{\mathrm{rat}}(F)^\times$ gives the standard embedding $\exp\colon \mathbb{Q}/\mathbb{Z} \to \mathbb{C}$. We endow X_F with its natural complex topology, cf. Definition 5.3.

One can check that X_F is one connected component of the complex points of the scheme Spec $\mathrm{W}_{\mathrm{rat}}(F)$. Actually, in the paper, X_F will denote a closely related space which is a deformation retract of the space considered here. This variant will be a compact Hausdorff space.

Theorem 1.5 (Theorem 5.2) *The functor taking a finite extension E of F to $X_E \to X_F$ induces an equivalence of categories between the category of finite extensions of F, and the category of connected finite covering spaces of X_F. In particular, the absolute Galois group of F agrees with the étale fundamental group of X_F.*

Here, the étale fundamental group of a connected topological space classifies, by definition, the finite covering spaces of the latter, cf. Definition 2.10. It is in general not directly related to the classical fundamental group defined in terms of paths. We also prove a version of this theorem in the world of schemes, replacing X_F by one connected component of $\mathrm{Spec}(\mathrm{W}_{\mathrm{rat}}(F) \otimes \mathbb{C})$, cf. Theorem 4.5.

Contrary to the results in p-adic Hodge theory cited above which reflect deep properties about ramification of local fields, this theorem is rather formal. In fact, the proof of the theorem is essentially an application of Hilbert's Satz 90 in its original form. Also, we cannot currently state a precise relationship between this theorem and the results in p-adic Hodge theory stated above. Still, we believe that there is such a relation, and that the theorem indicates that the ring of rational Witt vectors is an interesting object; in fact, we would go so far as to suggest to replace all occurences of the big Witt vectors by the rational Witt vectors.[2]

We warn the reader that the space X_F is highly infinite-dimensional, and in general far from path-connected. For example, if F is algebraically closed, its compact Hausdorff version can be (non-canonically) identified with an infinite product of copies of the solenoid (cf. Proposition/Definition 2.1)

$$\mathcal{S} = \varprojlim_{n \in \mathbb{N}} \mathbb{S}^1 = \mathrm{Hom}(\mathbb{Q}, \mathbb{S}^1) = \mathbb{A}/\mathbb{Q},$$

where $\mathbb{A} = \mathbb{A}_{\mathbb{Q}}$ is the ring of rational adèles.

Abstractly, it is clear that any group can be realised as the fundamental group of a topological space, by using the theory of classifying spaces. One may thus wonder what extra content Theorem 1.5 carries. We give several answers to this question. All are variants on the observation that our construction gives an actual topological space,

[2] An instance is the definition of a Λ-ring, which can be regarded as a commutative ring A with a map $A \to \mathrm{W}(A)$ satisfying certain properties. In most natural examples, including K_0 of a commutative ring, the map $A \to \mathrm{W}(A)$ actually factors through a map $A \to \mathrm{W}_{\mathrm{rat}}(A)$.

as opposed to a topological space up to homotopy; and in fact, it is not just any space, but a compact Hausdorff space. For example, in Proposition 7.10, we observe that properties of X_F formally imply that the absolute Galois group of F is torsion-free. Also, a compact Hausdorff space has certain finer homotopical and (co)homological invariants which give rise to non-profinitely completed structures on natural arithmetic invariants. From now on, let X_F denote the compact Hausdorff space defined in Sect. 5 below, which is a deformation retract of the space considered above.

Fundamental group. By design, the étale fundamental group of X_F agrees with the absolute Galois group of F. However, as a topological space, X_F also has a classical fundamental group, given by homotopy classes of loops; we denote it by $\pi_1^{\mathrm{path}}(X_F)$ (suppressing the choice of base point in the introduction). In general, $\pi_1^{\mathrm{path}}(X_F)$ could be trivial even when F is not algebraically closed; this happens whenever F is 'too big'.

However, for many examples of interest, the situation is better.

Theorem 1.6 (Sect. 6.3) *Assume that F is an abelian extension of a finite extension of $\mathbb{Q}$. Then X_F is path-connected, and the map $\pi_1^{\mathrm{path}}(X_F) \to \pi_1^{\text{ét}}(X_F) \cong \mathrm{Gal}(\overline{F}/F)$ is injective with dense image. Moreover, $\pi_1^{\mathrm{path}}(X_F)$ carries a natural topology, making it a complete topological group which can be written as an inverse limit of discrete infinite groups. The map $\pi_1^{\mathrm{path}}(X_F) \to \mathrm{Gal}(\overline{F}/F)$ is continuous for this topology, but $\pi_1^{\mathrm{path}}(X_F)$ does not carry the subspace topology.*

Remark 1.7 More precisely, $\pi_1^{\mathrm{path}}(X_F)$ is an inverse limit of discrete groups, each of which is an extension of a finite group by a free abelian group of finite rank. The essential difference between $\pi_1^{\mathrm{path}}(X_F)$ and $\mathrm{Gal}(\overline{F}/F)$ is that the Kummer map

$$F^\times \to \mathrm{Hom}(\mathrm{Gal}(\overline{F}/F), \hat{\mathbb{Z}})$$

lifts to a map

$$F^\times \to \mathrm{Hom}(\pi_1^{\mathrm{path}}(X_F), \mathbb{Z})\ .$$

One can characterise the image of $\pi_1^{\mathrm{path}}(X_F) \to \mathrm{Gal}(\overline{F}/F)$ as the stabiliser of the class in $\mathrm{Ext}(\overline{F}^\times, \mathbb{Z})$ coming by pullback along a fixed inclusion $\overline{F}^\times \hookrightarrow \mathbb{C}^\times$ from the exponential sequence

$$0 \to \mathbb{Z} \to \mathbb{C} \xrightarrow{\exp} \mathbb{C}^\times \to 0,$$

cf. Proposition 6.14. In particular, the group $\pi_1^{\mathrm{path}}(X_{\mathbb{Q}(\zeta_\infty)}) \subset \mathrm{Gal}(\overline{\mathbb{Q}}/\mathbb{Q}(\zeta_\infty))$ acts naturally on the group $\log(\overline{\mathbb{Q}}) \subset \mathbb{C}$ of logarithms of algebraic numbers.[3]

Cohomology. In general, the singular cohomology groups of X_F do not agree with the sheaf cohomology groups (as, e.g., path-connected and connected components

[3]The induced action on $2\pi i\mathbb{Z} \subset \mathbb{C}$ is trivial, as we are working over the cyclotomic extension. In fact, there can not be an action (except for complex conjugation) on $2\pi i\mathbb{Z}$, which presents an obstruction to extending this action beyond the cyclotomic extension.

do not agree), and sheaf cohomology behaves better. Thus, let $\mathrm{H}^i(X_F, A)$ denote the sheaf cohomology with coefficients in the constant sheaf A, for any abelian group A. The second part of the following theorem is a consequence of the Bloch–Kato conjecture, proved by Voevodsky, [5, Theorem 6.1].

Theorem 1.8 (Theorem 7.8, Proposition 7.9) *Let $i \geq 0$ and $n \geq 1$.*

(i) *There is a natural isomorphism*

$$\mathrm{H}^i(X_F, \mathbb{Z}/n\mathbb{Z}) \cong \mathrm{H}^i(\mathrm{Gal}(\overline{F}/F), \mathbb{Z}/n\mathbb{Z}) .$$

(ii) *The cohomology group $\mathrm{H}^i(X_F, \mathbb{Z})$ is torsion-free. In particular, using (i), there is a canonical isomorphism*

$$\mathrm{H}^i(X_F, \mathbb{Z})/n \cong \mathrm{H}^i(\mathrm{Gal}(\overline{F}/F), \mathbb{Z}/n\mathbb{Z}) .$$

Thus, one gets natural $\mathbb{Z}$-structures on the Galois cohomology groups. Note that we regard the choice ι of roots of unity as fixed throughout; in particular, all Tate twists are trivialised.

Remark 1.9 Recall that by the Bloch–Kato conjecture,

$$\mathrm{H}^i(\mathrm{Gal}(\overline{F}/F), \mathbb{Z}/n\mathbb{Z}) \cong \mathrm{K}_i^{\mathrm{M}}(F)/n,$$

where $\mathrm{K}_i^{\mathrm{M}}(F)$ denotes the Milnor K-groups of F. One might thus wonder whether $\mathrm{H}^i(X_F, \mathbb{Z}) = \mathrm{K}_i^{\mathrm{M}}(F)$. This cannot be true, as the latter contains torsion. However, it is known that all torsion in $\mathrm{K}_i^{\mathrm{M}}(F)$ comes via cup product by roots of unity

$$\mathbb{Q}/\mathbb{Z} \otimes \mathrm{K}_{i-1}^{\mathrm{M}}(F) \to \mathrm{K}_i^{\mathrm{M}}(F),$$

so that $\mathrm{K}_i^{\mathrm{M}}(F)_{\mathrm{tf}} \stackrel{\mathrm{def}}{=} \mathrm{K}_i^{\mathrm{M}}(F)/(\mathbb{Q}/\mathbb{Z} \cup \mathrm{K}_{i-1}^{\mathrm{M}}(F))$ is torsion-free. Also, as we are taking the quotient by a divisible subgroup, one still has

$$\mathrm{H}^i(\mathrm{Gal}(\overline{F}/F), \mathbb{Z}/n\mathbb{Z}) \cong \mathrm{K}_i^{\mathrm{M}}(F)_{\mathrm{tf}}/n.$$

One could then wonder whether

$$\mathrm{H}^i(X_F, \mathbb{Z}) = \mathrm{K}_i^{\mathrm{M}}(F)_{\mathrm{tf}}.$$

This is true for $i = 0, 1$, but not for $i > 1$, as the Steinberg relation $x \cup (1 - x) = 0$ for $x \neq 0, 1$ does not hold in $\mathrm{H}^2(X_F, \mathbb{Z})$. However, we regard this as a defect of X_F that should be repaired:

Question 1.10 Does there exist a topological space X_F^{M} mapping to X_F such that there are isomorphisms

$$\mathrm{H}^i(X_F^{\mathrm{M}}, \mathbb{Z}) \cong \mathrm{K}_i^{\mathrm{M}}(F)_{\mathrm{tf}}$$

for all $i \geq 0$, which are compatible with the isomorphisms in degrees $i = 0, 1$ for X_F?

For algebraically closed fields F, the space X_F^{M} would have to be constructed in such a way as to freely adjoin the Steinberg relation on its cohomology groups; the general case should reduce to this case by descent.

Descent along the cyclotomic extension. So far, all of our results were assuming that F contains all roots of unity. One may wonder whether the general case can be handled by a descent technique. This is, unfortunately, not automatic, as the construction for F involved the choice of roots of unity, so one cannot naïvely impose a descent datum. However, there are certain structures on X_F that we have not made use of so far.

First, X_F was defined as (one connected component of) the $\mathbb{C}$-valued points of some scheme defined over $\mathbb{Q}$ (or even $\mathbb{Z}$). In particular, by the machinery of étale homotopy types, its profinite homotopy type acquires an action of (a subgroup of) the absolute Galois group of $\mathbb{Q}$. This action should, in fact, factor over the Galois group of the cyclotomic extension of $\mathbb{Q}$, and allow one to define the descent datum. Unfortunately, this requires heavy technology, and does not play well with the purely topological considerations on cohomology and fundamental groups above; however, we record a version of this relationship on the level of cohomology as part of Theorem 8.9 below.

Second, X_F was defined in terms of the rational Witt vectors, and the rational Witt vectors carry extra endomorphisms, given by Frobenius operators.[4] Thus, one would expect to have Frobenius operators on X_F; however, the Frobenius operators exchange connected components, and it turns out that on the connected component X_F there are no remaining operators.[5] However, one can use a different connected component instead, at least in some situations. In this respect, we prove the following result.

Theorem 1.11 (Proposition 8.6(ii), Theorem 8.8 and 8.9) *Let ℓ be a fixed prime, and let F be a perfect field of characteristic different from ℓ (but possibly positive) whose absolute Galois group* $\mathrm{Gal}(\overline{F}/F)$ *is pro-ℓ. Let $n \leq \infty$ be maximal such that $\mu_{\ell^n} \subset F$; for simplicity, we assume $n \geq 2$ in case $\ell = 2$. Then there is a compact Hausdorff space $Y_{\ell^n,F}$ with an action of $U_{(\ell^n)} = 1 + \ell^n\mathbb{Z}_{(\ell)}$, with the following properties.*

(i) *Let $F(\zeta_{\ell^\infty})/F$ be the ℓ-cyclotomic extension. Then there is a natural isomorphism $\pi_1^{\text{ét}}(Y_{\ell^n,F}) \cong \mathrm{Gal}(\overline{F}/F(\zeta_{\ell^\infty}))$.*
(ii) *There is a natural isomorphism*

$$\mathrm{H}^i(Y_{\ell^n,F}, \mathbb{Z}/\ell^m\mathbb{Z}) \cong \mathrm{H}^i(\mathrm{Gal}(\overline{F}/F(\zeta_{\ell^\infty})), \mathbb{Z}/\ell^m\mathbb{Z}) .$$

---

[4] In fact, one can combine the first and second observation, which leads to the observation that $\mathrm{W}_{\mathrm{rat}}(F)$ is a Λ-ring; in fact, (almost tautologically) one for which the map $\mathrm{W}_{\mathrm{rat}}(F) \to \mathrm{W}(\mathrm{W}_{\mathrm{rat}}(F))$ factors over $\mathrm{W}_{\mathrm{rat}}(F) \to \mathrm{W}_{\mathrm{rat}}(\mathrm{W}_{\mathrm{rat}}(F))$.

[5] In fancy language, the 'dynamical system' of the connected components of $\mathrm{Spec}(\mathrm{W}_{\mathrm{rat}}(\mathbb{Q}(\zeta_\infty)) \otimes \mathbb{C})$ with its Frobenius operators is one form of the Bost–Connes system, [6].

Under this isomorphism, the action of $U_{(\ell^n)}$ on the left corresponds to the action of $1+\ell^n\mathbb{Z}_\ell \cong \mathrm{Gal}(F(\zeta_{\ell^\infty})/F)$ on the right via the tautological embedding $U_{(\ell^n)} \hookrightarrow 1+\ell^n\mathbb{Z}_\ell$.

We note that there is again an interesting difference between discrete and profinite groups: The Galois group of the cyclotomic extension is profinite, but the Frobenius operators live in a discrete subgroup. This is necessary, as the Frobenius operators will also act on $\mathrm{H}^i(Y_{\ell^n,F},\mathbb{Z})$.

Finally, let us give a brief summary of the different sections. In Sects. 2 and 3, we recall various basic facts about topological fundamental groups, and Pontrjagin duals, respectively. The material here is standard, but not always well-known. In Sect. 4, we prove Theorem 1.5 in the world of schemes, and in Sect. 5 we prove the version for topological spaces. Next, in Sect. 6, we prove Theorem 1.6; this relies on a careful analysis of the path-connected components of X_F and an analysis of the multiplicative groups of large extensions of number fields. In Sect. 7, we prove Theorem 1.8. Finally, in Sect. 8, we prove Theorem 1.11.

Notation. We denote the profinite completion of the integers by $\hat{\mathbb{Z}}$, the ring of finite (rational) adèles by $\mathbb{A}_\mathrm{f} = \hat{\mathbb{Z}} \otimes_\mathbb{Z} \mathbb{Q}$ and the full ring of adèles by $\mathbb{A} = \mathbb{A}_\mathrm{f} \times \mathbb{R}$.

2 Preliminaries on Fundamental Groups

In this section we assemble a number of results, some well-known, some less so, on different concepts of fundamental groups.

2.1 *Classical Fundamental Groups and Path Components*

For a topological space X with a base point $x \in X$ we let $\pi_1^{\mathrm{path}}(X,x)$ be the usual fundamental group defined in terms of loops. To be precise, a loop in X based at x is a continuous map $\gamma\colon [0,1] \to X$ with $\gamma(0)=\gamma(1)$, and for two loops γ, δ the product $\gamma * \delta$ is defined as 'run first through δ, then through γ',[6] i.e. as

$$\gamma * \delta\colon [0,1] \to X, \quad t \mapsto \begin{cases} \delta(2t) & \text{for } 0 \le t \le \frac{1}{2}, \\ \gamma(2t-1) & \text{for } \frac{1}{2} \le 1. \end{cases}$$

[6]Note that this convention is reverse to that prevalent in algebraic topology, but it is common in algebraic geometry and is more convenient when working with categories of covering spaces. Of course, these two conventions yield groups which are opposite groups of one another, hence related by a canonical isomorphism $[\gamma] \mapsto [\gamma]^{-1}$.

This composition induces a group structure on the set $\pi_1^{\text{path}}(X, x)$ of homotopy classes of loops in X based at x; we call $\pi_1^{\text{path}}(X, x)$ the *classical fundamental group* of X at x.

Path components. Recall that a space X is *path-connected* if for every two points $x, y \in X$ there is a path in X from x to y, i.e. a continuous map $\gamma\colon [0, 1] \to X$ with $\gamma(0) = x$ and $\gamma(1) = y$. More generally, introduce an equivalence relation on the points of a space X by declaring x and y equivalent if there is a path from x to y in X. Then the equivalence classes of this relation are called the *path components* of X; they can be characterised as the maximal path-connected subspaces of X. The set $\pi_0^{\text{path}}(X)$ of path components will be equipped with the quotient topology induced by the given topology on X.

Finally note that if $x \in X$ and $X^\circ \subseteq X$ is the path component containing x, then every loop in X based at x lies in X°, and similarly for every homotopy of loops. Hence the inclusion $X^\circ \hookrightarrow X$ induces an isomorphism of fundamental groups $\pi_1^{\text{path}}(X^\circ, x) \to \pi_1^{\text{path}}(X, x)$.

Since the interval $[0, 1]$ is connected, every path-connected space is connected, but the converse does not hold. The most well-known counterexample seems to be the *topologist's sine curve*

$$T = \{(0, y) \mid -1 \leq y \leq 1\} \cup \{(x, \sin \tfrac{1}{x}) \mid x > 0\} \subset \mathbb{R}^2$$

which is connected but has two path components, cf. Fig. 1. See [7, Example 117] for more details.

More instructive for our purposes is the following example, to which we will return several times in this section.

Proposition and Definition 2.1 *The following topological groups are all canonically isomorphic; each of them is called a (one-dimensional)* solenoid.

(i) *The Pontryagin dual $\mathbb{Q}^\vee$, i.e. the set of group homomorphisms $\mathbb{Q} \to \mathbb{S}^1$ endowed with the compact-open topology, where $\mathbb{Q}$ carries the discrete topology and $\mathbb{S}^1 \subset \mathbb{C}^\times$ is the unit circle;*

(ii) *the inverse limit $\varprojlim_{n \in \mathbb{N}} \mathbb{S}^1$, where the set $\mathbb{N}$ is partially ordered by divisibility, and for $m \mid n$ the transition map from the n-th to the m-th copy of $\mathbb{S}^1$ is $z \mapsto z^{n/m}$;*

(iii) *the inverse limit $\varprojlim_{n \in \mathbb{N}} \mathbb{R}/\frac{1}{n}\mathbb{Z}$, where the transition maps are induced by the identity $\mathbb{R} \to \mathbb{R}$;*

(iv) *the adèle class group $\mathbb{A}/\mathbb{Q}$, where $\mathbb{Q}$ is diagonally embedded in $\mathbb{A}$.*

Fig. 1 The topologist's sine curve

Proof We can write each of these groups as an inverse limit of certain topological groups, indexed by the partially ordered set $\mathbb{N}$. The constituents of index n are, respectively:

(i) the quotient $A_n = (\frac{1}{n}\mathbb{Z})^\vee$ of $\mathbb{Q}^\vee$ corresponding to the subgroup $\frac{1}{n}\mathbb{Z} \subset \mathbb{Q}$;
(ii) the n-th copy of $\mathbb{S}^1$, denoted by B_n;
(iii) the quotient $C_n = \mathbb{R}/\frac{1}{n}\mathbb{Z}$;
(iv) the double quotient $D_n = (\frac{1}{n}\hat{\mathbb{Z}})\backslash\mathbb{A}/\mathbb{Q} = \mathbb{A}/(\mathbb{Q} + \frac{1}{n}\hat{\mathbb{Z}})$.

We can write down some isomorphisms between these constituents:

- $B_n \to A_n$, $s \mapsto (\frac{1}{n}\mathbb{Z} \to \mathbb{S}^1,\ q \mapsto s^{nq})$;
- $C_n \to B_n$, $t + \frac{1}{n}\mathbb{Z} \mapsto e^{2\pi i n t}$;
- $C_n \to D_n$ induced by the inclusion $\mathbb{R} \hookrightarrow \mathbb{A} = \mathbb{A}_f \times \mathbb{R}$, $t \mapsto (0, t)$.

It is easy to see that these give three families of isomorphisms of topological groups $A_n \leftarrow B_n \leftarrow C_n \to D_n$, commuting with the structure maps of the inverse systems, hence defining isomorphisms between the limits. □

Proposition 2.2 *The solenoid $\mathcal{S}$ is a commutative compact Hausdorff group. It is connected, but not path-connected. The path component $\mathcal{S}^\circ$ containing the identity is a subgroup, hence the path components of $\mathcal{S}$ are precisely the cosets of $\mathcal{S}^\circ$. Therefore $\pi_0^{\text{path}}(\mathcal{S}) \cong \mathcal{S}/\mathcal{S}^\circ$ acquires the structure of a topological group.*

As such it is canonically isomorphic to $\mathbb{A}_f/\mathbb{Q} \cong \hat{\mathbb{Z}}/\mathbb{Z}$ with the quotient topology (which is indiscrete), where $\mathbb{Q}$ is embedded diagonally in $\mathbb{A}_f$.

Proof The claims in the third and fourth sentences of the proposition follow formally from $\mathcal{S}$ being a topological group.

We next show that $\mathcal{S}$ is connected, using the description (ii): $\mathcal{S} \cong \varprojlim B_n$ with each $B_n \cong \mathbb{S}^1$. If $f\colon \varprojlim B_n \to \{0, 1\}$ is continuous, then by construction of the inverse limit topology f must factor through some B_n, hence be constant.

It remains to determine $\pi_0^{\text{path}}(\mathcal{S})$. It is most convenient to do this using description (iv): $\mathcal{S} \cong \mathbb{A}/\mathbb{Q}$. It is not hard to see that the quotient map $\mathbb{A} \to \mathbb{A}/\mathbb{Q}$ has unique lifting of paths, i.e. if $\gamma\colon [0, 1] \to \mathbb{A}/\mathbb{Q}$ is continuous and $a \in \mathbb{A}$ is such that $\gamma(0) = a + \mathbb{Q}$, then there is a unique continuous $\tilde{\gamma}\colon [0, 1] \to \mathbb{A}$ inducing γ and satisfying $\tilde{\gamma}(0) = a$. From this we see that the neutral path component of $\mathbb{A}/\mathbb{Q}$ is precisely the image in $\mathbb{A}/\mathbb{Q}$ of $\{0\} \times \mathbb{R} \subset \mathbb{A}_f \times \mathbb{R} = \mathbb{A}$. Note that $\mathbb{Q}$ is embedded diagonally, hence it intersects trivially with $\{0\} \times \mathbb{R}$, and we obtain a group isomorphism (but not a homeomorphism, see below!) $\mathbb{R} \to \mathcal{S}^0$. Hence the topological group $\pi_0^{\text{path}}(\mathcal{S}) \cong \pi_0^{\text{path}}(\mathbb{A}/\mathbb{Q})$ is isomorphic to $(\mathbb{A}/\mathbb{R})/\mathbb{Q} \cong \mathbb{A}_f/\mathbb{Q}$. Since $\mathbb{Q}$ is dense in $\mathbb{A}_f$, this carries the indiscrete topology. Note that $\mathbb{Q} + \hat{\mathbb{Z}} = \mathbb{A}_f$ and $\mathbb{Q} \cap \hat{\mathbb{Z}} = \mathbb{Z}$, so we may also identify $\pi_0^{\text{path}}(\mathcal{S})$ with $\hat{\mathbb{Z}}/\mathbb{Z}$. □

Locally path-connected spaces. A topological space X is *locally path-connected* if for every $x \in X$ and every open neighbourhood $V \subseteq X$ of x there exists an open neighbourhood $U \subseteq V \subseteq X$ of x which is path-connected.

Let X be a topological space, and denote the given topology by $\mathfrak{O} = \{V \subseteq X \mid V \text{ open}\}$. For an open subset $V \in \mathfrak{O}$ and $x \in V$ set

$$U(V, x) = \{y \in X \mid \text{ there is a path } \gamma\colon [0, 1] \to V \text{ with } \gamma(0) = x \text{ and } \gamma(1) = y\};$$

i.e., $U(V, x)$ is the path component of V containing x. It is then clear that the $U(V, x)$ for varying $x \in X$ and $x \in V \in \mathfrak{O}$ form a basis of a topology $\mathfrak{O}^{\mathrm{lpc}}$ on the set X. We let X^{lpc} be the topological space with underlying space X and topology $\mathfrak{O}^{\mathrm{lpc}}$. Hence we obtain a continuous but not necessarily open bijection $X^{\mathrm{lpc}} \to X$. The following properties are easily checked:

Lemma 2.3 *Let X be a topological space, with topology $\mathfrak{O}$.*

(i) For every point $x \in X$ the sets $U(V, x)$, where V runs through all elements of $\mathfrak{O}$ containing x, is a basis of neighbourhoods of x in X^{lpc}.
(ii) The space X^{lpc} is locally path-connected.
(iii) The topology $\mathfrak{O}^{\mathrm{lpc}}$ is the coarsest topology on the set X which is finer than $\mathfrak{O}$ and locally path-connected.
(iv) The construction is functorial: if $f\colon Y \to X$ is continuous, then so is $f\colon Y^{\mathrm{lpc}} \to X^{\mathrm{lpc}}$.
(v) If Y is a locally path-connected space, then any continuous map $Y \to X$ factors uniquely as $Y \to X^{\mathrm{lpc}} \to X$. In other words, $X^{\mathrm{lpc}} \to X$ is universal among continuous maps from locally path-connected spaces to X.
This may be rephrased as follows: if **Top** *denotes the category of topological spaces with continuous maps and* **LPC** $\subset$ **Top** *denotes the full subcategory of locally path-connected spaces then the functor $(-)^{\mathrm{lpc}}$:* **Top** $\to$ **LPC** *is right adjoint to the inclusion functor* **LPC** $\hookrightarrow$ **Top**. □

Example 2.4 (i) If X is totally disconnected then X^{lpc} is discrete.
(ii) Let M be a smooth manifold and let $\mathscr{F}$ be a foliation on M, defined by a vector subbundle F of the tangent bundle $\mathrm{T}M$ such that the sections of F are stable under the Lie bracket (also known as an *involutive* or *integrable* subbundle). Recall that a *leaf* of the foliation is a smooth manifold L together with an injective immersion $i\colon L \hookrightarrow M$ which, for every $p \in L$, induces an identification of T_pL with $F_p \subseteq \mathrm{T}_{i(p)}M$, and which is maximal with respect to this property.
Since i is injective, we may identify L with a subset of M. In general, however, the topology of L will not be the subspace topology inherited from M. It is not too hard to show that if $i(L)$ is endowed with this subspace topology, then

$$i(L)^{\mathrm{lpc}} \cong L.$$

For instance, let $\vartheta \in \mathbb{R}$ and consider the *Kronecker foliation of slope ϑ*. This is the foliation $\mathscr{F}_\vartheta$ on the torus $M = \mathbb{R}^2/\mathbb{Z}^2$ given by the subbundle $F_\vartheta \subset \mathrm{T}M \cong \mathbb{R}^2 \times M$ with $F_{\vartheta,p} = \mathbb{R} \cdot \binom{\vartheta}{1}$ for every $p \in M$. Each leaf of $\mathscr{F}_\vartheta$ is then the image in M of an affine subspace in $\mathbb{R}^2$ parallel to $\mathbb{R} \cdot \binom{\vartheta}{1}$. If $\vartheta \in \mathbb{Q}$ then every leaf L

is homeomorphic to $\mathbb{S}^1$, and $L \to i(L)$ is a homeomorphism. If ϑ is irrational then all leaves are homeomorphic to $\mathbb{R}$ and have *dense image* in M. They are all translates, via the group structure on M, of the leaf through 0:

$$i\colon \mathbb{R} \to M, \quad t \mapsto \begin{pmatrix} t\vartheta \\ t \end{pmatrix} \bmod \mathbb{Z}^2.$$

The topology on $i(\mathbb{R})$ inherited from M defines a topology $\mathfrak{O}_\vartheta$ on $\mathbb{R}$. A basis of neighbourhoods of 0 for this topology is given by the sets

$$\{t \in \mathbb{R} \mid \text{both } t \text{ and } t\vartheta \text{ differ by less than } \varepsilon \text{ from an integer}\}$$

for varying $\varepsilon > 0$. Hence every $\mathfrak{O}_\vartheta$-neighbourhood of 0 is unbounded. Still, $\mathfrak{O}_\vartheta^{\mathrm{lpc}}$ is the Euclidean topology on $\mathbb{R}$.

Note that M is the completion of the topological group $(\mathbb{R}, \mathfrak{O}_\vartheta)$, and using this it is not hard to see that $(\mathbb{R}, \mathfrak{O}_{\vartheta_1}) \simeq (\mathbb{R}, \mathfrak{O}_{\vartheta_2})$ as topological groups if and only if ϑ_1 and ϑ_2 are in the same $\mathrm{GL}_2(\mathbb{Z})$-orbit in $\mathbb{P}^1(\mathbb{R}) \smallsetminus \mathbb{P}^1(\mathbb{Q})$.

(iii) There is a similar description of $\mathcal{S}^{\mathrm{lpc}}$, where $\mathcal{S} = \mathbb{A}/\mathbb{Q}$ is the solenoid. We have already determined the path components of $\mathcal{S}$ in the proof of Proposition 2.2. Since $\mathcal{S}$ is a topological group, they are all homeomorphic, and they are all dense in $\mathcal{S}$ by Proposition 2.2. One of them is the image of $\mathbb{R}$ under the obvious homomorphism $i\colon \mathbb{R} \to \mathbb{A} \to \mathcal{S}$. Again, i is injective and continuous, but not a homeomorphism onto its image. If $\mathfrak{O}$ denotes the topology on $\mathbb{R}$ corresponding to the subspace topology on $i(\mathbb{R}) \subset \mathcal{S}$, then a basis for $\mathfrak{O}$ is given by the open sets

$$x + \bigcup_{n\in\mathbb{Z}}]kn - \varepsilon, kn + \varepsilon[$$

for $x \in \mathbb{R}$, $k \in \mathbb{N}$ and $\varepsilon > 0$. This topology is not locally path-connected, but $\mathfrak{O}^{\mathrm{lpc}}$ is the Euclidean topology on $\mathbb{R}$.

From this we can determine the topology on $\mathcal{S}^{\mathrm{lpc}}$: it is the unique topology on $\mathcal{S}$ for which $\mathcal{S}$ is a topological group and $i\colon \mathbb{R} \to \mathcal{S}$ is a homeomorphism onto an open subgroup, where $\mathbb{R}$ has the Euclidean topology. Hence $\mathcal{S}^{\mathrm{lpc}}$ is (non-canonically) isomorphic to $\hat{\mathbb{Z}}/\mathbb{Z} \times \mathbb{R}$, with the discrete topology on $\hat{\mathbb{Z}}/\mathbb{Z}$.

Corollary 2.5 *Let X be a topological space.*

(i) The canonical map $\pi_0^{\mathrm{path}}(X^{\mathrm{lpc}}) \to \pi_0^{\mathrm{path}}(X)$ is a bijection.

(ii) For any $x \in X$ the canonical map $\pi_1^{\mathrm{path}}(X^{\mathrm{lpc}}, x) \to \pi_1^{\mathrm{path}}(X, x)$ is a group isomorphism, and similarly for higher homotopy groups defined in the usual way using spheres.

(iii) The complex $\mathrm{C}_\bullet(X)$ of singular simplices in X with integral coefficients is canonically isomorphic to $\mathrm{C}_\bullet(X^{\mathrm{lpc}})$. In particular $X^{\mathrm{lpc}} \to X$ induces isomorphisms on singular homology and cohomology, for any abelian coefficient group.

Proof Lemma 2.3 implies that a map of sets $[0, 1] \to X$ is continuous if and only if it is continuous when viewed as a map $[0, 1] \to X^{\mathrm{lpc}}$, and similarly for homotopies and singular simplices. □

Topologies on the classical fundamental group Let (X, x) be a pointed topological space. Then there exist several natural topologies on $\pi_1^{\mathrm{path}}(X, x)$.

- The *loop topology* on $\pi_1^{\mathrm{path}}(X, x)$ is the quotient topology defined by the surjective map $\Omega(X, x) \to \pi_1^{\mathrm{path}}(X, x)$, where $\Omega(X, x)$ is the loop space of (X, x), i.e. the set of all continuous pointed maps $(\mathbb{S}^1, 1) \to (X, x)$ endowed with the compact-open topology. While the loop topology is defined in a very natural way it does not always turn $\pi_1^{\mathrm{path}}(X, x)$ into a topological group, only into a *quasi-topological group*.
  Here a quasi-topological group is a group G together with a topology such that the inverse map $G \to G, g \mapsto g^{-1}$, and all multiplication maps $G \to G, g \mapsto gh$, and $G \to G, g \mapsto hg$, for $h \in G$ are continuous. These conditions do not imply that the multiplication map $G \times G \to G$, $(g, h) \mapsto gh$, is continuous (which would turn G into a topological group).
  For instance for the *Hawaiian earrings*
  $$H = \bigcup_{n \in \mathbb{N}} C_n$$
  where $C_n \subset \mathbb{R}^2$ is a circle of radius $\frac{1}{n}$ centered at $(0, \frac{1}{n})$, the fundamental group $\pi_1^{\mathrm{path}}(H, 0)$ with the loop topology is a quasi-topological group but not a topological group, see [8].
- Brazas [9] showed that for any pointed space (X, x) there is a finest topology on $\pi_1^{\mathrm{path}}(X, x)$ such that $\pi_1^{\mathrm{path}}(X, x)$ becomes a topological group and $\Omega(X, x) \to \pi_1^{\mathrm{path}}(X, x)$ is continuous. This topology is known as the *τ-topology*. Clearly it agrees with the loop topology if and only if the latter already turns $\pi_1^{\mathrm{path}}(X, x)$ into a topological group.
  Brazas introduces a generalised notion of covering spaces called *semicovering spaces*, cf. [10]. For a semicovering space $p: Y \to X$ the subspace topology on the fibre $p^{-1}(x)$ is discrete, and the monodromy action of $\pi_1^{\mathrm{path}}(X, x)$ on $p^{-1}(x)$ is continuous for the τ-topology. If X is path-connected and locally path-connected this construction provides an equivalence of categories between semicoverings of X and discrete sets with continuous $\pi_1^{\mathrm{path}}(X, x)$-action.
- For any monodromy action defined by a semicovering space the point stabilisers will be open subgroups of $\pi_1^{\mathrm{path}}(X, x)$ for the τ-topology. Hence it makes sense to define a new topology called the *σ-topology* on $\pi_1^{\mathrm{path}}(X, x)$ where a neighbourhood basis of the identity is given by the τ-open subgroups (rather than all τ-open neighbourhoods) of $\pi_1^{\mathrm{path}}(X, x)$.
- Finally we may consider the completion $\pi_1^{\mathrm{Gal}}(X, x)$ of $\pi_1^{\mathrm{path}}(X, x)$ with respect to the σ-topology (more precisely, with respect to the two-sided uniformity defined by

the σ-topology). This group is complete and has a basis of open neighbourhoods of the identity given by open subgroups. By [11, Proposition 7.1.5] it is then a *Noohi group*, i.e. the tautological map from $\pi_1^{\mathrm{Gal}}(X, x)$ to the automorphism group of the forgetful functor

$$\pi_1^{\mathrm{Gal}}(X, x)\text{–}\mathbf{Sets} \to \mathbf{Sets}$$

is an isomorphism. Here **Sets** is the category of sets and $\pi_1^{\mathrm{Gal}}(X, x)$-**Sets** is the category of (discrete) sets with a continuous left $\pi_1^{\mathrm{Gal}}(X, x)$-action.
In the case where X is path-connected and locally path-connected the category $\pi_1^{\mathrm{Gal}}(X, x)$-**Sets** is again equivalent to the category of semicovering spaces of X, and $\pi_1^{\mathrm{Gal}}(X, x)$ can be constructed from that category as the automorphism group of a fibre functor, see [12] for details.

2.2 *Etale Fundamental Groups of Topological Spaces*

Let X be a connected (but not necessarily path-connected!) topological space and $x \in X$. We shall construct a profinite group $\pi_1^{\text{ét}}(X, x)$ which classifies pointed finite coverings of (X, x), much like the étale fundamental group in algebraic geometry does. To do so we proceed analogously to the usual construction for schemes.

Categories of finite covering spaces. Recall that a continuous map of topological spaces $p \colon Y \to X$ is a trivial finite covering if there is a finite discrete space D and a homeomorphism $X \times D \to Y$ making the obvious diagram commute; more generally, $p \colon Y \to X$ is a *finite covering* if every point in X has an open neighbourhood $U \subseteq X$ such that the base change $p_U \colon Y_U = p^{-1}(U) \to U$ is a trivial finite covering. The map

$$X \to \mathbb{N}_0, \quad x \mapsto |p^{-1}(x)|,$$

is continuous. If X is connected, it is therefore constant; the unique value it assumes is called the *degree* of the covering.

Definition 2.6 Let X be a topological space. The category $\mathbf{FCov}(X)$ has as objects the pairs (Y, p) where Y is a topological space and $p \colon Y \to X$ is a finite covering, and as morphisms from (Y_1, p_1) to (Y_2, p_2) the continuous maps $f \colon Y_1 \to Y_2$ such that $p_1 = p_2 \circ f$.

For every point $x \in X$ we define a functor $\Phi_x \colon \mathbf{FCov}(X) \to \mathbf{FSet}$ (the target being the category of finite sets) by sending $p \colon Y \to X$ to the fibre $p^{-1}(x)$, with the obvious action on morphisms. For a continuous map $f \colon X \to X'$ we obtain a functor $f^* \colon \mathbf{FCov}(X') \to \mathbf{FCov}(X)$ by pullback. For $x \in X$ there is then a canonical isomorphism $\Phi_{f(x)} \cong \Phi_x \circ f^*$ of functors $\mathbf{FCov}(X') \to \mathbf{FSet}$.

Proposition 2.7 *Let X be a connected topological space and $x \in X$. Let $p \colon Y \to X$ be a finite covering of degree d. Then Y splits into finitely many connected components* *as* $Y = Y_1 \coprod \cdots \coprod Y_n$*, each* $Y_i \to X$ *is a finite covering of* X *of some degree* d_i*, and* $d = d_1 + \cdots + d_n$.

Proof To each open and closed subset $Z \subseteq Y$ we assign a counting function

$$c_Z : X \to \{0, 1, \ldots, d\}, \quad x \mapsto |p^{-1}(x) \cap Z|.$$

We claim this is continuous: let $U \subseteq X$ be an open subset over which p becomes trivial. Note we cannot assume U to be connected itself because we have not assumed X to be locally connected. Still, $Z \cap p^{-1}(U)$ is both open and closed in $p^{-1}(U)$, and we may assume the latter to be $U \times D$, where D is a discrete set of cardinality d. Hence for each $\delta \in D$ the locus of $u \in U$ with $(u, \delta) \in Z$ is both open and closed in U. Therefore c_Z is continuous on U. But any point in X is contained in a suitable U, therefore c_Z is continuous on X. But as X is connected, c_Z must be constant, equal to some $0 \leq d_Z \leq d$.

From this argument we also see that $Z \to X$ is a finite covering. The same applies to $Y \setminus Z$. The degrees of the two coverings thus obtained must be strictly smaller than d, hence after finitely many steps we arrive at a decomposition into connected finite coverings. □

Lemma 2.8 *Let* $p : Y \to X$ *be a finite covering, where* X *and* Y *are connected topological spaces. Let* $g \in \mathrm{Aut}(Y/X)$*, i.e.* g *is a homeomorphism* $Y \to Y$ *with* $p \circ g = g$*. If* g *has a fixed point, then it is the identity.*

Proof Similarly to the proof of Proposition 2.7, we show that the set $\{y \in Y \mid g(y) = y\}$ is both open and closed in Y. □

Proposition 2.9 *Let* X *be a connected topological space. Then* $\mathbf{FCov}(X)$ *is a Galois category in the sense of SGA 1, and for every* $x \in X$ *the functor* Φ_x *may serve as a fibre functor.*

Proof There are several equivalent characterisations of Galois categories, one being given as follows: an essentially small category $\mathscr{C}$ that admits a functor $\Phi : \mathscr{C} \to \mathbf{FSet}$ (called 'fibre functor') satisfying the following set of axioms (reproduced from cf. [13, Exposé V.4]).

- (G1) $\mathscr{C}$ has a final object, and arbitrary fibre products exist.
- (G2) $\mathscr{C}$ has finite coproducts and categorical quotients of objects by finite groups of automorphisms.
- (G3) Every morphism in $\mathscr{C}$ factors as $\iota \circ \pi$ where ι is the inclusion of a direct summand in a coproduct and π is a strict epimorphism.
- (G4) Φ commutes with fibre products and sends right units to right units.
- (G5) Φ commutes with finite coproducts, sends strict epimorphisms to strict epimorphisms and sends categorical quotients by finite groups to categorical quotients by finite groups.
- (G6) If $\Phi(f)$ is an isomorphism, then so is f.

To show that $\mathbf{FCov}(X)$ and Φ_x satisfy these axioms is mostly straightforward. The nontrivial parts are the existence of quotients in (G2), and (G6).

For the former let $p\colon Y \to X$ be an object in $\mathbf{FCov}(X)$ and let $G \subseteq \mathrm{Aut}(Y/X)$ be a finite subgroup. Endow $G\backslash Y$ with the quotient topology; we claim that $G\backslash Y \to X$ is an object of $\mathbf{FCov}(X)$, and it will follow formally that it is a categorical quotient for the group action. Take an open subset $U \subseteq X$ over which Y is trivialised; it suffices to show that $G\backslash p^{-1}(U) \to U$ is a finite covering. We may assume that $p^{-1}(U) = U \times \{1, 2, \ldots, d\}$. Then we obtain a continuous, hence locally constant, map

$$\alpha\colon U \to \mathrm{Hom}(G, \mathfrak{S}_d)$$

where $\alpha(u)\colon G \to \mathfrak{S}_d$ is the permutation action of G on the fibre $p^{-1}(u) \cong \{1, 2, \ldots, d\}$. From the fact that α is locally constant we deduce that the restriction of $G\backslash Y$ to U is a finite covering of U, as desired.

As for (G6) let $f\colon Y_1 \to Y_2$ be a morphism in $\mathbf{FCov}(X)$ which induces a bijection on the fibres over some $x \in X$. We need to show that f is a homeomorphism. First, by an argument analogous to the preceding, we show that f is in fact bijective. Then on any open subset $U \subseteq X$ trivialising both coverings we may assume that f takes the form

$$U \times \{1, 2, \ldots, d\} \to U \times \{1, 2, \ldots, d\}, \quad (u, \delta) \mapsto (u, \beta(u)(\delta))$$

for some finite sets D_i and some continuous map $\beta\colon U \to \mathfrak{S}_d$. It is then clear that f is also open. □

For a Galois category $\mathscr{C}$ with a fibre functor Φ the group $\pi = \mathrm{Aut}\,\Phi$ acquires a natural structure of a profinite group, as a projective limit over all the images of π in $\mathrm{Aut}(\Phi(Y))$ for $Y \in \mathrm{Ob}\,\mathscr{C}$. The functor Φ then factors through the category π-**FSet** of finite sets with a continuous left action by π, and in fact induces an equivalence between $\mathscr{C}$ and π-**FSet** by [13, Exposé V, Théorème 4.1]. In fact, Galois categories are precisely those that are equivalent to π-**FSet** for some profinite group π, cf. the remarks after [13, Exposé V, Définition 5.1].

Definition 2.10 Let X be a connected topological space and $x \in X$. The automorphism group of the fibre functor $\Phi_x\colon \mathbf{FCov}(X) \to \mathbf{FSet}$ is called the *étale fundamental group* of X at x and denoted by $\pi_1^{\text{ét}}(X, x)$.

It follows from the formalism of Galois categories that for two points $x, x' \in X$ the groups $\pi_1^{\text{ét}}(X, x)$ and $\pi_1^{\text{ét}}(X, x')$ are isomorphic, the isomorphism being canonical up to inner automorphisms, cf. [13, Exposé V, Corollary 5.7].

In a similar vein, let X be a connected topological space and $\gamma\colon [0, 1] \to X$ a path; write $x_0 = \gamma(0)$ and $x_1 = \gamma(1)$. Then γ induces an isomorphism of functors $\varphi_\gamma\colon \Phi_{x_0} \to \Phi_{x_1}$ as follows: for any finite covering $p\colon Y \to X$ the pullback $\gamma^* p\colon \gamma^* Y = Y \times_{X,\gamma} [0, 1] \to [0, 1]$ trivialises canonically, i.e. for any $t \in [0, 1]$ the composition

$$\Phi_{x_t}(Y) = p^{-1}(\gamma(t)) = (\gamma^* p)^{-1}(t) \hookrightarrow \gamma^* Y \to \pi_0(\gamma^* Y)$$

is a bijection, hence there is a canonical identification

$$\Phi_{x_0}(Y) \cong \pi_0(\gamma^* Y) \cong \Phi_{x_1}(Y). \tag{1}$$

We define the isomorphism of functors $\varphi_\gamma \colon \Phi_{x_0} \to \Phi_{x_1}$ applied to the object $p \colon Y \to X$ to be (1). By conjugation it induces an isomorphism of étale fundamental groups

$$\tau_\gamma \colon \pi_1^{\text{ét}}(X, x_0) = \operatorname{Aut} \Phi_{x_0} \to \operatorname{Aut} \Phi_{x_1} = \pi_1^{\text{ét}}(X, x_1), \quad \alpha \mapsto \varphi_\gamma \circ \alpha \circ \varphi_\gamma^{-1}. \tag{2}$$

The class of this isomorphism up to inner automorphisms is precisely the canonical class of isomorphisms for any two points of X mentioned above.

Continuity properties. Another important property of étale fundamental groups is their compatibility with cofiltered projective limits.

Proposition 2.11 *Let (X_α) be a cofiltered projective system of compact Hausdorff spaces and let $X = \varprojlim_\alpha X_\alpha$.*

(i) Let $p \colon Y \to X$ be a finite covering space. Then there exists some α_0, a finite covering $p_0 \colon Y_0 \to X_{\alpha_0}$ and a pullback diagram

$$\begin{array}{ccc} Y & \longrightarrow & Y_0 \\ {\scriptstyle p}\downarrow & & \downarrow{\scriptstyle p_0} \\ X & \longrightarrow & X_{\alpha_0}. \end{array}$$

(ii) Let $p_1 \colon Y_1 \to X_{\alpha_1}$ and $p_2 \colon Y_2 \to X_{\alpha_2}$ be finite covering spaces, and let $f \colon Y_1 \times_{X_{\alpha_1}} X \to Y_2 \times_{X_{\alpha_2}} X$ be a continuous map commuting with the projections to X. Then there exists some $\alpha_0 \geq \alpha_1, \alpha_2$ such that f is the base change along $X \to X_{\alpha_0}$ of a continuous map $Y_1 \times_{X_{\alpha_1}} X_{\alpha_0} \to Y_2 \times_{X_{\alpha_2}} X_{\alpha_0}$ commuting with the projections to X_{α_0}.

Proof We will only prove (i), the proof for (ii) being very similar.

Call a subset of X *basis-open* if it is the preimage of an open set in some X_α. As the name suggests, these form a basis of the topology on X.

Let $\mathfrak{U}$ be the set of basis-open subsets of X on which p becomes trivial. Then $\mathfrak{U}$ is an open cover of X, and since X is compact there exists a finite subcover, say $\{U_1, \ldots, U_n\}$. Then there exist finite sets $D_1, \ldots, D_n$ and continuous functions $\varphi_{ij} \colon U_{ij} = U_i \cap U_j \to \operatorname{Isom}(D_i, D_j)$ satisfying the cocycle condition

$$\varphi_{jk}(u) \circ \varphi_{ij}(u) = \varphi_{ik}(u) \qquad \text{for all } u \in U_{ijk} = U_i \cap U_j \cap U_k$$

such that Y is isomorphic to the union of the spaces $U_i \times D_i$, glued along the φ_{ij}.

Next, let $\mathfrak{V}$ be the set of all basis-open subsets $V \subseteq X$ such that V is contained in some U_{ij}, and for every i, j with $V \subseteq U_{ij}$ the restriction $\varphi_{ij}|_V$ is constant. Again,

$\mathfrak{V}$ is an open cover of X and hence has a finite subcover $\{V_1, \ldots, V_m\}$. Since the projective system (X_α) is cofiltered and the U_i and V_j are basis-open sets, there exists some α_0 such that all U_i and V_j are preimages of open sets in X_{α_0}. Then also the φ_{ij} are compositions with functions on X_{α_0}, and we see that $p\colon Y \to X$ is the pullback of a finite covering defined on X_{α_0}. □

Remark 2.12 With a little more effort we can show that in the situation of Proposition 2.11, for a compatible system of basepoints and under the assumption that the X_α are connected (and hence so is X), the natural map

$$\pi_1^{\text{ét}}(X, x) \to \varprojlim_\alpha \pi_1^{\text{ét}}(X_\alpha, x_\alpha)$$

is an isomorphism.

Lemma 2.13 *Let X be a connected compact Hausdorff space, and let $(p_\alpha\colon Y_\alpha \to X)_\alpha$ be a cofiltered projective system of connected finite covering spaces such that $\tilde{Y} = \varprojlim_\alpha Y_\alpha$ has trivial étale fundamental group.*

Then every connected finite covering space of X is dominated by some Y_α.

Proof Let $Z \to X$ be a connected finite covering. By assumption, the pullback covering $Z \times_X \tilde{Y} \to \tilde{Y}$ splits, i.e. $Z \times_X \tilde{Y}$ is isomorphic to $\tilde{Y} \times D$ as a $\tilde{Y}$-space, for some finite discrete set D. By Proposition 2.11 this splitting has to occur at a finite level, i.e. there has to be some α_0 such that $Z \times_X Y_{\alpha_0} \simeq Y_{\alpha_0} \times D$ as a Y_{α_0}-space. Choosing some $d \in D$ we obtain a commutative diagram

$$\begin{array}{ccccccc}
Y_{\alpha_0} & \xhookrightarrow{\mathrm{id}\times d} & Y_{\alpha_0} \times D & \xrightarrow{\simeq} & Z \times_X Y_{\alpha_0} & \longrightarrow & Z \\
\| & & \downarrow & & \downarrow & & \downarrow \\
Y_{\alpha_0} & = & Y_{\alpha_0} & = & Y_{\alpha_0} & \longrightarrow & X
\end{array}$$

The composition of the upper horizontal maps $Y_\alpha \to Z$ is a continuous map between finite covering spaces of X, respecting the projections to X, hence itself a finite covering, as desired. □

Homotopy invariance. We now show that étale fundamental groups are homotopy invariant. We will make extensive use of the following classical result:

Proposition 2.14 (Unique Homotopy Lifting Property) *Let X be a topological space, $p\colon Y \to X$ a finite covering, and S another topological space. Assume we are given a homotopy, i.e. a continuous map $H: S \times [0, 1] \to X$, together with a lift of $H(-, 0)$ to Y, i.e. a commutative diagram of continuous maps*

$$\begin{array}{ccc}
S \times \{0\} & \longrightarrow & Y \\
\downarrow & & \downarrow{\scriptstyle p} \\
S \times [0, 1] & \xrightarrow[H]{} & X.
\end{array}$$

Then there exists a unique continuous map $S \times [0, 1] \to Y$ making the resulting diagram

$$\begin{array}{ccc} S \times \{0\} & \longrightarrow & Y \\ \big\downarrow & \nearrow & \big\downarrow p \\ S \times [0,1] & \xrightarrow[H]{} & X. \end{array}$$

commute. □

This is of course well-known, see e.g. [14, Proposition 1.30]; however, we wish to explicitly stress that no (local) connectivity assumptions about X are made.

Proposition 2.15 *Let X and Z be connected topological spaces, let $f, g\colon X \to Z$ be continuous, let H be a homotopy between them (i.e. a continuous map $H\colon X \times [0, 1] \to Z$ such that $H(\xi, 0) = f(\xi)$ and $H(\xi, 1) = g(\xi)$ for all $\xi \in X$), and let $x \in X$ be a basepoint. These data determine a path $\gamma\colon [0, 1] \to Z$ by $\gamma(t) = H(x, t)$.*

Then the diagram

$$\begin{array}{ccc} & & \pi_1^{\text{ét}}(Z, f(x)) \\ & \overset{f_*}{\nearrow} & \\ \pi_1^{\text{ét}}(X, x) & & \cong \big\downarrow \tau_\gamma \\ & \underset{g_*}{\searrow} & \\ & & \pi_1^{\text{ét}}(Z, g(x)), \end{array}$$

where τ_γ is the map in (2), commutes.

To show this we first need a lemma.

Lemma 2.16 *Let X be a topological space and let $q\colon W \to X \times [0, 1]$ be a finite covering. For any $t \in [0, 1]$ consider the restriction $q_t\colon W_t = q^{-1}(X \times \{t\}) \to X \times \{t\} \cong X$. Then there is a canonical isomorphism $W_0 \cong W_1$ in* **FCov**(X)*; the construction of this isomorphism is functorial in W.*

Proof Applying Proposition 2.14 to the diagram

$$\begin{array}{ccc} W_0 & \hookrightarrow & W \\ {\scriptstyle \mathrm{id}_{W_0} \times 0} \big\downarrow & & \big\downarrow q \\ W_0 \times [0,1] & \xrightarrow[q_0 \times \mathrm{id}_{[0,1]}]{} & X \times [0,1], \end{array}$$

we deduce the existence of a unique continuous map $W_0 \times [0, 1] \to W$ making the resulting diagram commute. In particular this map induces an isomorphism on the fibres over any point in $X \times [0, 1]$ of the form $(x, 0)$. By Proposition 2.9 (or more precisely by axiom (G6) for Galois categories, mentioned in the proof thereof) it must be a homeomorphism $W_0 \times [0,1] \to W$. Functoriality is straightforward. □

Proof of Proposition 2.15 We first note that H induces an isomorphism of functors $f^* \Rightarrow g^*\colon \mathbf{FCov}(Z) \to \mathbf{FCov}(X)$, which by abuse of notation we call H^*, in the following way: for every finite covering $p\colon Y \to Z$ we consider the pullback $q = H^*p\colon H^*Y \to X \times [0,1]$. The natural isomorphism from Lemma 2.16 can then be rewritten as $f^*Y \xrightarrow{\simeq} g^*Y$, and it is easy to check that this indeed defines an isomorphism of functors $H^*\colon f^* \Rightarrow g^*$. The 'horizontal' composition of H^* with the identity on the fibre functor Φ_x induces an isomorphism of functors $\Phi_x \circ f^* \Rightarrow \Phi_x \circ g^*$; this isomorphism can be identified with

$$\Phi_x \circ f^* \cong \Phi_{f(x)} = \Phi_{\gamma(0)} \overset{\varphi_\gamma}{\underset{\cong}{\Rightarrow}} \Phi_{\gamma(1)} = \Phi_{g(x)} \cong \Phi_x \circ g^*,$$

where φ_γ is the isomorphism of functors from (1). This identity of isomorphisms between functors can be translated into the identity $\tau_y \circ f_* = g_*$ of maps between automorphism groups of fibre functors, i.e. the commutativity of the diagram under consideration. □

Two consequences are easily drawn:

Corollary 2.17 *Let (X,x) and (Z,z) be connected pointed topological spaces, and let $f,g\colon (X,x) \to (Z,z)$ be homotopic continuous maps in the pointed sense, that is, assume there exists a continuous map $H\colon X \times [0,1] \to Z$ with $H(\xi,0) = f(\xi)$ and $H(\xi,1) = g(\xi)$ for all $\xi \in X$, and also $H(x,t) = z$ for all $t \in [0,1]$.*

Then the group homomorphisms $f_*, g_*\colon \pi_1^{\text{ét}}(X,x) \to \pi_1^{\text{ét}}(Z,z)$ are equal.

Proof In this case the path $\gamma\colon [0,1] \to Z$ as in Proposition 2.15 is constant, hence induces the identity automorphism of $\pi_1^{\text{ét}}(Z,z)$. □

Corollary 2.18 *Let (X,x) and (Y,y) be pointed topological spaces, and let $f\colon (X,x) \to (Y,y)$ be a pointed homotopy equivalence, that is, assume there exists a pointed map $g\colon (Y,y) \to (X,x)$ such that $f \circ g$ and $g \circ f$ are homotopic in the pointed sense to the respective identities.*

Then X is connected if and only if Y is connected. Assuming this to be the case, $f^*\colon \mathbf{FCov}(Y) \to \mathbf{FCov}(X)$ is an equivalence of categories and $f_*\colon \pi_1^{\text{ét}}(X,x) \to \pi_1^{\text{ét}}(Y,y)$ is an isomorphism of topological groups.

Proof It suffices to show that connectedness is preserved under homotopy equivalence, then the remainder will follow formally from Corollary 2.17. So, let $f\colon X \to Y$ be a homotopy equivalence with quasi-inverse $g\colon Y \to X$. If Y is disconnected there exists a continuous surjection $c\colon Y \to \{0,1\}$. Then $c \circ f \circ g$ is homotopic to c; but since these two maps have discrete image, they must then be identical. Hence $c \circ f$ must be surjective, hence X is disconnected as well. □

The assumptions in Corollary 2.18 are met for the inclusion of a *deformation retract*. Recall that a subspace A of a topological space X is called a deformation retract if there exists a continuous map $H\colon X \times [0,1] \to X$ with the following properties:

(i) $H(x, 0) = x$ for all $x \in X$;
(ii) $H(x, 1) \in A$ for all $x \in X$;
(iii) $H(a, t) = a$ for all $a \in A$ and $t \in [0, 1]$.

Such a map H is then called a *defining homotopy* for the deformation retract $A \subseteq X$, and the map $r: X \to A$ sending x to $H(x, 1)$ is called a *deformation retraction*. Note that some authors do not require condition (iii), and call a deformation retraction in our sense a 'strong deformation retraction'.

The étale fundamental group as a limit of deck transformation groups. There is another way to view the étale fundamental group which will be useful; for proofs cf. [13, Exposé V.4]. A general fact about Galois categories is the existence of a fundamental pro-object representing a given fibre functor $\Phi: \mathscr{C} \to \mathbf{FSet}$; this is a cofiltered projective system (Y_α) of objects together with a functorial isomorphism

$$\Phi(T) \cong \varinjlim_{\alpha} \mathrm{Hom}_{\mathscr{C}}(Y_\alpha, T)$$

for objects T of $\mathscr{C}$. By passing to a cofinal subsystem we may assume that all Y_α are Galois objects, i.e. $\mathrm{Aut}(Y_\alpha)$ operates simply transitively on $\Phi(Y_\alpha)$. We then obtain identifications $\mathrm{Aut}(Y_\alpha) \cong \mathrm{Im}(\pi \to \mathrm{Aut}\ \Phi(Y_\alpha))$, and by passage to the limit

$$\varprojlim_{\alpha} \mathrm{Aut}(Y_\alpha) \cong \pi.$$

For $\mathscr{C} = \mathbf{FCov}(X)$ and $\Phi = \Phi_x$ a fundamental pro-object is a cofiltered projective system of connected finite coverings of X such that every connected finite covering is dominated by one of them. It serves as a replacement for a universal covering space of X, which may not even exist as a topological space. The Galois objects in $\mathbf{FCov}(X)$ are precisely the normal finite connected coverings, and so we obtain:

Proposition 2.19 *Let X be a connected topological space and $x \in X$. Then there exists a cofiltered projective system $(p_\alpha: Y_\alpha \to X)$ of finite connected normal coverings of X such that every finite connected covering of X is dominated by some Y_α, together with an isomorphism of functors*

$$\Phi_x \cong \varinjlim_{\alpha} \mathrm{Hom}_X(Y_\alpha, -).$$

For such a system there is a canonical isomorphism of profinite groups

$$\pi_1^{\text{ét}}(X, x) \cong \varprojlim_{\alpha} \mathrm{Aut}_X(Y_\alpha).$$

Given $x \in X$ there is a simple natural construction of this fundamental pro-object. We define a *pointed finite covering space* of (X, x) to be a continuous map of pointed spaces $p: (Y, y) \to (X, x)$ such that $p: Y \to X$ is an object of $\mathbf{FCov}(X)$; to put it another way, this is an object Y of $\mathbf{FCov}(X)$ together with an element of $\Phi_x(Y)$.

A morphism of pointed finite covering spaces, say from $p_1\colon (Y_1, y_1) \to (X, x)$ to $p_2\colon (Y_2, y_2) \to (X, x)$ is a continuous map of pointed spaces $(Y_1, y_1) \to (Y_2, y_2)$ that makes the obvious diagram commute. Then for each two given pointed finite covering spaces there is *at most one* morphism from the first to the second. In particular, if two pointed finite covering spaces are isomorphic, the isomorphism is unique.

It is easily seen that the isomorphism classes of pointed finite covering spaces of (X, x) form a set $P = P(X, x)$; it becomes a directed set when we define $(Y_1, y_1) \geq (Y_2, y_2)$ to mean that there exists a (necessarily unique) morphism of pointed direct covering spaces $(Y_1, y_1) \to (Y_2, y_2)$. We then define the *universal profinite covering space* of (X, x) to be the pair

$$(\tilde{X}, \tilde{x}) = \varprojlim_{(Y,y)\in P} (Y, y).$$

This is a pointed topological space coming with a continuous map $p\colon (\tilde{X}, \tilde{x}) \to (X, x)$, and (by Proposition 2.19) also with a continuous action by $\pi_1^{\text{ét}}(X, x)$ which preserves p; moreover, p is the quotient map for this action. The fibre $p^{-1}(x) \subseteq \tilde{X}$ is a principal homogeneous space for $\pi_1^{\text{ét}}(X, x)$, and the point $\tilde{x} \in p^{-1}(x)$ defines a canonical trivialisation.

We also note the following for later use:

Proposition 2.20 *Let (X, x) be a pointed connected topological space, and let $p\colon (\tilde{X}, \tilde{x}) \to (X, x)$ be its universal profinite covering space. Then $p\colon \tilde{X} \to X$ satisfies the unique homotopy lifting property.*

Proof This follows formally from Proposition 2.14 and the universal property of projective limits. □

Equivariant étale fundamental groups. Let X be a connected topological space and let Γ be a group acting on X from the left by homeomorphisms. Then we define a *Γ-equivariant finite covering space* of X as a finite covering $p\colon Y \to X$ together with a lift of the Γ-action to Y, i.e. an action of Γ by homeomorphisms on Y such that p becomes Γ-equivariant. If $p_i\colon Y_i \to X$ are Γ-equivariant finite covering spaces for $i = 1, 2$ then a morphism from Y_1 to Y_2 is a continuous Γ-equivariant map $f\colon Y_1 \to Y_2$ such that $p_2 \circ f = p_1$. We obtain a category $\mathbf{FCov}_\Gamma(X)$ of Γ-equivariant finite covering spaces of X. Essentially repeating the proof of Proposition 2.9 we see that $\mathbf{FCov}_\Gamma(X)$ is a Galois category, and for every $x \in X$ the functor $\Phi_x\colon \mathbf{FCov}_\Gamma(X) \to \mathbf{FSet}$ with $\Phi_x(p\colon Y \to X) = p^{-1}(Y)$ is a fibre functor.

Definition 2.21 Let X be a connected topological space endowed with a left action of a group Γ by homeomorphisms, and let $x \in X$. The automorphism group of the fibre functor $\Phi_x\colon \mathbf{FCov}_\Gamma(X) \to \mathbf{FSet}$ is called the *Γ-equivariant étale fundamental group* of X at x and denoted by $\pi_1^{\text{ét}}([\Gamma\backslash X], x)$.

The notation is purely symbolic at this point, though it is possible to define a stack $[\Gamma\backslash X]$ on a suitable site and extend the theory of étale fundamental groups to such stacks. For our purposes, however, the definition of $\pi_1^{\text{ét}}([\Gamma\backslash X], x)$ given above will suffice.

There is a forgetful functor $F\colon \mathbf{FCov}_\Gamma(X) \to \mathbf{FCov}(X)$ and also a functor $I\colon \Gamma$-**FSet** $\to \mathbf{FCov}_\Gamma(X)$ which is some sort of induction: it sends a finite Γ-set S to the topologically trivial covering $X \times S \to X$ with the diagonal Γ-action on $X \times S$ (note that as soon as the Γ-action on S is nontrivial this is nontrivial as an object of $\mathbf{FCov}_\Gamma(X)$). These two exact functors induce homomorphisms of fundamental groups $\pi_1^{\text{ét}}(X, x) \to \pi_1^{\text{ét}}([\Gamma\backslash X], x) \to \hat{\Gamma}$, where $\hat{\Gamma}$ is the profinite completion of Γ, which is canonically isomorphic to the fundamental group of Γ-**FSet** at the forgetful fibre functor Γ-**FSet** $\to$ **FSet**.

Proposition 2.22 *Let X be a connected topological space endowed with a left action of an abstract group Γ by homeomorphisms, and let $x \in X$. Then the sequence*

$$\pi_1^{\text{ét}}(X, x) \xrightarrow{F^*} \pi_1^{\text{ét}}([\Gamma\backslash X], x) \xrightarrow{I^*} \hat{\Gamma} \to 1 \tag{3}$$

is exact.

Proof The functor $I\colon \Gamma$-**FSet** $\to \mathbf{FCov}_\Gamma(X)$ is fully faithful, hence it induces a surjection on fundamental groups by [13, Exposé V, Proposition 6.9]. The functor $F \circ I$ sends every object of Γ-**FSet** to a completely decomposed object in $\mathbf{FCov}(X)$; by [13, Exposé V, Corollary 6.5] this implies that $I^* \circ F^* = (F \circ I)^* = 1$, or $\operatorname{im} F^* \subseteq \ker I^*$. For the reverse inclusion $\ker I^* \subseteq \operatorname{im} F^*$ we apply the criterion given in [13, Exposé V, Proposition 6.11].[7] Using said criterion we can reduce this inclusion to the following claim: if $Y \to X$ is a connected object in $\mathbf{FCov}_\Gamma(X)$ whose image under F admits a section (i.e. such that there is a continuous but not necessarily Γ-invariant section of $Y \to X$) then Y is in the essential image of I. Indeed, the tautological map $Y \to X \times \pi_0(Y)$ is then an isomorphism in $\mathbf{FCov}_\Gamma(X)$. □

Remark 2.23
(i) The homomorphism $\pi_1^{\text{ét}}(X, x) \to \pi_1^{\text{ét}}([\Gamma\backslash X], x)$ need not be injective. As a counterexample we may take $X = \mathbb{S}^1$ and $\Gamma = \mu_\infty$ acting by translations. Then for any $n > 1$ the homeomorphism of $\mathbb{S}^1$ given by a primitive n-th root of unity does not lift to a homeomorphism of the same order along the degree n covering $\mathbb{S}^1 \to \mathbb{S}^1$, which shows that a finite covering of $\mathbb{S}^1$ admitting a lift of the μ_∞-action must already be trivial. Hence the map $\hat{\mathbb{Z}} \cong \pi_1^{\text{ét}}(\mathbb{S}^1) \to \pi_1^{\text{ét}}([\mu_\infty\backslash\mathbb{S}^1])$ is trivial (therefore not injective), and consequently $\pi_1^{\text{ét}}([\mu_\infty\backslash\mathbb{S}^1]) \cong \hat{\mu}_\infty = 1$.
(ii) We will need to apply these constructions in the case where Γ operates from the right on X. This can be translated to an action from the left by setting $\gamma x = x\gamma^{-1}$ and $\pi_1^{\text{ét}}([X/\Gamma], x) = \pi_1^{\text{ét}}([\Gamma\backslash X], x)$. This way we still obtain an exact sequence

$$\pi_1^{\text{ét}}(X, x) \to \pi_1^{\text{ét}}([X/\Gamma], x) \to \hat{\Gamma} \to 1,$$

[7] Note, however, the misprint there: the two inclusions $\ker u \subset \operatorname{im} u'$ and $\ker u \supset \operatorname{im} u'$ must be exchanged.

but we have to bear in mind that the construction of the second map involves the inversion map $\gamma \mapsto \gamma^{-1}$.

2.3 Comparison Between Classical and Étale Fundamental Groups

Let X be a connected topological space and $x \in X$. Then there is a canonical homomorphism

$$\alpha\colon \pi_1^{\text{path}}(X, x) \to \pi_1^{\text{ét}}(X, x) = \operatorname{Aut}(\Phi_x) \tag{4}$$

constructed as follows. For $[\gamma] \in \pi_1^{\text{path}}(X, x)$ represented by a loop $\gamma\colon [0, 1] \to X$ and a finite covering $p\colon Y \to X$ we let $\tau([\gamma])$ operate on $\Phi_x(Y) = p^{-1}(x)$ by sending $y \in p^{-1}(x)$ to the end point $\tilde{\gamma}(1)$ of the unique continuous lift $\tilde{\gamma}\colon [0, 1] \to Y$ of γ with starting point $\tilde{\gamma}(0) = y$.

Proposition 2.24 *Let (X, x) be a connected pointed topological space. Then α as in (4) is continuous if $\pi_1^{\text{ét}}(X, x)$ is endowed with its profinite topology, and $\pi_1^{\text{path}}(X, x)$ is endowed with either of the loop, τ- and σ-topologies. It also extends uniquely to a continuous group homomorphism $\pi_1^{\text{Gal}}(X, x) \to \pi_1^{\text{ét}}(X, x)$.*

Proof We first show that α is continuous for the loop topology on $\pi_1^{\text{path}}(X, x)$. Since the open subgroups of $\pi_1^{\text{ét}}(X, x)$ form a basis of open neighbourhoods of the identity and since the loop topology turns $\pi_1^{\text{path}}(X, x)$ into a quasi-topological group, it suffices to show that the preimage of any open subgroup of $\pi_1^{\text{ét}}(X, x)$ under α is open in $\pi_1^{\text{path}}(X, x)$.

For such an open subgroup there is a pointed connected finite covering $p\colon (Y, y) \to (X, x)$ such that the subgroup is the image of $p_*\colon \pi_1^{\text{ét}}(Y, y) \to \pi_1^{\text{ét}}(X, x)$. There is a commutative diagram of continuous maps

$$\begin{array}{ccccc}
\Omega(Y, y) & \longrightarrow & \pi_1^{\text{path}}(Y, y) & \xrightarrow{\ \alpha\ } & \pi_1^{\text{ét}}(Y, y) \\
\Big\downarrow{\scriptstyle \Omega(p)} & & \Big\downarrow{\scriptstyle p_*} & & \Big\downarrow{\scriptstyle p_*} \\
\Omega(X, x) & \longrightarrow & \pi_1^{\text{path}}(X, x) & \xrightarrow[\ \alpha\ ]{} & \pi_1^{\text{ét}}(X, x).
\end{array}$$

By the unique homotopy lifting property for p the map $\Omega(p)$ defines a homeomorphism from $\Omega(Y, y)$ to an open and closed subset of $\Omega(X, x)$; in particular it is open as a map $\Omega(Y, y) \to \Omega(X, x)$. Hence $p_*\colon \pi_1^{\text{path}}(Y, y) \to \pi_1^{\text{path}}(X, x)$ is also open, and its image is equal to the preimage in $\pi_1^{\text{path}}(X, x)$ of $p_*(\pi_1^{\text{ét}}(Y, y)) \subseteq \pi_1^{\text{ét}}(X, x)$. Hence this preimage is open. Therefore α is continuous for the loop topology.

One way to construct the τ-topology from the loop topology is explained in [12, Sect. 7]: the forgetful functor from topological groups to quasi-topological groups has a left adjoint τ which preserves the underlying groups. Hence for any topological group G and any quasi-topological group π a group homomorphism $\pi \to G$ is continuous if and only if $\tau(\pi) \to G$ is continuous. Now τ applied to $\pi_1^{\mathrm{path}}(X, x)$ with the loop topology yields $\pi_1^{\mathrm{path}}(X, x)$ with the τ-topology, and therefore α remains continuous when $\pi_1^{\mathrm{path}}(X, x)$ is endowed with the τ-topology.

For the σ-topology we use again the fact that the open subgroups of $\pi_1^{\text{ét}}(X, x)$ form a basis of open neighbourhoods of the identity. Since the preimage of each of these under α is an open subgroup for the τ-topology, it is also an open subgroup for the σ-topology. Hence α is also continuous for the σ-topology.

Finally, because $\pi_1^{\text{ét}}(X, x)$ is complete, α extends to a continuous homomorphism $\pi_1^{\mathrm{Gal}}(X, x) \to \pi_1^{\text{ét}}(X, x)$. □

Proposition 2.25 *Let (X, x) be a pointed connected topological space, and let $p\colon (\tilde{X}, \tilde{x}) \to (X, x)$ be its universal profinite covering space. Then the sequence of groups*

$$1 \to \pi_1^{\mathrm{path}}(\tilde{X}, \tilde{x}) \xrightarrow{p_*} \pi_1^{\mathrm{path}}(X, x) \xrightarrow{\alpha} \pi_1^{\text{ét}}(X, x) \tag{5}$$

is exact.

Proof We first show that p_* is injective. Let $\tilde{\gamma}\colon [0, 1] \to \tilde{X}$ be a loop based at $\tilde{x}$ such that $\gamma = p \circ \tilde{\gamma}$ is nullhomotopic, say by a homotopy $H\colon [0, 1] \times [0, 1] \to X$ with $H(0, t) = H(1, t) = x$ and $H(t, 0) = \gamma(t)$ for all $t \in [0, 1]$. Then by Proposition 2.20 H lifts to a homotopy of paths $\tilde{H}\colon [0, 1] \times [0, 1] \to \tilde{X}$ with $\tilde{H}(0, t) = \tilde{x}$ and $\tilde{H}(t, 0) = \tilde{\gamma}(t)$ for all $t \in [0, 1]$. By construction, $\tilde{H}(1, t) \in p^{-1}(x)$ for all $t \in [0, 1]$, and $\tilde{H}(1, 0) = \tilde{x}$. Since $p^{-1}(x)$ is totally disconnected, $\tilde{H}(1, t)$ must be equal to $\tilde{x}$ for all $t \in [0, 1]$. Hence $\tilde{H}$ really defines a homotopy of loops, and not just of paths, from $\tilde{\gamma}$ to the constant loop. Therefore the class of $\tilde{\gamma}$ in $\pi_1^{\mathrm{path}}(\tilde{X}, \tilde{x})$ is trivial. This shows the injectivity of p_*.

For exactness at $\pi_1^{\mathrm{path}}(X, x)$, let $\gamma\colon [0, 1] \to X$ be a loop based at x. Then by Proposition 2.20 there exists a unique lift $\tilde{\gamma}\colon [0, 1] \to \tilde{X}$ which is continuous and satisfies $\tilde{\gamma}(0) = \tilde{x}$. The end point $\tilde{x}' = \tilde{\gamma}(1)$ is another element of the fibre $p^{-1}(x)$, not necessarily equal to $\tilde{x}$. Recall that $p^{-1}(x)$ is a principal homogeneous space for $\pi_1^{\text{ét}}(X, x)$, hence there is a unique element of $\pi_1^{\text{ét}}(X, x)$ that sends $\tilde{x}$ to $\tilde{x}'$. Unravelling of definitions shows that this element is equal to $\alpha([\gamma])$. Hence we find that the following conditions are equivalent:

(i) $[\gamma] \in p_*(\pi_1^{\mathrm{path}}(\tilde{X}, \tilde{x}))$;
(ii) $\tilde{\gamma}(1) = \tilde{\gamma}(0)$;
(iii) $\tilde{x} = \tilde{x}'$;
(iv) $\alpha([\gamma]) = 1$.

The equivalence of (i) and (iv) then shows exactness at $\pi_1^{\mathrm{path}}(X, x)$. □

We can also characterise the image of α. Since $\pi_1^{\text{ét}}(X, x)$ acts continuously on $\tilde{X}$ it permutes the path-components of that space.

Proposition 2.26 *Let (X, x) be a pointed connected space and let $p\colon (\tilde{X}, \tilde{x}) \to (X, x)$ be its universal profinite covering space. Let $\tilde{X}^\circ$ be the path-component of $\tilde{X}$ containing $\tilde{x}$.*

Then the image of $\alpha\colon \pi_1^{\text{path}}(X, x) \to \pi_1^{\text{ét}}(X, x)$ is the stabiliser of $\tilde{X}^\circ$ in $\pi_1^{\text{ét}}(X, x)$.

Proof Let $[\gamma] \in \pi_1^{\text{path}}(X, x)$ be represented by a loop $\gamma\colon [0, 1] \to X$ based at x, and let $\tilde{x}' \in p^{-1}(x) \subseteq \tilde{X}$. Let $\tilde{\gamma}\colon [0, 1] \to \tilde{X}$ be the unique continuous lift of γ with $\tilde{\gamma}(0) = \tilde{x}'$. Then $\alpha([\gamma])(\tilde{x}') = \tilde{\gamma}(1)$, hence $\tilde{x}'$ and its image under $\alpha([\gamma])$ lie in the same path component of $\tilde{X}$. Since $\tilde{x}' \in p^{-1}(x)$ was arbitrary this shows that $\alpha([\gamma])$ preserves all path components of $\tilde{X}$ meeting $p^{-1}(x)$, in particular $\tilde{X}^\circ$.

For the other inclusion let $\beta \in \pi_1^{\text{ét}}(X, x)$ be an element preserving $\tilde{X}^\circ$. Then $\beta(\tilde{x}) \in \tilde{X}^\circ$, hence there is a path $\tilde{\gamma}$ in $\tilde{X}$ from $\tilde{x}$ to $\beta(\tilde{x})$. Then $\gamma = p \circ \tilde{\gamma}$ is a closed loop in X based at x and therefore represents an element of $\pi_1^{\text{path}}(X, x)$. By construction, both $\alpha([\gamma]) \in \pi_1^{\text{ét}}(X, x)$ and β send $\tilde{x}$ to $\beta(\tilde{x})$, but since $\pi_1^{\text{ét}}(X, x)$ acts freely on $\tilde{X}$, they must be equal. Hence $\beta = \alpha([\gamma])$ is in the image of α. □

Hence we can rewrite (5) in a more precise way: the sequence

$$1 \to \pi_1^{\text{path}}(\tilde{X}, \tilde{x}) \xrightarrow{p_*} \pi_1^{\text{path}}(X, x) \xrightarrow{\alpha} \text{Stab}_{\pi_1^{\text{ét}}(X,x)}(\tilde{X}^\circ) \to 1 \tag{6}$$

is exact.

Let $\hat{\pi}_1^{\text{path}}(X, x)$ be the profinite completion of the group $\pi_1^{\text{path}}(X, x)$. By the universal property of profinite completions α induces a continuous group homomorphism

$$\hat{\alpha}\colon \hat{\pi}_1^{\text{path}}(X, x) \to \pi_1^{\text{ét}}(X, x). \tag{7}$$

Proposition 2.27 *Let X be path-connected, locally path-connected and semi-locally simply connected. Then $\hat{\alpha}$ as in (7) is an isomorphism of topological groups.*

Proof This follows from the classical theory of covering spaces: under the given circumstances, (X, x) has a universal (possibly infinite) covering space, and discrete sets with an operation by $\pi_1^{\text{path}}(X, x)$ are equivalent to covering spaces of X. Hence finite covering spaces of X are equivalent to finite sets with $\pi_1^{\text{path}}(X, x)$-action, which are in turn equivalent to finite sets with continuous $\pi_1^{\text{ét}}(X, x)$-action. □

For general path-connected spaces $\hat{\alpha}$ need not be an isomorphism; however, there is a weaker technical condition which still ensures that $\hat{\alpha}$ is surjective.

Definition 2.28 Let X be a topological space. We say that X is *stably path-connected* if X is path-connected, and for every finite covering $Y \to X$ with Y connected, Y is already path-connected.

Example 2.29 (i) If a topological space X is connected and locally path-connected then it is also stably path-connected. To see this, note that any finite covering space of X is then also locally path-connected, and a space which is connected and locally path-connected is also globally path-connected, cf. [15, Theorem 25.5].

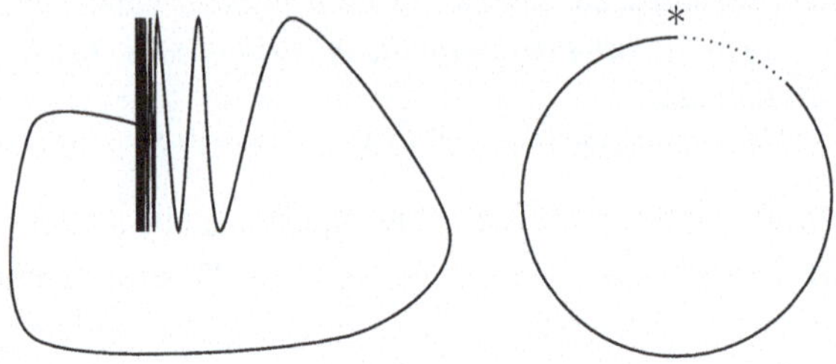

Fig. 2 The Warsaw circle (left) and the long circle (right)

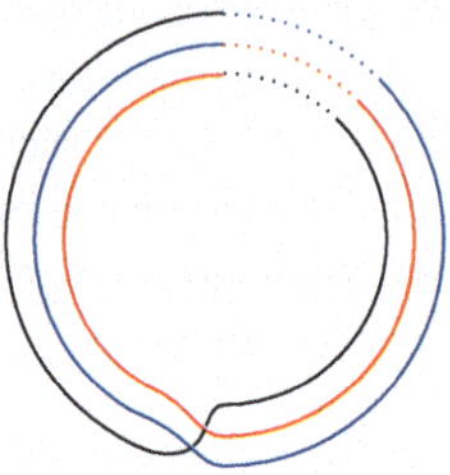

Fig. 3 A connected but not path-connected covering of the long circle

(ii) Let ω_1 be the first uncountable ordinal, and let $L = \omega_1 \times [0, 1)$ be the *long line*, equipped with the order topology (see [7, Example 45]). Then L is Hausdorff and locally homeomorphic to $\mathbb{R}$ but not paracompact. Every two points in L are contained in an open subset homeomorphic to $\mathbb{R}$, hence L is path-connected and locally path-connected. The one-point compactification $L^* = L \cup \{*\}$ is no longer path-connected.
We then define the *long circle* to be $C = L^*/(* \sim (0, 0))$, see Fig. 2. The dotted part is so long on the left that no path can enter it from the left, but every point in it can be reached by a path entering from the right, which shows that C is path-connected. C is not locally path-connected at $*$, but everywhere else.
For every $n \in \mathbb{N}$ the long circle admits a connected covering of degree n with n path components, hence it is not stably path-connected. See Fig. 3 for the case $n = 3$; the different colours encode the path components.

(iii) Similar remarks apply to the *Warsaw circle*. Consider the truncated topologist's sine curve

$$S' = \{(0, y) \mid -1 \le y \le 1\} \cup \left\{ \left(x, \sin\frac{1}{x}\right) \,\middle|\, 0 < x \le \frac{1}{\pi} \right\};$$

the Warsaw circle is defined as the quotient $W = S'/((0, 0) \sim (0, \frac{1}{\pi}))$, cf. Fig. 2. Like the long circle, W is path-connected but not stably path-connected.

Proposition 2.30 *Let X be a path-connected topological space and let $x \in X$. Then the following are equivalent.*

(i) X is stably path-connected.

(ii) For every finite Galois covering $p\colon Y \to X$ the map $\alpha_Y\colon \pi_1^{\text{path}}(X,x) \to \text{Aut}(Y/X)$ is surjective.
(iii) The map $\alpha\colon \pi_1^{\text{path}}(X,x) \to \pi_1^{\text{ét}}(X,x)$ has dense image.
(iv) The map $\hat{\alpha}\colon \hat{\pi}_1^{\text{path}}(X,x) \to \pi_1^{\text{ét}}(X,x)$ is surjective.

Proof The equivalence of (ii), (iii) and (iv) is easily seen.

We now show that (i) implies (ii). Assume that X is stably path-connected, let $\beta \in \text{Aut}(Y/X)$, and fix some $y \in p^{-1}(Y)$. By assumption Y is path-connected, hence there exists a path $\tilde{\gamma}\colon [0,1] \to Y$ from y to $\beta(y)$. Then $\gamma = p \circ \tilde{\gamma}$ is a closed loop in X, and as in the proof of Proposition 2.26 we find that $\alpha_Y([\gamma]) = \beta$. Hence (iv) holds.

Finally we assume (ii) and show that it implies (i). Since every connected finite covering of X is dominated by a finite Galois covering it suffices to show that all finite Galois covering spaces are path-connected. So, let $p\colon Y \to X$ be a finite Galois covering. First we note that every $y \in Y$ is in the same path-component as some element of the fibre $p^{-1}(x)$ (take a path from $p(y)$ to x and lift it). It then suffices to show that any two points y_1, y_2 in $p^{-1}(x)$ can be linked by a path in Y. Since $\text{Aut}(Y/X)$ operates transitively on p^{-1} and since α_Y is surjective, there is some loop γ in (X,x) such that $\alpha([\gamma])$ sends y_1 to y_2. There is then a lift of γ to a path in Y starting at y_1, and by our choice of γ this path must end at y_2. This shows (iv). □

Remark 2.31 By Proposition 2.30 the image of α cannot be dense for our examples of path-connected spaces which are not stably path-connected. Indeed, the long circle C from Example 2.29(ii) is path-connected, and C^{lpc} is homeomorphic to the long line, hence $\pi_1^{\text{path}}(C) \cong \pi_1^{\text{path}}(C^{\text{lpc}})$ is trivial. However, from the finite connected coverings of C mentioned in Example 2.29(ii) we see that $\pi_1^{\text{ét}}(C) \cong \hat{\mathbb{Z}}$. Similarly, the Warsaw circle W from Example 2.29(iii) has trivial classical fundamental group, and there exists a surjection $\pi_1^{\text{ét}}(W) \to \hat{\mathbb{Z}}$.

2.4 *Etale Fundamental Groups of Schemes*

We assume the classical theory of étale fundamental groups for schemes, as exposed in [13], to be known to the reader. Briefly, for a connected scheme $\mathscr{X}$ the category $\mathbf{FEt}(\mathscr{X})$ of étale coverings of $\mathscr{X}$ (i.e. schemes $\mathscr{Y}$ together with a finite étale morphism $\mathscr{Y} \to \mathscr{X}$) is a Galois category, and for every geometric point $\overline{x}\colon \text{Spec}\, \Omega \to \mathscr{X}$ the functor

$$\Phi_{\overline{x}}\colon \mathbf{FEt}(\mathscr{X}) \to \mathbf{FSet}$$

is a fibre functor, and the corresponding fundamental group $\text{Aut}(\Phi_{\overline{x}})$ is called the *étale fundamental group* of $\mathscr{X}$ at $\overline{x}$ and denoted by $\pi_1^{\text{ét}}(\mathscr{X},\overline{x})$.

If $\mathscr{X}$ is a connected scheme of finite type over $\mathbb{C}$, there is a canonical topology called the *complex topology* on $\mathscr{X}(\mathbb{C})$, turning it into a connected topological space. For an étale covering $\mathscr{Y} \to \mathscr{X}$ the map $\mathscr{Y}(\mathbb{C}) \to \mathscr{X}(\mathbb{C})$ is then a finite covering.

Hence we obtain a functor $\mathbf{FEt}(\mathscr{X}) \to \mathbf{FCov}(\mathscr{X}(\mathbb{C}))$ which is an equivalence of categories by *Riemann's Existence Theorem*, cf. [13, Exposé XII, Théorème 5.1]. In particular we obtain an isomorphism of profinite groups

$$\pi_1^{\text{ét}}(\mathscr{X}, \overline{x}) \cong \pi_1^{\text{ét}}(\mathscr{X}(\mathbb{C}), \overline{x}).$$

See Sect. 5.2 for a partial extension of these observations to schemes of infinite type over $\mathbb{C}$.

For a field F, the étale covers of Spec F are of the form Spec E where E is an étale F-algebra, i.e. a finite product of finite separable field extensions of F. Consequently, a universal profinite covering is given by Spec $\overline{F} \to$ Spec F where $\overline{F}$ is a separable closure of F. Note that the morphism $\overline{x}\colon$ Spec $\overline{F} \to$ Spec F is also a geometric point of Spec F, and we obtain an isomorphism of profinite groups

$$\pi_1^{\text{ét}}(\operatorname{Spec} F, \overline{x}) \cong \operatorname{Gal}(\overline{F}/F),$$

where $\operatorname{Gal}(\overline{F}/F)$ is endowed with the Krull topology.

We note a technical result on étale fundamental groups of schemes which is analogous to one for compact Hausdorff spaces mentioned before.

Lemma 2.32 *Let $\mathscr{X}$ be a connected quasi-compact quasi-separated scheme, and let $(p_\alpha\colon \mathscr{Y}_\alpha \to \mathscr{X})_\alpha$ be a cofiltered projective system of connected finite étale coverings such that $\tilde{\mathscr{Y}} = \varprojlim_\alpha \mathscr{Y}_\alpha$ is simply connected (in the sense that $\pi_1^{\text{ét}}(\tilde{\mathscr{Y}})$ is trivial).*

Then every connected finite étale covering space of $\mathscr{X}$ is dominated by some $\mathscr{Y}_\alpha$.

Proof The proof of Lemma 2.32 is strictly parallel to that of Lemma 2.13. □

3 Topological Invariants of Pontryagin Duals

We begin by briefly summarising the basic results about Pontryagin duals; for a systematic introduction see [16].

For a commutative, locally compact topological group M let $M^\vee$ be its *Pontryagin dual*, i.e. the set of continuous group homomorphisms $M \to \mathbb{S}^1$, endowed with the compact open topology. By Pontryagin duality, this is again a commutative, locally compact topological group, and the tautological map $M \to (M^\vee)^\vee$ is an isomorphism.

The Pontryagin dual $M^\vee$ is compact if and only if M is discrete. Moreover, $M^\vee$ is connected if and only if M is torsion-free, and $M^\vee$ is totally disconnected if and only if M is a torsion group.

3.1 Pontryagin Duals of Discrete Abelian Groups

For discrete abelian M there is a short exact sequence

$$0 \to M_{\text{tors}} \to M \to M_{\text{tf}} \to 0,$$

where M_{tors} is the torsion subgroup of M, and M_{tf} is the maximal torsion-free quotient of M. By duality we obtain a short exact sequence of compact topological groups

$$0 \to M_{\text{tf}}^{\vee} \to M^{\vee} \to M_{\text{tors}}^{\vee} \to 0$$

with $M_{\text{tf}}^{\vee}$ connected and $M_{\text{tors}}^{\vee}$ totally disconnected. In particular, the connected components of $M^{\vee}$ are precisely the translates of the subgroup $M_{\text{tf}}^{\vee}$. Note that M_{tf} can be written as a filtered inductive limit over free abelian groups of finite rank, hence $M_{\text{tf}}^{\vee}$ can be written as a cofiltered projective limit over finite-dimensional tori. We then also see that the canonical homomorphism $\pi_0(M^{\vee}) \to M_{\text{tors}}^{\vee}$ is an isomorphism of topological groups, in particular $\pi_0(M^{\vee})$ is compact and totally disconnected, therefore profinite.

The determination of the path components of $M^{\vee}$ is a bit more involved. Consider the short exact sequence

$$0 \to \mathbb{Z} \to \mathbb{R} \to \mathbb{S}^1 \to 0,$$

where $\mathbb{R} \to \mathbb{S}^1$ is the map $t \mapsto e^{2\pi i t}$, and apply the left-exact functor $\mathrm{Hom}(M, -)$ to it. This yields an exact sequence

$$\begin{aligned} 0 \to \mathrm{Hom}(M, \mathbb{Z}) \to \mathrm{Hom}(M, \mathbb{R}) \to \mathrm{Hom}(M, \mathbb{S}^1) \\ \xrightarrow{\delta} \mathrm{Ext}(M, \mathbb{Z}) \to \mathrm{Ext}(M, \mathbb{R}) \to \ldots ; \end{aligned}$$

here $\mathrm{Hom}(M, \mathbb{S}^1) = M^{\vee}$ (because M is discrete), and since $\mathbb{R}$ is a $\mathbb{Q}$-vector space, it is injective as an abelian group, hence $\mathrm{Ext}(M, \mathbb{R}) = 0$. That is, the interesting part of our sequence can be rewritten as

$$\mathrm{Hom}(M, \mathbb{R}) \to M^{\vee} \xrightarrow{\delta} \mathrm{Ext}(M, \mathbb{Z}) \to 0. \tag{8}$$

Recall that the connecting homomorphism δ can be described in a more explicit way: if $\chi \in M^{\vee} = \mathrm{Hom}(M, \mathbb{S}^1)$ then $\delta(\chi) \in \mathrm{Ext}(M, \mathbb{Z})$ is the class of the extension

$$0 \to \mathbb{Z} \to E_{\chi} \to M \to 0,$$

where

$$E_{\chi} = \mathbb{R} \times_{\mathbb{S}^1, \chi} M = \{(r, m) \in \mathbb{R} \times M \mid \exp r = \chi(m)\}.$$

Proposition 3.1 *Let M be a discrete abelian group. The path component of $M^\vee$ containing the trivial element is precisely the image of* $\mathrm{Hom}(M, \mathbb{R})$. *In particular, the connecting homomorphism δ in (8) induces a group isomorphism*

$$\pi_0^{\text{path}}(M^\vee) \xrightarrow{\cong} \mathrm{Ext}(M, \mathbb{Z}).$$

Proof Endow $\mathrm{Hom}(M, \mathbb{R})$ with the compact-open topology. Then it is clearly path-connected (if f is an element of this space, then so is tf for every $t \in [0, 1]$, and the assignment $t \mapsto tf$ is continuous). Furthermore, the map $\mathrm{Hom}(M, \mathbb{R}) \to M^\vee$ is continuous, therefore its image must be contained in the trivial path component of $M^\vee$.

For the other inclusion, consider a path beginning at the trivial element of $M^\vee$. Such a path is given by a family $(\chi_t)_{t\in[0,1]}$ of group homomorphisms $\chi_t : M \to \mathbb{S}^1$ such that for every $m \in M$, $\chi_0(m) = 1$ and the map $\gamma_m : [0, 1] \to \mathbb{S}^1, t \mapsto \chi_t(m)$, is continuous. We need to show that then χ_t is in the image of $\mathrm{Hom}(M, \mathbb{R})$, for every $t \in [0, 1]$.

Since $\mathbb{R}$ is the universal covering space of $\mathbb{S}^1$, there is a unique continuous lift $\tilde{\gamma}_m : [0, 1] \to \mathbb{R}$ with $\tilde{\gamma}_m(0) = 0$. Then for each $t \in [0, 1]$ we define a map

$$\tilde{\chi}_t : M \to \mathbb{R}, \quad m \mapsto \tilde{\gamma}_m(t).$$

By construction, the composition

$$M \xrightarrow{\tilde{\chi}_t} \mathbb{R} \xrightarrow{\exp} \mathbb{S}^1$$

is equal to χ_t; it remains to be shown that $\tilde{\chi}_t$ is indeed a group homomorphism.

For each $m, n \in M$ consider the map

$$f_{m,n} : [0, 1] \to \mathbb{R}, \quad t \mapsto \tilde{\chi}_t(m) + \tilde{\chi}_t(n) - \tilde{\chi}_t(m + n) = \tilde{\gamma}_m(t) + \tilde{\gamma}_n(t) - \tilde{\gamma}_{m+n}(t).$$

This is a continuous map, as can be seen from the second expression. Also it has image in $\mathbb{Z}$ because the $\chi_t = \exp \tilde{\chi}_t$ are group homomorphisms. Hence it is constant. Since $f_{m,n}(0) = 0$, it has to be zero. That this holds for every $m, n \in M$ precisely means that each $\tilde{\chi}_t$ is a group homomorphism. □

Remark 3.2 A *Whitehead group* is an abelian group M with $\mathrm{Ext}(M, \mathbb{Z}) = 0$. By Proposition 3.1 an abelian group is a Whitehead group if and only if its Pontryagin dual is path-connected. Clearly free abelian groups are Whitehead groups; their Pontryagin duals are products of circle groups. The *Whitehead problem* is the question whether the reverse implication holds, i.e. whether every Whitehead group is free abelian or, equivalently, whether every abelian path-connected compact Hausdorff topological group is a product of circles.

Stein [17] showed that every countable Whitehead group is indeed free abelian; Shelah [18, 19] showed that the statement 'every Whitehead group is free abelian' is independent of ZFC. More precisely, if Gödel's constructibility axiom $V = L$ holds, then every Whitehead group is free abelian; if $2^{\aleph_0} > \aleph_1$ and Martin's Axiom holds, then there is a Whitehead group of cardinality $\aleph_1$ which is not free abelian.

For a discrete abelian group M, the subspace topology on the path components of $M^\vee$ may be odd (cf. Example 2.4(iii)), but its minimal locally path-connected refinement admits a simple description, at least in the case relevant to us:

Proposition 3.3 *Let M be a $\mathbb{Q}$-vector space with discrete topology, and let $(M^\vee)^\circ$ be the neutral path component of its Pontryagin dual. Then there is a natural isomorphism of topological groups*

$$\mathrm{Hom}(M, \mathbb{R}) \cong ((M^\vee)^\circ)^{\mathrm{lpc}},$$

where $\mathrm{Hom}(M, \mathbb{R})$ *is endowed with the compact-open topology.*

Proof From the short exact sequence (8) we obtain a map $\mathrm{Hom}(M, \mathbb{R}) \to ((M^\vee)^\circ)^{\mathrm{lpc}}$. This map is a continuous bijection and an isomorphism of abstract groups. Arguing as in Example 2.4(iii) we find that it is in fact a homeomorphism. □

Corollary 3.4 *Let M be a $\mathbb{Q}$-vector space with discrete topology. Then $\pi_1^{\mathrm{path}}(M^\vee, 1)$ is the trivial group.*

Proof This follows from Corollary 2.5 and Proposition 3.3.

We will now determine the étale fundamental groups of Pontryagin duals.

Proposition 3.5 *Let M be a torsion-free discrete abelian group. Then the connected étale coverings of $M^\vee$ are precisely the $N^\vee$, where N runs through the subgroups of $V = M \otimes \mathbb{Q}$ that contain M with finite index.*

Proof Let us first assume that M is finitely generated; then $M^\vee$ is an n-dimensional torus, and the statement is well-known.

Now consider the general case. We first show that each N indeed defines a finite covering of $M^\vee$. Without loss of generality we may assume that $(N : M)$ is a prime p. Then there exists a finitely generated subgroup $N' \subseteq N$ which is not completely contained in M; the subgroup $M' = M \cap N'$ is then also finitely generated, and $(N' : M')$ must be p because it cannot be 1. Then $N = N' + M$ and thus

$$N^\vee \cong (N')^\vee \times_{(M')^\vee} M^\vee.$$

Hence the map $N^\vee \to M^\vee$ is a base change of the finite covering $(N')^\vee \to (M')^\vee$, hence itself a finite covering. Since N is also torsion-free, $N^\vee$ is connected.

For the other implication, let $Y \to M^\vee$ be a finite covering. By Proposition 2.11 this must be the base change of some finite covering space $Y_0 \to (M')^\vee$ for a finitely generated subgroup $M' \subseteq M$, and by what we have already shown, $Y_0 = (N')^\vee$ with $(N' : M')$ finite. Then $Y = (N' \oplus_{M'} M)^\vee$; since this is assumed to be connected, $N' \oplus_{M'} M$ must be torsion-free, hence embed via the obvious map into $V = M \otimes \mathbb{Q}$. □

The next corollary follows essentially formally from Proposition 3.5.

Corollary 3.6 *Let M be a torsion-free discrete abelian group. Then $\pi_1^{\text{ét}}(M^\vee, 0) \cong (M \otimes \mathbb{Q}/M)^\vee$ as topological groups.*

In particular, if M is a $\mathbb{Q}$-vector space, then $\pi_1^{\text{ét}}(M^\vee)$ is trivial. □

Applying this to $M = \mathbb{Q}$ we find that the solenoid $\mathcal{S} \cong \mathbb{Q}^\vee$ has trivial étale fundamental group.

Cohomology. For a topological space X and an abelian group A there are several different ways to define cohomology groups $\mathrm{H}^p(X, A)$, and they do not always agree.

(i) First there is the singular cohomology group $\mathrm{H}^p_{\text{sing}}(X, A)$, which is the cohomology of the singular cochain complex $\mathrm{C}^\bullet(X, A)$ with

$$\mathrm{C}^p(X, A) = \mathrm{Hom}(\mathrm{C}_p(X), A).$$

(ii) There is another construction using sheaf cohomology: the category $\mathrm{Sh}(X)$ of sheaves of abelian groups on X is an abelian category with enough injectives, and the global sections functor $\Gamma_X \colon \mathrm{Sh}(X) \to \mathbf{Ab}$ sending a sheaf $\mathscr{A}$ to $\Gamma(X, \mathscr{A})$ is left exact. Hence it admits right derived functors, and we set $\mathrm{H}^p(X, \mathscr{A}) = \mathrm{R}^p\Gamma_X(\mathscr{A})$. For each abelian group A there is a constant sheaf A_X modelled on A, and we set $\mathrm{H}^p(X, A) = \mathrm{H}^p(X, A_X)$.

(iii) Finally we can also consider Čech cohomology groups for sheaves:

$$\check{\mathrm{H}}^p(X, \mathscr{A}) = \varinjlim_{\mathfrak{U}} \check{\mathrm{H}}^p(\mathfrak{U}, \mathscr{A})$$

where the limits goes over ever finer open covers of X, and again we set $\check{\mathrm{H}}^p(X, A) = \check{\mathrm{H}}^p(X, A_X)$.

If X is a paracompact Hausdorff space, sheaf cohomology in terms of derived functors is always isomorphic to Čech cohomology by [20, Théorème 5.10.1]. However, sheaf cohomology and singular cohomology do not even agree on all compact Hausdorff spaces, see Remark 3.9 below. We prefer to work with sheaf cohomology because it behaves better with respect to projective limits of spaces:

Proposition 3.7 *Let A be an abelian group and let $(X_j)_{j \in J}$ be a cofiltered projective system of compact Hausdorff spaces and let $X = \varprojlim_j X_j$. Then the canonical map of sheaf cohomology groups*

$$\varinjlim_{j \in J} \mathrm{H}^p(X_j, A) \to \mathrm{H}^p(X, A)$$

is an isomorphism.

Proof As already remarked we may identify these groups with Čech cohomology groups; the corresponding statement for Čech cohomology groups is [21, Chap. X,

Theorem 3.1]. The proof is based on the observation that for a compact space Čech cohomology may be computed using only finite covers. □

Proposition 3.8 *Let V be a $\mathbb{Q}$-vector space.*

(i) If A is an abelian torsion group then $\mathrm{H}^0(V^\vee, A) = A$ and $\mathrm{H}^p(V^\vee, A) = 0$ for all $p > 0$.
(ii) For each $p \geq 0$ there is a canonical isomorphism

$$\mathrm{H}^p(V^\vee, \mathbb{Q}) \cong \textstyle\bigwedge_{\mathbb{Q}}^p V.$$

Under these isomorphisms wedge and cup products correspond to each other.
(iii) We have $\mathrm{H}^0(V^\vee, \mathbb{Z}) = \mathbb{Z}$, and for each $p > 0$ the inclusion $\mathbb{Z} \hookrightarrow \mathbb{Q}$ induces an isomorphism

$$\mathrm{H}^p(V^\vee, \mathbb{Z}) \cong \mathrm{H}^p(V^\vee, \mathbb{Q}).$$

Proof (i) First note that it suffices to prove this statement for $A = \mathbb{Z}/n\mathbb{Z}$: On compact Hausdorff spaces, sheaf cohomology commutes with direct sums and filtered limits of sheaves, and every abelian torsion group can be built from the groups $\mathbb{Z}/n\mathbb{Z}$ for $n \in \mathbb{N}$ using these constructions.
Let $\mathcal{M}$ be the set of all finitely generated free abelian subgroups of V, so that $V = \varinjlim_{M \in \mathcal{M}} M$ is a filtered limit. Then by Proposition 3.7 we can write

$$\mathrm{H}^p(V^\vee, \mathbb{Z}/n\mathbb{Z}) \cong \varinjlim_{M \in \mathcal{M}} \mathrm{H}^p(M^\vee, \mathbb{Z}/n\mathbb{Z}).$$

We need to show that the image of each $c \in \mathrm{H}^p(M^\vee, \mathbb{Z}/n\mathbb{Z})$ for $p > 0$ is trivial. Consider the subgroup $\frac{1}{n}M \in \mathcal{M}$; a short calculation using that $M^\vee$ and $(\frac{1}{n}M)^\vee$ are tori of the same dimension shows that the pullback of c to $\mathrm{H}^p((\frac{1}{n}M)^\vee, \mathbb{Z}/n\mathbb{Z})$ is zero (but only for $p > 0$).
(ii) Again we write $V = \varinjlim_{M \in \mathcal{M}} M$ and use that $\mathrm{H}^p(M^\vee, \mathbb{Q}) \cong \bigwedge_{\mathbb{Q}}^p(M \otimes \mathbb{Q})$.
(iii) This follows from (i) and (ii) using the long exact cohomology sequence induced by the short exact sequence $0 \to \mathbb{Z} \to \mathbb{Q} \to \mathbb{Q}/\mathbb{Z} \to 0$. □

Remark 3.9 Note that, by contrast, for any abelian group A the singular cohomology of $V^\vee$ with A-coefficients can be calculated as

$$\mathrm{H}^0_{\mathrm{sing}}(V^\vee, A) \cong A^{\mathrm{Ext}(V,\mathbb{Z})}, \quad \mathrm{H}^p_{\mathrm{sing}}(V^\vee, A) = 0 \text{ for all } p > 0$$

by Corollary 2.5(iii) and Proposition 3.1. So, for every $p \geq 0$ the groups $\mathrm{H}^p(V^\vee, \mathbb{Z})$ and $\mathrm{H}^p_{\mathrm{sing}}(V^\vee, \mathbb{Z})$ are not isomorphic. The same holds for rational coefficients.

3.2 Spectra of Group Algebras

We assemble some simple results on the spectra of group algebras; they will play the role in our scheme-theoretic considerations that is played by Pontryagin duals in the topological case.

Proposition 3.10 *Let M be an abelian group. Then* $\operatorname{Spec}\mathbb{C}[M]$ *is connected if and only if M is torsion-free.*

Proof Note that a $\mathbb{C}$-scheme $\mathscr{X}$ is connected if and only if every $\mathbb{C}$-morphism $\mathscr{X} \to \mathscr{S} = \operatorname{Spec}\mathbb{C} \coprod \operatorname{Spec}\mathbb{C}$ is constant.

First assume that M is torsion-free. Then M is a filtered limit $\varinjlim_{i\in I} M_i$ where the M_i are free abelian groups of finite rank. Then

$$\operatorname{Spec}\mathbb{C}[M] = \varprojlim_{i\in I} \operatorname{Spec}\mathbb{C}[M_i],$$

so each $\mathbb{C}$-morphism $\operatorname{Spec}\mathbb{C}[M] \to \mathscr{S}$ factors through some $\operatorname{Spec}\mathbb{C}[M_i]$. But $M_i \simeq \mathbb{Z}^n$ for some $n \in \mathbb{N}$, hence $\operatorname{Spec}\mathbb{C}[M_i] \simeq \mathbb{G}_m^n$ is connected.

Now assume that M contains a nontrivial finite subgroup M_0. Then $\operatorname{Spec}\mathbb{C}[M] \to \operatorname{Spec}\mathbb{C}[M_0]$ is surjective, but the target is a disjoint union of $|M_0|$ copies of $\operatorname{Spec}\mathbb{C}$, so $\operatorname{Spec}\mathbb{C}[M]$ cannot be connected. □

For any abelian group M the group algebra $\mathbb{C}[M]$ is a Hopf algebra: the comultiplication $\mathbb{C}[M] \to \mathbb{C}[M] \otimes_{\mathbb{C}} \mathbb{C}[M]$ is determined by $m \mapsto m \otimes m$ for all $m \in M$, and the other structure maps are even more obvious. This turns $\mathbb{C}[M]$ into a commutative group scheme over $\mathbb{C}$. From the short exact sequence

$$0 \to M_{\text{tors}} \to M \to M_{\text{tf}} \to 0,$$

where M_{tors} is the torsion subgroup and M_{tf} is the maximal torsion-free quotient, we obtain a short exact sequence of group schemes

$$0 \to \operatorname{Spec}\mathbb{C}[M_{\text{tf}}] \to \operatorname{Spec}\mathbb{C}[M] \to \operatorname{Spec}\mathbb{C}[M_{\text{tors}}] \to 0.$$

Here $\operatorname{Spec}\mathbb{C}[M_{\text{tf}}]$ is connected by Proposition 3.10, and it is easy to see that the topological group underlying $\operatorname{Spec}\mathbb{C}[M_{\text{tors}}]$ is isomorphic to $(M_{\text{tors}})^\vee$, in particular totally disconnected. From this we see:

Corollary 3.11 *Let M be an abelian group. Then* $\operatorname{Spec}\mathbb{C}[M] \to \operatorname{Spec}\mathbb{C}[M_{\text{tors}}]$ *induces a homeomorphism on $\pi_0(\cdot)$, and the identity component is isomorphic to* $\operatorname{Spec}\mathbb{C}[M_{\text{tf}}]$. *In particular $\pi_0(\operatorname{Spec}\mathbb{C}[M]))$ is canonically isomorphic to $(M_{\text{tors}})^\vee$ as a topological group, and the étale fundamental group of* $\operatorname{Spec}\mathbb{C}[M]$ *at any base point is isomorphic to that of* $\operatorname{Spec}\mathbb{C}[M_{\text{tf}}]$.

Proof Everything is clear from the preceding, except the statement about fundamental groups; but because $\operatorname{Spec}\mathbb{C}[M]$ is a group scheme and any connected component contains a $\mathbb{C}$-rational point, every two connected components are isomorphic as schemes. □

Proposition 3.12 *Let M be a torsion-free abelian group. Then the connected étale coverings of* $\operatorname{Spec}\mathbb{C}[M]$ *are precisely given by the* $\operatorname{Spec}\mathbb{C}[N]$, *where N runs through the subgroups of $V = M \otimes \mathbb{Q}$ that contain M with finite index.*

Proof The proof Proposition 3.12 is closely analogous to that of Proposition 3.5. We begin again by observing that the desired result is well-known in the finitely generated case.

For the general case, the argument that each N defines an étale covering $\operatorname{Spec}\mathbb{C}[N] \to \operatorname{Spec}\mathbb{C}[M]$ is directly parallel to the corresponding argument in the proof of Proposition 3.5, and we shall not repeat it.

For the other implication, let $\mathcal{Y} \to \operatorname{Spec}\mathbb{C}[M]$ be an étale covering. Then $\mathcal{Y}$ must be affine, say $\mathcal{Y} = \operatorname{Spec} B$ for some finite étale ring homomorphism $\operatorname{Spec}\mathbb{C}[M] \to B$. By [22, Tag 00U2, item (9)] this must be the base change of some étale ring homomorphism $\mathbb{C}[M'] \to B'$ for a finitely generated subgroup $M' \subseteq M$. Since B is finite over $\mathbb{C}[M]$ so must B' be over $\mathbb{C}[M']$, i.e. $\operatorname{Spec} B' \to \operatorname{Spec}\mathbb{C}[M']$ must be an étale covering, and by what we have already shown, $B' = \mathbb{C}[N']$ with $(N' : M')$ finite. Then $B = \mathbb{C}[N' \oplus_{M'} M]$; since the spectrum of this algebra must be connected, $N' \oplus_{M'} M$ must be torsion-free, hence embed via the obvious map into $V = M \otimes \mathbb{Q}$. □

Corollary 3.13 *Let M be a torsion-free abelian group. Then* $\operatorname{Spec}\mathbb{C}[M]$ *is connected, and for any geometric base point x we obtain a natural isomorphism of profinite groups*

$$\pi_1^{\text{ét}}(\operatorname{Spec}\mathbb{C}[M], x) \cong (M \otimes \mathbb{Q}/M)^\vee.$$

In particular, $\operatorname{Spec}\mathbb{C}[M]$ *is simply connected if and only if M is a $\mathbb{Q}$-vector space.* □

This corollary is again directly analogous to Corollary 3.6. Finally we note the following analogue of Proposition 3.8(i):

Proposition 3.14 *Let V be a $\mathbb{Q}$-vector space and let A be an abelian torsion group. Then the étale cohomology groups* $\mathrm{H}^p_{\text{ét}}(\operatorname{Spec}\mathbb{C}[V], A)$ *vanish for all $p > 0$.*

Proof The proof is analogous to that of Proposition 3.8(i). □

4 Galois Groups as Étale Fundamental Groups of $\mathbb{C}$-Schemes

4.1 Rational Witt Vectors

We begin by recalling rings of 'big' Witt vectors. Note that there are many different constructions of these rings; for more information see [23].

In the following, rings are always supposed commutative and unital. The ring $\mathrm{W}(A)$ of Witt vectors in A is defined for any ring A. Its underlying set is the set $1 + tA[\![t]\!]$ of formal power series in $A[\![t]\!]$ with constant coefficient one. Addition of Witt vectors is multiplication of power series: $f \oplus g = fg$. Multiplication of Witt vectors is more involved.

Proposition 4.1 *There is a unique system of binary operations $\odot$, consisting of one binary operation $\odot\colon \mathrm{W}(A) \times \mathrm{W}(A) \to \mathrm{W}(A)$ for each ring A, such that the following statements hold:*

(i) With $\oplus$ as addition and $\odot$ as multiplication, $\mathrm{W}(A)$ becomes a ring.
(ii) For any ring A and elements $a, b \in A$ the equation

$$(1 - at) \odot (1 - bt) = 1 - abt$$

holds in $\mathrm{W}(A)$.
(iii) The operation $\odot$ is functorial in A: for a ring homomorphism $\varphi\colon A \to B$ and elements $f, g \in \mathrm{W}(A)$ the equation $\mathrm{W}(\varphi)(f \odot g) = \mathrm{W}(\varphi)(f) \odot \mathrm{W}(\varphi)(g)$ holds.
Here $\mathrm{W}(\varphi)\colon \mathrm{W}(A) \to \mathrm{W}(B)$ is the obvious map that sends t to t and acts as φ on the coefficients.
(iv) The operation $\odot$ is continuous for the t-adic topology on $\mathrm{W}(A)$.

Hence $(\mathrm{W}(A), \oplus, \odot)$ becomes a complete topological ring, and W becomes a functor from rings to complete topological rings.

The proof of this result can be found in many sources, e.g. [23, Sect. 9].

Proposition 4.2 *Let A be a ring. The set*

$$\mathrm{W}_{\mathrm{rat}}(A) \stackrel{\mathrm{def}}{=} \left\{ f \in \mathrm{W(A)} \;\middle|\; f = \frac{1 + a_1 t + a_2 t^2 + \cdots + a_n t^n}{1 + b_1 t + b_2 t^2 + \cdots + b_m t^m},\ a_i, b_j \in A \right\}$$

is a subring of $\mathrm{W}(A)$.

It seems that this result first appeared explicitly in the literature as [4, Proposition 3.4]. The elements of $\mathrm{W}_{\mathrm{rat}}(A)$ are called *rational Witt vectors*. The rings $\mathrm{W}_{\mathrm{rat}}(A)$ occur naturally in some problems in K-theory, see [4, 24, 25].

Remark 4.3 In case $A = F$ is a field of characteristic zero, there is a more elementary description of $\mathrm{W}_{\mathrm{rat}}(F)$. First assume that F is algebraically closed. Then the set of all polynomials $1 - \alpha t$, where α runs through $F^\times$, is a basis of the abelian group underlying $\mathrm{W}_{\mathrm{rat}}(F)$, and the product of two basis elements corresponding to α and β, respectively, is the basis element corresponding to $\alpha\beta$. This means that $\mathrm{W}_{\mathrm{rat}}(F)$ is canonically isomorphic to the group ring $\mathbb{Z}[F^\times]$.

In the general case, choose an algebraic closure $\overline{F}/F$. Then there is a natural action of $\mathrm{Gal}(\overline{F}/F)$ on $\mathrm{W}_{\mathrm{rat}}(\overline{F})$, the ring of invariants being canonically isomorphic to $\mathrm{W}_{\mathrm{rat}}(F)$. The isomorphism $\mathrm{W}_{\mathrm{rat}}(\overline{F}) \cong \mathbb{Z}[\overline{F}^\times]$ is equivariant for this Galois action, hence $\mathrm{W}_{\mathrm{rat}}(F)$ is canonically isomorphic to the ring of $\mathrm{Gal}(\overline{F}/F)$-invariants in $\mathbb{Z}[\overline{F}^\times]$.

Now we can give the the construction of the schemes $\mathscr{X}_F$. So, let F be a field containing $\mathbb{Q}(\zeta_\infty) \stackrel{\text{def}}{=} \bigcup_n \mathbb{Q}(\zeta_n)$, where the latter is assumed embedded into $\mathbb{C}$. The group homomorphism

$$\mu_\infty \to \mathrm{W}_{\mathrm{rat}}(F)^\times, \quad \zeta \mapsto [\zeta] \stackrel{\text{def}}{=} 1 - \zeta t,$$

and the canonical inclusion

$$\mu_\infty \hookrightarrow \mathbb{C}^\times$$

define ring homomorphisms $\mathbb{Z}[\mu_\infty] \to \mathrm{W}_{\mathrm{rat}}(F)$ and $\mathbb{Z}[\mu_\infty] \to \mathbb{C}$, respectively. We then set

$$A_F = \mathrm{W}_{\mathrm{rat}}(F) \otimes_{\mathbb{Z}[\mu_\infty]} \mathbb{C} \quad \text{and} \quad \mathscr{X}_F = \operatorname{Spec} A_F.$$

Note that if $\overline{F}/F$ is an algebraic closure, $A_{\overline{F}}$ comes with an action by $\mathrm{Gal}(\overline{F}/F)$, and the ring of invariants is canonically isomorphic to A_F.

Remark 4.4 In fact we may define a $\mathbb{C}$-algebra A_E and a $\mathbb{C}$-scheme $\mathscr{X}_E = \operatorname{Spec} A_E$ for any $\mathbb{Q}(\zeta_\infty)$-algebra E by the same formula. We will only use this construction in the case where E is a finite product of fields, say $E = E_1 \times \cdots \times E_n$ with each E_ν a field extension of $\mathbb{Q}(\zeta_\infty)$; then there are natural isomorphisms

$$A_E \cong \prod_{\nu=1}^{n} A_{E_\nu} \quad \text{and} \quad \mathscr{X}_E \cong \coprod_{\nu=1}^{n} \mathscr{X}_{E_\nu}.$$

The following is the version of Theorem 1.5 for schemes.

Theorem 4.5 *For any field $F \supset \mathbb{Q}(\zeta_\infty)$, $\mathscr{X}_F$ is connected, and the étale fundamental group of $\mathscr{X}_F$ is isomorphic to the absolute Galois group of F. More precisely, the functor $E \mapsto \mathscr{X}_E$ induces a (degree-preserving) equivalence of categories between the category of finite étale F-algebras and the category of finite étale schemes over $\mathscr{X}_F$.*

A special case of this can be handled directly.

Proposition 4.6 *If F is algebraically closed, $\mathscr{X}_F$ is connected and simply connected.*

Proof We may identify $\mathscr{X}_F$ with

$$\operatorname{Spec}\big((\mathrm{W}_{\mathrm{rat}}(F) \otimes_{\mathbb{Z}} \mathbb{C}) \otimes_{\mathbb{C}[\mu_\infty]} \mathbb{C}\big) \cong \operatorname{Spec} \mathbb{C}[F^\times] \times_{\operatorname{Spec} \mathbb{C}[\mu_\infty]} \operatorname{Spec} \mathbb{C}.$$

The embedding $\mu_\infty \hookrightarrow F^\times$ induces an isomorphism on torsion subgroups, hence by Corollary 3.11 the morphism $\operatorname{Spec} \mathbb{C}[F^\times] \to \operatorname{Spec} \mathbb{C}[\mu_\infty]$ induces a homeomorphism on $\pi_0(\cdot)$. The morphism $\operatorname{Spec} \mathbb{C} \to \operatorname{Spec} \mathbb{C}[\mu_\infty]$ picks one connected component, hence the fibre product $\mathscr{X}_F$ must be connected, more precisely identified with one particular connected component of $\operatorname{Spec} \mathbb{C}[F^\times]$.

By Corollary 3.11, $\pi_1^{\text{ét}}(\mathscr{X}_F)$ is therefore isomorphic to $\pi_1^{\text{ét}}(\operatorname{Spec}\mathbb{C}[(F^\times)_{\text{tf}}])$. Since $F^\times$ is divisible, $(F^\times)_{\text{tf}}$ is a $\mathbb{Q}$-vector space, and by Corollary 3.13 the spectrum of its group algebra is simply connected. Hence so is $\mathscr{X}_F$. □

4.2 Recognising Properties of Scheme Morphisms on Geometric Points

We assemble a few well-known observations on morphisms of schemes which will be useful later.

Proposition 4.7 *Let $\mathscr{X} = \operatorname{Spec} A$ be an affine scheme and let G be a finite group operating on $\mathscr{X}$, hence on A. Assume that for any algebraically closed field k the group G operates freely on $\mathscr{X}(k)$.*

Then the natural morphism $\pi\colon \mathscr{X} \to \mathscr{Y} = \operatorname{Spec} A^G$ is a finite étale Galois covering with deck group G.

Proof We claim that the natural morphism $\mathscr{X} \times G \to \mathscr{X} \times \mathscr{X}$, $(x, g) \mapsto (x, xg)$ is a closed immersion. It is clearly finite, as the composite with each projection $\mathscr{X} \times \mathscr{X} \to \mathscr{X}$ is finite. Thus, by [26, Corollary 18.12.6], it remains to show that it is a monomorphism. Thus, let $\mathscr{S}$ be some scheme with two maps $a = (a_{\mathscr{X}}, a_G)$, $b = (b_{\mathscr{X}}, b_G)\colon \mathscr{S} \to \mathscr{X} \times G$ whose composites with $\mathscr{X} \times G \to \mathscr{X} \times \mathscr{X}$ agree. In particular, it follows that the two maps $a_{\mathscr{X}}, b_{\mathscr{X}} : \mathscr{S} \to \mathscr{X}$ agree. It remains to see that the two maps $a_G, b_G : \mathscr{S} \to G$ agree. As both maps are constant on connected components, it suffices to check this on geometric points, where it follows from the assumption.

Since the component maps $\mathscr{X} \times G \to \mathscr{X}$ are also finite étale, the morphism $\mathscr{X} \times G \to \mathscr{X} \times \mathscr{X}$ is a finite étale equivalence relation on $\mathscr{X}$. Hence its quotient $[\mathscr{X}/G]$ is an algebraic space; by [22, Tag 03BM] it must be representable by an affine scheme, and the morphism $\mathscr{X} \to [\mathscr{X}/G]$ is finite. This affine scheme represents the same functor as $\mathscr{Y}$, therefore $\mathscr{Y} = [\mathscr{X}/G]$ and $\mathscr{X} \to \mathscr{Y}$ is finite. It is also étale by [22, Tag 02WV]. □

Lemma 4.8 *Let $B \subseteq A$ be an integral ring extension, and let $f\colon \mathscr{X} = \operatorname{Spec} A \to \operatorname{Spec} B = \mathscr{Y}$ be the corresponding morphism of schemes. Then for every algebraically closed field k the map $\mathscr{X}(k) \to \mathscr{Y}(k)$ is surjective.*

Proof Let $\overline{y}\colon \operatorname{Spec} k \to \mathscr{Y}$ be a geometric point with image $y \in \mathscr{Y}$. Then $\overline{y}$ factors as $\operatorname{Spec} k \to \operatorname{Spec} \kappa(y) \hookrightarrow \mathscr{Y}$, defined by a field extension $\kappa(y) \hookrightarrow k$.

By [22, Tag 00GQ] f is surjective on topological points, hence there exists some $x \in \mathscr{X}$ with $f(x) = y$. By integrality, $\kappa(x)$ is an algebraic extension of $\kappa(y)$. Since k was assumed to be algebraically closed, we can find an embedding $\kappa(y) \hookrightarrow k$ making the diagram

$$\begin{array}{ccccc} & & \operatorname{Spec}\kappa(x) & \longrightarrow & \mathscr{X} \\ & \nearrow & \downarrow & & \downarrow \\ \operatorname{Spec} k & \longrightarrow & \operatorname{Spec}\kappa(y) & \longrightarrow & \mathscr{Y} \end{array}$$

commute. Then the composition $\overline{x}\colon \operatorname{Spec} k \to \operatorname{Spec} \kappa(x) \to \mathcal{X}$ is a preimage of $\overline{y}$ in $\mathcal{X}(k)$. □

Lemma 4.9 *Let A be a ring, and let G be a profinite group acting continuously on A. Let $A^G \subseteq A$ be the ring of invariants, and let $\varphi\colon \mathcal{X} = \operatorname{Spec} A \to \operatorname{Spec} A^G = \mathcal{Y}$ be the associated morphism of schemes. Then for every algebraically closed field k the induced map $\mathcal{X}(k)/G \to \mathcal{Y}(k)$ is bijective.*

Proof Note that the map $A^G \to A$ is integral: For any $a \in A$, the G-orbit $Ga = \{a_1, \ldots, a_n\}$ of a is finite, and then a is a root of the monic polynomial $P(X) = \prod_{i=1}^{n}(X - a_i) \in A^G[X]$. Surjectivity therefore follows from Lemma 4.8, and we only need to show injectivity.

First, we handle the case of finite G. Let p be the characteristic of k. First, we reduce to the case A is an algebra over the corresponding prime field. If $p = 0$, then $(A \otimes \mathbb{Q})^G = A^G \otimes \mathbb{Q}$ as invariants commute with filtered colimits, so we can replace A by $A \otimes \mathbb{Q}$. If $p > 0$, we first assume that A is flat over $\mathbb{Z}$, by replacing A by the free algebra $\mathbb{Z}[x_a | a \in A]$ on the elements of A, which admits a natural G-equivariant surjective map to A. Assuming now that A is flat over $\mathbb{Z}$, the map $A^G/p \to (A/p)^G$ is injective, but need not be an isomorphism. However, we claim that the map induces an isomorphism on perfections, i.e. on the filtered colimit over $a \mapsto a^p$; in particular, the k-valued points are the same. We need to show that whenever $a \in A/p$ is G-invariant, there is some n such that a^{p^n} lifts to an element of A^G. Note that there is a commutative diagram

$$\begin{array}{ccccccccc}
0 & \longrightarrow & A^G & \longrightarrow & A^G & \longrightarrow & (A/p)^G & \longrightarrow & \mathrm{H}^1(G, A) \\
 & & \downarrow & & \downarrow & & \downarrow & & \downarrow \\
0 & \longrightarrow & (A/p^n)^G & \longrightarrow & (A/p^{n+1})^G & \longrightarrow & (A/p)^G & \longrightarrow & \mathrm{H}^1(G, A/p^n)\,.
\end{array}$$

Choose n large enough that the p-part of the order of G divides p^n. Then $\mathrm{H}^1(G, A)$ is killed by p^n, and thus the map $\mathrm{H}^1(G, A) \to \mathrm{H}^1(G, A/p^n)$ is injective. Thus, if an element of $(A/p)^G$ can be lifted to $(A/p^{n+1})^G$, then it can be lifted all the way to A^G. But for any $a \in A/p$, the element a^{p^n} lifts canonically to A/p^{n+1}: Indeed, for any two lifts $\tilde{a}_1, \tilde{a}_2 \in A$ of a, one has $\tilde{a}_1^{p^n} = \tilde{a}_2^{p^n} \in A/p^{n+1}$. It follows that for any $a \in (A/p)^G$, a^{p^n} lifts to $(A/p^{n+1})^G$.

In particular, we can assume that A is defined over a field. Let $x_0, x_1 \in \mathcal{X}(k)$ be in different G-orbits. Then for every $g \in G$ the induced homomorphisms $x_\nu \circ g\colon A_k = A \otimes k \to k$ are surjective, hence their kernels are maximal ideals in A_k. By assumption, these ideals are all distinct, hence (by maximality) coprime. By the Chinese Remainder Theorem we then find some $f \in A_k$ which is sent to 0 by all $x_0 \circ g$ and to 1 by all $x_1 \circ g$. After possibly replacing f by

$$\prod_{g \in G} g(f)$$

we may assume that $f \in A_k^G = A^G \otimes k$ (as k is a free module over its prime field). Then $f(\varphi(x_0)) = 0$ and $f(\varphi(x_1)) = 1$, whence $\varphi(x_0) \neq \varphi(x_1)$.

This finishes the case that G is finite. In general, A is the filtered colimit of its subrings A^H, where $H \subset G$ runs through open subgroups. Let $\mathcal{Y}_H = \mathrm{Spec}(A^H)$, so that $\mathcal{X}(k) = \varprojlim_{H \subseteq G} \mathcal{Y}_H(k)$. By the case of finite G, we know that $\mathcal{Y}_H(k)/(G/H) = \mathcal{Y}(k)$. Therefore, if $x, y \in \mathcal{X}(k)$ map to the same element of $\mathcal{Y}(k)$, then their images in $\mathcal{Y}_H(k)$ lie in the same G/H-orbit, in particular in the same G-orbit. For each H, we get a nonempty closed subset of G of elements which carry the image of x in $\mathcal{Y}_H(k)$ to the image of y in $\mathcal{Y}_H(k)$. By the variant of Cantor's Intersection Theorem given as Lemma 4.10 below, their intersection is nonempty, which gives an element of G carrying x to y. □

Lemma 4.10 *Let X be a compact topological space and let $(A_i)_{i \in I}$ be a family of non-empty closed subspaces of X. Assume the family is cofiltered in the sense that for every $i, j \in I$ there is some $k \in I$ such that $A_k \subseteq A_i \cap A_j$.*

Then the intersection $\bigcap_{i \in I} A_i$ is non-empty.

Proof Assume that the intersection is empty. Then the union of the open subsets $U_i = X \smallsetminus A_i$ is all of X, that is, the U_i form an open cover of X. By compactness there exists some finite subcover, say $X = U_{i_1} \cup \cdots \cup U_{i_n}$. This means that $A_{i_1} \cap \cdots \cap A_{i_n} = \varnothing$. But by our assumption there exists some $k \in I$ such that $A_k \subseteq A_{i_1} \cap \cdots \cap A_{i_n} = \varnothing$, which contradicts our assumption that $A \neq \varnothing$.

4.3 Classification of Étale Covering Spaces of $\mathcal{X}_F$

Now let F be any field containing $\mathbb{Q}(\zeta_\infty) \subset \mathbb{C}$. Consider the $\mathbb{C}$-scheme $\mathcal{X}_F$ as defined before. By Remark 4.4 we obtain a contravariant functor $E \mapsto \mathcal{X}_E$ from F-algebras to $\mathcal{X}_F$-schemes, in other words, a covariant functor from affine (Spec F)-schemes to $\mathcal{X}_F$-schemes.

Theorem 4.11 *For any field $F \supseteq \mathbb{Q}(\zeta_\infty)$ the $\mathbb{C}$-scheme $\mathcal{X}_F$ is connected. If E/F is a finite étale F-algebra then $\mathcal{X}_E \to \mathcal{X}_F$ is a finite étale covering space. The resulting functor*

$$\mathbf{FEt}(\mathrm{Spec}\, F) \to \mathbf{FEt}(\mathcal{X}_F), \quad \mathrm{Spec}\, E \mapsto \mathcal{X}_E,$$

is an equivalence of categories.

The proof of this theorem rests on the generalities proved before, as well as the following observations.

Lemma 4.12 *Let $F \supseteq \mathbb{Q}(\zeta_\infty)$ be a field, let $\overline{F}/F$ be an algebraic closure and let $G = \mathrm{Gal}(\overline{F}/F)$ be the corresponding Galois group. Let k be a field. Then G operates freely on the set*

$$\mathcal{I}(\overline{F}, k) = \{\chi\colon (\overline{F})^\times \to k^\times \mid \chi \textit{ is a group homomorphism, injective on } \mu_\infty\}.$$

Proof Let $\sigma \in G \smallsetminus \{1\}$ and $\chi \in \mathscr{I}(\overline{F}, k)$. We need to show that $\sigma(\chi) \neq \chi$.

Without loss of generality we may assume that σ topologically generates G (otherwise we replace F by the fixed field of σ). Then G is a nontrivial procyclic group. We claim that there exists some prime p such that the closed subgroup generated by σ^p is a proper subgroup of G: otherwise, for every finite quotient H of G and every prime p the p-th power map $H \to H$ is surjective; since H is a finite abelian group this is only possible if H is trivial. But if every finite quotient of G is trivial, then so is G itself, contradiction.

Now pick some p such that $\overline{\langle \sigma^p \rangle}$ is properly contained in G, and let E be the fixed field of σ^p. Then E/F is a cyclic Galois extension of degree p.

Let $\zeta \in F$ be a primitive p-th root of unity. Note that since χ is injective on μ_∞ we must have $\chi(\zeta) \neq 1$. Furthermore $\mathrm{N}_{E/F}(\zeta) = \zeta^p = 1$, hence by the original form of Hilbert's 'Satz 90' ([27, Satz 90], see also [28, Chap. IV, Theorem 3.5]) there is some $\alpha \in E^\times$ with $\zeta = \sigma(\alpha)/\alpha$. But then $\chi(\sigma(\alpha)/\alpha) = \chi(\zeta) \neq 1$, hence $\chi(\sigma(\alpha)) \neq \chi(\alpha)$. Therefore $\sigma(\chi) \neq \chi$.

Lemma 4.13 *Let F, $\overline{F}$, G and k be as in Lemma 4.12. Then G operates freely on $\mathscr{X}_{\overline{F}}(k)$.*

Proof Just note that

$$\mathscr{X}_{\overline{F}}(k) = \mathrm{Hom}(\mathbb{Z}[\overline{F}^\times] \otimes_{\mathbb{Z}[\mu_\infty]} \mathbb{C}, k)$$

can be identified with the set of pairs (χ, g), where $\chi \colon \overline{F}^\times \to k^\times$ is a group homomorphism and $g \colon \mathbb{C} \to k$ is a field embedding such that χ and g agree on all roots of unity. The Galois action is given by $\sigma(\chi, g) = (\sigma(\chi), g)$. The χ occurring are all injective on μ_∞, and therefore the desired result follows from Lemma 4.12. □

Lemma 4.14 *The map $\mathscr{X}_{\overline{F}}(k) \to \mathscr{X}_F(k)$ is constant on G-orbits, and the induced map $\mathscr{X}_{\overline{F}}(k)/G \to \mathscr{X}_F(k)$ is a bijection.*

Proof Recall that $\mathscr{X}_{\overline{F}} = \mathrm{Spec}\, A_{\overline{F}}$ and $\mathscr{X}_F = \mathrm{Spec}\, A_{\overline{F}}^G$ (the ring of G-invariants), so the result follows from Lemma 4.9.

Proof of Theorem 4.11 Let E be a finite étale F-algebra. We will show that $\mathscr{X}_E \to \mathscr{X}_F$ is a finite étale covering.

First assume that E is a field, Galois over F. Combining Lemmas 4.13 and 4.14 we find that $\mathrm{Gal}(E/F)$ operates freely on $\mathscr{X}_E(k)$, for any algebraically closed field k. By Proposition 4.7 then $\mathscr{X}_E \to \mathscr{X}_F$ must be finite étale.

For a general finite field extension E/F let E'/F be the Galois closure of E in $\overline{F}$; then the composition $\mathscr{X}_{E'} \to \mathscr{X}_E \to \mathscr{X}_F$ is finite étale, as is the first component, hence the second has to be finite étale, too.

Finally, if E is an arbitrary finite étale F-algebra, there is a canonical isomorphism

$$E \xrightarrow{\cong} \prod_{\mathfrak{p} \in \mathrm{Spec}\, E} E/\mathfrak{p}$$

with the $E/\mathfrak{p}$ being finite field extensions of F. Therefore, by Remark 4.4, we obtain an isomorphism of $\mathscr{X}_F$-schemes

$$\mathscr{X}_E \overset{\cong}{\to} \coprod_{\mathfrak{p}\in \operatorname{Spec} E} \mathscr{X}_{E/\mathfrak{p}},$$

hence $\mathscr{X}_E \to \mathscr{X}_F$ is also finite étale.

We now have shown that the functor $\operatorname{Spec} E \mapsto \mathscr{X}_E$ really sends $\mathbf{FEt}(\operatorname{Spec} F)$ to $\mathbf{FEt}(\mathscr{X}_F)$. Note that if E/F is Galois, then $\mathscr{X}_E \to \mathscr{X}_F$ is a Galois covering with group $\operatorname{Gal}(E/F)$. From this we deduce that the functor is fully faithful. It remains to be shown that it is essentially surjective; this follows by applying Lemma 2.32 to the system of étale coverings $(\mathscr{X}_E \to \mathscr{X}_F)_{E/F \text{ finite}}$, using that the limit $\mathscr{X}_{\overline{F}} = \varprojlim_E \mathscr{X}_E$ is simply connected by Proposition 4.6. □

Corollary 4.15 *'The' étale fundamental group of $\mathscr{X}_F$ is naturally isomorphic to 'the' absolute Galois group of F.* □

The formulation of this corollary requires some explanation. Note that classically, to speak sensibly of the absolute Galois group of a field F one needs to fix an algebraic closure $\overline{F}$. However, we can get by with a slightly more general type of object. In Grothendieck's interpretation of the absolute Galois group, $\operatorname{Gal}(\overline{F}/F)$ is the étale fundamental group of $\operatorname{Spec} F$ at the geometric point $\operatorname{Spec} \overline{F} \to \operatorname{Spec} F$. This is the automorphism group of the fibre functor $\Psi_{\overline{F}}$ on the Galois category $\mathbf{FEt}(\operatorname{Spec} F)$, given by $\Psi_{\overline{F}}(X) = X(\overline{F})$. If Ψ is *any* fibre functor on $\mathbf{FEt}(\operatorname{Spec} F)$, then $\operatorname{Aut} \Psi$ is still isomorphic to $\operatorname{Gal}(\overline{F}/F)$, the isomorphism being canonical up to inner automorphisms. Hence we might call $\operatorname{Aut}(\Psi)$ 'the absolute Galois group of F at Ψ'.

Now, let $\overline{x}\colon \operatorname{Spec} \Omega \to \mathscr{X}_F$ be a geometric point. Then $\pi_1^{\text{ét}}(\mathscr{X}_F, \overline{x})$ is the automorphism group of the fibre functor $\Phi_{\overline{x}}$ on $\mathbf{FEt}(\mathscr{X}_F)$ with $\Phi_{\overline{x}}(Y) = Y_{\overline{x}}$. The composition

$$\Psi_{\overline{x}}\colon \mathbf{FEt}(\operatorname{Spec} F) \xrightarrow{\mathscr{X}_{(\cdot)}} \mathbf{FEt}(\mathscr{X}_F) \xrightarrow{\Phi_{\overline{x}}} \mathbf{Sets}$$

is still a fibre functor on $\mathbf{FEt}(\operatorname{Spec} F)$, and Corollary 4.15 says that the absolute Galois group of F at $\Psi_{\overline{x}}$ is canonically isomorphic to $\pi_1^{\text{ét}}(\mathscr{X}_F, \overline{x})$, for any geometric point $\overline{x}$ of $\mathscr{X}_F$.

Remark 4.16 Let A be any ring and B a finite étale A-algebra of constant degree d. Above, we have shown that if A is a field over $\mathbb{Q}(\zeta_\infty)$, then the map $\mathrm{W}_{\text{rat}}(A) \to \mathrm{W}_{\text{rat}}(B)$ becomes finite étale after base change from $\mathbb{Z}[\mu_\infty]$ to $\mathbb{C}$; by faithfully flat base change, this is already true after base change along $\mathbb{Z}[\mu_\infty] \to \mathbb{Q}(\zeta_\infty)$. It is not evident from our proof how general this result is, so we want to mention the following generalisation. We do not know whether the assumption $\operatorname{Pic}(A) = 0$ is necessary, and it would be nice to remove it.

Theorem 4.17 *Let A be a ring with* $\mathrm{Pic}(A) = 0$*, let S be a set of primes which is invertible in A, and let B be a finite étale A-algebra. For a prime $p \in S$, consider the element $\Phi_p \in \mathrm{W}_{\mathrm{rat}}(A)$ given by the cyclotomic polynomial $\prod_{i=1}^{p-1}(1 - \zeta_p^i t) \in \mathrm{W}_{\mathrm{rat}}(\mathbb{Z}) \to \mathrm{W}_{\mathrm{rat}}(A)$. Then the map*

$$\mathrm{W}_{\mathrm{rat}}(A)[(\Phi_p - (p-1))^{-1} \mid p \in S] \to \mathrm{W}_{\mathrm{rat}}(B)[(\Phi_p - (p-1))^{-1} \mid p \in S]$$

is finite étale in the following cases:

(i) *The algebra B is everywhere of degree $\leq d$ over A, and S contains the set of primes $p \leq d$.*
(ii) *The algebra B is Galois over A with Galois group G, and S contains the set of primes dividing the order of G.*

Note that if A contains a p-th root of unity $\zeta_p \in A$ and we denote $[\zeta_p] = 1 - \zeta_p t \in \mathrm{W}_{\mathrm{rat}}(A)$, then $\Phi_p = [\zeta_p] + [\zeta_p^2] + \ldots + [\zeta_p^{p-1}]$, and inverting $\Phi_p - (p-1)$ is equivalent to inverting $[\zeta_p] - 1$. We do not know to what extent it is necessary to invert $\Phi_p - (p-1)$ for all $p \in S$.

Proof One may reduce part (i) to part (ii) by passing to the Galois closure. Now $\mathrm{W}_{\mathrm{rat}}(A) = \mathrm{W}_{\mathrm{rat}}(B)^G$ and this passes to the filtered colimit, giving

$$\mathrm{W}_{\mathrm{rat}}(A)[(\Phi_p - (p-1))^{-1} \mid p \in S] = \mathrm{W}_{\mathrm{rat}}(B)[(\Phi_p - (p-1))^{-1} \mid p \in S]^G .$$

By Proposition 4.7, it is enough to check that G acts freely on geometric points of $\mathrm{Spec}(\mathrm{W}_{\mathrm{rat}}(B)[([\zeta_p] - 1)^{-1} \mid p \in S])$. Thus, for every $1 \neq g \in G$, we need to check that g acts freely; replacing g by a power, we can assume that the order of g is a prime p. We can also replace G by the cyclic subgroup generated by g (and A by the invariants of B under this subgroup), and assume that $G \cong \mathbb{Z}/p\mathbb{Z}$ is cyclic of prime order.

We make a further reduction to assume that A contains a p-th root of unity. Indeed, let $A_1 = A \otimes_{\mathbb{Z}} \mathbb{Z}[\zeta_p]$, which is a finite étale Galois cover with Galois group $G_1 = (\mathbb{Z}/p\mathbb{Z})^\times$. We claim that

$$\mathrm{W}_{\mathrm{rat}}(A)[(\Phi_p - (p-1))^{-1}] \to \mathrm{W}_{\mathrm{rat}}(A_1)[(\Phi_p - (p-1))^{-1}] = \mathrm{W}_{\mathrm{rat}}(A_1)[([\zeta_p] - 1)^{-1}]$$

is a finite étale Galois cover with Galois group G_1. To check this, as before it is enough to check that G_1 acts trivially on geometric points of $\mathrm{Spec}(\mathrm{W}_{\mathrm{rat}}(A_1)[([\zeta_p] - 1)^{-1}])$. But for any map $\mathrm{W}_{\mathrm{rat}}(A_1)[([\zeta_p] - 1)^{-1}] \to k$, the image of $[\zeta_p]$ in k will be a nontrivial p-th root of 1, so that no nontrivial element of G_1 fixes the image of $[\zeta_p]$ in k, and in particular G_1 acts freely on geometric points. A similar statement holds for $B \to B_1 \stackrel{\mathrm{def}}{=} B \otimes_A A_1$, and by faithfully flat base change it is enough to check the result for $A_1 \to B_1$.

Thus, we are reduced to the case that A is a $\mathbb{Z}[\frac{1}{p}, \zeta_p]$-algebra, and B is a finite étale $G = \mathbb{Z}/p\mathbb{Z}$-cover of A; we want to prove that the map

$$\mathrm{W}_{\mathrm{rat}}(A)[([\zeta_p] - 1)^{-1}] \to \mathrm{W}_{\mathrm{rat}}(B)[([\zeta_p] - 1)^{-1}]$$

is finite étale, for which it is enough to check that G acts freely on geometric points of $\mathrm{Spec}(\mathrm{W}_{\mathrm{rat}}(B)[([\zeta_p] - 1)^{-1}])$. In that case, by Kummer theory and our assumption $\mathrm{Pic}(A) = 0$, B is given by adjoining the p-th root $a^{1/p} \in B$ of some element $a \in A^\times$. But then $a^{1/p}$ gives an element $[a^{1/p}] \in \mathrm{W}_{\mathrm{rat}}(B)$ on which $1 \in G = \mathbb{Z}/p\mathbb{Z}$ acts by $[a^{1/p}] \mapsto [\zeta_p a^{1/p}] = [\zeta_p][a^{1/p}]$. For any geometric point $\mathrm{W}_{\mathrm{rat}}(B)[([\zeta_p] - 1)^{-1}] \to k$, the image of $[\zeta_p] - 1$ is invertible in k, so it follows that $[a^{1/p}]$ maps to 0 in k; but this is impossible, as $a^{1/p}$ and thus $[a^{1/p}]$ is a unit. □

5 Galois Groups as Étale Fundamental Groups of Topological Spaces

We now present the construction of a compact Hausdorff space X_F for every field F containing $\mathbb{Q}(\mu_\infty)$, with properties analogous to those of $\mathscr{X}_F$. In fact, the two constructions are closely related, see Theorem 5.7 below.

5.1 *The Spaces X_F*

Let F be a field containing $\mathbb{Q}(\zeta_\infty)$ and let $\overline{F}/F$ be an algebraic closure. We endow $\overline{F}^\times$ with the discrete topology and consider the Pontryagin dual $(\overline{F}^\times)^\vee = \mathrm{Hom}(\overline{F}^\times, \mathbb{S}^1)$. Letting $\iota \colon \mu_\infty \hookrightarrow \mathbb{S}^1$ be the obvious embedding, we then set

$$X_{\overline{F}} = \{\chi \in (\overline{F}^\times)^\vee \mid \chi|_{\mu_\infty} = \iota\},$$

endowed with the subspace topology from $(\overline{F}^\times)^\vee$. The absolute Galois group $G = \mathrm{Gal}(\overline{F}/F)$ operates from the left on $X_{\overline{F}}$ by homeomorphisms via $\sigma(f) = f \circ \sigma^{-1}$. Then we set

$$X_F = G \backslash X_{\overline{F}},$$

endowed with the quotient topology. Note that if $\overline{F}'/F$ is another algebraic closure of F, the version of X_F constructed from $\overline{F}$ and that constructed from $\overline{F}'$ are canonically homeomorphic.

Proposition 5.1 *Let $F \supseteq \mathbb{Q}(\zeta_\infty)$ be a field with algebraic closure $\overline{F}$ and absolute Galois group $G = \mathrm{Gal}(\overline{F}/F)$.*

(i) The G-action on $X_{\overline{F}}$ is proper and free.

(ii) The space X_F is nonempty, connected, compact and Hausdorff.
(iii) The étale fundamental group of $X_{\overline{F}}$ is trivial.

Proof We begin by showing that $X_{\overline{F}}$ is nonempty; it will follow that X_F is also nonempty. The group $\overline{F}^\times$ is divisible, and its torsion subgroup is equal to μ_∞. By general facts about divisible groups it can then be written as a direct sum $\overline{F}^\times = V \oplus \mu_\infty$ where V is a $\mathbb{Q}$-vector space. Then we can construct an element $\chi \in X_{\overline{F}} \subset \mathrm{Hom}(\overline{F}^\times, \mathbb{S}^1) = \mathrm{Hom}(V, \mathbb{S}^1) \times \mathrm{Hom}(\mu_\infty, \mathbb{S}^1)$ by declaring it to be ι on μ_∞ and any group homomorphism on V. Note, however, that V can in general not be chosen Galois invariant.

The space $X_{\overline{F}}$ is a translate of, and therefore homeomorphic to, the closed subgroup $\mathrm{Hom}(\overline{F}^\times/\mu_\infty, \mathbb{S}^1) \subset \mathrm{Hom}(\overline{F}^\times, \mathbb{S}^1)$. This is clearly a connected compact Hausdorff space, and it has trivial étale fundamental group by Corollary 3.6, which proves (iii).

For (i), note that the action being proper means that the map $G \times X_{\overline{F}} \to X_{\overline{F}} \times X_{\overline{F}}$ sending (g, x) to (gx, x) is proper; but since both G and $X_{\overline{F}}$ are compact, this follows automatically from continuity. That the action is free is a direct consequence of Lemma 4.12.

For (ii), note that we already know the corresponding statement for $X_{\overline{F}}$. From this and (i) we easily deduce (ii). □

Similarly to the scheme-theoretic case, let E be finite étale F-algebra. Recall that there is a canonical isomorphism

$$E \xrightarrow{\cong} \prod_{\mathfrak{p} \in \mathrm{Spec}\, E} E/\mathfrak{p};$$

we therefore set

$$X_E = \coprod_{\mathfrak{p} \in \mathrm{Spec}\, E} X_{E/\mathfrak{p}}.$$

This assignment extends to a functor from $\mathbf{FEt}(\mathrm{Spec}\, F)$ to the category $\mathbf{CH}/X_F$ of compact Hausdorff spaces over X_F. To give its action on morphisms, note that

$$\begin{aligned}
\mathrm{Hom}_F(\mathrm{Spec}\, E_1, \mathrm{Spec}\, E_2) &\cong \mathrm{Hom}_F\left(\coprod_{\mathfrak{p}_1 \in \mathrm{Spec}\, E_1} \mathrm{Spec}\, E_1/\mathfrak{p}_1, \coprod_{\mathfrak{p}_2 \in \mathrm{Spec}\, E_2} \mathrm{Spec}\, E_2/\mathfrak{p}_2\right) \\
&\cong \prod_{\mathfrak{p}_1} \coprod_{\mathfrak{p}_2} \mathrm{Hom}_F(\mathrm{Spec}\, E_1/\mathfrak{p}_1, \mathrm{Spec}\, E_2/\mathfrak{p}_2) \\
&\cong \prod_{\mathfrak{p}_1} \coprod_{\mathfrak{p}_2} \mathrm{Hom}_F(E_2/\mathfrak{p}_2, E_1/\mathfrak{p}_1)
\end{aligned}$$

and similarly

$$\begin{aligned}\mathrm{Hom}_{X_F}(X_{E_1}, X_{E_2}) &\cong \mathrm{Hom}_{X_F}\left(\coprod_{\mathfrak{p}_1 \in \mathrm{Spec}\, E_1} X_{E_1/\mathfrak{p}_1}, \coprod_{\mathfrak{p}_2 \in \mathrm{Spec}\, E_2} X_{E_2/\mathfrak{p}_2}\right) \\ &\cong \prod_{\mathfrak{p}_1} \coprod_{\mathfrak{p}_2} \mathrm{Hom}_{X_F}(X_{E_1/\mathfrak{p}_1}, X_{E_2/\mathfrak{p}_2}).\end{aligned}$$

By piecing together morphisms the obvious way, it therefore suffices to give a continuous map $X_E \to X_F$ for a finite field extension E/F. For this, choose an algebraic closure $\overline{E}/E$, which is then also an algebraic closure of F, and let $X_E \to X_F$ be the forgetful map

$$X_E = \mathrm{Gal}(\overline{E}/E)\backslash X_{\overline{E}} \to \mathrm{Gal}(\overline{E}/F)\backslash X_{\overline{E}} = X_F.$$

Theorem 5.2 *The functor* $\mathbf{FEt}(\mathrm{Spec}\, F) \to \mathbf{CH}/X_F$ *sending an étale covering space* $\mathrm{Spec}\, E \to \mathrm{Spec}\, F$ *to the map* $X_E \to X_F$ *in fact has image in* $\mathbf{FCov}(X_F)$ *and defines an equivalence of categories* $\mathbf{FEt}(\mathrm{Spec}\, F) \to \mathbf{FCov}(X_F)$.

Proof Let first E/F be a finite Galois extension. Then $X_E \to X_F$ is the quotient map for the action of the finite group $\mathrm{Gal}(E/F)$ on X_E, which is free by Proposition 5.1. Since the spaces involved are compact Hausdorff spaces, $X_E \to X_F$ is then a finite covering, and $\mathrm{Gal}(E/F) = \mathrm{Aut}_{\mathrm{Spec}\, F}(\mathrm{Spec}\, E) \to \mathrm{Aut}_{X_F}(X_E)$ is an isomorphism of finite groups.

By passing to a Galois closures, we find that for finite but not necessarily Galois field extensions E/F the map $X_E \to X_F$ is still a finite covering. From this, the same statement for finite étale F-algebras follows formally.

Hence the functor $\mathbf{FEt}(\mathrm{Spec}\, F) \to \mathbf{FCov}(X_F)$ is well-defined. That it is fully faithful can again be reduced to the case of automorphisms of a Galois object, which we have already seen. Finally, its essential surjectivity follows from the combination of Lemma 2.13 and 5.1 (iii). □

5.2 *The Relation Between X_F and $\mathscr{X}_F$*

Recall that for a scheme $\mathscr{X}$ of finite type over $\mathbb{C}$ there is a canonical topology on $\mathscr{X}(\mathbb{C})$, called the *complex topology*, cf. [29] and [13, Exposé XII]. Here $\mathscr{X}(\mathbb{C})$ designates the set of sections of the structural morphism $\mathscr{X} \to \mathrm{Spec}\, \mathbb{C}$ (rather than all scheme morphisms $\mathrm{Spec}\, \mathbb{C} \to \mathscr{X}$); we hope that no confusion with the usage in Sect. 4 will arise. The complex topology is characterised by the following properties:

(i) The complex topology on $\mathbb{A}^1(\mathbb{C}) \cong \mathbb{C}$ is the Euclidean topology, i.e. the metric topology on $\mathbb{C}$ induced by the metric $d(z, w) = |z - w|$.

(ii) The complex topology on $(\mathscr{X} \times_{\operatorname{Spec}\mathbb{C}} \mathscr{Y})(\mathbb{C}) \cong \mathscr{X}(\mathbb{C}) \times \mathscr{Y}(\mathbb{C})$ is the product topology defined by the complex topologies on $\mathscr{X}(\mathbb{C})$ and $\mathscr{Y}(\mathbb{C})$.
(iii) If $\mathscr{Y} \hookrightarrow \mathscr{X}$ is an open, resp. closed, embedding, then so is $\mathscr{Y}(\mathbb{C}) \hookrightarrow \mathscr{X}(\mathbb{C})$.

In particular, for a quasiprojective variety $\mathscr{X} \subseteq \mathbb{P}^n_{\mathbb{C}}$ the complex topology on $\mathscr{X}(\mathbb{C})$ is the subspace topology induced by the Euclidean topology on $\mathbb{P}^n(\mathbb{C})$.

The complex topology can easily be generalised to arbitrary $\mathbb{C}$-schemes; its description is facilitated when restricted to affine $\mathbb{C}$-schemes, which suffices for our purposes.

Let A be a $\mathbb{C}$-algebra, so that $\mathscr{X} = \operatorname{Spec} A$ becomes an affine $\mathbb{C}$-scheme. Note that $\mathscr{X}(\mathbb{C})$ can be identified with the set of $\mathbb{C}$-algebra homomorphisms $A \to \mathbb{C}$. We can interpret an element $f \in A$ as a function $f\colon \mathscr{X}(\mathbb{C}) \to \mathbb{C}$ by sending $\mathscr{X}(\mathbb{C}) \ni x\colon A \to \mathbb{C}$ to $x(f)$, i.e. by writing $f(x)$ for what formally is $x(f)$.

This way we define for every $f \in A$ a subset $U_f \subseteq \mathscr{X}(\mathbb{C})$ by

$$U_f = \{x \in \mathscr{X}(\mathbb{C}) \mid |f(x)| < 1\}.$$

Definition 5.3 Let $\mathscr{X} = \operatorname{Spec} A$ be an affine $\mathbb{C}$-scheme. The *complex topology* on $\mathscr{X}(\mathbb{C})$ is the unique topology for which $\{U_f \mid f \in A\}$ is a subbasis of open sets.

That is, a subset of $\mathscr{X}(\mathbb{C})$ is open if it can be written as a union of subsets of the form $U_{f_1,\dots,f_n} = U_{f_1} \cap \cdots \cap U_{f_n}$ (for possibly varying n).

Proposition 5.4 *(i) If $\mathscr{X}$ is an affine $\mathbb{C}$-scheme of finite type, the above definition of the complex topology on $\mathscr{X}(\mathbb{C})$ agrees with the classical one.*
(ii) If $\mathscr{X} = \varprojlim_\alpha \mathscr{X}_\alpha$ is a cofiltered limit of affine $\mathbb{C}$-schemes $\mathscr{X}_\alpha$ of finite type, then $\mathscr{X}(\mathbb{C}) = \varprojlim_\alpha \mathscr{X}_\alpha(\mathbb{C})$ as topological spaces.
(iii) The complex topology is compatible with fibre products: $(\mathscr{X} \times_{\mathscr{S}} \mathscr{Y})(\mathbb{C}) = \mathscr{X}(\mathbb{C}) \times_{\mathscr{S}(\mathbb{C})} \mathscr{Y}(\mathbb{C})$ as topological spaces.
(iv) If $\mathscr{Y} \to \mathscr{X}$ is a finite étale covering, then $\mathscr{Y}(\mathbb{C}) \to \mathscr{X}(\mathbb{C})$ is a finite covering.

Proof (i) Easy.
(ii) Note that we can write $\mathscr{X} = \operatorname{Spec} A$ and $\mathscr{X}_\alpha = \operatorname{Spec} A_\alpha$ where $A = \varinjlim_\alpha A_\alpha$. Then because the limit is filtered every finite subset of A is the image of a finite subset of some A_α. Hence every basic open set $U_{f_1,\dots,f_n} \subseteq \mathscr{X}(\mathbb{C})$ is the preimage of an open subset of some $\mathscr{X}_\alpha(\mathbb{C})$.
(iii) By (i) this holds if $\mathscr{X}$, $\mathscr{Y}$ and $\mathscr{S}$ are of finite type over $\mathbb{C}$. Apply (ii) to deduce the general case from this.
(iv) Again this is well-known if $\mathscr{X}$ (and then automatically also $\mathscr{Y}$) is of finite type over $\mathbb{C}$. It follows e.g. from [13, Exposé XII, Propositions 3.1 (iii) and 3.2 (vi)]. Every étale covering of an arbitrary $\mathscr{X}$ is the pullback from an étale covering of a $\mathbb{C}$-scheme of finite type, cf. [22, Tag 00U2, item (9)]. Apply this, (ii) and (iii) to deduce the general case. □

Remark 5.5 Note that for a nonempty affine $\mathbb{C}$-scheme $\mathscr{X}$ the space $\mathscr{X}(\mathbb{C})$ may well be empty, e.g. for $\mathscr{X} = \operatorname{Spec} \mathbb{C}(T)$.

For each field F containing $\mathbb{Q}(\zeta_\infty)$ we may consider $\mathscr{X}_F(\mathbb{C})$, and from now on we will tacitly assume it to be endowed with its complex topology.

Corollary 5.6 *Let F be a field extension of $\mathbb{Q}(\zeta_\infty)$. Then $\mathscr{X}_F(\mathbb{C})$ is connected. For a finite extension E/F the map $\mathscr{X}_E(\mathbb{C}) \to \mathscr{X}_F(\mathbb{C})$ is a covering map of degree $[E : F]$.*

Proof We only need to prove that $\mathscr{X}_F(\mathbb{C})$ is connected. This reduces to the case that F is algebraically closed, where it follows from Proposition 5.4 and Theorem 5.7(i). □

This suggests that also $\pi_1^{\text{ét}}(\mathscr{X}_F(\mathbb{C})) \cong \operatorname{Gal}(\overline{F}/F)$; the difficulty here is to show that every topological finite covering of $\mathscr{X}_F(\mathbb{C})$ comes from an étale covering of the scheme $\mathscr{X}_F$. An analogous statement in the case of $\mathbb{C}$-schemes of finite type is known under the name 'Riemann's existence theorem', cf. [13, Exposé XII, Théorème 5.1]. It is, however, not known to the authors in which generality this can hold for affine schemes of infinite type over $\mathbb{C}$.

However, we can now show in a roundabout way that indeed $\pi_1^{\text{ét}}(\mathscr{X}_F(\mathbb{C})) \cong \operatorname{Gal}(\overline{F}/F)$.

Theorem 5.7 *Let F be a field containing $\mathbb{Q}(\zeta_\infty)$, and let $\overline{F}/F$ be an algebraic closure.*

(i) There is a canonical $\operatorname{Gal}(\overline{F}/F)$*-equivariant homeomorphism*

$$\mathscr{X}_{\overline{F}}(\mathbb{C}) \xrightarrow{\cong} X_{\overline{F}} \times \operatorname{Hom}(\overline{F}^\times, \mathbb{R}),$$

where the second factor denotes the set of group homomorphisms $\overline{F}^\times \to \mathbb{R}$, endowed with the compact-open topology (the topology on $\overline{F}^\times$ being discrete).

(ii) The homeomorphism from (i) induces a continuous map $\mathscr{X}_F(\mathbb{C}) \to X_F$, which is a deformation retraction. Each fibre of this map is homeomorphic to $\operatorname{Hom}(\overline{F}^\times, \mathbb{R})$.

(iii) The diagram

$$\begin{array}{ccc} \mathbf{FEt}(\operatorname{Spec} F) & \xrightarrow{\Phi_{\text{sch}}} & \mathbf{FEt}(\mathscr{X}_F) \\ \downarrow{\scriptstyle \Phi_{\text{top}}} & & \downarrow{\scriptstyle \mathscr{Y} \mapsto \mathscr{Y}(\mathbb{C})} \\ \mathbf{FCov}(X_F) & \xrightarrow[\Psi]{} & \mathbf{FCov}(\mathscr{X}_F(\mathbb{C})) \end{array}$$

commutes up to isomorphism of functors, and all functors in it are equivalences of categories. Here the two functors Φ_{sch} and Φ_{top} are those from Theorems 4.11 and 5.2, respectively, and Ψ is induced by the map $\mathscr{X}_F(\mathbb{C}) \to X_F$ in (ii).

Proof (i) Note that there are canonical bijections

$$\begin{aligned}\mathscr{X}_{\overline{F}}(\mathbb{C}) &\cong \mathrm{Hom}_{\mathbb{C}\text{-algebras}}(\mathrm{W}_{\mathrm{rat}}(\overline{F}) \otimes_{\mathbb{Z}[\mu_\infty]} \mathbb{C}, \mathbb{C}) \\ &\cong \mathrm{Hom}_{\mathbb{C}\text{-algebras}}(\mathbb{Z}[\overline{F}^\times] \otimes_{\mathbb{Z}[\mu_\infty]} \mathbb{C}, \mathbb{C}) \\ &\cong \mathrm{Hom}_{\mathbb{C}\text{-algebras}}(\mathbb{C}[\overline{F}^\times] \otimes_{\mathbb{C}[\mu_\infty]} \mathbb{C}, \mathbb{C}) \\ &\cong \{\chi \in \mathrm{Hom}_{\mathrm{groups}}(\overline{F}^\times, \mathbb{C}^\times) \mid \chi|_{\mu_\infty} = \mathrm{id}_{\mu_\infty}\}.\end{aligned}$$

Now $\mathbb{C}^\times \cong \mathbb{S}^1 \times \mathbb{R}$ as Lie groups, and from this we obtain a product decomposition

$$\mathscr{X}_{\overline{F}}(\mathbb{C}) \cong \{\chi \in \mathrm{Hom}(\overline{F}^\times, \mathbb{S}^1) \mid \chi|_{\mu_\infty} = \mathrm{id}_{\mu_\infty}\} \times \mathrm{Hom}(\overline{F}^\times, \mathbb{R}).$$

The first factor is equal to $X_{\overline{F}}$. This bijection is clearly equivariant for $\mathrm{Gal}(\overline{F}/F)$, and it is straightforward to show that it is a homeomorphism.

(ii) There is a $\mathrm{Gal}(\overline{F}/F)$-equivariant deformation retraction $\mathscr{X}_{\overline{F}}(\mathbb{C}) \to X_{\overline{F}}$: using the description of $\mathscr{X}_{\overline{F}}(\mathbb{C})$ from (i) we can define it as

$$\begin{aligned} H\colon X_{\overline{F}} \times \mathrm{Hom}(\overline{F}^\times, \mathbb{R}) \times [0,1] &\to X_{\overline{F}} \times \mathrm{Hom}(\overline{F}^\times, \mathbb{R}) \\ (\chi, \lambda, t) &\mapsto (\chi, t\lambda).\end{aligned}$$

By equivariance, H descends to a deformation retraction $\mathscr{X}_F(\mathbb{C}) \to X_F$ (note that X_F is the quotient of $X_{\overline{F}}$ by the Galois group by construction, and $\mathscr{X}_F(\mathbb{C})$ is the quotient by the Galois group by Lemma 4.14). The statement about the fibres follows from the fact that $\mathrm{Gal}(\overline{F}/F)$ operates freely on $X_{\overline{F}}$.

(iii) That the diagram commutes up to isomorphism of functors is a direct calculation. We already know that three of the functors are equivalences: Φ_{sch} is an equivalence by Theorem 4.11, Φ_{top} is an equivalence by Theorem 5.2, and Ψ is an equivalence by (ii) and Corollary 2.18. Hence the fourth functor is also an equivalence. □

6 Classical Fundamental Groups Inside Galois Groups

6.1 Path Components of the Spaces X_F

First let $\overline{F}$ be an algebraically closed field. To determine the set of path components of $X_{\overline{F}}$ we need to contemplate a large commutative diagram.

Lemma 6.1 *Let $\overline{F}$ be an algebraically closed field containing $\mathbb{Q}(\zeta_\infty)$. Then there is a commutative diagram with exact rows and columns:*

$$\begin{array}{ccccccccc}
 & & 0 & & 0 & & 0 & & \\
 & & \downarrow & & \downarrow & & \downarrow & & \\
0 & \longrightarrow & \mathrm{Hom}(\overline{F}_{\mathrm{tf}}^{\times}, \mathbb{R}) & \longrightarrow & \mathrm{Hom}(\overline{F}_{\mathrm{tf}}^{\times}, \mathbb{S}^1) & \longrightarrow & \mathrm{Ext}(\overline{F}_{\mathrm{tf}}^{\times}, \mathbb{Z}) & \longrightarrow & 0 \\
 & & \downarrow & & \downarrow & & \downarrow & & \\
0 & \longrightarrow & \mathrm{Hom}(\overline{F}^{\times}, \mathbb{R}) & \longrightarrow & \mathrm{Hom}(\overline{F}^{\times}, \mathbb{S}^1) & \longrightarrow & \mathrm{Ext}(\overline{F}^{\times}, \mathbb{Z}) & \longrightarrow & 0 \\
 & & \downarrow & & \downarrow & & \downarrow & & \\
 & & 0 & \longrightarrow & \mathrm{Hom}(\mu_\infty, \mathbb{S}^1) & \longrightarrow & \mathrm{Ext}(\mu_\infty, \mathbb{Z}) & \longrightarrow & 0 \\
 & & & & \downarrow & & \downarrow & & \\
 & & & & 0 & & 0 & &
\end{array} \tag{9}$$

Here, all Hom *and* Ext *groups are understood to be in the category of abelian groups.*

Proof This diagram is essentially obtained by applying the bifunctor $\mathrm{Hom}(-,-)$ and its derivative $\mathrm{Ext}(-,-)$ to the short exact sequences

$$0 \to \mu_\infty \to \overline{F}^{\times} \to \overline{F}_{\mathrm{tf}}^{\times} \to 0$$

in the first variable and

$$0 \to \mathbb{Z} \to \mathbb{R} \to \mathbb{S}^1 \to 0$$

in the second variable. Hence it commutes by functoriality. We now show exactness in the rows and columns.

For the rows, let A be one of the groups μ_∞, $\overline{F}^{\times}$ and $\overline{F}_{\mathrm{tf}}^{\times}$. Then we obtain an exact sequence

$$\begin{aligned} &\mathrm{Hom}(A, \mathbb{Z}) \to \mathrm{Hom}(A, \mathbb{R}) \to \mathrm{Hom}(A, \mathbb{S}^1) \to \\ &\mathrm{Ext}(A, \mathbb{Z}) \to \mathrm{Ext}(A, \mathbb{R}). \end{aligned} \tag{10}$$

In all three cases A is divisible, therefore the first term $\mathrm{Hom}(A, \mathbb{Z})$ vanishes; the last term $\mathrm{Ext}(A, \mathbb{R})$ vanishes because $\mathbb{R}$ is divisible. In the case $A = \mu_\infty$ the term $\mathrm{Hom}(A, \mathbb{R})$ is also trivial. Hence the exact sequences (10) for these choices of A can be identified with the rows of (9).

As to the columns, the exactness of the first column is trivial. The second column is the exact sequence

$$0 \to \mathrm{Hom}(\overline{F}_{\mathrm{tf}}^{\times}, \mathbb{S}^1) \to \mathrm{Hom}(\overline{F}^{\times}, \mathbb{S}^1) \to \mathrm{Hom}(\mu_\infty, \mathbb{S}^1) \to \mathrm{Ext}(\overline{F}_{\mathrm{tf}}^{\times}, \mathbb{S}^1)$$

where the last term is zero because $\mathbb{S}^1$ is divisible. Finally the third column is the exact sequence

$$\mathrm{Hom}(\mu_\infty, \mathbb{Z}) \to \mathrm{Ext}(\overline{F}_{\mathrm{tf}}^{\times}, \mathbb{S}^1) \to \mathrm{Ext}(\overline{F}^{\times}, \mathbb{S}^1) \to \mathrm{Ext}(\mu_\infty, \mathbb{S}^1) \to 0,$$

which is exact because Ext^2 is zero on the category of abelian groups, and whose first member $\mathrm{Hom}(\mu_\infty, \mathbb{Z})$ is clearly trivial. □

Recall that by Proposition 3.1 there is a canonical bijection between $\pi_0^{\mathrm{path}}((\overline{F}^\times)^\vee)$ and $\mathrm{Ext}(\overline{F}^\times, \mathbb{Z})$. Denote by

$$\mathrm{Ext}_{\mathrm{exp}}(\overline{F}^\times, \mathbb{Z}) \subset \mathrm{Ext}(\overline{F}^\times, \mathbb{Z})$$

the subset of those extensions whose restriction to $\mu_\infty \subset \overline{F}^\times$ is isomorphic to the exponential sequence

$$0 \to \mathbb{Z} \to \mathbb{Q} \to \mu_\infty \to 0,$$

i.e. the preimage of the element in $\mathrm{Ext}(\mu_\infty, \mathbb{Z})$ encoding this extension in $\mathrm{Ext}(\overline{F}^\times, \mathbb{Z})$.

Proposition 6.2 *Let $\overline{F} \supset \mathbb{Q}(\zeta_\infty)$ be an algebraically closed field. The subset $X_{\overline{F}} \subset (\overline{F}^\times)^\vee$ is the union of those path components corresponding to the subset* $\mathrm{Ext}_{\mathrm{exp}}(\overline{F}^\times, \mathbb{Z}) \subset \mathrm{Ext}(\overline{F}^\times, \mathbb{Z}) \cong \pi_0^{\mathrm{path}}((\overline{F}^\times)^\vee)$.

Proof The middle column in (9) can be rewritten as

$$0 \to (\overline{F}_{\mathrm{tf}}^\times)^\vee \to (\overline{F}^\times)^\vee \to \mu_\infty^\vee \to 0,$$

and X_F is equal to the preimage of the inclusion $\iota \in \mathrm{Hom}(\mu_\infty, \mathbb{S}^1) = \mu_\infty^\vee$. The horizontal map $(\overline{F}^\times)^\vee = \mathrm{Hom}(\overline{F}^\times, \mathbb{S}^1) \to \mathrm{Ext}(\overline{F}^\times, \mathbb{Z})$ occurring in (9) is precisely the one taking a point in $(\overline{F}^\times)^\vee$ to its path component. The exactness and commutativity of (9) imply that X_F can equally be described as the preimage of the exponential sequence

$$[0 \to \mathbb{Z} \to \mathbb{Q} \to \mu_\infty \to 0] \in \mathrm{Ext}(\mu_\infty, \mathbb{Z})$$

in $(\overline{F}^\times)^\vee$. □

Corollary 6.3 *There is a canonical* $\mathrm{Gal}(\overline{F}/F)$*-equivariant bijection between the sets $\pi_0^{\mathrm{path}}(X_{\overline{F}})$ and* $\mathrm{Ext}_{\mathrm{exp}}(\overline{F}^\times, \mathbb{Z})$. □

By Theorem 5.7(i) there is then also a $\mathrm{Gal}(\overline{F}/F)$-equivariant bijection between $\pi_0^{\mathrm{path}}(\mathscr{X}_{\overline{F}}(\mathbb{C}))$ and $\mathrm{Ext}_{\mathrm{exp}}(\overline{F}^\times, \mathbb{Z})$.

6.2 Multiplicatively Free Fields

We will now investigate conditions for X_F, and therefore also $\mathscr{X}_F(\mathbb{C})$, to be path-connected. First we note that for a large class of fields it cannot be path-connected.

Proposition 6.4 *Let $F \supseteq \mathbb{Q}(\zeta_\infty)$ be a field, and assume that there exists an element $\alpha \in F^\times$ which is not a root of unity, such that for infinitely many $n \in \mathbb{N}$ there exists an n-th root of α in F.*

Then X_F has uncountably many path components.

For example, X_F has uncountably many path components for $F = \mathbb{Q}(\zeta_\infty, p^{1/\infty})$ with p a rational prime.

Proof Let $V \subset F^\times$ be the smallest saturated subgroup of $F^\times$ containing α; then $\mu_\infty \subset V$, and $M = V_{\mathrm{tf}} = V/\mu_\infty$ is a torsion-free abelian group of rank one, hence it can be embedded into the additive group of $\mathbb{Q}$. We fix such an embedding $\psi\colon M \hookrightarrow \mathbb{Q}$, which amounts to a short exact sequence

$$0 \to \mu_\infty \to V \xrightarrow{\psi} \mathbb{Q}.$$

By construction $M \cong \psi(V) \subseteq \mathbb{Q}$ has unbounded denominators, i.e. it is not contained in $\frac{1}{n}\mathbb{Z}$ for any $n \in \mathbb{N}$.

The inclusion $M \hookrightarrow \overline{F}^\times$ defines a continuous surjection

$$X_{\overline{F}} \to \mathrm{Hom}_{\exp}(V, \mathbb{S}^1) \overset{\text{def}}{=} \{\chi \in \mathrm{Hom}(V, \mathbb{S}^1) \mid \chi(\zeta) = \zeta \text{ for all } \zeta \in \mu_\infty\}$$

which is clearly $\mathrm{Gal}(\overline{F}/F)$-equivariant, hence we obtain a continuous surjection $X_{\overline{F}} \to \mathrm{Hom}_{\exp}(V, \mathbb{S}^1)$, where the target is (non-canonically) homeomorphic to $M^\vee$. We therefore obtain a surjection $\pi_0^{\text{path}}(X_F) \to \pi_0^{\text{path}}(M^\vee)$.

By Proposition 3.1 there is a bijection $\pi_0^{\text{path}}(M^\vee) \cong \mathrm{Ext}(M, \mathbb{Z})$. Hence we need to show that $\mathrm{Ext}(M, \mathbb{Z})$ is uncountable.

We will from now on identify M with its image $\psi(M) \subseteq \mathbb{Q}$; moreover, we may assume that $\mathbb{Z} \subset M$. Then there is a short exact sequence

$$0 \to \mathbb{Z} \to M \to M/\mathbb{Z} \to 0$$

and hence a long exact sequence

$$\underbrace{\mathrm{Hom}(M, \mathbb{Z})}_{=0} \to \underbrace{\mathrm{Hom}(\mathbb{Z}, \mathbb{Z})}_{=\mathbb{Z}} \to \mathrm{Ext}(M/\mathbb{Z}, \mathbb{Z}) \to \mathrm{Ext}(M, \mathbb{Z}) \to \underbrace{\mathrm{Ext}(\mathbb{Z}, \mathbb{Z})}_{=0}. \quad (11)$$

Letting $\mathfrak{N}$ be the set of cyclic subgroups $N \subset M$ containing $\mathbb{Z}$, we can then write $M = \varinjlim_{N \in \mathfrak{N}} N$ and $M/\mathbb{Z} = \varinjlim_{N \in \mathfrak{N}} N/\mathbb{Z}$. These are filtered limits, and we obtain a spectral sequence with

$$E_2^{p,q} = \mathrm{R}^p \varprojlim_{N \in \mathfrak{N}} \mathrm{Ext}^q(N/\mathbb{Z}, \mathbb{Z}) \Rightarrow \mathrm{Ext}^{p+q}(M/\mathbb{Z}, \mathbb{Z}).$$

Here the quotients $N/\mathbb{Z}$ are finite cyclic groups, therefore the groups $\mathrm{Ext}^q(N/\mathbb{Z}, \mathbb{Z})$ can be computed as follows:

- $\mathrm{Ext}^0(N/\mathbb{Z}, \mathbb{Z}) = \mathrm{Hom}(N/\mathbb{Z}, \mathbb{Z}) = 0$,
- $\mathrm{Ext}^1(N/\mathbb{Z}, \mathbb{Z}) = \mathrm{Ext}(N/\mathbb{Z}, \mathbb{Z})$ is a finite cyclic group of the same order as $N/\mathbb{Z}$, and
- $\mathrm{Ext}^q(N/\mathbb{Z}, \mathbb{Z}) = 0$ for all $q > 1$ because these Ext groups are taken over $\mathbb{Z}$, which has homological dimension one.

Moreover, note that if $C' \subset C$ are finite cyclic groups, the nontrivial part of the long exact sequence of $\mathrm{Ext}(-, \mathbb{Z})$ groups for the short exact sequence $0 \to C' \to C \to C/C' \to 0$ reads

$$\underbrace{\mathrm{Hom}(C', \mathbb{Z})}_{=0} \to \mathrm{Ext}(C/C', \mathbb{Z}) \to \mathrm{Ext}(C, \mathbb{Z}) \to \mathrm{Ext}(C', \mathbb{Z}) \to \underbrace{\mathrm{Ext}^2(C/C', \mathbb{Z})}_{=0},$$

hence the natural map $\mathrm{Ext}(C, \mathbb{Z}) \to \mathrm{Ext}(C', \mathbb{Z})$ is surjective.

From these observations we conclude: if $N, N' \in \mathfrak{N}$ with $N' \subseteq N$, then for every $q \geq 0$ the natural map $\mathrm{Ext}^q(N/\mathbb{Z}, \mathbb{Z}) \to \mathrm{Ext}^q(N'/\mathbb{Z}, \mathbb{Z})$ is surjective. In technical parlance, the $\mathrm{Ext}^q(N, \mathbb{Z})$ satisfy the *Mittag–Leffler condition*. Therefore

$$\mathrm{R}^p \varprojlim_{N \in \mathfrak{N}} \mathrm{Ext}^q(N/\mathbb{Z}, \mathbb{Z}) = 0 \qquad \text{for all } p > 0,\ q \geq 0,$$

(cf. [30, Chap. III, Proposition 10.3] or [31, Chap. 0, Proposition 13.2.2]) and the spectral sequence degenerates at E_2. We therefore obtain an isomorphism

$$\mathrm{Ext}(M/\mathbb{Z}, \mathbb{Z}) \cong \varprojlim_{N \in \mathfrak{N}} \mathrm{Ext}(N/\mathbb{Z}, \mathbb{Z}).$$

This is an inverse limit over infinitely many nontrivial finite groups with surjective transition maps, hence an uncountable profinite group. By (11) $\mathrm{Ext}(M, \mathbb{Z}) \cong \pi_0^{\mathrm{path}}(M^\vee)$ must then also be uncountable. □

Here is a class of fields where an α as in Proposition 6.4 cannot occur.

Definition 6.5 Let F be a field containing $\mathbb{Q}(\zeta_\infty)$. Then F is called *multiplicatively free* if $F^\times_{\mathrm{tf}} = F^\times/\mu_\infty$ is a free abelian group. F is called *stably multiplicatively free* if every finite extension of F is multiplicatively free.

Remark 6.6 (i) It is unknown to the authors whether there exists a field which is multiplicatively free but not stably multiplicatively free.

(ii) Recall that a subgroup B of an abelian group A is *saturated* if whenever $a \in A$ and there is some $n \in \mathbb{N}$ with $na \in B$, then $a \in B$. By a result of Pontryagin [32, Lemma 16], a torsion-free abelian group A is free if and only if every finitely generated subgroup of A is contained in a saturated finitely generated subgroup of A. Hence a field $F \supseteq \mathbb{Q}(\zeta_\infty)$ is multiplicatively free if and only if every finite subset of $F^\times$ is contained in a saturated subgroup of $F^\times$ generated (as a group) by all roots of unity and possibly finitely many additional elements.

(iii) The condition in Proposition 6.4 and the property of being multiplicatively free are mutually exclusive. It is again unknown to the authors whether always one or the other holds. For general abelian torsion-free groups, not necessarily isomorphic to $F^\times_{\mathrm{tf}}$ for a field F, both can be false.
More precisely, there exists an abelian group A with the following properties: it is torsion-free; it has rank two (i.e., $A \hookrightarrow A \otimes_{\mathbb{Z}} \mathbb{Q} \cong \mathbb{Q}^2$); for any $a \in A \setminus \{0\}$ there are only finitely many $n \in \mathbb{N}$ for which there exists $b \in A$ with $a = nb$; it is not free, in fact, every rank one quotient of A is divisible. This group A is constructed in [33, Lemma 2].[8]

Proposition 6.7 *Let F be an algebraic extension of $\mathbb{Q}(\zeta_\infty)$ which can be written as an abelian extension of a finite extension of $\mathbb{Q}$. Then F is stably multiplicatively free. In particular, $\mathbb{Q}(\zeta_\infty)$ is stably multiplicatively free.*

Proof We first show that F is multiplicatively free.

For a finite set S of rational primes let $\mathfrak{o}_{F,S}$ be the ring of S-integers in F. By a result of May [34, Theorem] the group $\mathfrak{o}^\times_{F,S}/\mu_\infty$ is then free abelian. Note that the free abelian subgroups $\mathfrak{o}^\times_{F,S}/\mu_\infty$ are saturated in $F^\times/\mu_\infty$, and every finitely generated subgroup of $F^\times/\mu_\infty$ is contained in one of them. By [34, Lemma] $F^\times/\mu_\infty$ must then be free abelian itself. Hence F is multiplicatively free.

Now let E/F be a finite extension; we can write $E = F(\alpha)$ for some $\alpha \in E$. By assumption there exists a subfield $K \subset F$ which is finite over $\mathbb{Q}$ such that the extension F/K is abelian. Then $E/K(\alpha)$ is abelian, too. By what we have just shown E is therefore multiplicatively free. □

The Kummer pairing. For a field F containing $\mathbb{Q}(\zeta_\infty)$ and an algebraic closure $\overline{F}/F$, let $F^\times_{\mathrm{sat}}$ be the saturation of $F^\times$ in $\overline{F}^\times$, i.e. the group of all $\alpha \in \overline{F}^\times$ such that there exists some $n \in \mathbb{N}$ with $\alpha^n \in F^\times$. Note that $F^\times_{\mathrm{sat}}$ is divisible, hence $F^\times_{\mathrm{sat}}/\mu_\infty = (F^\times_{\mathrm{sat}})_{\mathrm{tf}}$ is a $\mathbb{Q}$-vector space. Then there is a canonical biadditive pairing

$$\langle\cdot,\cdot\rangle\colon \operatorname{Gal}(\overline{F}/F) \times F^\times_{\mathrm{sat}} \to \mu_\infty, \quad (\sigma,\alpha) \mapsto \langle\sigma,\alpha\rangle = \frac{\sigma(\alpha)}{\alpha}.$$

It clearly factors through $\operatorname{Gal}(F^{\mathrm{ab}}/F) \times F^\times_{\mathrm{sat}}/F^\times$. By Kummer theory, cf. [30, Sect. VI.8], the maximal abelian extension F^{ab} is obtained by adjoining all elements of $F^\times_{\mathrm{sat}}$ to F, and the resulting homomorphism

$$\kappa\colon \operatorname{Gal}(F(F^\times_{\mathrm{sat}})/F) \to \operatorname{Hom}(F^\times_{\mathrm{sat}}/F^\times, \mu_\infty), \quad \sigma \mapsto \langle\sigma,\cdot\rangle \tag{12}$$

is an isomorphism.

Now assume in addition that F is stably multiplicatively free. We wish to understand the action of $\operatorname{Gal}(\overline{F}/F)$ on $\operatorname{Ext}_{\exp}(\overline{F}^\times, \mathbb{Z})$.

---

[8] Here is the construction. Let $\varphi \in \operatorname{End}(\mathbb{Q}/\mathbb{Z}) \cong \operatorname{End}(\hat{\mathbb{Z}}) \cong \hat{\mathbb{Z}} \cong \prod_{p \text{ prime}} \mathbb{Z}_p$ be such that the component $\varphi_p \in \mathbb{Z}_p$ at each p is transcendental over $\mathbb{Q}$. Then we set $A = \{(a,b) \in \mathbb{Q}^2 \mid \varphi(a \bmod \mathbb{Z}) = b \bmod z\}$.

We begin by considering a saturated subgroup $V \subset \overline{F}^{\times}$ such that $\mu_\infty \subseteq V$, $V_{\mathrm{tf}} = V/\mu_\infty$ is a $\mathbb{Q}$-vector space of finite rank, V is stable under the Galois action, and V is the saturation of $\Lambda = V \cap F^{\times}$. Note that then $\Lambda_{\mathrm{tf}} = \Lambda/\mu_\infty$ is a $\mathbb{Z}$-lattice in V.

Then the Galois action on V admits the following description: for $\alpha \in V$ and $\sigma \in \mathrm{Gal}(\overline{F}/F)$ we obtain

$$\sigma(\alpha) = \langle \sigma, \alpha \rangle \cdot \alpha.$$

Then there is also a natural action of $\mathrm{Gal}(\overline{F}/F)$ on $\mathrm{Hom}(V, \mu_\infty)$, namely $\sigma(\chi) = \chi \circ \sigma^{-1}$, that is,

$$\sigma(\chi)(\alpha) = \chi(\sigma^{-1}\alpha) = \chi(\langle \sigma^{-1}, \alpha \rangle \cdot \alpha) = \chi(\langle \sigma, \alpha \rangle)^{-1} \cdot \chi(\alpha).$$

Hence for the subset

$$\mathrm{Hom}_{\mathrm{exp}}(V, \mu_\infty) = \{\chi \in \mathrm{Hom}(V, \mu_\infty) \mid \chi|_{\mu_\infty} = \mathrm{id}_{\mu_\infty}\},$$

which is a translate of the subgroup $\mathrm{Hom}(V_{\mathrm{tf}}, \mu_\infty) \subset \mathrm{Hom}(V, \mu_\infty)$, we obtain a particularly simple description of the Galois action: for $\chi \in \mathrm{Hom}_{\mathrm{exp}}(V, \mu_\infty)$ and $\sigma \in \mathrm{Gal}(\overline{F}/F)$ we have

$$\sigma(\chi) = \langle \sigma, \cdot \rangle^{-1} \cdot \chi = \kappa_V(\sigma)^{-1} \cdot \chi, \tag{13}$$

where κ_V is the isomorphism $\mathrm{Gal}(F(V)/F) \to \mathrm{Hom}(V/\Lambda, \mu_\infty)$ induced by (12).

Lemma 6.8 *Let $F \supseteq \mathbb{Q}(\zeta_\infty)$ be a stably multiplicatively free field, let $\overline{F}/F$ be an algebraic closure, and let $V \subset \overline{F}^{\times}$ be a Galois-stable saturated subgroup of finite rank.*

Then there exist

- *an open subgroup $H \subseteq \mathrm{Gal}(\overline{F}/F)$,*
- *an open compact subgroup $L \subseteq \mathrm{Hom}(V_{\mathrm{tf}}, \mu_\infty) \cong \mathrm{Hom}(V, \mathbb{A}_{\mathrm{f}})$ and*
- *a surjective continuous group homomorphism $\kappa \colon H \to L$*

such that $H \subseteq \mathrm{Gal}(\overline{F}/F)$ operates on $\mathrm{Hom}_{\mathrm{exp}}(V, \mu_\infty)$ by $\tau(\chi) = \chi + \kappa(\tau)$ (where the group structure on μ_∞ is written additively).

Proof Let $B \subset V$ be a finite set which maps to a $\mathbb{Q}$-basis of V_{tf}, and let E be the subfield of $\overline{F}$ generated by B. Then E/F is a finite extension, hence E is multiplicatively free. Let $\Lambda = V \cap E^{\times}$; this is a subgroup of V containing μ_∞, and the quotient $\Lambda_{\mathrm{tf}} = \Lambda/\mu_\infty$ is a full $\mathbb{Z}$-lattice in V_{tf}.

We set $L = \mathrm{Hom}(V/\Lambda, \mu_\infty) \subset \mathrm{Hom}(V_{\mathrm{tf}}, \mu_\infty)$; under the isomorphism

$$\mathrm{Hom}(V_{\mathrm{tf}}, \mu_\infty) \cong \mathrm{Hom}(V, \mathbb{A}_{\mathrm{f}}) \cong \mathrm{Hom}(\Lambda, \mathbb{A}_{\mathrm{f}})$$

it corresponds to the subgroup $\mathrm{Hom}(\Lambda, \hat{\mathbb{Z}})$, which is clearly open and compact.

We then set $H = \mathrm{Gal}(\overline{F}/E)$ and let $\kappa\colon \mathrm{Gal}(\overline{F}/E) \to \mathrm{Hom}(V/\Lambda, \mu_\infty)$ be the map induced by the Kummer pairing, as in (12). From (13) we see that the H-action on $\mathrm{Hom}_{\mathrm{exp}}(V, \mu_\infty)$ is indeed as described. □

Lemma 6.9 *Let $F \supseteq \mathbb{Q}(\zeta_\infty)$ be a stably multiplicatively free field, let $\overline{F}/F$ be an algebraic closure and let $V \subset \overline{F}^\times$ be a Galois-stable saturated subgroup of finite rank.*

Then $\mathrm{Gal}(\overline{F}/F)$ *operates transitively on* $\mathrm{Ext}_{\mathrm{exp}}(V, \mathbb{Z})$.

Proof Using the short exact sequences $0 \to \mathbb{Z} \to \mathbb{Q} \to \mu_\infty \to 0$ and $0 \to \mu_\infty \to V \to V_{\mathrm{tf}} \to 0$ we obtain a commutative diagram with exact rows and columns analogous to (9):

$$\begin{array}{ccccccccc} & & 0 & & 0 & & 0 & & \\ & & \downarrow & & \downarrow & & \downarrow & & \\ 0 & \longrightarrow & \mathrm{Hom}(V_{\mathrm{tf}}, \mathbb{Q}) & \longrightarrow & \mathrm{Hom}(V_{\mathrm{tf}}, \mu_\infty) & \longrightarrow & \mathrm{Ext}(V_{\mathrm{tf}}, \mathbb{Z}) & \longrightarrow & 0 \\ & & \downarrow & & \downarrow & & \downarrow & & \\ 0 & \longrightarrow & \mathrm{Hom}(V, \mathbb{Q}) & \longrightarrow & \mathrm{Hom}(V, \mu_\infty) & \longrightarrow & \mathrm{Ext}(V, \mathbb{Z}) & \longrightarrow & 0 \\ & & \downarrow & & \downarrow & & \downarrow & & \\ & & 0 & \longrightarrow & \mathrm{Hom}(\mu_\infty, \mu_\infty) & \longrightarrow & \mathrm{Ext}(\mu_\infty, \mathbb{Z}) & \longrightarrow & 0. \\ & & & & \downarrow & & \downarrow & & \\ & & & & 0 & & 0 & & \end{array}$$

Here $\mathrm{Ext}_{\mathrm{exp}}(V, \mathbb{Z})$ is the preimage of the exponential extension $\varepsilon_{\mathrm{exp}} \in \mathrm{Ext}(\mu_\infty, \mathbb{Z})$ in $\mathrm{Ext}(V, \mathbb{Z})$, hence a translate of the subgroup $\mathrm{Ext}(V_{\mathrm{tf}}, \mathbb{Z})$. It can therefore also be described as the quotient of $\mathrm{Hom}_{\mathrm{exp}}(V, \mu_\infty)$, a translate of $\mathrm{Hom}(V_{\mathrm{tf}}, \mu_\infty)$ in $\mathrm{Hom}(V, \mu_\infty)$, by the subgroup $\mathrm{Hom}(V, \mathbb{Q})$.

From Lemma 6.8 we see that there is an open subgroup of $\mathrm{Hom}(V_{\mathrm{tf}}, \mu_\infty)$ on whose translates in $\mathrm{Hom}_{\mathrm{exp}}(V, \mu_\infty)$ a suitable open subgroup H of the Galois group acts transitively. Since the subgroup $\mathrm{Hom}(V_{\mathrm{tf}}, \mathbb{Q})$ is dense in $\mathrm{Hom}(V_{\mathrm{tf}}, \mu_\infty) \cong \mathrm{Hom}(V_{\mathrm{tf}}, \mathbb{A}_{\mathrm{f}})$ this implies that H, and therefore also $\mathrm{Gal}(\overline{F}/F)$, operates transitively on $\mathrm{Ext}_{\mathrm{exp}}(V, \mathbb{Z})$. □

Lemma 6.10 *Let $V \subseteq W \subset \overline{F}^\times$ be Galois-stable saturated subgroups of finite rank. Let also $\varepsilon_1, \varepsilon_2 \in \mathrm{Ext}_{\mathrm{exp}}(\overline{F}^\times, \mathbb{Z})$, and let $\chi_1^{(V)}, \chi_2^{(V)} \in \mathrm{Hom}(V, \mu_\infty)$ satisfying the following conditions:*

- *(i) the connecting homomorphism $\delta\colon \mathrm{Hom}(V, \mu_\infty) \to \mathrm{Ext}(V, \mathbb{Z})$ induced by the exponential sequence sends $\chi_i^{(V)}$ to $\varepsilon_i|_V$, for $i = 1, 2$;*
- *(ii) $\chi_1^{(V)}$ and $\chi_2^{(V)}$ lie in the same $\mathrm{Gal}(\overline{F}/F)$-orbit.*

Then there exist elements $\chi_1^{(W)}, \chi_2^{(W)} \in \mathrm{Hom}(W, \mu_\infty)$ such that

(iii) *the connecting homomorphism* $\delta\colon \operatorname{Hom}(W, \mu_\infty) \to \operatorname{Ext}(W, \mathbb{Z})$ *sends* $\chi_i^{(W)}$ *to* $\varepsilon_i|_W$, *for* $i = 1, 2$;

(iv) $\chi_i^{(W)}|_V = \chi_i^{(V)}$, *for* $i = 1, 2$;

(v) $\chi_1^{(W)}$ *and* $\chi_2^{(W)}$ *lie in the same* $\operatorname{Gal}(\overline{F}/F)$*-orbit.*

Proof Again we contemplate a large commutative diagram, obtained from $0 \to V \to W \to W/V \to 0$ and the exponential sequence:

$$
\begin{array}{ccccccccc}
 & & 0 & & 0 & & 0 & & \\
 & & \downarrow & & \downarrow & & \downarrow & & \\
0 & \longrightarrow & \operatorname{Hom}(W/V, \mathbb{Q}) & \longrightarrow & \operatorname{Hom}(W/V, \mu_\infty) & \longrightarrow & \operatorname{Ext}(W/V, \mathbb{Z}) & \longrightarrow & 0 \\
 & & \downarrow & & \downarrow & & \downarrow & & \\
0 & \longrightarrow & \operatorname{Hom}(W, \mathbb{Q}) & \longrightarrow & \operatorname{Hom}(W, \mu_\infty) & \longrightarrow & \operatorname{Ext}(W, \mathbb{Z}) & \longrightarrow & 0 \\
 & & \downarrow & & \downarrow & & \downarrow & & \\
0 & \longrightarrow & \operatorname{Hom}(V, \mathbb{Q}) & \longrightarrow & \operatorname{Hom}(V, \mu_\infty) & \longrightarrow & \operatorname{Ext}(V, \mathbb{Z}) & \longrightarrow & 0 \\
 & & \downarrow & & \downarrow & & \downarrow & & \\
 & & 0 & & 0 & & 0 & &
\end{array}
$$

First we consider the problem of lifting an individual character. So, let $\varepsilon \in \operatorname{Ext}(\overline{F}^\times, \mathbb{Z})$ and $\chi^{(V)} \in \operatorname{Hom}(V, \mu_\infty)$ with $\delta(\chi^{(V)}) = \varepsilon|_V$. A short diagram chase then shows that there exists some $\chi^{(W)} \in \operatorname{Hom}(W, \mu_\infty)$ with $\delta(\chi^{(W)}) = \varepsilon|_W$ and $\chi^{(W)}|_V = \chi^{(V)}$, and that moreover the set of all such $\chi^{(W)}$ is a translate of $\operatorname{Hom}(W/V, \mathbb{Q})$ (which can be considered a subgroup of $\operatorname{Hom}(W, \mu_\infty)$).

With this preparation we find an element $\chi_1^{(W)}$ satisfying (iii) and (iv). We choose some $\sigma \in \operatorname{Gal}(\overline{F}/F)$ with $\sigma(\chi_1^{(V)}) = \chi_2^{(V)}$, and let $\psi = \sigma(\chi_1^{(V)})$. Then ψ is our first approximation to $\chi_2^{(W)}$, and it clearly satisfies (iv) and (v), but not necessarily (iii).

From Lemma 6.8 we deduce the existence of the following objects:

- a closed subgroup $H \subseteq \operatorname{Gal}(\overline{F}/F)$,
- an open compact subgroup $L \subseteq \operatorname{Hom}(W/V, \mu_\infty) \cong \operatorname{Hom}(W/V, \mathbb{A}_\mathrm{f})$ and
- a surjective continuous group homomorphism $\kappa\colon H \to L$

such that on $\operatorname{Hom}_{\exp}(V, \mu_\infty)$ the group $H \subseteq \operatorname{Gal}(\overline{F}/F)$ operates by $\tau(\chi) = \chi + \kappa(\tau)$ (here we write the group structure in μ_∞ additively).

Since $\varepsilon_2|_W$ and $\delta(\psi)$ both restrict to $\varepsilon_2|_V \in \operatorname{Ext}(V, \mathbb{Z})$ their difference lies in $\operatorname{Ext}(W/V, \mathbb{Z})$. Hence there exists an element $\alpha \in \operatorname{Hom}(W/V, \mu_\infty) \cong \operatorname{Hom}(W/V, \mathbb{A}_\mathrm{f})$ such that $\delta(\alpha) = \varepsilon_2|_W - \delta(\psi)$. We let further $\tau \in H \subseteq \operatorname{Gal}(\overline{F}/F)$ with $\kappa(\tau) = \alpha$.

Finally we set $\chi_2^{(W)} = \psi + \alpha$. We check that all desired conditions are met:

(iii) $\chi_1^{(W)}$ was chosen such that $\delta(\chi_1^{(W)}) = \varepsilon_2|_W$, and

$$\delta(\chi_2^{(W)}) = \delta(\psi) + \delta(\alpha) = \delta(\psi) + \varepsilon_2|_W - \delta(\psi) = \varepsilon_2|_W.$$

(iv) $\chi_1^{(W)}$ was chosen such that $\chi_1^{(W)}|_V = \chi_1^{(V)}$, and

$$\chi_2^{(W)}|_V = \psi|_V + \alpha|_V = \chi_2^{(V)} + 0 = \chi_2^{(V)}.$$

(v) $\chi_1^{(W)}$ and $\chi_2^{(W)}$ are in the same Galois orbit because

$$\tau\sigma(\chi_1^{(W)}) = \tau(\psi) = \psi + \kappa(\tau) = \psi + \alpha = \chi_2^{(W)}. \qquad \square$$

Proposition 6.11 *Let* $F \supseteq \mathbb{Q}(\zeta_\infty)$ *be a* countable *stably multiplicatively free field, and let* $\overline{F}/F$ *be an algebraic closure.*

Then $\mathrm{Gal}(\overline{F}/F)$ *operates transitively on* $\mathrm{Ext}_{\exp}(\overline{F}^\times, \mathbb{Z})$.

Proof By countability we can find an ascending chain (indexed by $\mathbb{N}$) of Galois-stable subgroups $\mu_\infty \subset V_1 \subset V_2 \subset \cdots$ of finite rank, whose union is $\overline{F}^\times$. Hence there is a spectral sequence with

$$E_2^{p,q} = \mathrm{R}^p \varprojlim_n \mathrm{Ext}^q(V_n, \mathbb{Z}) \Rightarrow \mathrm{Ext}^{p+q}(\overline{F}^\times, \mathbb{Z}). \tag{14}$$

Since these Ext groups are taken in the category of abelian groups, the entries with $q > 1$ vanish. Likewise, it is easy to see that the structure maps $\mathrm{Hom}(V_n, \mathbb{Z}) \to \mathrm{Hom}(V_m, \mathbb{Z})$ and $\mathrm{Ext}(V_n, \mathbb{Z}) \to \mathrm{Ext}(V_m, \mathbb{Z})$ for $n \geq m$ are surjective, hence the inverse systems in (14) satisfy a Mittag-Leffler condition, and the higher direct images also vanish. Therefore the spectral sequence degenerates at E_2, and the natural map

$$\mathrm{Ext}(\overline{F}^\times, \mathbb{Z}) \to \varprojlim_n \mathrm{Ext}(V_n, \mathbb{Z})$$

is an isomorphism. We deduce that the restriction

$$\mathrm{Ext}_{\exp}(\overline{F}^\times, \mathbb{Z}) \to \varprojlim_n \mathrm{Ext}_{\exp}(V_n, \mathbb{Z}) \tag{15}$$

is a bijection.

Let $\varepsilon_1, \varepsilon_2 \in \mathrm{Ext}_{\exp}(\overline{F}^\times, \mathbb{Z})$. We need to show that there is a $\sigma \in \mathrm{Gal}(\overline{F}/F)$ with $\sigma(\varepsilon_1) = \varepsilon_2$; by what we have just seen this is equivalent to $\sigma(\varepsilon_1|_{V_n}) = \varepsilon_2|_{V_n}$ for all $n \in \mathbb{N}$. Using Lemmas 6.9 and 6.10 we inductively produce elements $\chi_1^{(n)}, \chi_2^{(n)} \in \mathrm{Hom}(V_n, \mu_\infty)$ such that the following conditions hold:

(iii) the connecting homomorphism $\delta\colon \mathrm{Hom}(V_n, \mu_\infty) \to \mathrm{Ext}(V_n, \mathbb{Z})$ sends $\chi_i^{(n)}$ to $\varepsilon_i|_{V_n}$, for $i = 1, 2$ and $n \in \mathbb{N}$;

(iv) $\chi_i^{(n+1)}|_{V_n} = \chi_i^{(n)}$ for $i = 1, 2$ and $n \in \mathbb{N}$;

(v) $\chi_1^{(n)}$ and $\chi_2^{(n)}$ lie in the same $\mathrm{Gal}(\overline{F}/F)$-orbit.

We let

$$T_n = \{\sigma \in \mathrm{Gal}(\overline{F}/F) \mid \sigma(\chi_1^{(n)}) = \chi_2^{(n)}\}.$$

By (v) each T_n is a nonempty subset of $\mathrm{Gal}(\overline{F}/F)$, and by (iv) the sequence of subsets (T_n) is descending, i.e. $T_n \supseteq T_{n+1}$ for all $n \in \mathbb{N}$. Moreover, $\mathrm{Gal}(\overline{F}/F)$ operates continuously on $\mathrm{Hom}(V_n, \mu_\infty)$ when the latter is endowed with the compact-open topology; since $\mathrm{Hom}(V_n, \mu_\infty)$ is Hausdorff, points in this space are closed, hence the T_n are closed subsets of $\mathrm{Gal}(\overline{F}/F)$.

Now we can apply Cantor's Intersection Theorem (Lemma 4.10 above) to conclude that

$$\bigcap_{n \in \mathbb{N}} T_n \neq \varnothing.$$

By construction, any element in this intersection sends ε_1 to ε_2. □

Remark 6.12 The reader may wonder why we do not simply proceed as follows to prove Proposition 6.11. For any Galois-stable saturated subgroup $V \subset \overline{F}^\times$ of finite rank we set $U_V = \{\sigma \in \mathrm{Gal}(\overline{F}/F) \mid \sigma(\varepsilon_1|_V) = \varepsilon_2|_V\}$; this is a coset of the stabiliser of $\varepsilon_1|_V$, and it is nonempty by Lemma 6.9. Then an application of Lemma 4.10 should show that the intersection of all U_V is nonempty.

The problem with this argument is that U_V is not closed, only an F_σ-subset (a countable union of closed subsets). Rigidifying the situation by adding the auxiliary conditions that the lifts $\chi_i^{(n)} \in \mathrm{Hom}(V, \mu_\infty)$ also be fixed replaces the U_V by the closed subsets T_V, which allows us to apply Lemma 4.10.

Corollary 6.13 *Let F be a countable stably multiplicatively free field. Then X_F and $\mathscr{X}_F(\mathbb{C})$ are path-connected.*

Proof By Corollary 6.3 and Proposition 6.11, $\mathrm{Gal}(\overline{F}/F)$ operates transitively on $\pi_0^{\mathrm{path}}(X_{\overline{F}})$, hence $X_F = \mathrm{Gal}(\overline{F}/F)\backslash X_{\overline{F}}$ is path-connected. By Theorem 5.7(ii) then also $\mathscr{X}_F(\mathbb{C})$ is path-connected. □

6.3 Classical Fundamental Groups of the Spaces X_F

It will be convenient to fix a basepoint $\tilde{\chi} \in X_{\overline{F}}$ and denote its image in X_F by χ. Then $p\colon (X_{\overline{F}}, \tilde{\chi}) \to (X_F, \chi)$ is a universal profinite covering space. Recall that there is then a short exact sequence of abstract groups (6) which in our case becomes

$$1 \to \pi_1^{\mathrm{path}}(X_{\overline{F}}, \tilde{\chi}) \xrightarrow{p_*} \pi_1^{\mathrm{path}}(X_F, \chi) \xrightarrow{\alpha} \mathrm{Stab}_{\pi_1^{\text{ét}}(X_F,\chi)} X_{\overline{F}}^{\circ} \to 1.$$

Since $X_{\overline{F}}$ is homeomorphic to the Pontryagin dual of a $\mathbb{Q}$-vector space its classical fundamental group $\pi_1^{\mathrm{path}}(X_{\overline{F}}, \tilde{\chi})$ is trivial by Corollary 3.4, hence α maps

$\pi_1^{\mathrm{path}}(X_F, \chi)$ isomorphically to the stabiliser. The latter can be rewritten: the set $\pi_0^{\mathrm{path}}(X_{\overline{F}})$ is in canonical bijection with $\mathrm{Ext}_{\exp}(\overline{F}^\times, \mathbb{Z})$, and this bijection is equivariant for the isomorphism $\pi_1^{\text{ét}}(X_F, \chi) \cong \mathrm{Gal}(\overline{F}/F)$. Hence we have shown the following:

Proposition 6.14 *Let $F \supseteq \mathbb{Q}(\zeta_\infty)$ be a field, and let $\tilde{\chi} \in X_{\overline{F}}$. Denote the image of $\tilde{\chi}$ in X_F by χ, and let $\varepsilon \in \mathrm{Ext}_{\exp}(\overline{F}^\times, \mathbb{Z})$ be the pullback of the extension*

$$[0 \to \mathbb{Z} \to \mathbb{R} \to \mathbb{S}^1 \to 0] \in \mathrm{Ext}(\mathbb{S}^1, \mathbb{Z})$$

along $\tilde{\chi}$.

Then $\pi_1^{\mathrm{path}}(X_F, \chi)$ is canonically isomorphic to the stabiliser of ε in Gal $(\overline{F}/F)$. □

This stabiliser seems to be hard to determine in general. However, in the countable stably multiplicatively free case we can at least say that it is large.

Proposition 6.15 *Let $F \supseteq \mathbb{Q}(\zeta_\infty)$ be a countable stably multiplicatively free field, and let $\chi \in X_F$ be any basepoint. Then the image of $\pi_1^{\mathrm{path}}(X_F, \chi)$ is a dense subgroup of* $\mathrm{Gal}(\overline{F}/F)$.

Proof If F is countable and stably multiplicatively free then so is any finite extension E of F. Hence all coverings X_E defined by finite extensions E of F are path-connected by Corollary 6.13. By Theorem 5.2 every connected covering space of X_F is of this form, therefore X_F is stably path-connected. Hence by Proposition 2.30 the image of $\pi_1^{\mathrm{path}}(X_F, \chi)$ in $\pi_1^{\text{ét}}(X_F, \chi) \cong \mathrm{Gal}(\overline{F}/F)$ is dense. □

This applies in particular to $F = \mathbb{Q}(\zeta_\infty)$.

The fundamental group as an inverse limit. Let $F \supseteq \mathbb{Q}(\zeta_\infty)$ be a countable stably multiplicatively free field. We shall write $\pi_1^{\mathrm{path}}(X_F)$ as an inverse limit of discrete groups which are extensions of finite groups by free abelian groups of finite rank. In particular, $\pi_1^{\mathrm{path}}(X_F)$ will be endowed with a non-discrete topology.

Fix an algebraic closure $\overline{F}/F$, and let $\mathfrak{L}(\overline{F}/F)$ denote the set of all subgroups $\Lambda < \overline{F}^\times$ satisfying the following conditions:

(i) Λ contains μ_∞, and $\Lambda_{\mathrm{tf}} = \Lambda/\mu_\infty$ is a free abelian group of finite rank.
(ii) Λ is stable under $\mathrm{Gal}(\overline{F}/F)$.
(iii) Set $E = F(\Lambda)$; this is a finite Galois extension of F by (i) and (ii). Also, let V be the saturation of Λ in $\overline{F}^\times$. Then $E^\times \cap V = \Lambda$.
(iv) For every $\sigma \in \mathrm{Gal}(E/F)$ there exists a $\lambda \in \Lambda$ such that $\sigma(\lambda)/\lambda = \zeta_n$, where n is the order of σ in $\mathrm{Gal}(E/F)$.

Lemma 6.16 *Let $F \supseteq \mathbb{Q}(\zeta_\infty)$ be a stably multiplicatively free field with algebraic closure $\overline{F}$. Then $\overline{F}^\times$ can be written as the filtered union*

$$\overline{F}^\times = \bigcup_{\Lambda \in \mathfrak{L}(\overline{F}/F)} \Lambda. \tag{16}$$

Proof For any finite collection of elements of $\overline{F}^{\times}$, say $\alpha_1, \dots, \alpha_n \in \overline{F}^{\times}$, we will construct a $\Lambda \in \mathfrak{L}(\overline{F}/F)$ with $\alpha_1, \dots, \alpha_n \in \Lambda$. This will then ensure that not only (16) holds, but also the union is filtered (note that all Λ are assumed finitely generated up to μ_∞, see condition (i) above).

So, assume that $\alpha_1, \dots, \alpha_n \in \overline{F}^{\times}$, and let E be the Galois closure of $F(\alpha_1, \dots, \alpha_n)$ in $\overline{F}$. By Hilbert's Theorem 90 in the form already used in the proof of Lemma 4.12, for every $\sigma \in \mathrm{Gal}(E/F)$ there exists some $\lambda_\sigma \in E^{\times}$ such that $\sigma(\lambda_\sigma)/\lambda_\sigma = \zeta_n$. Let V be the smallest saturated subgroup of $\overline{F}^{\times}$ containing the λ_σ and all Galois conjugates of $\alpha_1, \dots, \alpha_n$. Then V_{tf} is a $\mathbb{Q}$-vector space of finite rank, and $\Lambda = E^{\times} \cap V$ will be an element of $\mathfrak{L}(\overline{F}/F)$ containing $\alpha_1, \dots, \alpha_n$. □

For $\Lambda \in \mathfrak{L}(\overline{F}/F)$ set $X_{\overline{F}}(\Lambda) = \mathrm{Hom}_{\mathrm{exp}}(\Lambda, \mathbb{S}^1)$. There is a natural continuous action of $\mathrm{Gal}(\overline{F}/F)$ on $X_{\overline{F}}(\Lambda)$, and we set $X_F(\Lambda) = \mathrm{Gal}(\overline{F}/F)\backslash X_{\overline{F}}(\Lambda)$.

Lemma 6.17 *Let $F \supseteq \mathbb{Q}(\zeta_\infty)$ be a stably multiplicatively free field, and let $\overline{F}/F$ be an algebraic closure.*

(i) For every $\Lambda \in \mathfrak{L}(\overline{F}/F)$ the space $X_{\overline{F}}(\Lambda)$ is homeomorphic to a torus of dimension rank Λ. *For any basepoint $\chi \in X_{\overline{F}}(\Lambda)$ there is a canonical isomorphism*

$$\pi_1^{\mathrm{path}}(X_{\overline{F}}(\Lambda), \chi) \cong \mathrm{Hom}(\Lambda, \mathbb{Z}).$$

(ii) Let $\Lambda \in \mathfrak{L}(\overline{F}/F)$ and let E be the subfield of $\overline{F}$ generated by Λ, a finite Galois extension of F. Then the action of $\mathrm{Gal}(\overline{F}/F)$ *on $X_{\overline{F}}(\Lambda)$ factors through* $\mathrm{Gal}(E/F)$, *and the induced action of* $\mathrm{Gal}(E/F)$ *is free. Hence $X_{\overline{F}}(\Lambda) \to X_F(\Lambda)$ is a finite covering.*

(iii) For any basepoint $\chi \in X_F(\Lambda)$ there is a natural exact sequence

$$1 \to \mathrm{Hom}(\Lambda, \mathbb{Z}) \to \pi_1^{\mathrm{path}}(X_F(\Lambda), \chi) \to \mathrm{Gal}(E/F) \to 1.$$

Proof For (i) note that $X_{\overline{F}}(\Lambda)$ is a translate of the subgroup $\Lambda_{\mathrm{tf}}^{\vee}$ in $\Lambda^{\vee}$; this subgroup is a torus whose classical fundamental group is canonically isomorphic to $\mathrm{Hom}(\Lambda, \mathbb{Z})$.

As to (ii), it is clear that the Galois action factors through $\mathrm{Gal}(E/F)$. We will now show that the induced action of this finite group is free. Let $1 \neq \sigma \in \mathrm{Gal}(E/F)$ be an element of order $n > 1$, and assume that there is some $\chi \in X_{\overline{F}}(\Lambda)$ with $\sigma(\chi) = \chi$. By condition (iv) in the definition of $\mathfrak{L}(\overline{F}/F)$ there exists some $\lambda \in \Lambda$ with $\sigma(\lambda)/\lambda = \zeta_n$. But then

$$e^{2\pi i/n} = \chi(\zeta_n) = \frac{\chi(\sigma(\lambda))}{\chi(\lambda)} = \frac{\sigma(\chi)(\lambda)}{\chi(\lambda)} = \frac{\chi(\lambda)}{\chi(\lambda)} = 1,$$

a contradiction. Therefore σ cannot have a fixed point in $X_{\overline{F}}(\Lambda)$, and the action is free.

Part (iii) then follows easily (note that these spaces are path-connected and locally path-connected, in fact manifolds, so the classical theory of fundamental groups and covering spaces applies). □

Proposition 6.18 *Let $F \supseteq \mathbb{Q}(\zeta_\infty)$ be a stably multiplicatively free field with algebraic closure $\overline{F}/F$. Then the canonical map*

$$X_F \to \varprojlim_{\Lambda \in \mathfrak{L}(\overline{F}/F)} X_F(\Lambda) \tag{17}$$

is a homeomorphism.

Proof Consider first the map $X_{\overline{F}} \to \varprojlim X_F(\Lambda) = \varprojlim \mathrm{Gal}(\overline{F}/F)\backslash X_{\overline{F}}(\Lambda)$. This is clearly surjective, and if two elements of $X_{\overline{F}}$ have the same image they must be in the same Galois orbit, by an argument using Lemma 4.10 (Cantor's Intersection Theorem) similar to that used in the proof of Proposition 6.11. Hence the map (17) is bijective. It is also continuous, and domain and target are compact Hausdorff spaces. Therefore it is a homeomorphism.

Proposition 6.19 *Let $F \supseteq \mathbb{Q}(\zeta_\infty)$ be a countable stably multiplicatively free field with algebraic closure $\overline{F}$. Choose a basepoint $\chi \in X_F$, and for each $\Lambda \in \mathfrak{L}(\overline{F}/F)$ denote its image in $X_F(\Lambda)$ by χ_Λ. Then the natural map*

$$\pi_1^{\mathrm{path}}(X_F, \chi) \to \varprojlim_{\Lambda \in \mathfrak{L}(\overline{F}/F)} \pi_1^{\mathrm{path}}(X_F(\Lambda), \chi_\Lambda) \tag{18}$$

is an isomorphism.

Proof Since F (hence also $\overline{F}$) is countable, there exists a cofinal sequence $(\Lambda_n)_{n\in\mathbb{N}}$ in $\mathfrak{L}(\overline{F}/F)$. To see this, choose an enumeration $\overline{F} = \{a_1, a_2, a_3, \ldots\}$ and choose the Λ_n inductively in such a way that $\Lambda_n \subseteq \Lambda_{n+1}$ and $a_1, \ldots, a_n \in \Lambda_n$. Hence $X_F \to \varprojlim_n X_F(\Lambda_n)$ is is also a homeomorphism, and it suffices to show that

$$\pi_1^{\mathrm{path}}(X_F, \chi) \to \varprojlim_{n\in\mathbb{N}} \pi_1^{\mathrm{path}}(X_F(\Lambda_n), \chi_{\Lambda_n})$$

is an isomorphism.

For each $n \in \mathbb{N}$ there is a commutative diagram

$$\begin{array}{ccc} X_{\overline{F}}(\Lambda_{n+1}) & \xrightarrow{\tilde{p}} & X_{\overline{F}}(\Lambda_n) \\ \downarrow & & \downarrow \\ X_F(\Lambda_{n+1}) & \xrightarrow[p]{} & X_F(\Lambda_n) \end{array}$$

where $\tilde{p}$ is a fibration and the vertical maps are finite coverings. Therefore p is also a fibration.

In general if $(X_n)_{n\in\mathbb{N}}$ is a projective system of pointed topological spaces where the transition maps are fibrations, then there is a short exact sequence

$$1 \to \mathrm{R}^1 \varprojlim_{n\in\mathbb{N}} \pi_2^{\mathrm{path}}(X_n) \to \pi_1^{\mathrm{path}}(X) \to \varprojlim_{n\in\mathbb{N}} \pi_1^{\mathrm{path}}(X_n) \to 1$$

of abstract groups (compatible choice of basepoints understood), see [35] (see also [36, Theorem 2.1] for a more elementary exposition). Hence in our case there is a short exact sequence

$$1 \to \mathrm{R}^1 \varprojlim_{n\in\mathbb{N}} \pi_2^{\mathrm{path}}(X_F(\Lambda_n)) \to \pi_1^{\mathrm{path}}(X_F) \to \varprojlim_{n\in\mathbb{N}} \pi_1^{\mathrm{path}}(X_F(\Lambda_n)) \to 1.$$

Since the $X_F(\Lambda_n)$ admit finite covering spaces which are tori, their second homotopy groups vanish. □

Proposition 6.20 *The loop topology turns $\pi_1^{\mathrm{path}}(X_F, \chi)$ into a topological group with a basis of open neighborhoods of the identity given by open subgroups, hence it is equal to the τ- and σ-topologies. If we endow each $\pi_1^{\mathrm{path}}(X(\Lambda), \chi_\Lambda)$ with the discrete topology, then (18) becomes an isomorphism of topological groups.*

Moreover, $\pi_1^{\mathrm{path}}(X_F, \chi)$ is complete for this topology. Therefore $\pi_1^{\mathrm{path}}(X_F, \chi) \cong \pi_1^{\mathrm{Gal}}(X_F, \chi)$ is a Noohi group.

Proof Since the $X_F(\Lambda)$ are manifolds, their classical fundamental groups are discrete for the loop topology. Consider the commutative diagram

$$\begin{array}{ccc}
\Omega(X_F, \chi) & \longrightarrow & \varprojlim_{\Lambda\in\mathfrak{L}(\overline{F}/F)} \Omega(X_F(\Lambda), \chi_\Lambda) \\
\downarrow & & \downarrow \\
\pi_1^{\mathrm{path}}(X_F, \chi) & \xrightarrow[(18)]{} & \varprojlim_{\Lambda\in\mathfrak{L}(\overline{F}/F)} \pi_1^{\mathrm{path}}(X_F(\Lambda), \chi_\Lambda).
\end{array}$$

Here the upper horizontal map is a homeomorphism and the vertical maps are open. Hence (18) is a bijection which is continuous and open, hence also a homeomorphism.

A quasi-topological group which is a projective limit of topological groups is itself a topological group, and hence the loop and τ-topologies on $\pi_1^{\mathrm{path}}(X_F, \chi)$ agree. Since $\pi_1^{\mathrm{path}}(X, \chi)$ is a projective limit of discrete groups its τ- and σ-topologies agree, and it is complete. □

In particular we find that the loop topology turns $\pi_1^{\mathrm{path}}(X_F)$, which we may identify with a subgroup of $\mathrm{Gal}(\overline{F}/F)$, into a complete topological group whose topology is strictly finer than the subspace topology induced from the Krull topology on $\mathrm{Gal}(\overline{F}/F)$, because it has infinite discrete quotients.

7 Cohomology

We will next show how to realise Galois cohomology groups with constant coefficients as suitable cohomology groups of the spaces X_F and the schemes $\mathscr{X}_F$.

7.1 The Cartan–Leray Spectral Sequence

Consider the following situation: X is a compact Hausdorff space and G is a profinite group operating freely and continuously on X, with quotient space $Y = G\backslash X$. We will construct a spectral sequence relating the (sheaf) cohomologies of X and Y with the continuous group cohomology of G.

Continuous group cohomology. Let G be a profinite group. A *continuous G-module* is an abelian group A with a G-operation which becomes continuous when A is endowed with the discrete topology. The continuous G-modules form an abelian category G-**Mod** in an obvious way; this category has enough injectives. The functor $(-)^G\colon\ G\text{-}\mathbf{Mod}\rightarrow \mathbf{Ab}$ sending a G-module A to its invariant submodule A^G is left exact. Hence we obtain a total derived functor between derived categories $\mathrm{R}(-)^G : \mathscr{D}(\mathrm{G}\text{-}\mathbf{Mod}) \to \mathscr{D}(\mathbf{Ab})$, and derived functors in the classical sense which we call *continuous group cohomology*:

$$\mathrm{H}^p(G, A) = \mathrm{R}^p(-)^G(A).$$

Note that this may well differ from the classical group cohomology $\mathrm{H}^p(G^\delta, A)$ where G^δ is G as an abstract group. However, there is a canonical isomorphism

$$\varinjlim_{H} \mathrm{H}^p(G/H, A^H) \cong \mathrm{H}^p(G, A),$$

where the limit is over all normal open subgroups H of G, cf. the discussion in [37, Sect. 2.2].

For a field F with separable closure $\overline{F}/F$ and a continuous $\mathrm{Gal}(\overline{F}/F)$-module A we write shortly

$$\mathrm{H}^m(F, A) = \mathrm{H}^m(\mathrm{Gal}(\overline{F}/F), A);$$

these groups are called *Galois cohomology groups*. Note that if A is an abelian group interpreted as a constant module for the Galois group, then $\mathrm{H}^m(F, A)$ does not depend on the choice of a separable closure of F, i.e. for another separable closure $\overline{F}'/F$ there is a canonical isomorphism $\mathrm{H}^m(\mathrm{Gal}(\overline{F}'/F), A) \cong \mathrm{H}^m(\mathrm{Gal}(\overline{F}/F), A)$; this justifies the notation $\mathrm{H}^m(F, A)$.

Equivariant sheaves and their cohomology. Let X be a compact Hausdorff space and G a profinite group operating continuously and freely on X. There are several ways to define the category $\mathrm{Sh}_G(X)$ of G-equivariant abelian sheaves on X.

For instance, a sheaf of abelian groups $\mathscr{A}$ on X corresponds to an *espace étalé* $\pi\colon A \to X$, which is a topological space A with a local homeomorphism $\pi\colon A \to X$ and an abelian group structure in the category of X-spaces. Here $\pi^{-1}(x) \cong \mathscr{A}_x$ (the stalk), and for an open set $U \subseteq X$ sections of $\mathscr{A}$ on U are the same as continuous sections of the map $\pi^{-1}(U) \to U$. Then we define a *G-equivariant sheaf* on X to be a sheaf $\mathscr{A}$ of abelian groups on X together with a lift of the G-action on X to a continuous G-action on A. For a more 'modern' definition that is more amenable to generalisations see e.g. [38, Sect. 1].

Again the G-equivariant sheaves on X form an abelian category $\mathrm{Sh}_G(X)$ with enough injectives; there is a canonical equivalence $\mathrm{Sh}_G(*) \simeq G\text{-}\mathbf{Mod}$, where $*$ denotes the one-point space.

For a G-equivariant sheaf $\mathscr{A}$ the group $\Gamma(X, \mathscr{A})$ of global sections comes naturally with a continuous G-action. Hence we obtain a left exact functor

$$\Gamma_{X,G}\colon \mathrm{Sh}_G(X) \to G\text{-}\mathbf{Mod}, \quad \mathscr{A} \mapsto \Gamma(X, \mathscr{A}).$$

For a G-equivariant sheaf $\mathscr{A}$ on X we obtain a complex $\mathrm{R}\Gamma_{X,G}(\mathscr{A}) \in \mathscr{D}(G\text{-}\mathbf{Mod})$; but we may also forget its G-structure and apply the derived functor of the usual global sections functor $\Gamma_X\colon \mathrm{Sh}(X) \to \mathbf{Ab}$ to it.

Lemma 7.1 *For $\mathscr{A} \in \mathrm{Sh}_G(X)$ the complex of abelian groups underlying $\mathrm{R}\Gamma_{X,G}(\mathscr{A})$ (i.e. its image in $\mathscr{D}(\mathbf{Ab})$) is canonically isomorphic to the complex $\mathrm{R}\Gamma_X(\mathscr{A})$.*

In particular, the cohomology groups of either of these complexes become continuous G-modules whose underlying abelian groups are the ordinary sheaf cohomology groups $\mathrm{H}^p(X, \mathscr{A})$.

Proof This follows directly from the fact, proved in [38, Corollary 3], that the forgetful functor $\mathrm{Sh}_G(X) \to \mathrm{Sh}(X)$ sends injective objects to soft sheaves, hence sends an injective resolution of $\mathscr{A}$ in $\mathrm{Sh}_G(X)$ to an acyclic resolution of $\mathscr{A}$ in $\mathrm{Sh}(X)$. □

Sheaves on the quotient. Let X and G as before, and consider the quotient map $p\colon X \to G\backslash X = Y$. There is a canonical equivalence of abelian categories between $\mathrm{Sh}_G(X)$ and $\mathrm{Sh}(Y)$, which can again be described rather simply in terms of espaces étalés:

If $\mathscr{B}$ is a sheaf of abelian groups on Y with espace étalé B, then $\pi^{-1}\mathscr{B}$ has a natural G-structure since its espace étalé is the fibre product $B \times_Y X$, where G operates on the second factor. Vice versa, if $\mathscr{A}$ is a G-equivariant sheaf on X with espace étalé A, we may form the quotient $G\backslash A \to Y$ which is the espace étalé of a sheaf on Y. It is not hard to see that these two constructions are mutually inverse.

Proposition 7.2 *Let X be a compact Hausdorff space and let G be a profinite group acting continuously and freely on X, with quotient $Y = G\backslash X$. Let A be an abelian group, and denote the constant sheaves on X and Y modelled on A by A_X and A_Y, respectively; endow A_X with the tautological G-operation. Then there is a natural isomorphism in $\mathscr{D}(\mathbf{Ab})$:*

$$\mathrm{R}(-)^G\left(\mathrm{R}\Gamma_{X,G}(A_X)\right) \cong \mathrm{R}\Gamma_Y(A_Y).$$

Proof Consider the following diagram of left exact functors between abelian categories:

$$\begin{array}{ccc} \mathrm{Sh}(Y) & \xrightarrow{\cong} & \mathrm{Sh}_G(X) \\ {\scriptstyle \Gamma_Y}\downarrow & & \downarrow{\scriptstyle \Gamma_{X,G}} \\ \mathbf{Ab} & \xleftarrow[(-)^G]{} & G\text{-}\mathbf{Mod} \end{array}$$

It is easy to see that it is commutative up to isomorphism of functors.

The equivalence on the upper horizontal line sends A_Y to A_X, as can be seen on their espaces étalés, which are simply $A \times Y$ and $A \times X$ with G operating trivially on A. The claim then follows by the chain rule for derived functors $\mathrm{R}(F \circ G) \cong \mathrm{R}F \circ \mathrm{R}G$. □

Corollary 7.3 *With the same assumptions as in Proposition 7.2 there is a spectral sequence with*

$$E_2^{p,q} = \mathrm{H}^p(G, \mathrm{H}^q(X, A)) \Rightarrow \mathrm{H}^{p+q}(Y, A),$$

where $\mathrm{H}^p(G, -)$ *denotes continuous group cohomology, and* $\mathrm{H}^q(X, A)$ *and* $\mathrm{H}^{p+q}(Y, A)$ *denote sheaf cohomology.* □

Proposition 7.2 and Corollary 7.3 have analogues in étale cohomology. We content ourselves with stating the analogue of the latter.

Proposition 7.4 *Let* $\mathscr{X} \to \mathscr{Y}$ *be a pro-étale Galois covering of schemes, with a profinite deck transformation group G, and let A be an abelian group. Then there is a natural spectral sequence with*

$$E_2^{p,q} = \mathrm{H}^p(G, \mathrm{H}^q_{\text{ét}}(\mathscr{X}, A)) \Rightarrow \mathrm{H}^{p+q}_{\text{ét}}(\mathscr{Y}, A).$$

Proof This is shown in [39, Chap. III, Remark 2.21.(b)]. Here is a brief summary of the proof.

First we assume that G is finite. Then $\mathscr{X}$ is an object of the small étale site of $\mathscr{Y}$ on which G acts by automorphism, hence the functor

$$\mathrm{Sh}(\mathscr{Y}_{\text{ét}}) \to G\text{-}\mathbf{Mod}, \quad \mathscr{F} \mapsto \Gamma(\mathscr{X}, \mathscr{F}),$$

is well-defined. Its composition with the forgetful functor G-**Mod**→ **Ab** is the usual global sections functor. Hence we obtain a spectral sequence relating the derived functors of these functors.

We deduce the general case by passing to the limit over all coverings $H\backslash\mathscr{X} \to \mathscr{Y}$ with $H \subseteq G$ an open normal subgroup. □

7.2 The Cohomology of X_F and $\mathscr{X}_F$

We can now compute some cohomology groups of these spaces using the Cartan–Leray spectral sequence. We discuss the topological case in detail, the étale case for torsion coefficients is analogous.

We begin by computing the cohomology of X_F when F is algebraically closed.

Proposition 7.5 *Let $\overline{F}$ be an algebraically closed field containing $\mathbb{Q}(\zeta_\infty)$.*

(i) Let A be an abelian torsion group. Then $\mathrm{H}^0(X_{\overline{F}}, A) = A$ and $\mathrm{H}^p(X_{\overline{F}}, A) = 0$ for all $p > 0$.

(ii) There is a canonical isomorphism of graded algebras

$$\bigoplus_{p\geq 0} \mathrm{H}^p(X_{\overline{F}}, \mathbb{Q}) \cong \bigoplus_{p\geq 0} \bigwedge\nolimits_{\mathbb{Q}}^{p} \overline{F}_{\mathrm{tf}}^{\times}.$$

(iii) $\mathrm{H}^0(X_{\overline{F}}, \mathbb{Z}) = \mathbb{Z}$, and for each $p > 0$ the inclusion $\mathbb{Z} \hookrightarrow \mathbb{Q}$ induces an isomorphism $\mathrm{H}^p(X_{\overline{F}}, \mathbb{Z}) \cong \mathrm{H}^p(X_{\overline{F}}, \mathbb{Q})$.

Proof Recall that $X_{\overline{F}}$ is a translate of the subgroup $(\overline{F}_{\mathrm{tf}}^{\times})^\vee \subset (\overline{F}^{\times})^\vee$. Hence for any $\chi \in X_{\overline{F}}$ we obtain a homeomorphism $t_\chi : (\overline{F}_{\mathrm{tf}}^{\times})^\vee \to X_{\overline{F}}$ by $t_\chi(\omega) = \chi\omega$, and therefore an isomorphism of cohomology groups $t_\chi^* : \mathrm{H}^\bullet(X_{\overline{F}}, A) \to \mathrm{H}^\bullet((\overline{F}_{\mathrm{tf}}^{\times})^\vee, A)$. These depend continuously on $\chi \in X_{\overline{F}}$, but since $X_{\overline{F}}$ is connected, they must be independent of χ. Hence we obtain a canonical isomorphism $\mathrm{H}^\bullet(X_{\overline{F}}, A) \cong \mathrm{H}^\bullet((\overline{F}_{\mathrm{tf}}^{\times})^\vee, A)$, and the statements follow from Proposition 3.8. □

Theorem 7.6 *Let F be a field containing $\mathbb{Q}(\zeta_\infty)$.*

(i) For every abelian torsion group A and every $m \geq 0$ there is a natural isomorphism

$$\mathrm{H}^m(X_F, A) \cong \mathrm{H}^m(F, A).$$

(ii) For each $m \geq 0$ there are natural isomorphisms

$$\mathrm{H}^m(X_F, \mathbb{Q}) \cong \left(\bigwedge\nolimits_{\mathbb{Q}}^{m} \overline{F}_{\mathrm{tf}}^{\times}\right)^{\mathrm{Gal}(\overline{F}/F)}.$$

In low degrees this simplifies to

$$\mathrm{H}^0(X_F, \mathbb{Q}) = \mathbb{Q} \quad \textit{and} \quad \mathrm{H}^1(X_F, \mathbb{Q}) \cong F^\times \otimes_{\mathbb{Z}} \mathbb{Q}.$$

(iii) The cohomology groups with integral coefficients begin with

$$\mathrm{H}^0(X_F, \mathbb{Z}) = \mathbb{Z} \quad \textit{and} \quad \mathrm{H}^1(X_F, \mathbb{Z}) \cong F_{\mathrm{tf}}^{\times}.$$

Proof We consider the Cartan–Leray spectral sequence as in Corollary 7.3 for $X = X_{\overline{F}}$, $G = \mathrm{Gal}(\overline{F}/F)$ and $Y = X_F$:

$$E_2^{p,q} = \mathrm{H}^p(\mathrm{Gal}(\overline{F}/F), \mathrm{H}^q(X_{\overline{F}}, A)) \Rightarrow \mathrm{H}^{p+q}(X_F, A). \tag{19}$$

(i) If A is torsion then $\mathrm{H}^q(X_{\overline{F}}, A) = 0$ for all $q > 0$ by Proposition 7.5, hence the spectral sequence (19) degenerates at E_2 and we obtain an isomorphism

$$\mathrm{H}^p(\mathrm{Gal}(\overline{F}/F), A) \cong \mathrm{H}^p(X_F, A).$$

(ii) Consider the spectral sequence (19) for $A = \mathbb{Q}$. All the cohomology groups $\mathrm{H}^q(X_{\overline{F}}, \mathbb{Q})$ are $\mathbb{Q}$-vector spaces, hence have trivial Galois cohomology, so $E_2^{p,q} = 0$ for $p \neq 0$. Again the spectral sequence (19) degenerates at E_2 and we obtain isomorphisms

$$\mathrm{H}^0(\mathrm{Gal}(\overline{F}/F), \mathrm{H}^q(X_{\overline{F}}, \mathbb{Q})) \cong \mathrm{H}^q(X_F, \mathbb{Q}).$$

Using Proposition 7.5(ii) we can rewrite this in the desired form. It is clear that $\mathrm{H}^0(X_F, \mathbb{Q}) = \mathbb{Q}$; for the calculation of $\mathrm{H}^1(X_F, \mathbb{Q})$ we need that the Galois invariants in $\overline{F}_{\mathrm{tf}}^\times$ are isomorphic to $F^\times \otimes \mathbb{Q}$. To see this consider the short exact sequence of Galois modules

$$0 \to \mu_\infty \to \overline{F}^\times \to \overline{F}_{\mathrm{tf}}^\times \to 0$$

and the associated long exact Galois cohomology sequence

$$0 \to \mu_\infty \to F^\times \to (\overline{F}_{\mathrm{tf}}^\times)^{\mathrm{Gal}(\overline{F}/F)} \to \mathrm{Hom}(\mathrm{Gal}(\overline{F}/F), \mu_\infty) \to \cdots$$

which shows that the cokernel of the inclusion $F_{\mathrm{tf}}^\times \to (\overline{F}_{\mathrm{tf}}^\times)^{\mathrm{Gal}(\overline{F}/F)}$ is a torsion group. Hence $F^\times \otimes \mathbb{Q} \to (\overline{F}_{\mathrm{tf}}^\times)^{\mathrm{Gal}(\overline{F}/F)}$ must be an isomorphism.

(iii) Consider the long exact cohomology sequence for the short exact sequence of coefficient groups $0 \to \mathbb{Z} \to \mathbb{Q} \to \mu_\infty \to 0$:

$$\cdots \to \mathbb{Q} \twoheadrightarrow \mu_\infty \xrightarrow{0} \mathrm{H}^1(X_F, \mathbb{Z}) \to \mathrm{H}^1(X_F, \mathbb{Q}) \to \mathrm{H}^1(X_F, \mu_\infty) \to \cdots.$$

We see that $\mathrm{H}^1(X_F, \mathbb{Z})$ is the kernel of the map $\mathrm{H}^1(X_F, \mathbb{Q}) \to \mathrm{H}^1(X_F, \mu_\infty)$. By (ii) the domain of this map is isomorphic to $F^\times \otimes \mathbb{Q} \cong (F_{\mathrm{sat}}^\times)_{\mathrm{tf}}$, by (i) the target is isomorphic to $\mathrm{H}^1(\mathrm{Gal}(\overline{F}/F), \mu_\infty) = \mathrm{Hom}(\mathrm{Gal}(\overline{F}/F), \mu_\infty)$. A tedious but straightforward calculation shows that this map

$$(F_{\mathrm{sat}}^\times)_{\mathrm{tf}} \to \mathrm{Hom}(\mathrm{Gal}(\overline{F}/F), \mu_\infty)$$

is given by $\alpha \mapsto \langle -, \alpha \rangle$, where $\langle -, - \rangle$ is the Kummer pairing discussed in Sect. 6.2. Hence its kernel is precisely $F_{\mathrm{tf}}^\times$. □

Remark 7.7 Even for $A = \mathbb{Z}$ the spectral sequence (19) gets somewhat simplified, namely then $E_2^{p,q} = 0$ whenever $p \neq 0$ and $q \neq 0$. This is because then $\mathrm{H}^q(X_{\overline{F}}, \mathbb{Z})$ is a $\mathbb{Q}$-vector space by Proposition 7.5(iii), hence all higher Galois cohomology groups for this space vanish.

In a similar vein we can identify Galois cohomology with constant torsion coefficients with étale cohomology of $\mathscr{X}_F$:

Theorem 7.8 *Let $F \supseteq \mathbb{Q}(\zeta_\infty)$ be a field and let A be an abelian torsion group, viewed as a trivial Galois module. Then for every $m \geq 0$ there is a canonical isomorphism*

$$\mathrm{H}^m_{\text{ét}}(\mathscr{X}_F, A) \cong \mathrm{H}^m(F, A).$$

Proof Consider the Cartan–Leray spectral sequence as in Proposition 7.4:

$$E_2^{p,q} = \mathrm{H}^p(\mathrm{Gal}(\overline{F}/F), \mathrm{H}^q_{\text{ét}}(\mathscr{X}_{\overline{F}}, A)) \Rightarrow \mathrm{H}^{p+q}_{\text{ét}}(\mathscr{X}_F, A).$$

Since $\mathscr{X}_{\overline{F}} \cong \mathrm{Spec}\, \mathbb{C}[\overline{F}^\times_{\mathrm{tf}}]$ as a scheme, $\mathrm{H}^q(\mathscr{X}_{\overline{F}}, A) = 0$ for all $q > 0$ by Proposition 3.14. Hence the spectral sequence degenerates at E_2, and the claim follows. □

The Galois symbol. Recall that for a field F and an integer $m \geq 0$ the m-th *Milnor K-group* $\mathrm{K}^{\mathrm{M}}_m(F)$ is defined as the quotient of the m-th exterior power $\bigwedge^m_{\mathbb{Z}} F^\times$ by the subgroup generated by all expressions of the form $\alpha \wedge (1-\alpha) \wedge \beta_2 \wedge \cdots \wedge \beta_m$ for $\alpha \in F \smallsetminus \{0, 1\}$ and $\beta_i \in F^\times$. In other words, the graded algebra $\bigoplus_{m \geq 0} \mathrm{K}^{\mathrm{M}}_m(F)$ is the quotient of the exterior algebra $\bigwedge^\bullet F^\times$ by the two-sided homogeneous ideal generated by all $\alpha \wedge (1-\alpha)$ with $\alpha \in F \smallsetminus \{0, 1\}$. See [40] for more information.

The Milnor K-groups are related to the more universal and well-known *Quillen K-groups* $\mathrm{K}_m(F)$ as follows. There are canonical isomorphisms $\mathrm{K}_0(F) \cong \mathbb{Z}$ and $\mathrm{K}_1(F) \cong F^\times$. There is therefore a unique multiplicative extension $\bigwedge^\bullet_{\mathbb{Z}} F^\times \to \mathrm{K}_\bullet(F)$; it factors degreewise through a homomorphism $\mathrm{K}^{\mathrm{M}}_m(F) \to \mathrm{K}_m(F)$. This is an isomorphism for $m = 0, 1$ by construction and for $m = 2$ by Matsumoto [41], see also [40, §12].

For n prime to the characteristic of F there is a canonical homomorphism

$$\partial\colon F^\times \to \mathrm{H}^1(F, \mathbb{Z}/n\mathbb{Z}(1)) = \mathrm{H}^1(\mathrm{Gal}(\overline{F}/F), \mu_n)$$

that sends $\alpha \in F^\times$ to the cohomology class $\partial\alpha$ defined by the crossed homomorphism $\mathrm{Gal}(\overline{F}/F) \to \mu_n$ sending σ to $\sigma(\sqrt[n]{\alpha})/\sqrt[n]{\alpha}$. Taking cup products, this extends to a homomorphism of graded rings

$$\bigoplus_{m\geq 0} \bigwedge^m F^\times \to \bigoplus_{m \geq 0} \mathrm{H}^m(F, \mathbb{Z}/n\mathbb{Z}(m)), \quad \alpha_1 \wedge \cdots \wedge \alpha_m \mapsto \partial\alpha_1 \cup \cdots \cup \partial\alpha_m. \tag{20}$$

For any $\alpha \in F \smallsetminus \{0, 1\}$ the relation $\partial\alpha \cup \partial(1-\alpha) = 0$ holds in $\mathrm{H}^2(F, \mathbb{Z}/n\mathbb{Z}(2))$. Hence (20) factors through the Milnor K-groups of F, defining the *Galois symbols*

$$\partial^m \colon \mathrm{K}^{\mathrm{M}}_m(F) \to \mathrm{H}^m(F, \mathbb{Z}/n\mathbb{Z}(m)).$$

The *Bloch–Kato conjecture* [42, p. 118], now a theorem due to Voevodsky [5, Theorem 6.1], asserts that for every field F, every integer $m \geq 0$ and every $n \in \mathbb{N}$ prime to the characteristic of F the induced group homomorphism

$$\mathrm{K}^{\mathrm{M}}_m(F) \otimes \mathbb{Z}/n\mathbb{Z} \to \mathrm{H}^m(F, \mathbb{Z}/n\mathbb{Z}(m))$$

is an isomorphism.

Assume now that F contains $\mathbb{Q}(\zeta_\infty)$; then we may ignore Tate twists. By Theorem 7.6(i) we obtain therefore an isomorphism

$$\partial^m_n \colon \mathrm{K}^{\mathrm{M}}_m(F) \otimes \mathbb{Z}/n\mathbb{Z} \to \mathrm{H}^m(X_F, \mathbb{Z}/n\mathbb{Z}); \tag{21}$$

taking an inductive limit over all $n \in \mathbb{N}$ we can also construct an isomorphism

$$\partial^m_\infty \colon \mathrm{K}^{\mathrm{M}}_m(F) \otimes \mathbb{Q}/\mathbb{Z} \to \mathrm{H}^m(X_F, \mathbb{Q}/\mathbb{Z}).$$

Note also that ∂^1_n lifts canonically to an isomorphism $\partial^1 \colon \mathrm{K}^{\mathrm{M}}_1(F) = F^\times \to \mathrm{H}^1(X_F, \mathbb{Z})$ by Theorem 7.6(iii). However, for $\alpha \in F \smallsetminus \{0, 1\}$ the element $\partial^1(\alpha) \cup \partial^1(1-\alpha) \in \mathrm{H}^2(X_F, \mathbb{Z})$ is nonzero since its image in $\mathrm{H}^2(X_F, \mathbb{Q})$ is nonzero by Theorem 7.6(ii). In particular the resulting homomorphism

$$\bigwedge\nolimits^m_{\mathbb{Z}} F^\times \to \mathrm{H}^m(X_F, \mathbb{Z}), \quad \alpha_1 \wedge \cdots \wedge \alpha_m \mapsto \partial^1(\alpha_1) \cup \cdots \cup \partial^1(\alpha_m)$$

does not factor through $\mathrm{K}^{\mathrm{M}}_m(F)$ for any $m \geq 2$ and any field $F \supseteq \mathbb{Q}(\zeta_\infty)$.

Proposition 7.9 *Let $F \supseteq \mathbb{Q}(\zeta_\infty)$ be a field.*

(i) For every $m \geq 0$ the homomorphism

$$\mathrm{H}^m(X_F, \mathbb{Q}) \to \mathrm{H}^m(X_F, \mathbb{Q}/\mathbb{Z})$$

is surjective.

(ii) For every $m \geq 0$ the group $\mathrm{H}^m(X_F, \mathbb{Z})$ is torsion-free.

Proof Consider the commutative diagram

$$\begin{array}{ccc} \bigwedge^m_{\mathbb{Q}}(F^\times \otimes \mathbb{Q}) & \longrightarrow & \mathrm{K}^{\mathrm{M}}_m(F) \otimes \mathbb{Q}/\mathbb{Z} \\ \downarrow & & \downarrow \\ \mathrm{H}^m(X_F, \mathbb{Q}) & \longrightarrow & \mathrm{H}^m(X_F, \mathbb{Q}/\mathbb{Z}). \end{array}$$

Here the upper horizontal map is surjective by construction, and the right vertical map is surjective by the Bloch–Kato conjecture. Hence the lower horizontal map has to be surjective as well, which proves (i). The map from (i) is part of a long exact sequence

$$\cdots \to \mathrm{H}^m(X_F,\mathbb{Z}) \to \mathrm{H}^m(X_F,\mathbb{Q}) \to \mathrm{H}^m(X_F,\mathbb{Q}/\mathbb{Z}) \xrightarrow{\delta^m}$$
$$\mathrm{H}^{m+1}(X_F,\mathbb{Z}) \to \mathrm{H}^{m+1}(X_F,\mathbb{Q}) \to \mathrm{H}^{m+1}(X_F,\mathbb{Q}/\mathbb{Z}) \xrightarrow{\delta^{m+1}} \cdots .$$

By (i) the connecting homomorphisms δ^m have to vanish, hence the short sequences

$$0 \to \mathrm{H}^m(X_F,\mathbb{Z}) \to \mathrm{H}^m(X_F,\mathbb{Q}) \to \mathrm{H}^m(X_F,\mathbb{Q}/\mathbb{Z}) \to 0$$

are also exact. Therefore $\mathrm{H}^m(X_F,\mathbb{Z})$ injects into a $\mathbb{Q}$-vector space, which proves (ii). □

The cup product. Consider the following axiomatisation of the situation in Theorem 7.6(i). Assume that X is a connected compact Hausdorff space such that $\mathrm{H}^n(X, A) = 0$ for all finite abelian groups A and $n > 0$, and that a profinite group G operates continuously and freely on X. Then arguing as above we obtain canonical isomorphisms $\mathrm{H}^n(G, A) \cong \mathrm{H}^n(X/G, A)$ for all $n \geq 0$. The existence of such a space X can thus be used to relate continuous group cohomology with the cohomology of a compact Hausdorff space, and one may ask whether such a space always exists. The answer is no:

Proposition 7.10 *Let X be a connected compact Hausdorff space with $\mathrm{H}^n(X, A) = 0$ for $n > 0$, for all finite abelian groups A. Let G be a profinite group acting continuously and freely on X. Then G is torsion-free.*

It would be interesting to see whether one can deduce further properties of G.

Proof Assume the contrary; by passing to a suitable subgroup we may then assume that G is cyclic of order p for some prime p. By the Cartan–Leray spectral sequence we obtain isomorphisms $\mathrm{H}^n(X/G,\mathbb{F}_p) \cong \mathrm{H}^n(G,\mathbb{F}_p)$. It is not hard to see that these isomorphisms commute with the cup product.

By Lemma 7.11 below, every class in $\mathrm{H}^2(X/G,\mathbb{F}_p)$ is nilpotent. However, there exists a non-nilpotent class in $\mathrm{H}^2(G,\mathbb{F}_p)$ (see e.g. [43, Chap. II, Corollary 4.2]), a contradiction. □

Lemma 7.11 *Let X be a compact Hausdorff space, let A be a ring and let $n > 0$. Then every class in $\mathrm{H}^n(X, A)$ is nilpotent.*

Proof There is a canonical isomorphism of graded rings

$$\mathrm{H}^*(X, A) \cong \varinjlim_{\mathfrak{U}} \check{\mathrm{H}}^*(\mathfrak{U}, A)$$

where the limit runs over all *finite* open covers $\mathfrak{U}$ of X. By construction, every class of positive degree in $\check{\mathrm{H}}^*(\mathfrak{U}, A)$ is nilpotent. □

Corollary 7.12 *Let $F \supseteq \mathbb{Q}(\zeta_\infty)$ be a field. Then the absolute Galois group of F is torsion-free.* □

Of course, Corollary 7.12 also follows from the Theorem of Artin–Schreier ([44, Satz 4], see also [45, Theorem 11.14]) which says that the absolute Galois group of a field is finite and nontrivial if and only if that field is real closed. Clearly, no field containing $\mathbb{Q}(\zeta_\infty)$ can be real closed.

8 The Cyclotomic Character

In this section we develop a variant of the preceding constructions that works also for fields which do not contain all roots of unity, provided their absolute Galois groups are pro-ℓ-groups. We will summarise the necessary structural results, the proofs being very similar to the case treated before, and then discuss in more detail some actions on cohomology groups that only become nontrivial in this new case.

8.1 A Variant With (Some) Roots of Unity

Throughout this section we fix a rational prime ℓ and a perfect field F with algebraic closure $\overline{F}$ whose characteristic is not equal to ℓ (but may well be positive) such that $\mathrm{Gal}(\overline{F}/F)$ is a pro-ℓ-group.

We write μ_n for the group of all n-th roots of unity in $\overline{F}$, where $n \in \mathbb{N}$. We also set

$$\mu_{\ell^\infty} = \bigcup_{n\in\mathbb{N}} \mu_{\ell^n} \quad \text{and} \quad \mu_{\ell'} = \bigcup_{n\in\mathbb{N}\smallsetminus\ell\mathbb{N}} \mu_n.$$

The groups μ_{ℓ^n} for $n < \infty$ are cyclic of order ℓ^n. There is a continuous group homomorphism

$$\chi_{\ell,F}\colon \mathrm{Gal}(\overline{F}/F) \to \mathrm{Aut}\,\mu_{\ell^\infty} \xrightarrow{\cong} \mathbb{Z}_\ell^\times$$

called the *ℓ-adic cyclotomic character* and characterised by $\sigma(\zeta) = \zeta^{\chi_{\ell,F}(\sigma)}$ for all $\sigma \in \mathrm{Gal}(\overline{F}/F)$ and $\zeta \in \mu_{\ell^\infty}$. Its kernel is the group $\mathrm{Gal}(\overline{F}/F(\zeta_{\ell^\infty}))$, and the possibilities for its image are rather restricted.

Proposition 8.1 *Let n be maximal such that $\mu_{\ell^n} \subset F$. If ℓ is odd or if $n > 1$, the image of* $\mathrm{Gal}(\overline{F}/F)$ *under the ℓ-adic cyclotomic character is equal to the subgroup*

$$U_{\ell^n} \stackrel{\text{def}}{=} 1 + \ell^n\mathbb{Z}_\ell \subset \mathbb{Z}_\ell^\times.$$

If $\ell = 2$ and $n = 1$, there is some $m \in \{2, 3, 4, \dots, \infty\}$ such that the image is generated by $U_{2^m} \subset \mathbb{Z}_2^\times$ and $-1 \in \mathbb{Z}_2^\times$, where we set $U_{2^\infty} = \{1\}$.

Proof By assumption, this image $H = \operatorname{im} \chi_{\ell,F}$ is a closed subgroup of $\mathbb{Z}_\ell^\times$ which is contained in U_{ℓ^n} but not contained in $U_{\ell^{n+1}}$. If ℓ is odd or $n > 1$, the ℓ-adic exponential series defines an isomorphism of topological groups $\ell^n\mathbb{Z}_\ell \to U_{\ell^n}$. Any closed subgroup of $\mathbb{Z}_\ell$ is an ideal, hence of the form $\ell^m\mathbb{Z}_\ell$ for $0 \le m \le \infty$, and an ideal contained in $\ell^n\mathbb{Z}_\ell$ but not contained in $\ell^{n+1}\mathbb{Z}_\ell$ must be equal to $\ell^n\mathbb{Z}_\ell$.

The case $\ell = 2$ and $n = 1$ remains. Here the exponential series defines an isomorphism $4\mathbb{Z}_2 \to U_4$, hence by the previous argument we find that $H \cap U_4 = U_{2^m}$ for some $2 \le m \le \infty$. From the short exact sequence

$$0 \to 4\mathbb{Z}_2 \overset{\exp}{\to} U_4 \to \{\pm 1\} \to 0$$

and the assumption that $H \not\subseteq U_4$ we conclude that $H = \pm U_{2^m}$. □

By considering algebraic extensions of finite fields we see that all of these possibilities do occur. For another example, if F is a real closed field, then $\ell = 2$ and $n = 1$, and the image of $\chi_{2,F}\colon \operatorname{Gal}(\overline{F}/F) \hookrightarrow \mathbb{Z}_2^\times$ is simply $\{\pm 1\} \subset \mathbb{Z}_2^\times$.

We will only formulate our main results in this section for the case where $\operatorname{im} \chi_{\ell,F} = U_{\ell^n}$, the case $\operatorname{im} \chi_{2,F} = \pm U_{2^m}$ being similar.

The spaces. $Y_{\ell^n,F}$ For each $n \in \mathbb{N} \cup \{\infty\}$ and each perfect field F of characteristic different from ℓ with $\mu_{\ell^n} \subset F$ and each injective character $\iota\colon \mu_{\ell^n} \to \mathbb{S}^1$ we will now define a topological space $Y_{\ell^n,F}(\iota)$. We start with the algebraically closed case and let

$$Y_{\ell^n,\overline{F}}(\iota) = \{\chi \in \operatorname{Hom}(\overline{F}^\times/\mu_{\ell'}, \mathbb{S}^1) \mid \chi|_{\mu_{\ell^n}} = \iota\}$$

where 'Hom' denotes group homomorphisms; this space is endowed with the compact-open topology. The Galois group $\operatorname{Gal}(\overline{F}/F)$ operates continuously on $Y_{\ell^n,\overline{F}}(\iota)$, and we set

$$Y_{\ell^n,F}(\iota) = \operatorname{Gal}(\overline{F}/F)\backslash Y_{\ell^n,\overline{F}}(\iota).$$

Proposition 8.2 *Let ℓ be a rational prime and F a perfect field of characteristic different from ℓ such that* $\operatorname{Gal}(\overline{F}/F)$ *is a pro-ℓ-group. Let $n \in \mathbb{N} \cup \{\infty\}$ such that $\mu_{\ell^n} \subset F$ and let $\iota\colon \mu_{\ell^n} \to \mathbb{S}^1$ be an injective character.*

(i) $Y_{\ell^n,F}(\iota)$ is a nonempty compact Hausdorff space.
(ii) $\operatorname{Gal}(\overline{F}/F)$ *operates freely and properly on $Y_{\ell^n,\overline{F}}(\iota)$.*

Proof The proof is essentially analogous to that of Proposition 5.1(i) and (ii). For the freeness in (ii) we use Lemma 8.3 below, which is similar to Lemma 4.12 above. □

Lemma 8.3 *Let ℓ and F be as in Proposition 8.2, and let k be a field. Then* $\operatorname{Gal}(\overline{F}/F)$ *operates freely on the set $\mathcal{J}_\ell(\overline{F}, k)$ of all group homomorphisms $\overline{F}^\times \to k^\times$ which are injective on μ_{ℓ^∞} and trivial on $\mu_{\ell'}$.*

Proof Let $\sigma \in \operatorname{Gal}(\overline{F}/F)$ be different from the identity element, and let $\chi \in \mathcal{J}_\ell(\overline{F}, k)$. We need to show that $\sigma(\chi) \neq \chi$.

Replacing F by the fixed field of σ we may assume that σ topologically generates $\mathrm{Gal}(\overline{F}/F)$. Since $\mathrm{Gal}(\overline{F}/F)$ is a pro-ℓ-group, it is then either finite cyclic or isomorphic to $\mathbb{Z}_\ell$.

In the first case it has to be cyclic of order 2 by the Theorem of Artin–Schreier [44, Satz 4], and F has to be real closed. Then $\chi(\zeta_4)$ is a primitive fourth root of unity in k, and $\sigma(\chi)(\zeta_4) = \chi(\sigma(\zeta_4)) = \chi(\zeta_4^{-1}) = \chi(\zeta_4)^{-1} \neq \chi(\zeta_4)$, hence $\chi \neq \sigma(\chi)$.

In the second case we let E be the fixed field of σ^ℓ, so that E/F is a cyclic extension of degree ℓ. By Hilbert's Theorem 90 we find some $\alpha \in E$ with $\sigma(\alpha)/\alpha = \zeta_\ell$, and since $\chi(\zeta_\ell) \neq 1$ we then find $\sigma(\chi)(\alpha) \neq \chi(\alpha)$, i.e. $\chi \neq \sigma(\chi)$. □

It is here that the assumption that $\mathrm{Gal}(\overline{F}/F)$ is a pro-ℓ-group is used.

Lemma 8.4 *Let ℓ and F as before, and assume that the image of $\chi_{\ell,F}\colon \mathrm{Gal}(\overline{F}/F) \to \mathbb{Z}_\ell^\times$ is equal to U_{ℓ^n} for some $n \in \mathbb{N} \cup \{\infty\}$. Let $\iota\colon \mu_{\ell^n} \to \mathbb{S}^1$ be an injective character and let $\tilde{\iota}\colon \mu_{\ell^\infty} \to \mathbb{S}^1$ be an injective character with $\tilde{\iota}|_{\mu_{\ell^n}} = \iota$.*

Then the spaces $Y_{\ell^n,F}(\iota)$ and $Y_{\ell^\infty,F(\zeta_{\ell^\infty})}(\tilde{\iota})$ are canonically homeomorphic.

Proof There is a tautological inclusion $Y_{\ell^\infty,\overline{F}}(\tilde{\iota}) \hookrightarrow Y_{\ell^n,\overline{F}}(\iota)$, which is equivariant for the group inclusion $\mathrm{Gal}(\overline{F}/F(\zeta_\infty)) \hookrightarrow \mathrm{Gal}(\overline{F}/F)$. It therefore descends to a continuous map $Y_{\ell^\infty,F(\zeta_{\ell^\infty})}(\tilde{\iota}) \to Y_{\ell^n,F}(\iota)$, and since the spaces under consideration are compact Hausdorff spaces, it suffices to show that this map is a bijection.

Note that $Y_{\ell^n,\overline{F}}(\iota)$ is the disjoint union (as sets, not as topological spaces!)

$$\bigcup_{\substack{j \in \mathrm{Hom}(\mu_{\ell^\infty},\mathbb{S}^1) \\ j|_{\mu_{\ell^n}} = \iota}} Y_{\ell^\infty,\overline{F}}(j),$$

and the subgroup $\mathrm{Gal}(\overline{F}/F(\zeta_{\ell^\infty})) \subset \mathrm{Gal}(\overline{F}/F)$ preserves each summand while the quotient $\mathrm{Gal}(F(\zeta_{\ell^\infty})/F) \cong U_{\ell^n}$ permutes the summands simply transitively. Therefore the quotient of each summand by its stabiliser $\mathrm{Gal}(\overline{F}/F(\zeta_{\ell^\infty}))$ maps bijectively to the quotient of the whole set by $\mathrm{Gal}(\overline{F}/F)$. □

Proposition 8.5 *Let ℓ and F be as before with $\mathrm{im}\,\chi_{\ell,F} = U_{\ell^n}$ and let $\iota\colon \mu_{\ell^n} \hookrightarrow \mathbb{S}^1$. Then $Y_{\ell^n,F}(\iota)$ is connected.*

Proof By Lemma 8.4 we may assume that $n = \infty$. Then $Y_{\ell^\infty,\overline{F}}(\iota)$ is homeomorphic to the Pontryagin dual of the torsion-free group $\overline{F}^\times/\mu_{\ell'}\mu_{\ell^\infty} = \overline{F}^\times/\mu_\infty$, hence it is connected. Therefore its quotient $Y_{\ell^\infty,F}(\iota)$ is also connected. □

Proposition 8.6 *Let ℓ, F, n, ι and $\tilde{\iota}$ as before.*

(i) The étale fundamental group of $Y_{\ell^\infty,\overline{F}}(\tilde{\iota})$ is trivial.
(ii) The étale fundamental group of $Y_{\ell^n,F}(\iota)$ is isomorphic to $\mathrm{Gal}(\overline{F}/F(\zeta_{\ell^\infty}))$; this isomorphism is canonical up to inner automorphisms.

Proof For (i) note that $Y_{\ell^\infty,\overline{F}}(\tilde{\iota})$ is homeomorphic to the Pontryagin dual of $\overline{F}^\times/\mu_\infty$, a $\mathbb{Q}$-vector space. Hence $Y_{\ell^\infty,\overline{F}}(\tilde{\iota})$ is the universal profinite covering space of $Y_{\ell^\infty,F(\zeta_{\ell^\infty})}$

by Proposition 8.2. It follows that the deck transformation group $\mathrm{Gal}(\overline{F}/F(\zeta_{\ell^\infty}))$ is isomorphic to the étale fundamental group of $Y_{\ell^\infty,F(\zeta_{\ell^\infty})}(\tilde{\iota})$; by Lemma 8.4 this space is canonically homeomorphic to $Y_{\ell^n,F}(\iota)$. □

Theorem 8.7 *Let ℓ, F, n and ι as before, and let A be an abelian torsion group. Then there are canonical isomorphisms*

$$\mathrm{H}^m(Y_{\ell^n,F}, A) \cong \mathrm{H}^m(F(\zeta_{\ell^\infty}), A)$$

and

$$\mathrm{H}^m(Y_{\ell^n,F}, \mathbb{Q}) \cong \left(\bigwedge\nolimits_{\mathbb{Q}}^m \overline{F}_{\mathrm{tf}}^\times\right)^{\mathrm{Gal}(\overline{F}/F(\zeta_{\ell^\infty}))}$$

for each $m \geq 0$.

Proof The proof is similar to that of Theorem 7.6. □

There is also again a scheme-theoretic version of these constructions. For a perfect field F of characteristic different from ℓ with algebraic closure $\overline{F}$ such that $\mathrm{Gal}(\overline{F}/F)$ is a pro-ℓ-group, an $n \in \mathbb{N} \cup \{\infty\}$ with $\mu_{\ell^n} \subset F$ and an embedding $\iota\colon \mu_{\ell^n}(F) \hookrightarrow \mathbb{S}^1$ we set

$$B_{\ell^n,F} = \left(\mathbb{C}[\overline{F}^\times/\mu_{\ell'}]/I\right)^{\mathrm{Gal}(\overline{F}/F)},$$

where $I \subset \mathbb{C}[\overline{F}\times/\mu_{\ell'}]$ is the ideal generated by all $[\zeta] - \iota(\zeta)\cdot[1]$ with $\zeta \in \mu_{\ell^n}(F)$. Note that if $n < \infty$ then this ideal is generated by a single element $[\zeta_{\ell^n}] - \iota(\zeta_{\ell^n})\cdot[1]$ with ζ_{ℓ^n} a primitive ℓ^n-th root of unity.

Then we set $\mathcal{Y}_{\ell^n,F} = \operatorname{Spec} B_{\ell^n,F}$. Note that this still depends on ι, but we suppress this to lighten the notation. In complete analogy to the schemes $\mathcal{X}_F$ we obtain the following properties:

Theorem 8.8 *Let ℓ be a rational prime and let F be a perfect field of characteristic other than ℓ. Assume that* $\mathrm{Gal}(\overline{F}/F)$ *is a pro-ℓ-group and that the image of* $\chi_{\ell,n}\colon \mathrm{Gal}(\overline{F}/F) \to \mathbb{Z}_\ell^\times$ *is equal to U_{ℓ^n} for some $n \in \mathbb{N} \cup \{\infty\}$.*

Then the scheme $\mathcal{Y}_{\ell^n,F}$ is connected, and its étale fundamental group is isomorphic to $\mathrm{Gal}(\overline{F}/F(\zeta_{\ell^\infty}))$, *the isomorphism being canonical up to inner automorphisms. For abelian torsion groups A we obtain natural isomorphisms*

$$\mathrm{H}^m_{\text{ét}}(\mathcal{Y}_{\ell^n,F}, A) \cong \mathrm{H}^m(F(\zeta_{\ell^\infty}), A).$$

The space $Y_{\ell^n,F}$ can be identified with a subspace in $\mathcal{Y}_{\ell^n,F}(\mathbb{C})$ with the complex topology; it is a strong deformation retract.

Proof In analogy to Lemma 8.4 there is a canonical isomorphism $\mathcal{Y}_{\ell^n,F} \cong \mathcal{Y}_{\ell^\infty,F(\zeta_{\ell^\infty})}$, hence we may assume that $n = \infty$. The proof is then analogous to those of Corollary 4.15, Theorem 5.7(ii) and Theorem 7.8. □

8.2 Three Actions on Cohomology

We assume that ℓ is a prime, F a perfect field of characteristic other than ℓ with algebraic closure $\overline{F}$ such that $\mathrm{Gal}(\overline{F}/F)$ is a pro-ℓ-group and such that the image of $\chi_{\ell,F}\colon \mathrm{Gal}(\overline{F}/F) \to \mathbb{Z}_\ell^\times$ is equal to U_{ℓ^n} for some (finite!) $n \in \mathbb{N}$. We have seen that then for each abelian torsion group A and each $m \geq 0$ there are canonical isomorphisms

$$\mathrm{H}^m(Y_{\ell^n,F}, A) \cong \mathrm{H}^m_{\text{ét}}(\mathcal{Y}_{\ell^n,F}, A) \cong \mathrm{H}^m(F(\zeta_{\ell^\infty}), A). \tag{22}$$

On each of the groups in (22) there is a natural action from the left by a certain group; we will show below that these three actions are compatible.

The topological action. By construction $Y_{\ell^n,\overline{F}}$ is a closed subset of the Pontryagin dual $(\overline{F}^\times/\mu_{\ell'})^\vee$, more precisely a translate of $(\overline{F}^\times_{\mathrm{tf}})^\vee$. The group $\overline{F}^\times/\mu_{\ell'}$ is divisible and has no torsion elements of order prime to ℓ, therefore it is a $\mathbb{Z}_{(\ell)}$-module in a unique way (here $\mathbb{Z}_{(\ell)} \subset \mathbb{Q}$ is the ring of rational numbers whose denominators are prime to ℓ). Therefore the group of units $\mathbb{Z}_{(\ell)}^\times$ acts on $(\overline{F}^\times/\mu_{\ell'})$ by group automorphisms from the left: $u \cdot (\alpha \bmod \mu_{\ell'}) = (\alpha^u \bmod \mu_{\ell'})$. Hence it acts on the Pontryagin dual $(\overline{F}^\times/\mu_{\ell'})^\vee$ from the right: $(\chi \cdot u)(\alpha) = \chi(\alpha^u)$.

This action does not preserve the subspace $Y_{\ell^n,\overline{F}} \subset (\overline{F}^\times/\mu_{\ell'})^\vee$, but its restriction to the subgroup

$$U_{(\ell^n)} = 1 + \ell^n \mathbb{Z}_{(\ell)} \subset \mathbb{Z}_{(\ell)}^\times$$

will, because elements of $U_{(\ell^n)}$ operate trivially on μ_{ℓ^n}.

This action of $U_{(\ell^n)}$ on $Y_{\ell^n,\overline{F}}$ commutes with that of $\mathrm{Gal}(\overline{F}/F)$, hence it descends to a right action of $U_{(\ell^n)}$ on $Y_{\ell^n,F}$.

Then by functoriality we obtain a left action of $U_{(\ell^n)}$ on the cohomology group $\mathrm{H}^m(Y_{\ell^n,F}, A)$ for any abelian torsion group A and any $m \geq 0$.

The arithmetic action. The scheme $\mathcal{Y}_{\ell^n,F}$ admits a natural model over the ring of cyclotomic integers $\mathbb{Z}[\zeta_{\ell^n}]$. To be precise, let

$$\mathcal{Y}_{F,\mathrm{int}} = \operatorname{Spec} B_{F,\mathrm{int}}$$

with

$$B_{F,\mathrm{int}} = \left(\mathbb{Z}[\overline{F}^\times/\mu_{\ell'}][([\zeta_\ell]-1)^{-1}]\right)^{\mathrm{Gal}(\overline{F}/F)}$$

where $\zeta_\ell \in F$ is an ℓ-th root of unity. Each embedding $\iota\colon \mu_{\ell^n}(F) \hookrightarrow \mathbb{S}^1$ induces a ring embedding

$$e_\iota\colon \mathbb{Z}[\zeta_{\ell^n}] = \mathbb{Z}[\mu_{\ell^n}(\mathbb{C})] \hookrightarrow B_{F,\mathrm{int}}$$

with $e_\iota(\iota(\zeta)) = [\zeta]$ for each $\zeta \in \mu_{\ell^n}(F)$. This ring embedding turns $\mathcal{Y}_{F,\mathrm{int}}$ into a $\mathbb{Z}[\zeta_{\ell^n}]$-scheme, and there is a natural isomorphism

$$\mathcal{Y}_{F,\mathrm{int}} \times_{e_\iota, \operatorname{Spec}\mathbb{Z}[\zeta_{\ell^n}]} \operatorname{Spec}\mathbb{C} \cong \mathcal{Y}_{\ell^n,F}(\iota).$$

There are then also a natural isomorphisms

$$\mathrm{H}^m_{\text{ét}}(\mathcal{Y}_{\ell^n,F}, A) \cong \mathrm{H}^m_{\text{ét}}(\mathcal{Y}_{F,\text{int}} \times_{e_\iota, \operatorname{Spec}\mathbb{Z}[\zeta_{\ell^n}]} \operatorname{Spec}\overline{\mathbb{Q}}, A).$$

Now $\mathrm{Gal}(\overline{\mathbb{Q}}/\mathbb{Q}(\zeta_{\ell^n}))$ operates from the left on $B_{F,\text{int}} \otimes_{e_\iota, \mathbb{Z}[\zeta_{\ell^n}]} \overline{\mathbb{Q}}$ (trivially on the first factor and tautologically on the second factor), hence from the right on the spectrum of this algebra, hence from the left on the cohomology of the latter. Thus we obtain a left action of $\mathrm{Gal}(\overline{\mathbb{Q}}/\mathbb{Q}(\zeta_{\ell^n}))$ on $\mathrm{H}^m_{\text{ét}}(\mathcal{Y}_{\ell^n,F}, A)$.

The group-theoretic action. This is the simplest to describe: from the short exact sequence of profinite groups

$$1 \to \mathrm{Gal}(\overline{F}/F(\zeta_{\ell^\infty})) \to \mathrm{Gal}(\overline{F}/F) \overset{\chi_{\ell,F}}{\to} U_{\ell^n} \to 1$$

we obtain a right action of $\mathrm{Gal}(\overline{F}/F)$ on its normal subgroup $\mathrm{Gal}(\overline{F}/F(\zeta_{\ell^\infty}))$ by $h^g = g^{-1}hg$, hence a left action on $\mathrm{H}^m(F(\zeta_{\ell^\infty}), A) = \mathrm{H}^m(\mathrm{Gal}(\overline{F}/F(\zeta_{\ell^\infty})), A)$. Since inner group automorphisms act trivially on cohomology, this descends to a left action by $U_{\ell^n} \cong \mathrm{Gal}(F(\zeta_{\ell^\infty})/F)$.

Theorem 8.9 *Let ℓ be a rational prime, F a perfect field of characteristic other than ℓ with algebraic closure $\overline{F}$ such that* $\mathrm{Gal}(\overline{F}/F)$ *is a pro-ℓ-group and such that* $\operatorname{im}\chi_{\ell,F} = U_{\ell^n}$ *for some $n \in \mathbb{N}$. Let A be an abelian torsion group. Then the following claims hold for each $m \geq 0$:*

(i) The arithmetic action of $\mathrm{Gal}(\overline{\mathbb{Q}}/\mathbb{Q}(\zeta_{\ell^n}))$ *on* $\mathrm{H}^m_{\text{ét}}(\mathcal{Y}_{\ell^n,F}, A)$ *factors through the ℓ-adic cyclotomic character*

$$\chi_\ell = \chi_{\ell,\mathbb{Q}(\zeta_{\ell^n})} \colon \mathrm{Gal}(\overline{\mathbb{Q}}/\mathbb{Q}(\zeta_{\ell^n})) \twoheadrightarrow U_{\ell^n} \subset \mathbb{Z}_\ell^\times.$$

Hence it defines an action by U_{ℓ^n} on $\mathrm{H}^m_{\text{ét}}(\mathcal{Y}_{\ell^n,F}, A)$ which we also call arithmetic.

(ii) The topological action of $U_{(\ell^n)}$ on $\mathrm{H}^m(Y_{\ell^n,F}, A)$ extends uniquely to a continuous action of $U_{\ell^n} = 1 + \ell^n\mathbb{Z}_\ell$ on the same space. Here continuity refers to the ℓ-adic topology on U_{ℓ^n}.

Moreover the isomorphisms in (22) are U_{ℓ^n}-equivariant up to a sign. More precisely, the isomorphism

$$\mathrm{H}^m_{\text{ét}}(\mathcal{Y}_{\ell^n,F}, A) \cong \mathrm{H}^m(F(\zeta_{\ell^\infty}), A)$$

is equivariant for the identity $U_{\ell^n} \to U_{\ell^n}$, while the other two isomorphisms

$$\mathrm{H}^m(Y_{\ell^n,F}, A) \cong \mathrm{H}^m_{\text{ét}}(\mathcal{Y}_{\ell^n,F}, A) \quad \text{and} \quad \mathrm{H}^m(Y_{\ell^n,F}, A) \cong \mathrm{H}^m(F(\zeta_{\ell^\infty}), A)$$

are equivariant for the inverse map $U_{\ell^n} \to U_{\ell^n}$, $u \mapsto u^{-1}$.

Proof Broken up into smaller pieces, this is proved below in Proposition 8.13, Corollary 8.16 and Proposition 8.25 below. □

The proof of Theorem 8.9 will fill up the remainder of this section. More precisely, for each two of the three actions we will establish equivariance for these two actions, and prove along the way that the actions factor through U_{ℓ^n}. From a strictly logical point of view this is redundant, but we believe each of the three proofs reveals something particular about the objects under consideration.

8.3 Compatibility of the Arithmetic and Group-Theoretic Actions

For each field F satisfying the conditions of Theorem 8.9 and each 'coefficient field' k we set $\mathcal{Y}_{F,k} = \mathcal{Y}_{F,\mathrm{int}} \times_{\operatorname{Spec}\mathbb{Z}} \operatorname{Spec} k$. The scheme $\mathcal{Y}_{F,k}$ is not necessarily connected, but for $k = \mathbb{Q}$ it is. We will now determine the étale fundamental group of $\mathcal{Y}_{F,\mathbb{Q}}$.

There is a pro-étale but possibly disconnected normal covering $\mathcal{Y}_{\overline{F},\overline{\mathbb{Q}}}$. Its deck transformation group can be identified with $\operatorname{Gal}(\overline{F}/F) \times \operatorname{Gal}(\overline{\mathbb{Q}}/\mathbb{Q})$ which operates in the obvious way on $\mathcal{Y}_{\overline{F},\overline{\mathbb{Q}}}$ (from the right, however). The space of connected components $\pi_0(\mathcal{Y}_{\overline{F},\overline{\mathbb{Q}}})$ is canonically homeomorphic to $\operatorname{Isom}(\mu_{\ell^\infty}(\overline{F}), \mu_{\ell^\infty}(\overline{\mathbb{Q}}))$, with its obvious left action by $\operatorname{Gal}(\overline{\mathbb{Q}}/\mathbb{Q})$ and its obvious right action by $\operatorname{Gal}(\overline{F}/F)$. The space $\operatorname{Isom}(\mu_{\ell^\infty}(\overline{F}), \mu_{\ell^\infty}(\overline{\mathbb{Q}}))$ is a two-sided principal homogeneous space for the abelian group $\mathbb{Z}_\ell^\times$, and the Galois actions respect this structure.

Lemma 8.10 *Let* $\operatorname{Gal}(\overline{F}/F) \times \operatorname{Gal}(\overline{\mathbb{Q}}/\mathbb{Q})$ *act on* $\operatorname{Isom}(\mu_{\ell^\infty}(\overline{F}), \mu_{\ell^\infty}(\overline{\mathbb{Q}}))$ *by* $(\sigma_F, \sigma_{\mathbb{Q}}) \cdot \iota = \chi_\ell(\sigma_F)\chi_\ell(\sigma_{\mathbb{Q}})^{-1} \cdot \iota$. *Then the homeomorphism* $\pi_0(\mathcal{Y}_{\overline{F},\overline{\mathbb{Q}}}) \to \operatorname{Isom}(\mu_{\ell^\infty}(\overline{F}), \mu_{\ell^\infty}(\overline{\mathbb{Q}}))$ *is equivariant for* $\operatorname{Gal}(\overline{F}/F) \times \operatorname{Gal}(\overline{\mathbb{Q}}/\mathbb{Q})$.

Proof This follows from the previous discussion. As to the different signs, note that $\operatorname{Gal}(\overline{\mathbb{Q}}/\mathbb{Q})$ operates most naturally from the left on $\operatorname{Isom}(\mu_{\ell^\infty}(\overline{F}), \mu_{\ell^\infty}(\overline{\mathbb{Q}}))$ and from the right on $\mathcal{Y}_{\overline{F},\mathbb{Q}}$, whereas $\operatorname{Gal}(\overline{F}/F)$ operates most naturally from the right on both spaces. □

Corollary 8.11 *Fix some group isomorphism* $\tilde{\iota}\colon \mu_{\ell^\infty}(\overline{F}) \to \mu_{\ell^\infty}(\overline{\mathbb{Q}})$*, and denote the corresponding component of* $\mathcal{Y}_{\overline{F},\overline{\mathbb{Q}}}$ *by* $\mathcal{Y}^\circ_{\overline{F},\overline{\mathbb{Q}}}$. *Then* $\mathcal{Y}^\circ_{\overline{F},\overline{\mathbb{Q}}} \to \mathcal{Y}_{F,\mathbb{Q}}$ *is a universal profinite covering space, and its deck transformation group is*

$$G_{F,\mathbb{Q}} \stackrel{\text{def}}{=} \{(\sigma_F, \sigma_{\mathbb{Q}}) \in \operatorname{Gal}(\overline{F}/F) \times \operatorname{Gal}(\overline{\mathbb{Q}}/\mathbb{Q}) \mid \chi_{\ell,F}(\sigma_F) = \chi_\ell(\sigma_{\mathbb{Q}})\}.$$

Hence there is a natural isomorphism $\pi_1^{\text{ét}}(\mathcal{Y}_{F,\mathbb{Q}}) \cong G_{F,\mathbb{Q}}$*, canonical up to inner automorphisms.* □

Now take an isomorphism $\iota\colon \mu_{\ell^n}(F) \to \mu_{\ell^n}(\overline{\mathbb{Q}})$, and extend it to an isomorphism $\tilde{\iota}\colon \mu_{\ell^\infty}(\overline{F}) \to \mu_{\ell^\infty}(\overline{\mathbb{Q}})$. We wish to find our space $\mathcal{Y}_{\ell^n,F}$ and its model over $\mathbb{Q}(\zeta_{\ell^n})$ back as a quotient of $\mathcal{Y}^\circ_{\overline{F},\overline{\mathbb{Q}}}$.

We construct two intermediate coverings of $\mathcal{Y}^\circ_{\overline{F},\overline{\mathbb{Q}}} \to \mathcal{Y}_{F,\mathbb{Q}}$.

- First note that there is a continuous epimorphism

$$\chi_\ell\colon G_{F,\mathbb{Q}} \to \mathbb{Z}_\ell^\times, \quad (\sigma_F, \sigma_\mathbb{Q}) \mapsto \chi_\ell(\sigma_F) = \chi_\ell(\sigma_\mathbb{Q}).$$

  Then $G_{F,\mathbb{Q}}(\ell^n) = \chi_\ell^{-1}(U_{\ell^n})$ is an open normal subgroup of $G_{F,\mathbb{Q}} \cong \pi_1^{\text{ét}}(\mathcal{Y}_{F,\mathbb{Q}})$, and the corresponding intermediate covering is equal to $\mathcal{Y}^\circ_{F,\mathbb{Q}(\zeta_{\ell^n})}$, the connected component of $\mathcal{Y}_{F,\mathbb{Q}(\zeta_{\ell^n})}$ determined by ι. This is precisely the model of $\mathcal{Y}_{\ell^n,F}$ over $\mathbb{Q}(\zeta_{\ell^n})$ used to define the $\mathrm{Gal}(\overline{\mathbb{Q}}/\mathbb{Q}(\zeta_{\ell^n}))$-action on its étale cohomology.
- The closed normal subgroup $\mathrm{Gal}(\overline{F}/F(\zeta_{\ell^\infty})) \times \{1\} \subset G_{F,\mathbb{Q}} \subset \mathrm{Gal}(\overline{F}/F) \times \mathrm{Gal}(\overline{\mathbb{Q}}/\mathbb{Q})$ defines the normal profinite covering space

$$\mathcal{Y}^\circ_{F(\zeta_{\ell^\infty}),\overline{\mathbb{Q}}} \to \mathcal{Y}^\circ_{F,\mathbb{Q}(\zeta_{\ell^n})} \to \mathcal{Y}_{F,\mathbb{Q}};$$

  as a $\mathcal{Y}^\circ_{F,\mathbb{Q}(\zeta_{\ell^n})}$-scheme this is isomorphic to $\mathcal{Y}_{F,\mathbb{Q}(\zeta_{\ell^n})} \otimes_{\mathbb{Q}(\zeta_{\ell^n})} \overline{\mathbb{Q}}$.

Lemma 8.12 *There are natural isomorphisms of cohomology groups*

$$\mathrm{H}^m(F(\zeta_{\ell^\infty}), A) \overset{\cong}{\leftarrow} \mathrm{H}^m_{\text{ét}}(\mathcal{Y}^\circ_{F(\zeta_{\ell^\infty}),\overline{\mathbb{Q}}}, A) \overset{\cong}{\to} \mathrm{H}^m_{\text{ét}}(\mathcal{Y}_{\ell^n,F,\overline{\mathbb{Q}}}, A)$$

equivariant for the group homomorphisms (natural projections)

$$\mathrm{Gal}(\overline{F}/F) \leftarrow G_{F,\mathbb{Q}}(\ell^n) \to \mathrm{Gal}(\overline{\mathbb{Q}}/\mathbb{Q}(\zeta_{\ell^n})).$$

Proof The isomorphism $\mathrm{H}^m_{\text{ét}}(\mathcal{Y}^\circ_{F(\zeta_{\ell^\infty}),\overline{\mathbb{Q}}}, A) \to \mathrm{H}^m(F(\zeta_{\ell^\infty}), A)$ is obtained from the Cartan–Leray spectral sequence applied to the universal covering $\mathcal{Y}^\circ_{\overline{F},\overline{\mathbb{Q}}} \to \mathcal{Y}^\circ_{F(\zeta_{\ell^\infty}),\overline{\mathbb{Q}}}$; note that all the higher cohomology groups of $\mathcal{Y}^\circ_{\overline{F},\overline{\mathbb{Q}}}$ with torsion coefficients vanish.

The rest follows from the preceding discussion. □

Proposition 8.13 *Let* $\sigma_F \in \mathrm{Gal}(\overline{F}/F)$ *and* $\sigma_\mathbb{Q} \in \mathrm{Gal}(\overline{\mathbb{Q}}/\mathbb{Q}(\zeta_{\ell^n}))$ *be such that* $\chi_\ell(\sigma_F) = \chi_\ell(\sigma_\mathbb{Q})$. *Then under the isomorphism* $\mathrm{H}^m(F(\zeta_{\ell^\infty}), A) \cong \mathrm{H}^m_{\text{ét}}(\mathcal{Y}_{\ell^n,F,\overline{\mathbb{Q}}}, A)$ *in (22) the actions of* σ_F *and* $\sigma_\mathbb{Q}$ *correspond to each other.*

Proof This follows from Lemma 8.12: just note how the element $(\sigma_F, \sigma_\mathbb{Q}) \in G_{F,\mathbb{Q}}(\ell^n)$ acts.

8.4 Compatibility of the Group-Theoretic and Topological Actions

We shall consider diverse Galois categories and exact functors between them:

$$\begin{array}{ccccc} U_{\ell^n}\text{-}\mathbf{FSet} & \longrightarrow & \mathbf{FEt}(\operatorname{Spec} F) & \longrightarrow & \mathbf{FEt}(\operatorname{Spec} F(\zeta_{\ell^\infty})) \\ \downarrow & & \downarrow & & \downarrow \\ U_{(\ell^n)}\text{-}\mathbf{FSet} & \longrightarrow & \mathbf{FCov}_{U_{(\ell^n)}}(Y_{\ell^n,F}) & \longrightarrow & \mathbf{FCov}(Y_{\ell^n,F}) \end{array} \tag{23}$$

Though we have suppressed this in the notation, this diagram will depend on a choice of $\iota\colon \mu_{\ell^n} \hookrightarrow \mathbb{S}^1$ and of an extension $\tilde{\iota}\colon \mu_{\ell^\infty} \hookrightarrow \mathbb{S}^1$. The functors in (23) are as follows:

- U_{ℓ^n}–**FSet** $\to$ **FEt**(Spec F) is the composition of two functors

$$U_{\ell^n}\text{-}\mathbf{FSet} \to \operatorname{Gal}(\overline{F}/F)\text{-}\mathbf{FSet} \to \mathbf{FEt}(\operatorname{Spec} F),$$

  the first of which is induced by the group homomorphism $\chi_{\ell,F}\colon \operatorname{Gal}(\overline{F}/F) \to U_{\ell^n}$ and the second of which is 'Grothendieck's Galois theory'. For an explicit description, let S be a finite set with a continuous left action of $\operatorname{Gal}(\overline{F}/F)$, then $\operatorname{Gal}(\overline{F}/F)$ acts from the left on the F-algebra $\overline{F}^S$ by

$$\sigma((\alpha_s)_{s\in S}) = (\alpha_{\sigma^{-1}s})_{s\in S},$$

  and the ring of invariants $E(S) = (\overline{F}^S)^{\operatorname{Gal}(\overline{F}/F)}$ is a finite étale F-algebra. Then the functor can be described as $S \mapsto \operatorname{Spec} E(S)$.
- **FEt**(Spec F) $\to$ **FEt**(Spec $F(\zeta_{\ell^\infty})$) is the functor $X \mapsto X \times_{\operatorname{Spec} F} \operatorname{Spec} F(\zeta_{\ell^\infty})$.
- The two functors in the lower horizontal line are obtained from the $U_{(\ell^n)}$-action on $Y_{\ell^n,F}$ as on p. 22, that is, the first one sends a finite $U_{(\ell^n)}$-set S to the product $Y_{\ell^n,F} \times S$ with the diagonal $U_{(\ell^n)}$-action, and the second one is the obvious forgetful functor.
- U_{ℓ^n}–**FSet** $\to$ $U_{(\ell^n)}$–**FSet** is induced by the ℓ-adic completion map $U_{(\ell^n)} \to U_{\ell^n}$.
- **FEt**(Spec F) $\to$ $\mathbf{FCov}_{U_{(F)}}(Y_{\ell^n,F})$ sends Spec E for a finite extension E/F to $Y_{\ell^n,E}$, and more generally for an étale F-algebra E we set

$$Y_{\ell^n,E} = \coprod_{\mathfrak{p}\in \operatorname{Spec} E} Y_{\ell^n,E/\mathfrak{p}}.$$

  Note that in the basic case where E is a field we need to choose an algebraic closure $\overline{E}/E$ to even define $Y_{\ell^n,E}$, and an isomorphism $\overline{E} \to \overline{F}$ to obtain a map $Y_{\ell^n,E} \to Y_{\ell^n,F}$. However, as for the spaces Y_F we check that $Y_{\ell^n,F}$ and $Y_{\ell^n,E} \to Y_{\ell^n,F}$ are independent up to canonical isomorphism from the choices of $\overline{F}$, $\overline{E}$ and $\overline{E} \to \overline{F}$.
- **FEt**(Spec $F(\zeta_{\ell^\infty})$) $\to$ **FCov**($Y_{\ell^n,F}$) is the composition

$$\mathbf{FEt}(\operatorname{Spec} F(\zeta_{\ell^\infty})) \to \mathbf{FCov}(Y_{\ell^\infty,F(\zeta_{\ell^\infty})}) \to \mathbf{FCov}(Y_{\ell^n,F})$$

  where the first functor sends Spec E to $Y_{\ell^\infty,E}$ and the second functor is induced by the homeomorphism $Y_{\ell^n,F} \cong Y_{\ell^\infty,F(\zeta_{\ell^\infty})}$.

Lemma 8.14 *The diagram of exact functors between Galois categories (23) commutes up to isomorphism of functors.*

Proof The commutativity of the right square is straightforward but tedious.

For the commutativity of the left hand side, let S be a finite set endowed with a continuous left action by U_{ℓ^n}. We will construct a natural isomorphism

$$Y_{\ell^n,F} \times S \cong Y_{\ell^n,E(S)}$$

of $U_{(\ell^n)}$-equivariant finite covering spaces of $Y_{\ell^n,F}$.

First we may assume that the U_{ℓ^n}-action on S is transitive, because all the functors involved respect finite direct sums. Hence $E(S)$ is a finite field extension of F. As explained above, to construct the covering $Y_{\ell^n,E(S)} \to Y_{\ell^n,F}$ we need to choose an embedding of $E(S)$ into $\overline{F}$, and later check that the choice of this embedding changes everything by canonical isomorphisms only (we omit that later part). By definition, $E(S) = (\overline{F}^S)^{\mathrm{Gal}(\overline{F}/F)}$, and hence

$$\mathrm{Hom}_F(E(S), \overline{F}) \cong \mathrm{Hom}_{\overline{F}}(\overline{F}^S, \overline{F}) \cong S.$$

Therefore the choice we need to make is that of a particular element $s_0 \in S$, which then allows us to trivialise the $\mathrm{Gal}(\overline{F}/F)$-set S as $\mathrm{Gal}(\overline{F}/F)/H$, where $H = \mathrm{Gal}(\overline{F}/E(S))$ is the stabiliser of s_0.

Note that

$$\mathrm{Hom}_\iota(\mu_{\ell^\infty}, \mathbb{S}^1) = \{j : \mu_{\ell^\infty} \to \mathbb{S}^1 \mid j|_{\mu_{\ell^n}} = \iota\}$$

is a U_{ℓ^n}-torsor, and it is trivialised by the choice of $\tilde{\iota} \in \mathrm{Hom}_\iota(\mu_{\ell^\infty}, \mathbb{S}^1)$ (which is implicit in the construction of the rightmost vertical functor in (23)). There is then a unique U_{ℓ^n}-equivariant map

$$q : \mathrm{Hom}_\iota(\mu_{\ell^\infty}, \mathbb{S}^1) \twoheadrightarrow S$$

with $q(\tilde{\iota}) = s_0$. We define a continuous map

$$Y_{\ell^n,\overline{F}} \to Y_{\ell^n,\overline{F}} \times S, \quad \chi \mapsto (\chi, q(\chi|_{\mu_{\ell^\infty}}))$$

which is equivariant for the group inclusion $\mathrm{Gal}(\overline{F}/E(S)) \hookrightarrow \mathrm{Gal}(\overline{F}/F)$ (trivial Galois action on S), hence it descends to a continuous map

$$Y_{\ell^n,E(S)} \to Y_{\ell^n,F} \times S.$$

It is straightforward to check that this last map is a bijection, hence a homeomorphism, and that it is $U_{(\ell^n)}$-equivariant. □

Choosing a point in $Y_{\ell^\infty,\overline{F}}(\tilde{\iota})$ we obtain a compatible family of fibre functors on all the categories in (23), hence a commutative diagram of profinite groups.

Proposition 8.15 *The diagram of profinite groups and continuous group homomorphisms*

$$\begin{array}{ccccccccc}
1 & \longrightarrow & \pi_1^{\text{ét}}(Y_{\ell^n,F}) & \longrightarrow & \pi_1^{\text{ét}}([Y_{\ell^n,F}/U_{(\ell^n)}]) & \longrightarrow & \hat{U}_{(\ell^n)} & \longrightarrow & 1 \\
 & & \downarrow{\cong} & & \downarrow & & \downarrow & & \\
1 & \longrightarrow & \mathrm{Gal}(\overline{F}/F(\zeta_{\ell^\infty})) & \longrightarrow & \mathrm{Gal}(\overline{F}/F) & \xrightarrow[\chi_{\ell,F}]{} & U_{\ell^n} & \longrightarrow & 1
\end{array}$$

(where $\hat{U}_{(\ell^n)}$ denotes the profinite completion of the abstract group $U_{(\ell^n)}$) commutes, and the rows are exact. The leftmost vertical map is an isomorphism, the other two vertical maps are surjective but not injective.

Proof The commutativity follows from Lemma 8.14. The lower row is exact by Galois theory and our assumptions on F; the upper row is exact except possibly at $\pi_1^{\text{ét}}(Y_{\ell^n,F})$ by Proposition 2.22. The map $\pi_1^{\text{ét}}(Y_{\ell^n,F}) \to \mathrm{Gal}(\overline{F}/F(\zeta_{\ell^\infty}))$ is an isomorphism by Proposition 8.6(ii). Combining the last two observations we also obtain exactness of the upper row at $\pi_1^{\text{ét}}(Y_{\ell^n,F})$. Finally, the rightmost vertical map is clearly surjective but not injective, hence the same holds for the middle vertical map. □

Corollary 8.16 *For each abelian torsion group A the natural isomorphisms of cohomology groups*

$$\mathrm{H}^m(Y_{\ell^n,F}, A) \overset{\cong}{\to} \mathrm{H}^m(\pi_1^{\text{ét}}(Y_{\ell^n,F}), A) \overset{\cong}{\to} \mathrm{H}^m(F(\zeta_{\ell^\infty}), A)$$

are equivariant for the canonical group homomorphisms

$$U_{(\ell^n)} \to \hat{U}_{(\ell^n)} \to U_{\ell^n},$$

up to an exponent which is 1 for the first map and -1 for the second map.

In particular the $U_{(\ell^n)}$-action on $\mathrm{H}^m(Y_{\ell^n,F})$ extends uniquely to a continuous U_{ℓ^n}-action.

Proof This follows from Proposition 8.15. The exponent -1, i.e. inversion, on the acting groups for $\mathrm{H}^m(\pi_1^{\text{ét}}(Y_{\ell^n,F}), A) \to \mathrm{H}^m(F(\zeta_{\ell^\infty}), A)$ occurs because $U_{(\ell^n)}$ operates on $Y_{\ell^n,F}$ from the right, see Remark 2.23 (ii). □

8.5 Compatibility of the Topological and Arithmetic Actions

This argument is technically much more difficult than the other two, so considering its logical redundancy we will be very sketchy here.

The essential ingredient is the fact that $\mathscr{Y}_{F,\mathrm{int}}$ is a Λ-scheme in the sense of Borger [46]. The general definition of Λ-schemes is rather involved, but a naïve variant suffices for our purposes.

Definition 8.17 Let $\mathscr{X}$ be a scheme which is flat over $\mathbb{Z}$. A Λ-structure on $\mathscr{X}$ is a family of mutually commuting endomorphisms $\varphi_p : \mathscr{X} \to \mathscr{X}$, indexed by the rational primes $p = 2, 3, 5, 7, \dots$, such that for all p the base change

$$\varphi_p \times \mathrm{id}\colon \mathscr{X} \times \operatorname{Spec} \mathbb{F}_p \to \mathscr{X} \times \operatorname{Spec} \mathbb{F}_p$$

is the absolute Frobenius of $\mathscr{X}_p = \mathscr{X} \times \operatorname{Spec} \mathbb{F}_p$, i.e. the morphism which is the identity on the topological space underlying $\mathscr{X}_p$ and which is $f \mapsto f^p$ on sections of the structure sheaf.

For example let M be an abelian group in multiplicative notation. Then there is a canonical Λ-structure on $\operatorname{Spec} \mathbb{Z}[M]$ given by

$$\varphi_p^\sharp\colon \mathbb{Z}[M] \to \mathbb{Z}[M], \qquad [m] \mapsto [m^p]. \tag{24}$$

We will now construct a Λ-structure on $\mathscr{Y}_{F,\mathrm{int}} = \operatorname{Spec} B_{F,\mathrm{int}}$ by writing down the corresponding endomorphisms $\varphi_p^\sharp$ of $B_{F,\mathrm{int}}$. Recall that

$$B_{F,\mathrm{int}} = \left(\mathbb{Z}[\overline{F}^\times/\mu_{\ell'}][([\zeta_\ell]-1)^{-1}]\right)^{\mathrm{Gal}(\overline{F}/F)}$$

We begin by setting $\varphi_\ell^\sharp = \mathrm{id}$ (or anything)[9], and next consider the case $p \neq \ell$.

Here we define $\varphi_p^\sharp$ on $\mathbb{Z}[\overline{F}^\times/\mu_{\ell'}]$ by (24). This descends to an endomorphism

$$\varphi_p^\sharp\colon \mathbb{Z}[\overline{F}^\times/\mu_{\ell'}][([\zeta_\ell]-1)^{-1}] \to \mathbb{Z}[\overline{F}^\times/\mu_{\ell'}][([\zeta_\ell]-1)^{-1}].$$

Lemma 8.18 *Let $p \neq \ell$. The canonical ring homomorphism*

$$B_{F,\mathrm{int}} \otimes \mathbb{F}_p \to \left(\mathbb{F}_p[\overline{F}^\times/\mu_{\ell'}][([\zeta_\ell]-1)^{-1}]\right)^{\mathrm{Gal}(\overline{F}/F)}$$

is an isomorphism.

Proof For $F = \overline{F}$ this is clear. For the general case, we use that $B_{F,\mathrm{int}} \to B_{\overline{F},\mathrm{int}}$ is a pro-finite étale $\mathrm{Gal}(\overline{F}/F)$-cover (cf. Theorem 4.17), so that forming quotients under $\mathrm{Gal}(\overline{F}/F)$ commutes with base change. □

Corollary 8.19 *The family of endomorphisms $\varphi_p\colon \mathscr{Y}_{F,\mathrm{int}} \to \mathscr{Y}_{F,\mathrm{int}}$ given on rings by the $\varphi_p^\sharp$ as constructed above define a Λ-structure on $\mathscr{Y}_{F,\mathrm{int}}$.* □

Note that for $p \neq \ell$ the φ_p are even automorphisms of $\mathscr{Y}_{F,\mathrm{int}}$. They can be described in a different way as follows:

The group $\overline{F}^\times/\mu_{\ell'}$ is a $\mathbb{Z}_{(\ell)}$-module, hence $\mathbb{Z}_{(\ell)}^\times$ acts by automorphisms from the left on $\overline{F}^\times/\mu_{\ell'}$ and also on $\mathbb{Z}[\overline{F}^\times/\mu_{\ell'}][([\zeta_\ell]-1)^{-1}]$. Since this action commutes with $\mathrm{Gal}(\overline{F}/F)$, it restricts to a left $\mathbb{Z}_{(\ell)}^\times$-action on $B_{F,\mathrm{int}}$; hence a right $\mathbb{Z}_{(\ell)}^\times$-action on $\mathscr{Y}_{F,\mathrm{int}}$.

[9]Note that inverting $[\zeta_\ell]-1$ in particular inverts ℓ, so the condition that $\varphi_\ell^\sharp$ lifts Frobenius is vacuous.

Lemma 8.20 *Consider the $\mathbb{Z}_{(\ell)}^\times$-action on $\mathcal{Y}_{F,\mathrm{int}}$ as just described. For each prime $p \neq \ell$ the element $p \in \mathbb{Z}_{(\ell)}^\times$ acts by the automorphism φ_p, and the element $-1 \in \mathbb{Z}_{(\ell)}^\times$ acts by the automorphism $[a] \mapsto [a^{-1}]$ on $\mathbb{Z}[\overline{F}^\times/\mu_{\ell'}]$.* □

Note that $\mathbb{Z}_{(\ell)}^\times$ is generated by -1 and the primes different from ℓ, hence the $\mathbb{Z}_{(\ell)}^\times$-action on $\mathcal{Y}_{F,\mathrm{int}}$ is uniquely determined by Lemma 8.20.

Extending scalars from $\mathbb{Z}$ to $\mathbb{C}$ we then obtain a right $\mathbb{Z}_{(\ell)}^\times$-action on

$$\mathcal{Y}_{F,\mathrm{int}} \times \operatorname{Spec} \mathbb{C} \cong \coprod_{\iota\colon \mu_{\ell^n} \hookrightarrow \mathbb{S}^1} \mathcal{Y}_{\ell^n,F}(\iota).$$

The stabiliser of each component is $U_{(\ell^n)}$, and the quotient $\mathbb{Z}_{(\ell)}^\times/U_{(\ell^n)} \cong (\mathbb{Z}/\ell^n\mathbb{Z})^\times$ operates simply transitively on the set of components. Hence there is a canonical $\mathbb{Z}_{(\ell)}^\times$-equivariant isomorphism of cohomology groups

$$\mathrm{H}^m(\mathcal{Y}_{F,\mathrm{int}} \otimes \overline{\mathbb{Q}}, A) \cong \mathrm{H}^m(\mathcal{Y}_{F,\mathrm{int}} \otimes \mathbb{C}, A) \cong \bigoplus_\iota \mathrm{H}^m(\mathcal{Y}_{\ell^n,F}(\iota), A),$$

and each summand on the right hand side is stable under $U_{(\ell^n)}$. By unravelling definitions we see that the $U_{(\ell^n)}$-action on each summand corresponds to the 'topological action' described above under the canonical isomorphism $\mathrm{H}^m(\mathcal{Y}_{\ell^n,F}(\iota), A) \cong \mathrm{H}^m(Y_{\ell^n,F}, A)$.

We now draw some consequences from the fact that the φ_p constitute a Λ-structure on $\mathcal{Y}_{F,\mathrm{int}}$.

Proposition 8.21 *For each $p \neq \ell$ and each abelian torsion group A the automorphism of $\mathrm{H}^m_{\text{ét}}(\mathcal{Y}_{F,\mathrm{int}} \times \operatorname{Spec} \overline{\mathbb{F}}_p, A)$ induced by $\varphi_p \bmod p$ is the inverse of the 'arithmetic' automorphism induced by the canonical generator $\sigma_p \in \operatorname{Gal}(\overline{\mathbb{F}}_p/\mathbb{F}_p)$ with $\sigma_p(a) = a^p$.*

Sketch of proof This follows from the fact that $\sigma_p^\flat$ (the scheme automorphism of $\mathcal{Y}_{F,\mathrm{int}} \times \operatorname{Spec} \overline{\mathbb{F}}_p$ which is the identity on the first factor and the automorphism defined by σ_p on the second factor) and $\varphi_p \bmod p$ commute, their product is the absolute Frobenius endomorphism of the $\mathbb{F}_p$-scheme $\mathcal{Y}_{F,\mathrm{int}} \times \operatorname{Spec} \overline{\mathbb{F}}_p$, and this absolute Frobenius acts trivially on étale cohomology. The latter is clear for $\mathrm{H}^0_{\text{ét}}$ and then follows formally for $\mathrm{H}^m_{\text{ét}}$ by the universal property of sheaf cohomology. See [47, Rapport, Sect. 1.8] for a detailed discussion in the finite type case. □

Proposition 8.22 *Fix an algebraic closure $\overline{\mathbb{Q}}_p$ of $\mathbb{Q}_p$ and an embedding $\overline{\mathbb{Q}} \hookrightarrow \overline{\mathbb{Q}}_p$. Then for any abelian torsion group A and any $m \geq 0$ the natural maps*

$$\mathrm{H}^m_{\text{ét}}(\mathcal{Y}_{F,\mathrm{int}} \otimes \overline{\mathbb{F}}_p, A) \leftarrow \mathrm{H}^m_{\text{ét}}(\mathcal{Y}_{F,\mathrm{int}} \otimes \mathcal{O}_{\overline{\mathbb{Q}}_p}, A) \rightarrow \mathrm{H}^m_{\text{ét}}(\mathcal{Y}_{F,\mathrm{int}} \otimes \overline{\mathbb{Q}}_p, A) \leftarrow \mathrm{H}^m_{\text{ét}}(\mathcal{Y}_{F,\mathrm{int}} \otimes \overline{\mathbb{Q}}, A)$$

are isomorphisms, and they are equivariant for the action of $\mathbb{Z}_{(\ell)}^\times$ and the homomorphisms of absolute Galois groups

$$\mathrm{Gal}(\overline{\mathbb{F}}_p/\mathbb{F}_p) \leftarrow \mathrm{Gal}(\overline{\mathbb{Q}}_p/\mathbb{Q}_p) = \mathrm{Gal}(\overline{\mathbb{Q}}_p/\mathbb{Q}_p) \rightarrow \mathrm{Gal}(\overline{\mathbb{Q}}/\mathbb{Q}).$$

Sketch of proof This can be shown identifying each of the étale cohomology groups with $\mathrm{H}^m(F(\zeta_{\ell^\infty}), A)$. □

The next proposition is a variant of a result of Borger, [48, Theorem 6.1].

Proposition 8.23 *The action of* $\mathrm{Gal}(\overline{\mathbb{Q}}/\mathbb{Q})$ *on* $\mathrm{H}^m_{\text{ét}}(\mathcal{Y}_{F,\mathrm{int}} \otimes \overline{\mathbb{Q}}, A)$ *factors through the* ℓ*-adic cyclotomic character* $\chi_{\ell,\mathbb{Q}}\colon \mathrm{Gal}(\overline{\mathbb{Q}}/\mathbb{Q}) \rightarrow \mathbb{Z}_\ell^\times$.

Proof By Proposition 8.22 and Proposition 8.21, the action of any two Frobenius elements commute, as they can be identified with the action of the commuting operators φ_p. By Chebotarev, we see that the action of $\mathrm{Gal}(\overline{\mathbb{Q}}/\mathbb{Q})$ factors through its maximal abelian quotient; also, it is unramified at all primes different from ℓ by Proposition 8.22. This implies that it factors through the quotient $\mathrm{Gal}(\mathbb{Q}(\zeta_{\ell^\infty})/\mathbb{Q})$, which is precisely the quotient defined by the ℓ-adic cyclotomic character. □

Proposition 8.24 *(i) Let* $p \neq \ell$ *be a rational prime, and let* $\sigma_p \in \mathrm{Gal}(\overline{\mathbb{Q}}/\mathbb{Q})$ *be such that* $\chi_\ell(\sigma_p) = p \in \mathbb{Z}_\ell^\times$. *Then* σ_p *operates on* $\mathrm{H}^m_{\text{ét}}(\mathcal{Y}_{F,\mathrm{int}} \otimes \overline{\mathbb{Q}}, A)$ *through the inverse of* φ_p.

(ii) Let $\sigma_{-1} \in \mathrm{Gal}(\overline{\mathbb{Q}}/\mathbb{Q})$ *be such that* $\chi_\ell(\sigma_{-1}) = -1 \in \mathbb{Z}_\ell^\times$. *Then* σ_{-1} *operates on* $\mathrm{H}^m_{\text{ét}}(\mathcal{Y}_{F,\mathrm{int}} \otimes \overline{\mathbb{Q}}, A)$ *through the involution induced by the involution* $[a] \mapsto [a^{-1}]$ *on* $\mathbb{Z}[\overline{F}^\times/\mu_{\ell'}]$.

Proof (i) follows from the conjunction of Propositions 8.21, 8.22 and 8.23. For (ii), by Proposition 8.23 we may assume that σ_{-1} is complex conjugation, whose action on cohomology is easily determined by contemplating the isomorphisms

$$\mathrm{H}^m_{\text{ét}}(\mathcal{Y}_{F,\mathrm{int}} \otimes \overline{\mathbb{Q}}, A) \cong \mathrm{H}^m_{\text{ét}}(\mathcal{Y}_{F,\mathrm{int}} \otimes \mathbb{C}, A) \cong \mathrm{H}^m(\mathcal{Y}_{F,\mathrm{int}}(\mathbb{C}), A) \cong \bigoplus_{\iota} \mathrm{H}^m(Y_{\ell^n,F}(\iota), A).$$ □

Proposition 8.25 *The canonical isomorphism*

$$\mathrm{H}^m(Y_{\ell^n,F}(\iota), A) \rightarrow \mathrm{H}^m_{\text{ét}}(\mathcal{Y}_{F,\mathrm{int}} \times_{e_\iota, \operatorname{Spec}\mathbb{Z}[\zeta_{\ell^n}]} \operatorname{Spec}\overline{\mathbb{Q}}, A)$$

is equivariant for the inverse *inclusion* $U_{(\ell^n)} \rightarrow U_{\ell^n}$, $u \mapsto u^{-1}$, *in the sense that for any* $u \in U_{(\ell^n)}$ *and any* $\sigma \in \mathrm{Gal}(\overline{\mathbb{Q}}/\mathbb{Q}(\zeta_{\ell^n}))$ *with* $\chi_{\ell,\mathbb{Q}}(\sigma) = u^{-1}$ *the action by* u *on the left hand side agrees with the action by* σ *on the right hand side.*

Proof It suffices to show that the canonical isomorphism

$$\bigoplus_{\iota\colon \mu_{\ell^n} \hookrightarrow \mathbb{S}^1} \mathrm{H}^m(Y_{\ell^n,F}(\iota), A) \rightarrow \mathrm{H}^m_{\text{ét}}(\mathcal{Y}_{F,\mathrm{int}} \times_{\operatorname{Spec}\mathbb{Z}} \operatorname{Spec}\overline{\mathbb{Q}}, A)$$

is equivariant for the inverse inclusion $\mathbb{Z}_{(\ell)}^\times \rightarrow \mathbb{Z}_\ell^\times$, $u \mapsto u^{-1}$, where the $\mathbb{Z}_\ell^\times$-action on the right hand side is given by Proposition 8.23. Since $\mathbb{Z}_{(\ell)}^\times$ is generated by -1 and the primes different from ℓ it is sufficient to check this for these elements. The combination of Lemma 8.20 and Proposition 8.24 yields the desired result. □

Acknowledgements Part of this work was done while the second author was a Clay Research Fellow. All of it was done while the first author was supported by the Swiss National Science Foundation. The authors wish to thank Lennart Meier for asking a very helpful question, Markus Land and Thomas Nikolaus for a discussion about Proposition 7.10, and Eric Leichtnam for pointing out some typographical errors in an earlier version.

References

1. Fontaine, J.-M., Wintenberger, J.-P.: Extensions algébrique et corps des normes des extensions APF des corps locaux. C. R. Acad. Sci. Paris Sér. A-B **288**(8), A441–A444 (1979). MR 527692
2. Weinstein, J.: Gal($\bar{\mathbb{Q}}_p/\mathbb{Q}_p$) as a geometric fundamental group (2014). arXiv:1404.7192
3. Scholze, P.: Perfectoid spaces. Publ. Math. Inst. Hautes Études Sci. **116**, 245–313 (2012). MR 3090258
4. Almkvist, G.: Endomorphisms of finitely generated projective modules over a commutative ring. Ark. Mat. **11**, 263–301 (1973). MR 0424786
5. Voevodsky, V.: On motivic cohomology with $\mathbf{Z}/l$ -coefficients. Ann. Math. **174**(1), 401–438 (2011). MR 2811603
6. Bost, J.-B., Connes, A.: Hecke algebras, type III factors and phase transitions with spontaneous symmetry breaking in number theory. Selecta Math. (N.S.) **1**(3), 411–457 (1995). MR 1366621
7. Steen, L.A., Arthur Seebach, J.: Counterexamples in Topology, 2nd edn. Springer, New York, Heidelberg (1978). MR 507446
8. Fabel, P.: Multiplication is discontinuous in the Hawaiian earring group (with the quotient topology). Bull. Pol. Acad. Sci. Math. **59**(1), 77–83 (2011). MR 2810974
9. Brazas, J.: The fundamental group as a topological group. Topol. Appl. **160**(1), 170–188 (2013). MR 2995090
10. Brazas, J.: Semicoverings: a generalization of covering space theory. Homol. Homotopy Appl. **14**(1), 33–63 (2012). MR 2954666
11. Bhatt, B., Scholze, P.: The pro-étale topology for schemes. Astérisque (369), 99–201 (2015). MR 3379634
12. Klevdal, C.: A correspondence, Galois, with generalized covering spaces. Undergraduate honors Thesis. University of Colorado, Boulder (2015)
13. Revêtements étales et groupe fondamental (SGA 1), Documents Mathématiques (Paris) [Mathematical Documents (Paris)], 3, Société Mathématique de France, Paris, 2003, Séminaire de géométrie algébrique du Bois Marie 1960–61. [Algebraic geometry seminar of Bois Marie 1960–61], Directed by Grothendieck, A. With two papers by Raynaud, M. Updated and annotated reprint of the 1971 original. Lecture Notes in Mathematics, vol. 224, Springer, Berlin; MR0354651 (50 #7129). MR 2017446
14. Hatcher, A.: Algebraic Topology. Cambridge University Press, Cambridge (2002). MR 1867354
15. Munkres, J.R.: Topology, 2nd edn. Prentice Hall, Upper Saddle River (2000)
16. Rudin, W.: Fourier Analysis on Groups. Interscience tracts in pure and applied mathematics, vol. 12. Interscience Publishers (a division of John Wiley and Sons), New York, London (1962). MR 0152834
17. Stein, K.: Analytische Funktionen mehrerer komplexer Veränderlichen zu vorgegebenen Periodizitätsmoduln und das zweite Cousinsche Problem. Math. Ann. **123**, 201–222 (1951). MR 0043219
18. Shelah, S.: Infinite abelian groups, whitehead problem and some constructions. Isr. J. Math. **18**, 243–256 (1974). MR 0357114
19. Shelah, S.: A compactness theorem for singular cardinals, free algebras, whitehead problem and transversals. Isr. J. Math. **21**(4), 319–349 (1975). MR 0389579

20. Godement, R.: Topologie algébrique et théorie des faisceaux. Hermann, Paris (1973). Troisième édition revue et corrigée, Publications de l'Institut de Mathématique de l'Université de Strasbourg, XIII, Actualités Scientifiques et Industrielles, vol. 1252. MR 0345092
21. Eilenberg, S., Steenrod, N.: Foundations of Algebraic Topology. Princeton University Press, Princeton, New Jersey(1952). MR 0050886
22. The stacks project, website, available at http://stacks.math.columbia.edu/
23. Hazewinkel, M.: Witt vectors. I. In: Handbook of Algebra, vol. 6, pp. 319–472. Elsevier/North-Holland, Amsterdam (2009). MR 2553661
24. Almkvist, G.: K-theory of endomorphisms. J. Algebra **55**(2), 308–340 (1978). MR 523461
25. Kelley, J.L., Spanier, E.H.: Euler characteristics. Pac. J. Math. **26**, 317–339 (1968). MR 0260842
26. Grothendieck, A.: Éléments de géométrie algébrique. IV. Étude locale des schémas et des morphismes de schémas IV. Inst. Hautes Études Sci. Publ. Math. **32**, 5–361 (1967). MR 0238860
27. Hilbert, D.: Die Theorie der algebraischen Zahlkörper. Jahresber. Deutsch. Math.-Verein. **4**, 175–546 (1897)
28. Neukirch, J.: Algebraic Number Theory. Grundlehren der Mathematischen Wissenschaften [Fundamental principles of mathematical sciences], vol. 322, Springer, Berlin (1999): Translated from the 1992 German original and with a note by Schappacher, N. With a foreword by Harder, G. MR 1697859
29. Serre, J.-P.: Géométrie algébrique et géométrie analytique. Ann. Inst. Fourier **6**, 1–42 (1955–1956). MR 0082175 (18,511a)
30. Lang, S.: Algebra, 3rd edn., Graduate Texts in Mathematics, vol. 211, Springer, New York (2002). MR 1878556
31. Grothendieck, A.: Éléments de géométrie algébrique. III. Étude cohomologique des faisceaux cohérents I. Inst. Hautes Études Sci. Publ. Math. **11**, 5–167 (1961). MR0217085
32. Pontrjagin, L.: The theory of topological commutative groups. Ann. Math. **35**(2), 361–388 (1934). MR 1503168
33. Fuchs, L., Loonstra, F.: On the cancellation of modules in direct sums over Dedekind domains. Nederl. Akad. Wetensch. Proc. Ser. A 74 = Indag. Math. **33**, 163–169 (1971). MR 0289476
34. May, W.: Unit groups of infinite abelian extensions. Proc. Am. Math. Soc. **25**, 680–683 (1970). MR 0258786
35. Cohen, J.M.: Homotopy groups of inverse limits. Proceedings of the advanced study institute on algebraic topology (Aarhus Univ., Aarhus, 1970), vol. I, Mat. Inst., Aarhus Univ., Aarhus, 1970, pp. 29–43. Various Publ. Ser., No. 13. MR 0346781
36. Hirschhorn, P.S.: The homotopy groups of the inverse limit of a tower of fibrations (2015), preprint. http://www-math.mit.edu/~psh/notes/limfibrations.pdf
37. Serre, J-P.: Cohomologie galoisienne, Cours au Collège de France, vol. 1962, Springer, Berlin, Heidelberg, New York (1962/1963). MR 0180551 (31#4785)
38. Schneider, P.: Equivariant homology for totally disconnected groups. J. Algebra **203**(1), 50–68 (1998). MR 1620705
39. Milne, J.S.: Étale Cohomology. Princeton mathematical series. Princeton University Press, Princeton (1980). MR 559531
40. Milnor, J.: Algebraic K -theory and quadratic forms. Invent. Math. **9**, 318–344 (1969/1970). MR 0260844
41. Matsumoto, H.: Sur les sous-groupes arithmétiques des groupes semi-simples déployés. Ann. Sci. École Norm. Sup. **2**(4), 1–62 (1969). MR 0240214
42. Bloch, S., Kato, K.: p-adic étale cohomology. Inst. Hautes Études Sci. Publ. Math. (63), 107–152 (1986). MR 849653
43. Adem, A., James Milgram, R.: Cohomology of finite groups, 2nd edn., Grundlehren der Mathematischen Wissenschaften [Fundamental principles of mathematical sciences], vol. 309, Springer, Berlin (2004). MR 2035696
44. Artin, E., Schreier, O.: Algebraische Konstruktion reeller Körper. Abh. Math. Sem. Univ. Hambg. **5**(1), 85–99 (1927). MR 3069467
45. Jacobson, N.: Basic Algebra II, 2nd edn. W. H. Freeman and Company, New York (1989). MR 1009787

46. Borger, J.: The basic geometry of Witt vectors, I: the affine case. Algebra Number Theory **5**(2), 231–285 (2011). MR 2833791
47. Deligne, P.: Cohomologie étale. Lecture Notes in Mathematics, vol. 569. Springer, Berlin, New York (1977)
48. Borger, J.: Λ-rings and the field with one element, (2009). arXiv:0906.3146

Arithmeticity of Some Monodromy Groups

T. N. Venkataramana

Abstract Deligne and Mostow used families of cyclic coverings of the projective line to obtain non-arithmetic monodromy and non-arithmetic lattices. In their construction, the number of ramification points is less than the degree of the cover. In contrast, we show that if the number of ramification points exceeds the degree of the cover, the monodromy group is almost always arithmetic.

Keywords Monodromy · Thin groups · Arithmetic groups · Hermitian forms

1991 Mathematics Subject Classification Primary: 22E40 · Secondary: 20F36

1 Introduction

In their work [5] on variation of Hodge structure, Griffiths and Schmid considered monodromy groups associated to algebraic families of smooth projective varieties. These monodromy groups arise naturally as subgroups of the integral linear group $GL_n(\mathbb{Z})$ and Griffiths and Schmid ask if such a monodromy group is always an *arithmetic group* (i.e. has finite index in its integral Zariski closure in GL_n). Following Sarnak [10] we will say that a subgroup $\Gamma \subset GL_n(\mathbb{Z})$ is *thin* if it is not arithmetic.

In [2], Deligne and Mostow exhibited families of cyclic coverings (of degree $d \geq 2$, say) of the projective line for which the monodromy group is not arithmetic (i.e. is thin). One obtains families of cyclic coverings of $\mathbb{P}^1$ as follows. Let $n \geq 3$ and $d \geq 2$. Fix integers $k_1, k_2, \ldots, k_n$ with $1 \leq k_i \leq d-1$ for all i. We assume that $(d, k_1, \ldots, k_n)$ are co-prime i.e. $\mathbb{Z}d + \sum \mathbb{Z}k_i = \mathbb{Z}$. Let $S \subset \mathbb{C}^n$ be the open subset of n-tuples $z = (z_1, \ldots, z_n)$ with all entries distinct. We obtain a family of finite covers of $\mathbb{P}^1$ described by the equation

$$y^d = (x - z_1)^{k_1} \cdots (x - z_n)^{k_n},$$

T. N. Venkataramana (✉)
School of Mathematics, Tata Institute of Fundamental Research, Homi Bhabha Road, Colaba 400005, Mumbai, India
e-mail: venky@math.tifr.res.in

J. W. Cogdell et al. (eds.), *Cohomology of Arithmetic Groups*,
Springer Proceedings in Mathematics & Statistics 245,
https://doi.org/10.1007/978-3-319-95549-0_9

as z varies in S. We then get the monodromy representation of $\pi_1(S, p)$ (for some fixed base-point $p \in S$) on H^1 of the curve.

Deligne and Mostow show that for some special choices of the numbers $(n < 4)$ d and the k_i, the monodromy group is thin; they in fact show that (1) the Zariski closure G of the monodromy group is (in general) a product of unitary groups (the unitary structure coming from the intersection form on $H^1(C, \mathbb{R})$ of the foregoing curve C, extended as a Hermitian form on $H^1(C, \mathbb{C})$), (2) one of the factors of $G(\mathbb{R})$ is of the form $U(n-2, 1)$ and (3) the projection of the monodromy group to the factor $U(n-2, 1)$ is a *non-arithmetic* lattice (a discrete subgroup with finite co-volume, which is not commensurable to an arithmetic group). For example, they show, for the family

$$y^{18} = (x - z_1)(x - z_2)(x - z_3)(x - z_4),$$

that the Zariski closure is $U(2, 1) \times U(1, 2) \times U(3)$, the monodromy group is thin (and that the projection of the monodromy group to the factor $U(2, 1)$ is a non-arithmetic lattice). We refer to Theorem 9 for a more detailed description of the Deligne–Mostow construction.

In the cases considered by Deligne–Mostow, the presence of a rank one factor $U(n-2, 1)$ forces the inequality $n \leq 2d$.

In contrast, it is shown in [13, 14] that if $n \geq 2d + 1$ and we have the family of cyclic covers of degree d of $\mathbb{P}^1$ described by the equation

$$y^d = (x - z_1)^{k_1}(x - z_2)^{k_2} \cdots (x - z_n)^{k_n}$$

where $(z_1, \ldots z_n)$ varies over the set of n tuples in $\mathbb{C}^n$ with distinct entries, and if k_i are all co-prime to d, then the monodromy group is an arithmetic group. It is naturally a subgroup of a product of unitary groups of the form $\prod_{e|d} U(h)(O_e)$; here, for each divisor e of d, we denote by $E_e = \mathbb{Q}(\omega_e)$ the eth cyclotomic extension, and by $K_e = \mathbb{Q}(\omega_e + \omega_e^{-1})$ the maximal totally real subfield of E_e. The form h is Hermitian with respect to the quadratic extension E_e/K_e; the group $U(h)$ is the unitary group of h and is an algebraic group defined over K_e. Denote by O_e is the ring of integers in K_e for a more detailed description of the unitary groups involved see Sect. 2.3. We show that if $n \geq 2d + 1$, then the monodromy group has finite index in $\Gamma^* = \prod_{e|d} U(h)(O_e)$; by restricting scalars to $\mathbb{Q}$, it can be shown that Γ^* is an arithmetic group in the sense described in the first paragraph.

The bound $n > 2d$ can be improved to $n > d$ using the results of [2, 7]. For most divisors e of d, that the image of the monodromy group projects to a finite index subgroup of $U(h)(O_e)$, follows from the methods of [13, 14]. But, if the divisor $e \in \{2, 3, 4, 6\}$ one has to use the results of [2, 7]. However, we do not prove this in the present expository paper since the proof is rather long.

This article is a brief overview of the construction of Deligne and Mostow, the results in [7, 13, 14] and various related results.

2 Monodromy: Definitions and Examples

2.1 Fibrations

Suppose $\pi : X \to S$ is a locally trivial fibration of (smooth) manifolds with fibre F. Assume that S is connected. We assume that the integral cohomology of F is a finitely generated abelian group. Write $\mathbb{Z}^n$ for the group $H^*(F, \mathbb{Z})/torsion$. The local triviality of the fibration implies that the sheaf of abelian groups $\{H^*(F_s, \mathbb{Z})/torsion; s \in S\}$ forms a local system over S. Hence we get an action of the fundamental group $\pi_1(S, p)$ (with a fixed base-point $p \in S$) on the integral cohomology $\mathbb{Z}^n$ and hence a representation $\pi_1(S) \to GL_n(\mathbb{Z})$. This is called the monodromy representation of the fibration and the image $\Gamma \subset GL_n(\mathbb{Z})$ of the monodromy representation is called the **monodromy group**.

Denote by G the Zariski closure of the monodromy group Γ in GL_n. We say that Γ is an **arithmetic group** if Γ has finite index in $G \cap GL_n(\mathbb{Z}) = G(\mathbb{Z})$. Otherwise Γ is said to be **thin** [10].

2.2 Algebraic Families

Assume in addition to the preceding hypotheses, that the locally trivial fibration $\pi : X \to S$ is an algebraic morphism of smooth varieties whose fibres $F_s; s \in S$ are smooth projective varieties. In [5], Griffiths and Schmid ask whether the foregoing monodromy group is always an arithmetic group.

In [2], Deligne and Mostow reinterpreted the work of H.A. Schwartz and Picard as instances of non-arithmeticity of the monodromy; thus the answer to the preceding question of Griffiths and Schmid is not always yes. In the examples of Deligne–Mostow (and Schwartz and Picard), the fibres are (smooth projective) curves obtained as cyclic covers of $\mathbb{P}^1$. Later, we will give a more detailed description of the Deligne–Mostow construction. We first look at examples concerning the question of Griffiths and Schmid.

Example 1 The simplest example of a family of cyclic covers of $\mathbb{P}^1$ is the **Legendre** family of elliptic curves. Let S be the set $\mathbb{P}^1 \setminus \{0, 1, \infty\}$. The fundamental group of S (based at, say, 1/2) is the free group on two generators, namely a small loop h_0 around 0 (going in clockwise direction) and a small loop around infinity h_∞ (going in the anticlockwise direction). For each $\lambda \in S$, consider the elliptic curve $E = E_\lambda$ given by the (Legendre form of the) equation

$$y^2 = x(x-1)(x-\lambda).$$

The monodromy action on $H^1(E, \mathbb{Z}) = \mathbb{Z}^2$ is given by

$$h_\infty \mapsto \begin{pmatrix} 1 & 2 \\ 0 & 1 \end{pmatrix}, h_0 \mapsto \begin{pmatrix} 1 & 0 \\ 2 & 1 \end{pmatrix}.$$

The monodromy group is generated by the images of h_0, h_∞, and is the principal congruence subgroup of $SL_2(\mathbb{Z})$ of level 2 (up to ± 1). Hence the monodromy is an arithmetic group.

On the other hand we have the following (closely related) example where the monodromy group is thin.

Example 2 In [8], Nori constructed an example of thin monodromy as follows. Consider the space $U = \mathbb{P}^1 \setminus \{0, 1, \infty\}$ of the preceding example, and fix $c \in U$. Take for S the open set $U = \mathbb{P}^1 \setminus \{0, 1, \infty, c^{-1}\}$. For each $\lambda \in S$, consider the abelian surface (in fact a product of elliptic curves) $A_\lambda = E_\lambda \times E_{c\lambda}$. We have the family X of these abelian surfaces A_λ as λ varies over S. The fundamental group of S is the free group on three generators and under the monodromy representation, maps into $SL_2(\mathbb{Z}) \times SL_2(\mathbb{Z})$. It can easily be shown that the monodromy group is Zariski dense in $SL_2 \times SL_2$. However, Nori shows that the monodromy group has infinite index in the product $SL_2(\mathbb{Z}) \times SL_2(\mathbb{Z})$:

Theorem 1 (Nori) *The monodromy group is Zariski dense in $SL_2 \times SL_2$ and has infinite index in $SL_2(\mathbb{Z}) \times SL_2(\mathbb{Z})$. In fact, the monodromy group is infinitely presented.*

Proof We give a brief sketch of Nori's proof. The space S is $\mathbb{P}^1$ punctured at the four points $0, 1, c^{-1}, \infty$. For a puncture p, denote by h_p a small loop going around p counterclockwise. The fundamental group of S (with a suitable base-point fixed once and for all) is easily seen to be the quotient of the free group on $h_0, h_1, h_{c^{-1}}, h_\infty$ by the relation $h_0 h_1 h_{c^{-1}} h_\infty = 1$. We write ($F$ denotes the free group on the relevant letters)

$$\pi_1(S) = \frac{F\{h_0, h_1, h_{c^{-1}}, h_\infty\}}{(h_0 h_1 h_{c^{-1}} h_\infty = 1)}.$$

Similarly, if $V = \mathbb{P}^1 \setminus \{0, 1, \infty\}$,

$$\pi_1(V) = \frac{F\{h_0, h_1, h_\infty\}}{(h_0 h_1 h_\infty = 1)} \subset SL_2(\mathbb{Z}),$$

as in Example 1. The map $S \to V \times V$ given by $\lambda \mapsto (\lambda, c\lambda)$ induces a map of $\pi_1(S)$ into $\pi_1(V) \times \pi_1(V)$. In terms of the generators h_p it is easily seen that this map is

$$h_0 \mapsto (h_0, h_0), \quad h_\infty \mapsto (h_\infty, h_\infty), \quad h_1 \mapsto (h_1, 1) \quad \text{and} \quad h_{c^{-1}} \mapsto (1, h_1).$$

Since the image of $\pi_1(V)$ has finite index in $SL_2(\mathbb{Z})$, it follows that the image under the monodromy representation (call it ρ), of the normal subgroup generated by the element $h_1 \in \pi_1(S)$ (resp the element $h_{c^{-1}}$), is all of $SL_2 \times 1$ (resp $1 \times SL_2$). Consequently, the Zariski closure of the monodromy group is $SL_2 \times SL_2$.

Moreover, the image under ρ of every conjugate of h_1 commutes with the image $\rho(h_{c^{-1}})$, since a conjugate of $\rho(h_1)$ lies in $SL_2 \times 1$ and $\rho(h_{c^{-1}})$ lies in $1 \times SL_2$. This gives an infinite set of elements in the kernel of ρ: a commutator of gh_1g^{-1} with $h_{c^{-1}}$ is in the kernel for every $g \in \pi_1(S)$. It is known that the relations $[gh_1g^{-1}, h_{c^{-1}}] = 1$ give an infinite presentation of the image of $\pi_1(S)$ in $\pi_1(V) \times \pi_1(V)$. By the preceding example, $\pi_1(V) \times \pi_1(V)$ is a subgroup of $SL_2(\mathbb{Z}) \times SL_2(\mathbb{Z})$. Hence the monodromy group is infinitely presented and in particular, cannot have finite index in $SL_2(\mathbb{Z}) \times SL_2(\mathbb{Z})$. □

Example 3 Consider the family of hyperelliptic curves of a fixed genus. Thus, let S be the space of monic polynomials P of degree $2g + 1$ with distinct roots, and for each $P \in S$, consider the curve

$$y^2 = P(x), x \in \mathbb{C} \setminus \{z \in \mathbb{C} : P(z) = 0\}.$$

This is a compact Riemann surface C_P with finitely many punctures, and by filling in the punctures, we get the family of smooth projective curves C_P as P varies in S. The action of the fundamental group on $H^1(C_P, \mathbb{Z})$ is easily seen to preserve the intersection form. Hence the monodromy representation gives a map $\pi_1(S) \to Sp_{2g}(\mathbb{Z})$ where g is the genus of the curves C_P. A result of A'Campo [1] says that the monodromy group has finite index in $Sp_{2g}(\mathbb{Z})$; in particular, it is arithmetic.

It can be shown ([4]) that the fundamental group of S with S as in A'Campo's theorem is the braid group on $2g + 1$ generators and (see Sect. 3) that the monodromy representation is simply the integral Burau representation.

2.3 Cyclic Covers of the Projective Line

Example 3 was a family of degree two covers of the projective line ramified at $2g + 1$ points. Deligne and Mostow ([2]) considered families of degree $d \geq 3$ cyclic covers of the projective line, ramified at n points in the complex plane (and possibly at infinity). Let $k_1, k_2, \ldots, k_n$ be integers with $1 \leq k_i \leq d - 1$, and let $S = \{z \in \mathbb{C}^n : z_i \neq z_j \quad \forall i \neq j\}$. Thus S in a Zariski open subset of $\mathbb{C}^n$. For each $z \in S$, the set

$$\{(x, y) \in \mathbb{C}^2 : x \neq z_i \quad \forall i, \quad y^d = (x - z_1)^{k_1} \cdots (x - z_n)^{k_n}\} \tag{1}$$

is the complement of a finite set in a compact Riemann surface C_z of genus g. The collection $\{C_z; z \in S\}$ defines a family of Riemann surfaces of genus g. One may show, using the Riemann Hurwitz Formula, that the genus is given by the formula

$$g = 1 + \frac{1}{2}(nd - \sum_{i=1}^{n}(d, k_i) - (d, k_\infty)),$$

where $k_\infty = \sum_{i=1}^{n} k_i$ and (a, b) denotes the g.c.d. of the two positive integers a and b.

The collection $C_z : z \in S$ defines a family of Compact Riemann surfaces of genus g and hence the monodromy representation yields a map $\pi_1(S) \to Sp_{2g}(\mathbb{Z})$ since the monodromy group preserves the intersection form ($(v, w) \mapsto v \wedge w$) on $H^1(C_z, \mathbb{Z}) \simeq \mathbb{Z}^{2g}$.

The symplectic form $v \wedge w$ on $H^1(C, \mathbb{R})$ extends to a *Hermitian form* h : $(v, w) \mapsto v \wedge \overline{w}$ on $V = H^1(C, \mathbb{C})$. Now, the map $(x, y) \mapsto (x, \omega y)$ gives an action on the curve C_z in (1) of the group $\mathbb{Z}/d\mathbb{Z}$, where the group $\mathbb{Z}/d\mathbb{Z}$ is viewed as the group of dth roots of unity $\{\omega \in \mathbb{C}^* : \omega^d = 1\}$. Fix a generator ω of $\mathbb{Z}/d\mathbb{Z}$ and let V_m denote the ω^m-eigenspace of $V = H^1(C, \mathbb{C})$ for the action of ω. Since the monodromy group patently commutes with the action of $\mathbb{Z}/d\mathbb{Z}$, it follows that the monodromy group leaves stable the subspace V_m and the Hermitian form on V_m has signature (p_m, q_m) say. Therefore, the monodromy representation maps $\pi_1(S)$ into the product $G(\mathbb{R}) \subset U(h) = \prod_{m=1}^{d} U(p_m, q_m)$ where, as before, G is the Zariski closure of the monodromy group.

3 The Braid Group and the Burau Representation

3.1 The Braid Group and the Burau Representation

Definition 1 Let B_n denote the quotient of the free group on $n-1$ generators denoted $s_1, s_2, \ldots, s_{n-1}$, by the relations: for all $i, j \leq n-1$ with $\mid i - j \mid \geq 2$,

$$s_i s_j = s_j s_i,$$

and for $i \leq n-2$ the braiding relation

$$s_i s_{i+1} s_i = s_{i+1} s_i s_{i+1}.$$

Definition 2 Let q be a variable and consider $R = \mathbb{Z}[q, q^{-1}]$ the ring of Laurent polynomials. The braid group representation $\rho(q) : B_n \to GL_{n-1}(R) = GL(R^{n-1})$ acts naturally on R^{n-1} by R linear maps: we describe the action of the elements s_i on the standard basis elements $e_1, e_2, \ldots, e_{n-1}$ of R^{n-1} as follows. For each generator s_i of B_n, one has

$$s_i(e_j) = e_j \quad \forall j \quad with \quad \mid i - j \mid \geq 2$$

$$s_i(e_{i-1}) = e_{i-1} + q e_j, \quad s_i(e_{i+1}) = e_{i+1} + e_i (i \leq n-2), \quad s_i(e_i) = -q e_i.$$

This extends to a representation $\rho(n) : B_n \to GL_{n-1}(R)$ called the "Reduced Burau Representation". In this article, we are always concerned with the reduced

Burau representation, and therefore we will drop the phrase "reduced Burau representation" and call it the **Burau representation**.

Definition 3 For each integer d, let $\Phi_d(q)$ be the dth cyclotomic polynomial in q. Consider the ring homomorphism (quotient map) $R \to \mathbb{Z}[q]/(\Phi_d(q))$ (the latter is the ring of integers A_d in the dth cyclotomic extension $\mathbb{Q}[\omega_d]$ of $\mathbb{Q}$ where ω is a primitive dth root of unity). We then get a corresponding representation $\rho_d(n) : B_n \to GL_{n-1}(A_d)$, called the (reduced) *Burau representation evaluated at the (primitive) d-th root of unity* ω_d.

Example 4 Recall that the group S_n of permutations on n letters is generated by the involutions $\sigma_i = (i, i+1)$ with relations $\sigma_i\sigma_j = \sigma_j\sigma_i$ if $| i - j | \geq 2$, $\sigma_i\sigma_{i+1}\sigma_i = \sigma_{i+1}\sigma_i\sigma_{i+1}$ and $\sigma_i^2 = 1$. Thus there is a surjective homomorphism $p : B_n \to S_n$ given by $s_i \mapsto \sigma_i$. The group S_n has the reflection representation $r : S_n \to GL_{n-1}(\mathbb{Z})$ on $\mathbb{Z}^n/\mathbb{Z}(1, 1, \ldots, 1)$. It is easily seen that the composite $r \circ p : B_n \to GL_{n-1}(\mathbb{Z})$ is the (reduced) Burau representation evaluated at $q = 1$.

Example 5 If $n = 2g + 1$, consider the representation $\rho_{-1}(n) : B_n \to GL_{n-1}(\mathbb{Z})$, the Burau representation evaluated at $q = -1$. One can show that the image lies in the symplectic group: $\rho_{-1}(2g + 1) : B_{2g+1} \to Sp_{2g}(\mathbb{Z})$. It is shown in [1] that this is exactly the monodromy representation considered in the Example 3.

3.2 Properties of the Burau Representation

In [11], a Hermitian form on $\mathbb{C}^{n-1}$ preserved by the braid group under the composite $\rho(n) : B_n \to GL_{n-1}(R) \to GL_n(\mathbb{C})$ is constructed. However, the explicit form of the Hermitian form constructed in [11] does not behave well under reduction modulo ideals of R, and hence we use the hermitian form as in [13]. The ring $R = \mathbb{Z}[q, q^{-1}]$ has an involution given by $q \mapsto q^{-1}$. On R^{n-1} define a Hermitian form h with respect to this involution, by setting $h(e_i, e_i) = \frac{(q+1)^2}{q}$, $h(e_i, e_{i+1}) = -q - 1$ and $h(e_i, e_j) = 0$ for $| i - j | \geq 2$. It is shown in [13] that the braid group B_n preserves this hermitian form.

Let $R_0 = \mathbb{Z}[q + q^{-1}]$ be the ring of invariants in the ring $R = \mathbb{Z}[q, q^{-1}]$ for the action of the involution $q \mapsto q^{-1}$. On R^{n-1} we have the hermitian form defined in the preceding paragraph. The unitary group $U(h)$ of the hermitian form is then an affine algebraic group scheme defined over the ring R_0 of invariants, and the image of the Burau representation lies in the R_0 points of this algebraic group i.e. $\rho_q(n) : B_n \to U(h)(R_0) \subset GL_{n-1}(R)$.

Let $A_d \simeq \mathbb{Z}[q, q^{-1}]/(\Phi_d(q)) = R/(\Phi_d(q))$ where Φ_d, as before, is the dth cyclotomic polynomial. Then A_d is the ring of integers in the cyclotomic extension $E = \mathbb{Q}(\omega_d)$. Let $K = \mathbb{Q}(\omega_d + \omega_d^{-1})$ be its maximal totally real subfield; hence E/K is a CM extension. The specialisation of q to ω_d maps B_n into $U(h)(O_d)$ where O_d is the ring of integers in K:

$$\rho_d(n) : B_n \to U(h)(O_d) \subset GL_{n-1}(\mathbb{Z}[\omega_d]).$$

It can be shown (for example, see [13]) that unless the image of $\rho_d(n)$ is finite (which happens rarely), the image of ρ_d is Zariski dense in $U(h)$.

Remark 1 If $q = -1$, the above hermitian form vanishes identically. However, if, in the general hermitian form h as above we divide by $q + 1$ and let q tend to -1, we get a non-degenerate alternating form in the limit and the symplectic form in Example 3 is precisely this form (up to a multiple).

Lemma 2 *If d does not divide n, then the Burau representation $\rho_d(n)$ is absolutely irreducible. The Hermitian form on $A_d^{n-1} = (\mathbb{Z}[\omega_d])^{n-1}$ obtained by setting $q \mapsto \omega_g$ for a primitive dth root of unity is non-degenerate.*

If d divides n, then the Hermitian form h on A_d^{n-1} is degenerate and the kernel is a one dimensional line of the form $A_d v$ for some nonzero $v \in A_d^{n-1}$; the vector v is fixed by B_n and the quotient representation on A_d^{n-2}, denoted $\overline{\rho_d(n)}$, is absolutely irreducible.

For the (easy) proof, we refer to Proposition 14 in [13]. By an abuse of notation, we denote by $\overline{\rho_d(n)}$ the representation $\rho_d(n)$ if the latter is irreducible (by the preceding lemma, this happens if and only if d does not divide n).

3.3 Restriction of Scalars

Before we relate monodromy to the Burau representation, we describe a construction in algebraic groups which converts groups over larger fields into groups over smaller fields. Consider a linear algebraic group $\mathcal{G}$ defined over a number field K. Let us recall that an algebraic group G (say, over $\mathbb{Q}$) associates to each commutative $\mathbb{Q}$-algebra A a group $G(A)$, namely the group of A valued points of G). Consider, for any commutative $\mathbb{Q}$-algebra A, the assignment $A \mapsto \mathcal{G}(A \otimes_{\mathbb{Q}} K)$. The theory of Weil-restriction of scalars says that this assignment is (uniquely) represented by a linear algebraic group G over $\mathbb{Q}$. We write $G = R_{K/Q}(\mathcal{G})$, and refer to it as the group over $\mathbb{Q}$ obtained from $\mathcal{G}$ by restricting scalars from K to $\mathbb{Q}$. It follows that $G(\mathbb{R}) = \mathcal{G}(\mathbb{R} \otimes_{\mathbb{Q}} K) = \prod_v \mathcal{G}(K_v)$ where v runs over all archimedean completions of the number field K. One may show that under this identification of $G(\mathbb{R})$ with $\mathcal{G}(\mathbb{R} \otimes_{\mathbb{Q}} K)$, the groups $G(\mathbb{Z})$ and $\mathcal{G}(O_K)$ are commensurable. Thus the arithmetic group $G(\mathbb{Z})$, is up to finite index the same as the group $\mathcal{G}(O_K)$. In this sense, the group $\mathcal{G}(O_K)$ is an arithmetic group as described in the introduction.

For example, suppose $\mathcal{G} = SL_n$ over K, and K has degree d over $\mathbb{Q}$. then K^n may be viewed as $\mathbb{Q}^{dn}$ on which the algebra K acts by $\mathbb{Q}$-endomorphisms. Then $R_{K/Q}(SL_n)$ may be identified with the subgroup of SL_{nd} which commutes with the $\mathbb{Q}$-endomorphism algebra K. From now on, we will freely refer to groups of the form $\mathcal{G}(O_K)$ with K a number field, as arithmetic groups as in the introduction.

3.4 Burau and Monodromy

Consider the collection S of monic polynomials of degree n with $\mathbb{C}$ coefficients which have *distinct* roots. It is well known ([4]) that the fundamental group of the affine variety S is exactly the braid group B_n. For each $P \in S$ consider the set

$$\{(x, y) \in \mathbb{C}^2 : y^d = P(x), P(x) \neq 0\}. \tag{2}$$

This is the complement of a finite set in a compact Riemann surface C_P of genus g and the family C_P of these smooth projective genus g curves fibers over S, and one has the associated monodromy representation $\theta_d(n)$.

We will now state a result (see [7] for a reference) relating the Burau representation to the monodromy representation of the foregoing algebraic family (2) of cyclic coverings of $\mathbb{P}^1$ analogous to Example 3.

Lemma 3 *The monodromy representation on $H^1(C_P, \mathbb{Q})$ of this family is the direct sum $\bigoplus_{e|d} \rho_e(n)$ of the Burau representations evaluated at primitive eth roots of unity, where e runs through the divisors of d.*

Using this lemma (and certain well known results of Margulis on normal subgroups of higher rank arithmetic groups) one can reduce the arithmeticity or not, of the monodromy of the family considered in (2) to the arithmeticity or otherwise, of the image of the representation $\rho_d(n)$.

In [2], Deligne and Mostow exhibit (essentially) the first examples of non-arithmetic lattices in $U(2, 1)$. Their construction implies in particular, that certain monodromy groups are "thin" (i.e. not arithmetic). One of the examples is a special case of the family (2).

Example 6 Consider the family (2) with $d = 18$, $n = 4$. Then the monodromy group is not arithmetic.

To see this, it is "enough" to show (by Lemma 3) that the image of the representation $\rho_{18}(4) : B_4 \to U(h)(O_{18})$ is not arithmetic. As before, O_{18} is the ring of integers in the (totally real) field $\mathbb{Q}(cos(\frac{2\pi}{18}))$ and the field K has degree $\phi(18)/2 = 3$ over $\mathbb{Q}$. Therefore, K has three embeddings into $\mathbb{R}$. Under these embeddings it can be shown that the Hermitian form has signatures (3,0), (2,1) and (1,2). Under one of these embeddings, Deligne–Mostow (in [2]) show that the image in $U(2, 1)$ is *discrete*. It is not hard to show that the image of the braid group B_4 is Zariski dense in $PU(h)$ (the projective unitary group). Consequently, the image of B_4 is not of finite index in $U(h)(A_d)$. For, by strong approximation, the projection of a finite index subgroup of $U(h)(A_d)$ to $PU(2, 1)$ is dense and cannot be discrete.

As was mentioned before, the main focus of Deligne–Mostow is actually to prove that the projection gives a **non-arithmetic lattice** in $U(2, 1)$.

4 The Arithmeticity

In this section, we describe the result of [13] which says that

Theorem 4 [13] *If $n \geq 2d+1$, then the image of the Burau representation evaluated at primitive dth roots, namely the representation $\rho_d(n)$, is an arithmetic group. More precisely, the image has finite index in the integral unitary group $U(h)(O_d)$, where O_d is the ring of integers in the totally real field $\mathbb{Q}(\omega_d+\omega_d^{-1})$.*

(As explained in Sect. 3.3, the group $U(h)(O_d)$ may be viewed an arithmetic group as in the introduction, by an argument involving restriction of scalars from K to $\mathbb{Q}$).

It follows from Lemma 3 that

Corollary 1 [13] *The monodromy group of the family (2) is an arithmetic group.*

The proof is by induction on the integer $n \geq 2d+1$. The crucial case in the course of the proof is the case $n = 2d+1$; the other cases can be proved using a "bootstrapping" technique (see (5.1.2) of [13] for details). We describe the proof in the case $n = 2d+1$. The proof is by analysing the restriction of the Burau representation $\rho_d(n)$ to certain subgroups.

Remark 2 The range in Theorem 4 can be improved to $n \geq d+1$; for most d, the proof is the same as that of Theorem 4. One will have to use results of [7] or [2], in case one of the divisors of d is 4 or 6. We will provide details in a forthcoming paper.

4.1 Restriction of the Burau Representation to Smaller Braid Groups

Denote temporarily by G the group $B_n = B_{2d+1}$ generated by $s_1, \ldots, s_{n-1}$, by P the group $B_{n-1} = B_{2d}$ the group generated by $s_1, \ldots, s_{n-2}$ and by M the subgroup B_{n-2} the group generated by $s_2, \ldots, s_{n-2}$ (the suffices start from 2 and end in $n-2$). It follows from definitions that the representation $\rho_d(n)$ restricted to P contains the representation $\rho_d(n-1)$. By Lemma 2 it follows that the representation $\rho_d(2d) = \rho_d(n-1)$ of the smaller braid group $B_{2d} = P$ is not irreducible, but contains the trivial representation, with quotient being irreducible (we denoted this $\overline{\rho_d(2d)}$). It can be shown that the exact sequence of $P = B_{2d}$ modules

$$0 \rightarrow A_d v \rightarrow \rho_d(2d) \rightarrow \overline{\rho_d(2d)} \rightarrow 1 \tag{3}$$

does not split. However, its restriction to the braid group $M = B_{n-2}$ does split: $\rho_d(2d) = \rho_d(n-1)$ restricted to M is the direct sum of $\rho_d(n-2)$ and the trivial representation $A_d v$.

Consider the ordered basis $v, e_2, \dots, e_{n-2}, e_{n-1}$ of the representation $\rho_d(n) = \rho_d(2d+1)$. By Lemma 2, the representation $\rho_d(n)$ is irreducible. The preceding paragraph says that (with respect to the above basis) the restriction of $\rho_d(n)$ to $p \in P = B_{n-1}$ has the matrix form

$$\rho_d(n)(p) = \begin{pmatrix} 1 & * & * \\ 0 & p' & * \\ 0 & 0 & 1 \end{pmatrix}$$

where p' is the matrix of the transformation $\overline{\rho_d(n-1)(p)}$ with respect to the basis $e_2, \dots, e_{n-2}$. The entries of the above matrix marked $*$ are in general non-zero since the exact sequence (3) does not split. We now restrict the representation $\rho_d(n)$ to the subgroup $M = B_{n-2}$ and use the remarks after Eq. (3). Thus, every $m \in M$ has the matrix form

$$\begin{pmatrix} 1 & 0 & 0 \\ 0 & m' & 0 \\ 0 & 0 & 1 \end{pmatrix},$$

where m' is the matrix of the transformation $\rho_d(n-2)(m)$.

4.2 A Central Element of $M = B_{n-2}$

Consider the braid group B_n generated by $s_1, \dots, s_{n-1}$. Let Π_i be the product $s_1 s_2 \dots s_i$ and write Δ for the product $\Pi_{n-1}\Pi_{n-2}\dots\Pi_2\Pi_1$. The braid relations show immediately that

$$s_i \Delta = \Delta s_{n-i}$$

and hence Δ^2 lies in the centre of the braid group B_n. (It can be shown that this generates the centre of B_n but we do not use this fact). If d does not divide n, the irreducibility of the representation $\rho_d(n)$ (Lemma 2) implies that the element Δ^2 acts by a scalar λ. In [14] this scalar is computed to be $\omega_d^n \neq 1$ where ω_d is a primitive dth root of unity. We now write c for the corresponding central element of the smaller braid group $M = B_{n-2}$. The conclusion of the preceding paragraph implies that the scalar z by which c acts on $\rho_d(n-2)$ is not 1. Hence the matrix of element c with respect to the basis $v, e_2, \dots, e_{n-1}$ is

$$\rho_d(n)(c) = \begin{pmatrix} 1 & 0 & 0 \\ 0 & z & 0 \\ 0 & 0 & 1 \end{pmatrix}, z \neq 1.$$

We now compute the image of the commutator element $[p, c]$. This can be computed to be

$$\rho_d(n)([p,c]) = \begin{pmatrix} 1 & * & * \\ 0 & 1 & * \\ 0 & 0 & 1 \end{pmatrix}$$

where for sufficiently general $p \in P$, the entries $*$ are non-zero (in view of the fact that $z \neq 1$). We note that (the image of) the group generated by the conjugates $m[p,c]m^{-1}$ (using the irreducibility of the representation $\rho_d(n-2)$) has finite index in $U(O_K)$ where $U \subset U(h)$ is the subgroup which fixes the vector v, therefore stabilises the flag $Av \subset v^{\perp} \subset A^{n-1}$, and acts trivially on successive quotients of this flag. Thus U, with respect to the basis $v, e_2, \ldots, e_{n-1}$, consists of matrices of the form $\begin{pmatrix} 1 & * & * \\ 0 & 1 & * \\ 0 & 0 & 1 \end{pmatrix}$. Hence U is the unipotent radical of a parabolic subgroup of the unitary group G defined over $\mathbb{Q}(\omega_d + \omega_d^{-1})$. We now use the following result (due to Tits and various others; see [9, 12] for references).

Theorem 5 *Let G be a $\mathbb{Q}$-simple linear algebraic group with $\mathbb{R}-rank(G) \geq 2$. Suppose P is a non-trivial parabolic $\mathbb{Q}$-subgroup with unipotent radical U. Suppose $\Gamma \subset G(\mathbb{Z})$ is Zariski dense and intersects the unipotent radical $U(\mathbb{Z})$ in a finite index subgroup. Then Γ has finite index.*

This result immediately implies the arithmeticity of the Burau image for the representation $\rho_d(2d+1)$ in view of the above matrix computations.

5 The Gassner Representation and the Deligne–Mostow Theory

The preceding section dealt with the class of cyclic coverings (2) consisting of the cyclic coverings (1) for which all the integers k_i are 1; this yields a representation of the whole braid group (essentially the Burau representation at roots of unity). In this section, we consider the case when the integers k_i may not be equal to each other (see (4) below). This makes the representation more complicated.

5.1 The Pure Braid Group

Consider the homomorphism $B_n \to S_n$ described in Example 4. The kernel P_n of this homomorphism is called the "pure braid group". This can easily be identified with the fundamental group of the space $S = \{z \in \mathbb{C}^n : z_i \neq z_j \quad \forall \quad i \neq j\}$.

5.2 Gassner Representation

In this section, we denote by R the ring $\mathbb{Z}[X_1^{\pm 1}, \ldots, X_n^{\pm 1}]$ of Laurent polynomials in n variables $X_1, \ldots, X_n$ with $\mathbb{Z}$-coefficients. There is a homomorphism $G : P_n \to GL_{n-1}(R)$ called the *Gassner representation* (see [14] for precise definitions). Fix integers $k_1, \ldots, k_n$ with $1 \le k_i \le d-1$ and consider the homomorphism $\pi : R \to \mathbb{Z}[\omega_d]$ given by $X_i \mapsto \omega_d^{k_i}$ for each i. The resulting representation $\pi \circ G : P_n \to GL_{n-1}(\mathbb{Z}[\omega_d])$ is called the specialization of the Gassner representation at the n-tuple $(k_i/d)_{1 \le i \le n}$ and is denoted $g_d(k, n)$. One has a result analogous to Lemma 2 (for a reference see Lemma 20 of [14]).

Lemma 6 *(1) The specialisations $g_d(k, n)$ are irreducible unless*

$$\sum k_i \equiv 0 \pmod{d}.$$

(2) If $\sum k_i \equiv 0 \pmod{d}$ then $g_d(k, n)$ has a one dimensional space of invariant vectors $A_d v$ with quotient $\overline{g_d(k, n)}$ being irreducible.

By an abuse of notation, we denote by $\overline{g_d(k, n)}$ the representation $g_d(k, n)$ even when it is irreducible.

5.3 Monodromy and Gassner Representation

Fix n integers $k_1, \ldots, k_n$ with $1 \le k_i \le d-1$. Assume that $\sum \mathbb{Z}k_i + \mathbb{Z}d = \mathbb{Z}$. Let $S = \{z \in \mathbb{C}^n : z_i \ne z_j \quad \forall i \ne j\}$. Consider the family C_z' of curves parametrised by S given by

$$y^d = (x - z_1)^{k_1} \cdots (x - z_n)^{k_n}, \quad x \ne z_i \quad \forall \ i. \tag{4}$$

Then C_z' is the complement of a finite set in a compact Riemann surface C_z and the family $X \to S$ of these Riemann surfaces C_z is a family of smooth projective curves fibering over S. We have therefore the monodromy representation $m_d(n)$ of $P_n = \pi_1(S)$ on $H^1(C_z, \mathbb{Z}) = \mathbb{Z}^{2g}$ and for reasons as in the Burau case, the image lies in $U(h)(O_d)$ where h is the hermitian intersection form on $H^1(C_z, \mathbb{C})$. We have then a result analogous to Lemma 3.

Lemma 7 *The monodromy representation $m_d(n)$ is the direct sum of the specialised Gassner representations:*

$$m_d(n) = \bigoplus_{e|d} \overline{g_e(k, n)}$$

where the sum is over all the divisors e of d.

Thus arithmeticity or not of the monodromy of (4) amounts to arithmeticity or not, of the image of the specialised Gassner representation $g_d(k, n)$. In [14], the case when all the k_i are co-prime to d, with $n > 2d$ is considered. In this case, the monodromy is indeed arithmetic:

Theorem 8 [14] *Suppose* $n \geq 2d+1$ *and* $gcd(k_i, d) = 1 \ \forall \ i$. *Then the image of the specialised Gassner representation* $g_d(k, n)$ *is arithmetic.*

The proof is by induction on $n \geq 2d+1$ and quite analogous to that for Theorem 4 but the computations are more complicated. In the range $n < d$ the monodromy group is not always arithmetic, as was shown in [2].

5.4 Deligne–Mostow

We now give a very brief description of the results of [2]. Assume that $\sum \mathbb{Z}k_i + \mathbb{Z}d = \mathbb{Z}$. Fix $s \in (\mathbb{Z}/d\mathbb{Z})^*$. For a real number x let $[x]$ denote its integral part; i.e. it is the integer uniquely defined by $[x] \leq x < [x]+1$. Let $\{x\} = x - [x]$ denote its fractional part. For each suffix i, write $\mu_i = \{\frac{k_i s}{d}\}$ for the fractional part of $\frac{k_i s}{d}$.

Now, the image of the representation $g_d(k, n)$ lies in a unitary group $U(h)(O_d)$ where O_d is the ring of integers in the totally real subfield $K_d = \mathbb{Q}(\omega_d + \omega_d^{-1})$ of the dth cyclotomic extension $\mathbb{Q}(\omega_d)$. The latter is an irreducible lattice in the product $\prod_{t \in (\mathbb{Z}/d\mathbb{Z})^*/\pm 1} U(p_t, q_t)$ where for each archimedean (real) embedding of K_d in $\mathbb{R} \simeq (K_d)_t$, the hermitian form h over $(K_d)_t^{n-1}$ has signature (p_t, q_t).

Suppose, in the foregoing product $\prod U(p_t, q_t)$ there are *two* suffices $t = s, t = s'$ such that $U(p_t, q_t)$ is non-compact. Being an irreducible lattice, the image of $U(h)(O_d)$ in $U(p_t, q_t)$ for each of these t is dense in $PU(p_t, q_t)$ (the projectivised unitary group) by strong approximation. Consequently, if one can show that the monodromy Γ (the image of P_n) is *discrete* for $t = s$, then it follows that Γ has infinite index in $U(h)(O_d)$. It can be shown without much difficulty that Γ is Zariski dense in $U(h)$. Therefore, it proves that the monodromy group is thin (i.e. not arithmetic). This is the strategy used in [2] to exhibit non-arithmetic monodromy.

Deligne and Mostow prove discreteness of Γ, in some special cases. In all these cases, the group $U(p_s, q_s)$ is a real rank one group. Thus, fix $s \in (\mathbb{Z}/d\mathbb{Z})^*$ and let μ_i be as before. Write $\mu_\infty = 2 - \sum_{i=1}^n \mu_i$. *Assume that* $0 < \mu_\infty < 1$. Consider $i \neq j$ such that either $i \leq n$ or $i = \infty$ (similarly j). Consider the following **(semi-)integrality conditions** satisfied by the μ_i:

$$\frac{1}{1-\mu_i-\mu_j} \in \mathbb{Z} \ \ (\text{if} \ \ \mu_i \neq \mu_j), \quad \frac{1}{1-\mu_i-\mu_j} \in \frac{1}{2}\mathbb{Z} \ \ (\text{if} \ \ \mu_i = \mu_j). \tag{5}$$

We recall the famous theorem of Deligne–Mostow:

Theorem 9 (Deligne–Mostow) *(1) Suppose* $s \in (\mathbb{Z}/d\mathbb{Z})^*$. *The group* $U(p_s, q_s)$ *is a rank one group (isomorphic to* $U(n-2, 1)$ *if and only if* $0 < \mu_\infty < 1$.

(2) Suppose $0 < \mu_\infty < 1$ *and the integrality conditions (5) hold. Then the projection* Γ_s *of the monodromy group* P_n *into the* s*th factor* $U(p_s, q_s) = U(n-2, 1)$ *is discrete and has finite co-volume in* $U(p_s, q_s)$.

(3) Suppose $0 < \mu_\infty < 1$. *If there is another* $s' \neq \pm s \in (\mathbb{Z}/d\mathbb{Z})^*$ *such that* $U(p_{s'}, q_{s'})$ *is non-compact, then the projection* Γ_s *in* $U(n-2,1)$ *is a non-arithmetic lattice. In particular, the image of* P_n *in* $U(h)(O_d)$ *has infinite index.*

Corollary 2 *If there is an* $s \in (\mathbb{Z}/d\mathbb{Z})^*$ *satisfying the integrality conditions (5) and (3) of Theorem 9, the monodromy group of the family (4) is not an arithmetic group.*

5.5 Non-arithmetic Lattices in $U(2,1)$

By the Theorem 9, (finitely) many examples of non-arithmetic lattices in $U(2,1)$ can be constructed. Consider the case $d = 18, n = 4\ k_i = 1\ \ \forall\ \ i$ (cf. Example 6). Take $s = 7$. Then $\mu_i = \frac{7}{18}$ and $\mu_\infty = 2 - \sum_{i=1}^{4} \mu_i = \frac{8}{18}$. Then for finite i, j we have

$$\frac{1}{1-\mu_i-\mu_j} = \frac{1}{1-\frac{14}{18}} = \frac{9}{2}$$

is a half integer and if i is finite and $j = \infty$ we have

$$\frac{1}{1-\mu_i-\mu_\infty} = \frac{1}{1-\frac{15}{18}} = \frac{18}{3} \in \mathbb{Z}.$$

Thus the integrality conditions (5) are indeed satisfied and thus the projection of the braid group B_4 to $U(p_s, q_s) = U(2,1)$ is discrete.

Moreover, it can be checked that if $s' = 5$ then $U(p_5, q_5) = U(1,2)$ is non-compact and hence the projection of B_4 to $U(p_s, q_s)$ is a non-arithmetic lattice.

Deligne and Mostow give a large list of non-arithmetic lattices in $U(2,1)$ and some in $U(3,1)$. It is not known (see [3]) whether there exist non-arithmetic lattices in $U(n-2,1)$ for $n \geq 6$.

Remark 3 If there is a rank one factor as in Theorem 9, then we must have $0 \leq 2 - \sum_{i=1}^{n} \mu_i = \mu_\infty (< 1)$. Therefore

$$\frac{n}{d} = \sum_{i=1}^{n} \frac{1}{d} \leq \sum_{i=1}^{n} \{\frac{k_i s}{d}\} \leq 2.$$

Therefore $n \leq 2d$. Thus, if $n \geq 2d+1$, the Deligne Mostow strategy for obtaining non-arithmetic monodromy cannot be implemented. Indeed, if all the k_i are co-prime to d, then Theorem 8 says that the monodromy is arithmetic.

Acknowledgements My hearty thanks to the organisers of the conference in Bonn (in honour of Joachim Schwermer) for their kind invitation to take part in the conference, to speak there and contribute to the conference proceedings. I also thank the Max Planck Institute and the organisers for their warm hospitality. My best wishes to Joachim Schwermer.

It is a pleasure to thank the referee for a careful reading of the manuscript and for very helpful suggestions which have made a considerable improvement in the exposition of the paper.

The support of the J.C Bose fellowship for the period 2013-2018 is gratefully acknowledged.

References

1. A'Campo: Tresses, monodromie et groupes symplectique. Comment. Math. Helv. **54**, 318–327 (1979)
2. Deligne, P., Mostow, G.D.: Monodromy of hypergeometric functions and non-lattice integral monodromy. I.H.E.S Publ. Math. **63**, 5–89 (1986)
3. Deraux, M., Parker, J., Paupert, J.: New nonarithmetic complex hyperbolic lattices. Invent. Math. **203**, 681–773 (2016)
4. Fox, R., Neuwirth, L.: The braid groups. Math. Scand. **10**, 119–126 (1952)
5. Griffiths, P., Schmid, W.: Recent developments in Hodge theory: a discussion of techniques and results. Discrete Subgroups and Applications to Moduli, pp. 31–127. International Colloquium, Bombay (1973)
6. Looijenga, E.: Uniformization by Lauricella functions- an overview of the theory of Deligne and Mostow, arithmetic and geometry around hypergeometric functions. Progress in Mathematics, vol. 260, pp. 207–244. Birkhauser, Basel (2007)
7. Mcmullen, C.: Braid groups and Hodge theory. Math. Ann. **355**, 893–946 (2013)
8. Nori, M.V.: A non-arithmetic monodromy group. C.R Acad. Sci. Ser. I Math. **302**, 71–72 (1986)
9. Raghunathan, M.S.: Generators for arithmetic groups. Pac. J. Math. **152**, 365–373 (1992)
10. Sarnak, P.: Notes on thin matrix groups, thin groups and super-strong approximation. MSRI, vol. 61, pp. 343–362. Cambridge University Press, Cambridge (2014)
11. Squier, C.C.: The Burau representation is unitary. Proc. Amer. Math. Soc. **90**, 199–202 (1984)
12. Venkataramana, T.N.: On systems of generators of higher rank arithmetic subgroups. Pac. J. Math. **166**(1), 193–212 (1994)
13. Venkataramana, T.N.: Image of the Burau representation at dth roots of unity. Ann. Math. **179**, 1041–1083 (2014)
14. Venkataramana, T.N.: Monodromy of cyclic coverings of the projective line. Invent. Math. **197**, 1–45 (2014)

The manufacturer's authorised representative in the EU is Springer Nature Customer Service Centre GmbH, Europaplatz 3, 69115 Heidelberg, Germany. If you have any concerns regarding our products, please contact ProductSafety@springernature.com

Printed and bound by CPI Group (UK) Ltd, Croydon, CR0 4YY
29/07/2026
02180578-0007